Learning Modern Algebra

From Early Attempts to Prove
Fermat's Last Theorem

Print ISBN: 978-1-93951-201-7

Electronic ISBN: 978-1-61444-612-5

Printed in the United States of America

Current Printing (last digit):
10 9 8 7 6 5 4 3 2 1

Learning Modern Algebra

From Early Attempts to Prove
Fermat's Last Theorem

Al Cuoco

EDC, Waltham MA

and

Joseph J. Rotman

University of Illinois at Urbana–Champaign

Published and distributed by
The Mathematical Association of America

MAA TEXTBOOKS

Bridge to Abstract Mathematics, Ralph W. Oberste-Vorth, Aristides Mouzakitis, and Bonita A. Lawrence

Calculus Deconstructed: A Second Course in First-Year Calculus, Zbigniew H. Nitecki

Combinatorics: A Guided Tour, David R. Mazur

Combinatorics: A Problem Oriented Approach, Daniel A. Marcus

Complex Numbers and Geometry, Liang-shin Hahn

A Course in Mathematical Modeling, Douglas Mooney and Randall Swift

Cryptological Mathematics, Robert Edward Lewand

Differential Geometry and its Applications, John Oprea

Elementary Cryptanalysis, Abraham Sinkov

Elementary Mathematical Models, Dan Kalman

An Episodic History of Mathematics: Mathematical Culture Through Problem Solving, Steven G. Krantz

Essentials of Mathematics, Margie Hale

Field Theory and its Classical Problems, Charles Hadlock

Fourier Series, Rajendra Bhatia

Game Theory and Strategy, Philip D. Straffin

Geometry Revisited, H. S. M. Coxeter and S. L. Greitzer

Graph Theory: A Problem Oriented Approach, Daniel Marcus

Knot Theory, Charles Livingston

Learning Modern Algebra: From Early Attempts to Prove Fermat's Last Theorem, Al Cuoco and and Joseph J. Rotman

Lie Groups: A Problem-Oriented Introduction via Matrix Groups, Harriet Pollatsek

Mathematical Connections: A Companion for Teachers and Others, Al Cuoco

Mathematical Interest Theory, Second Edition, Leslie Jane Federer Vaaler and James W. Daniel

Mathematical Modeling in the Environment, Charles Hadlock

Mathematics for Business Decisions Part 1: Probability and Simulation (electronic textbook), Richard B. Thompson and Christopher G. Lamoureux

Mathematics for Business Decisions Part 2: Calculus and Optimization (electronic textbook), Richard B. Thompson and Christopher G. Lamoureux

Mathematics for Secondary School Teachers, Elizabeth G. Bremigan, Ralph J. Bremigan, and John D. Lorch

The Mathematics of Choice, Ivan Niven

The Mathematics of Games and Gambling, Edward Packel

Math Through the Ages, William Berlinghoff and Fernando Gouvea

Noncommutative Rings, I. N. Herstein

Non-Euclidean Geometry, H. S. M. Coxeter

Number Theory Through Inquiry, David C. Marshall, Edward Odell, and Michael Starbird

A Primer of Real Functions, Ralph P. Boas

A Radical Approach to Lebesgue's Theory of Integration, David M. Bressoud

A Radical Approach to Real Analysis, 2nd edition, David M. Bressoud

Real Infinite Series, Daniel D. Bonar and Michael Khoury, Jr.

Topology Now!, Robert Messer and Philip Straffin

Understanding our Quantitative World, Janet Andersen and Todd Swanson

MAA Service Center
P.O. Box 91112
Washington, DC 20090-1112
1-800-331-1MAA   FAX: 1-301-206-9789

Per Micky: Tutto quello che faccio, lo faccio per te.

לרעייתי מרגנית באהבה

Contents

Preface   **xiii**

  Some Features of This Book xiv

  A Note to Students xv

  A Note to Instructors xv

Notation   **xvii**

1 Early Number Theory   **1**

  1.1 Ancient Mathematics 1

  1.2 Diophantus 7

        Geometry and Pythagorean Triples 8

        The Method of Diophantus 11

        Fermat's Last Theorem 14

        Connections: Congruent Numbers 16

  1.3 Euclid 20

        Greek Number Theory 21

        Division and Remainders 22

        Linear Combinations and Euclid's Lemma 24

        Euclidean Algorithm 30

  1.4 Nine Fundamental Properties 36

  1.5 Connections 41

        Trigonometry 41

        Integration 42

2 Induction   **45**

  2.1 Induction and Applications 45

        Unique Factorization 53

        Strong Induction 57

        Differential Equations 60

  2.2 Binomial Theorem 63

        Combinatorics 69

  2.3 Connections 73

        An Approach to Induction 73

        Fibonacci Sequence 75

3 Renaissance   **81**

  3.1 Classical Formulas 82

  3.2 Complex Numbers 91

		Algebraic Operations	92
		Absolute Value and Direction	99
		The Geometry Behind Multiplication	101
	3.3	Roots and Powers	106
	3.4	Connections: Designing Good Problems	116
		Norms	116
		Pippins and Cheese	118
		Gaussian Integers: Pythagorean Triples Revisited	119
		Eisenstein Triples and Diophantus	122
		Nice Boxes	123
		Nice Functions for Calculus Problems	124
		Lattice Point Triangles	126

4 Modular Arithmetic   **131**

	4.1	Congruence	131
	4.2	Public Key Codes	149
	4.3	Commutative Rings	154
		Units and Fields	160
		Subrings and Subfields	166
	4.4	Connections: Julius and Gregory	169
	4.5	Connections: Patterns in Decimal Expansions	177
		Real Numbers	177
		Decimal Expansions of Rationals	179
		Periods and Blocks	182

5 Abstract Algebra   **191**

	5.1	Domains and Fraction Fields	192
	5.2	Polynomials	196
		Polynomial Functions	204
	5.3	Homomorphisms	206
		Extensions of Homomorphisms	213
		Kernel, Image, and Ideals	216
	5.4	Connections: Boolean Things	221
		Inclusion-Exclusion	227

6 Arithmetic of Polynomials   **233**

	6.1	Parallels to $\mathbb{Z}$	233
		Divisibility	233
		Roots	239
		Greatest Common Divisors	243
		Unique Factorization	248
		Principal Ideal Domains	255
	6.2	Irreducibility	259
		Roots of Unity	264
	6.3	Connections: Lagrange Interpolation	270

7 Quotients, Fields, and Classical Problems   **277**

	7.1	Quotient Rings	277
	7.2	Field Theory	287
		Characteristics	287
		Extension Fields	289

Algebraic Extensions 293
Splitting Fields 300
Classification of Finite Fields 305
7.3 Connections: Ruler–Compass Constructions 308
Constructing Regular *n*-gons 320
Gauss's construction of the 17-gon 322

8 Cyclotomic Integers   **329**
8.1 Arithmetic in Gaussian and Eisenstein Integers 330
Euclidean Domains 333
8.2 Primes Upstairs and Primes Downstairs 337
Laws of Decomposition 339
8.3 Fermat's Last Theorem for Exponent 3 349
Preliminaries 350
The First Case 351
Gauss's Proof of the Second Case 354
8.4 Approaches to the General Case 359
Cyclotomic integers 360
Kummer, Ideal Numbers, and Dedekind 365
8.5 Connections: Counting Sums of Squares 371
A Proof of Fermat's Theorem on Divisors 373

9 Epilog   **379**
9.1 Abel and Galois 379
9.2 Solvability by Radicals 381
9.3 Symmetry 384
9.4 Groups 389
9.5 Wiles and Fermat's Last Theorem 396
Elliptic Integrals and Elliptic Functions 397
Congruent Numbers Revisited 400
Elliptic Curves 404

A Appendices   **409**
A.1 Functions 409
A.2 Equivalence Relations 420
A.3 Vector Spaces 424
Bases and Dimension 427
Linear Transformations 435
A.4 Inequalities 441
A.5 Generalized Associativity 442
A.6 A Cyclotomic Integer Calculator 444
Eisenstein Integers 445
Symmetric Polynomials 446
Algebra with Periods 446

References   **449**

Index   **451**

About the Authors   **459**

Preface

This book is designed for college students who want to teach mathematics in high school, but it can serve as a text for standard abstract algebra courses as well. First courses in abstract algebra usually cover number theory, groups, and commutative rings. We have found that the first encounter with groups is not only inadequate for future teachers of high school mathematics, it is also unsatisfying for other mathematics students. Hence, we focus here on number theory, polynomials, and commutative rings. We introduce groups in our last chapter, for the earlier discussion of commutative rings allows us to explain how groups are used to prove Abel's Theorem: there is no generalization of the quadratic, cubic, and quartic formulas giving the roots of the general quintic polynomial. A modest proposal: undergraduate abstract algebra should be a sequence of two courses, with number theory and commutative rings in the first course, and groups and linear algebra (with scalars in arbitrary fields) in the second.

We invoke an historically accurate organizing principle: Fermat's Last Theorem (in Victorian times, the title of this book would have been *Learning Modern Algebra by Studying Early Attempts, Especially Those in the Nineteenth Century, that Tried to Prove Fermat's Last Theorem Using Elementary Methods*). To be sure, another important problem at that time that contributed to modern algebra was the search for formulas giving the roots of polynomials. This search is intertwined with the algebra involved in Fermat's Last Theorem, and we do treat this part of algebra as well. The difference between our approach and the standard approach is one of emphasis: the natural direction for us is towards algebraic number theory, whereas the usual direction is towards Galois theory.

Four thousand years ago, the quadratic formula and the Pythagorean Theorem were seen to be very useful. To teach them to new generations, it was best to avoid square roots (which, at the time, were complicated to compute), and so problems were designed to have integer solutions. This led to Pythagorean triples: positive integers a, b, c satisfying $a^2 + b^2 = c^2$. Two thousand years ago, all such triples were found and, when studying them in the seventeenth century, Fermat wondered whether there are positive integer solutions to $a^n + b^n = c^n$ for $n > 2$. He claimed in a famous marginal note that there are no solutions, but only his proof of the case $n = 4$ is known. This problem, called Fermat's Last Theorem, intrigued many of the finest mathematicians, but it long resisted all attempts to solve it. Finally, using sophisticated techniques of algebraic geometry developed at the end of the twentieth century, Andrew Wiles proved Fermat's Last Theorem in 1995.

Before its solution, Fermat's Last Theorem was a challenge to mathematicians (as climbing Mount Everest was a challenge to mountaineers). There are no dramatic applications of the result, but it is yet another triumph of human intellect. What is true is that, over the course of 350 years, much of contemporary mathematics was invented and developed in trying to deal with it. The number theory recorded in Euclid was shown to have similarities with the behavior of polynomials, and generalizations of prime numbers and unique factorization owe their initial study to attempts at proving Fermat's Last Theorem. But these topics are also intimately related to what is actually taught in high school. Thus, abstract algebra is not merely beautiful and interesting, but it is also a valuable, perhaps essential, topic for understanding high school mathematics.

Some Features of This Book

We include sections in every chapter, called *Connections*, in which we explicitly show how the material up to that point can help the reader understand and implement the mathematics that high school teachers use in their profession. This may include the many ways that results in abstract algebra connect with core high school ideas, such as solving equations or factoring. But it may also include mathematics for teachers themselves, that may or may not end up "on the blackboard;" things like the use of abstract algebra to make up good problems, to understand the foundations of topics in the curriculum, and to place the topics in the larger landscape of mathematics as a scientific discipline.

Many students studying abstract algebra have problems understanding proofs; even though they can follow each step of a proof, they wonder how anyone could have discovered its argument in the first place. To address such problems, we have tried to strike a balance between giving a logical development of results (so the reader can see how everything fits together in a coherent package) and discussing the messier kinds of thinking that lead to discovery and proofs. A nice aspect of this sort of presentation is that readers participate in *doing* mathematics as they learn it.

One way we implement this balance is our use of several design features, such as the *Connections* sections described above. Here are some others.

- Sidenotes provide advice, comments, and pointers to other parts of the text related to the topic at hand. What could be more fitting for a book related to Fermat's Last Theorem than to have large margins?
- Interspersed in the text are boxed "callouts," such as **How to Think About It**, which suggest how ideas in the text may have been conceived in the first place, how we view the ideas, and what we guess underlies the formal exposition. Some other callouts are:

  Historical Note, which provides some historical background. It often helps to understand mathematical ideas if they are placed in historical context; besides, it's interesting. The biographies are based on those in the MacTutor History of Mathematics Archive of the School of Mathematics and Statistics, University of St. Andrews, Scotland. It can be found on the internet: its URL is

  www-history.mcs.st-andrews.ac.uk

  Etymology, which traces out the origin of some mathematical terms. We believe that knowing the etymology of terms often helps to understand the ideas they name.

Etymology. The word *mathematics* comes from classical Greek; it means "knowledge," "something learned." But in ancient Rome through the thirteenth century, it meant "astronomy" and "astrology." From the Middle Ages, it acquired its present meaning.

The word *arithmetic* comes from the Greek word meaning "the art of counting." The word *geometry*, in classical Greek, meant "science of measuring;" it arose from an earlier term meaning "land survey."

It is a pleasure to acknowledge those who have contributed valuable comments, suggestions, ideas, and help. We thank Don Albers, Carol Baxter, Bruce Berndt, Peter Braunfeld, Keith Conrad, Victoria Corkery, Don DeLand, Ben Fischer, Andrew Granville, Heini Halberstam, Zaven Karian, Tsit-Yuen Lam, Paul Monsky, Beverly Ruedi, Glenn Stevens, and Stephen Ullom.

Conrad's website
`www.math.uconn.edu/`
`~kconrad/blurbs/`
is full of beautiful ideas.

A Note to Students

The heart of a mathematics course lies in its problems. We have tried to orchestrate them to help you build a solid understanding of the mathematics in the sections. Everything afterward will make much more sense if you work through as many exercises as you can, especially those that appear difficult. Quite often, you will learn something valuable from an exercise even if you don't solve it completely. For example, a problem you can't solve may show that you haven't fully understood an idea you thought you knew; or it may force you to discover a fact that needs to be established to finish the solution.

There are two special kinds of exercises.

* Those labeled **Preview** may seem to have little to do with the section at hand; they are designed to foreshadow upcoming topics, often with numerical experiments.

* Those labeled **Take it Further** develop interesting ideas that are connected to the main themes of the text, but are somewhat off the beaten path. They are not essential for understanding what comes later in the text.

An exercise marked with an asterisk, such as 1.8*, means that it is either used in some proof or it is referred to elsewhere in the text. For ease of finding such exercises, all references to them have the form "Exercise 1.8 on page 6" giving both its number and the number of the page on which it occurs.

A Note to Instructors

We recommend giving reading assignments to preview upcoming material. This contributes to balancing experience and formality as described above, and it saves time. Many important pages can be read and understood by students, and they should be discussed in class only if students ask questions about them.

It is possible to use this book as a text for a three hour one-semester course, but we strongly recommend that it be taught four hours per week.

—Al Cuoco and Joe Rotman

Notation

$\Delta(a, b, c)$	4	triangle with sides of lengths a, b, c
ΔABC	4	triangle with vertices A, B, C
$\mathbb{N}$	21	natural numbers
$\mathbb{Z}$	21	integers
$a \mid b$	21	a is a divisor of b
$\gcd(a, b)$	24	greatest common divisor
$\lfloor x \rfloor$	29	greatest integer in x
$\mathbb{Q}$	36	rational numbers
$\mathbb{R}$	36	real numbers
$\Rightarrow$	46	implies
$\text{lcm}(a, b)$	55	least common multiple
$\binom{n}{r}$	63	binomial coefficient
$\Re(z)$	92	real part of complex number z
$\Im(z)$	92	imaginary part of complex number z
$\mathbb{C}$	92	complex numbers
$\overrightarrow{PQ}$	93	arrow from P to Q
$\bar{z}$	96	conjugate of z
$\lvert z \rvert$	99	modulus of z
$\arg(z)$	100	argument of z
e^z	108	complex exponential
$\phi(n)$	111	Euler ϕ-function
$N(z)$	116	norm of z
$\mathbb{Z}[i]$	119	Gaussian integers
$\mathbb{Z}[\omega]$	120	Eisenstein integers
$a \equiv b \bmod m$	132	a is congruent to b modulo m
$m_1 \cdots \widehat{m_i} \cdots m_r$	147	expression with m_i deleted
$[a]$	154	congruence class of integer a
$\mathbb{Z}_m$	154	integers mod m
$\mathbb{Z}[\zeta]$	157	cyclotomic integers
R^S	157	ring of functions $R \to S$
$C(X)$	157	ring of continuous functions $X \to \mathbb{R}$

$\mathrm{Fun}(R)$	157	ring of functions $R \to R$
$\mathbb{F}_4$	165	field with 4 elements
2^X	167	Boolean ring of subsets of set X
$j(m)$	172	calendar month function
$\mathrm{Frac}(D)$	194	fraction field of domain D
a/b	195	element of $\mathrm{Frac}(D)$
$\deg(f)$	198	degree of polynomial f
$R[[x]]$	198	all power series over R
$R[x]$	198	all polynomials over R
x	200	indeterminate in $R[x]$
$f'(x)$	202	derivative of $f(x) \in R[x]$
$f^{\#}$	204	associated polynomial function of f
$\mathrm{Poly}(R)$	204	all polynomials functions over R
$k(x)$	205	field of rational functions over k
$\mathbb{F}_q$	205	finite field with exactly q elements
$R[x_1, \ldots, x_n]$	205	polynomials in several variables over R
$D(x_1, \ldots, x_n)$	206	rational functions in several variables over domain D
$R \cong S$	207	rings R and S are isomorphic
$\ker \varphi$	217	kernel of homomorphism φ
$\mathrm{im}\,\varphi$	217	image of homomorphism φ
$(b_1, \ldots, b_n)$	218	ideal generated by $b_1, \ldots, b_n$
(a)	218	principal ideal generated by a
(0)	219	zero ideal $= \{0\}$
IJ	220	product of ideals I and J
$I + J$	220	sum of ideals I and J
$R \times S$	221	direct product of rings R and S
$a \vee b$	223	binary operation in Boolean ring
$|A|$	227	number of elements in finite set A
PID	255	principal ideal domain
UFD	258	unique factorization domain
$\Phi_d(x)$	265	cyclotomic polynomial
$a + I$	278	coset of element a mod ideal I
$a \equiv b \bmod I$	279	congruent mod ideal I
R/I	280	quotient ring R mod I
$\langle X \rangle$	293	subfield generated by subset X
$[K : k]$	291	degree of extension field K/k
$k(z_1, \ldots, z_n)$	294	extension field adjoining $z_1, \ldots, z_n$ to k
$\mathrm{irr}(z, k)$	296	minimal polynomial of z over k
$\overline{PQ}$	310	line segment with endpoints P, Q
PQ	310	length of segment $\overline{PQ}$

$L(P, Q)$	309	line determined by points P, Q
$C(P, Q)$	309	circle with center P, radius PQ
∂	333	size function on Euclidean domain
λ	348	$\lambda = 1 - \omega$
ν	350	valuation
$r(n)$	371	number of non-associate $z \in \mathbb{Z}[i]$ of norm n
Q_1	372	first quadrant
$\zeta(s)$	374	Riemann zeta function
$\chi(n)$	375	a multiplicative function on $\mathbb{Z}[i]$
$\mathrm{Gal}(f)$	386	Galois group of polynomial f
$\mathrm{Gal}(E/k)$	387	Galois group of field extension E/k
S_n	389	symmetric group on n letters
G/N	392	quotient group
$a \in A$	409	a is an element of set A
1_X	411	identity function on set X
$f : a \mapsto b$	411	$f(a) = b$
$U \subseteq V$	410	U is a subset of set V
$U \subsetneqq V$	410	U is a proper subset of V
$\varnothing$	410	empty set
$g \circ f$	414	composite f followed by g
$[a]$	421	equivalence class of element a
$\mathrm{Span}\langle X \rangle$	427	subspace spanned by subset X
$\dim(V)$	433	dimension of vector space V
V^*	437	dual space of vector space V
$A^\top$	438	transpose of matrix A

1

Early Number Theory

Algebra, geometry, and number theory have been used for millennia. Of course, numbers are involved in counting and measuring, enabling commerce and architecture. But reckoning was also involved in life and death matters such as astronomy, which was necessary for navigation on the high seas (naval commerce flourished four thousand years ago) as well as to predict the seasons, to apprise farmers when to plant and when to harvest. Ancient texts that have survived from Babylon, China, Egypt, Greece, and India provide evidence for this. For example, the Nile River was the source of life in ancient Egypt, for its banks were the only arable land in the midst of desert. Mathematics was used by the priestly class to predict flooding as well as to calculate area (taxes were assessed according to the area of land, which changed after flood waters subsided). And their temples and pyramids are marvels of engineering.

1.1 Ancient Mathematics

The quadratic formula was an important mathematical tool, and so it was taught to younger generations training to be royal scribes. Here is a problem from an old Babylonian cuneiform text dating from about 1700 BCE. We quote from van der Waerden [35], p. 61 (but we write numbers in base 10 instead of in base 60, as did the Babylonians). We also use modern algebraic notation that dates from the fifteenth and sixteenth centuries (see Cajori [6]).

> *I have subtracted the side of the square from the area, and it is* 870. *What is the side of my square?*

The text rewrites the data as the quadratic equation $x^2 - x = 870$; it then gives a series of steps showing how to find the solution, illustrating that the Babylonians knew the quadratic formula.

Historians say that teaching played an important role in ancient mathematics (see van der Waerden [35], pp. 32–33). To illustrate, the coefficients of the quadratic equation were chosen wisely: the discriminant $b^2 - 4ac = 1 - 4(-870) = 3481 = 59^2$ is a perfect square. Were the discriminant not a perfect square, the problem would have been much harder, for finding square roots was not routine in those days. Thus, the quadratic in the text is well-chosen for teaching the quadratic formula; a good teaching prize would not be awarded for $x^2 - 47x = 210$.

The number 59 may have been chosen because the Babylonians wrote numbers in base 60, and $59 = 60 - 1$.

The Babylonians were not afraid of cubics. Another of their problems from about the same time is

Solve $12x^3 = 3630$,

and the answer was given. The solution was, most likely, obtained by using tables of approximations of cube roots.

A standard proof of the quadratic formula is by "completing the square." This phrase can be taken literally. Given a quadratic $x^2 + bx = c$ with b and c positive, we can view $x^2 + bx$ as the shaded area in Figure 1.1. Complete the

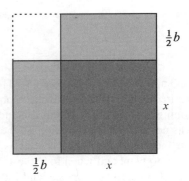

Figure 1.1. Completing the Square.

figure to a square by attaching the corner square having area $\frac{1}{2}b \times \frac{1}{2}b = \frac{1}{4}b^2$; the new square has area

$$c + \tfrac{1}{4}b^2 = x^2 + bx + \tfrac{1}{4}b^2 = (x + \tfrac{1}{2}b)^2.$$

Thus, $x + \frac{1}{2}b = \sqrt{c + \frac{1}{4}b^2}$, which simplifies to the usual formula giving the roots of $x^2 + bx - c$. The algebraic proof of the validity of the quadratic formula works without assuming that b and c are positive, but the idea of the proof is geometric.

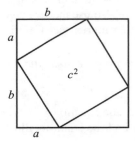 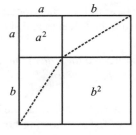

Figure 1.2. Pythagorean Theorem.

In [35], pp. 26–35, van der Waerden considers the origin of proofs in mathematics, suggesting that they arose in Europe and Asia in Neolithic (late Stone Age) times, 4500 BCE–2000 BCE.

Exercise 1.4 on page 5 asks you to show that the rhombus in Figure 1.2 with sides of length c is a square.

The Babylonians were aware of the Pythagorean Theorem. Although they believed it, there is no evidence that the Babylonians had proved the Pythagorean Theorem; indeed, no evidence exists that they even saw a need for a proof. Tradition attributes the first proof of this theorem to Pythagoras, who lived around 500 BCE, but no primary documents extant support this. An elegant proof of the Pythagorean Theorem is given on page 354 of Heath's 1926 translation [16] of Euclid's *The Elements*; the theorem follows from equality of the areas of the two squares in Figure 1.2.

Here is an ancient application of the Pythagorean Theorem. Aristarchus (ca. 310 BCE–250 BCE) saw that the Moon and the Sun appear to be about the same size, and he wondered how far away they are. His idea was that at the time of the half-moon, the Earth E, Moon M, and Sun S form a right triangle with right angle $\angle M$ (that is, looking up at the Moon, the line of sight seems to be perpendicular to the Sun's rays). The Pythagorean Theorem gives

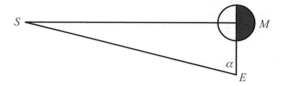

Figure 1.3. Earth, Moon, and Sun.

$|SE|^2 = |SM|^2 + |ME|^2$. Thus, the Earth is farther from the Sun than it is from the Moon. Indeed, at sunset, $\alpha = \angle E$ seems to be very close to 90°: if we are looking at the Moon and we wish to watch the Sun dip below the horizon, we must turn our head all the way to the left. Aristarchus knew trigonometry; he reckoned that $\cos\alpha$ was small, and he concluded that the Sun is very much further from the Earth than is the Moon.

Example 1.1. Next, we present a geometric problem from a Chinese collection of mathematical problems, *Nine Chapters on the Mathematical Art*, written during the Han Dynasty about two thousand years ago. Variations of this problem still occur in present day calculus books!

There is a door whose height and width are unknown, and a pole whose length p is also unknown. Carried horizontally, the pole does not fit by 4 *ch'ih; vertically, it does not fit by* 2 *ch'ih; slantwise, it fits exactly. What are the height, width, and diagonal of the door?*

There are similar problems from the Babylonians and other ancient cultures.

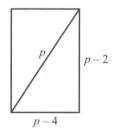

Figure 1.4. Door Problem.

The data give a right triangle with sides $p - 4$, $p - 2$, and p, and the Pythagorean Theorem gives the equation $(p - 4)^2 + (p - 2)^2 = p^2$, which simplifies to $p^2 - 12p + 20 = 0$. The discriminant $b^2 - 4ac$ is $144 - 80 = 64$, a perfect square, so that $p = 10$ and the door has height 8 and width 6 (the other root of the quadratic is $p = 2$, which does not fit the physical data). The sides of the right triangle are 6, 8, 10, and it is similar to the triangle with sides 3, 4, 5. Again, the numbers have been chosen wisely. The idea is to teach students how to use the Pythagorean Theorem and the quadratic formula. As we have already remarked, computing square roots was then quite difficult, so that the same problem for a pole of length $p = 12$ would not have been very bright because there is no right triangle with sides of integral length that has hypotenuse 12.   ▲

Are there right triangles whose three sides have integral length that are not similar to the $3, 4, 5$ triangle? You are probably familiar with the $5, 12, 13$ triangle. Let's use $\triangle(a, b, c)$ (lower case letters) to denote the triangle whose sides have length a, b, and c; if $\triangle(a, b, c)$ is a right triangle, then c denotes the length of its hypotenuse, while a and b are its **legs**. Thus, the right triangle with side-lengths 5, 12, 13 is denoted by $\triangle(5, 12, 13)$. (We use the usual notation, $\triangle ABC$, to denote a triangle whose vertices are A, B, C.)

Definition. A triple (a, b, c) of positive integers with $a^2 + b^2 = c^2$ is called a *Pythagorean triple*.

If (a, b, c) is a Pythagorean triple, then the triangles $\triangle(a, b, c)$ and $\triangle(b, a, c)$ are the same. Thus, we declare that the Pythagorean triples (a, b, c) and (b, a, c) are the same.

Historical Note. Pythagorean triples are the good choices for problems teaching the Pythagorean Theorem. There are many of them: Figure 1.5 shows a Babylonian cuneiform tablet dating from the dynasty of Hammurabi, about 1800 BCE, whose museum name is Plimpton 322, which displays fifteen Pythagorean triples (translated into our number system).

b	a	c
120	119	169
3456	3367	4825
4800	4601	6649
13500	12709	18541
72	65	97
360	319	481
2700	2291	3541
960	799	1249
600	481	769
6480	4961	8161
60	45	75
2400	1679	2929
240	161	289
2700	1771	3229
90	56	106

Figure 1.5. Plimpton 322.

It is plain that the Babylonians had a way to generate large Pythagorean triples. Here is one technique they might have used. Write

$$a^2 = c^2 - b^2 = (c+b)(c-b).$$

If there are integers m and n with

$$c + b = m^2$$
$$c - b = n^2,$$

then

$$a = \sqrt{(c+b)(c-b)} = mn. \qquad (1.1)$$

We can also solve for b and c:

$$b = \tfrac{1}{2}\left(m^2 - n^2\right) \qquad (1.2)$$
$$c = \tfrac{1}{2}\left(m^2 + n^2\right). \qquad (1.3)$$

Summarizing, here is what we call the **Babylonian method**. Choose *odd* numbers m and n (forcing $m^2 + n^2$ and $m^2 - n^2$ to be even, so that b and c are integers), and define a, b, and c by Eqs. (1.1), (1.2), and (1.3). For example, if $m = 7$ and $n = 5$, we obtain 35, 12, 37. If we choose $m = 179$ and $n = 71$, we obtain 13500, 12709, 18541, the largest triple on Plimpton 322.

The Babylonian method does not give all Pythagorean triples. For example, $(6, 8, 10)$ is a Pythagorean triple, but there are no *odd* numbers $m > n$ with $6 = mn$ or $8 = mn$. Of course, $(6, 8, 10)$ is not signifcantly different from $(3, 4, 5)$, which arises from $3 > 1$. In the next section, we will show, following Diophantus, ca. 250 CE, how to find *all* Pythagorean triples. But now we should recognize that practical problems involving applications of pure mathematics (e.g., surveying) led to efforts to teach this mathematics effectively, which led to more pure mathematics (Pythagorean triples) that seems at first to have no application outside of teaching. The remarkable, empirical, fact is that pure mathematics yields new and valuable applications. For example, we shall see in the next section that classifying Pythagorean triples leads to simplifying the verification of some trigonometric identities as well as the solution of certain integration problems (for example, we will see a natural way to integrate $\sec x$).

After all, what practical application does the Pythagorean triple $(13500, 12709, 18541)$ have?

Exercises

1.1 Prove the quadratic formula for the roots of $ax^2 + bx + c = 0$ whose coefficients a, b, and c may not be positive.

1.2 Give a geometric proof that $(a+b)^2 = a^2 + 2ab + b^2$ for a, b positive.

1.3 * Let $f(x) = ax^2 + bx + c$ be a quadratic whose coefficients a, b, c are rational. Prove that if $f(x)$ has one rational root, then its other root is also rational.

1.4 *

    (i) Prove that the rhombus with side lengths c in the left square of Figure 1.2 is a square.

    (ii) Prove the Pythagorean Theorem in a way suggested by Figure 1.2.

    (iii) Give a proof of the Pythagorean Theorem different from the one suggested by Figure 1.2.

The book by Loomis [20] contains 370 different proofs of the Pythagorean Theorem, by the author's count.

1.5 Here is another problem from *Nine Chapters on the Mathematical Art*. A pond is 10 ch'ih square. A reed grows at its center and extends 1 ch'ih out of the water. If the reed is pulled to the side of the pond, it reaches the side precisely. What are the depth of the water and the length of the reed?

Answer. Depth = 12 ch'ih and length = 13 ch'ih.

1.6 *

(i) Establish the algebraic identity

$$\left(\frac{a+b}{2}\right)^2 - \left(\frac{a-b}{2}\right)^2 = ab.$$

(ii) Use (i) to establish the **Arithmetic–Geometric Mean Inequality**: if a and b are positive reals, then

$$\sqrt{ab} \leq \tfrac{1}{2}(a+b).$$

When is there equality?

(iii) Show how to dissect an $a \times b$ rectangle so that it fits inside a square with side-length $(a+b)/2$. How much is "left over?"

Hint. Try it with numbers. Cut an 8×14 rectangle to fit inside an 11×11 square.

(iv) Show that a rectangle of maximum area with fixed perimeter is a square.

(v) The **hyperbolic cosine** is defined by

$$\cosh x = \tfrac{1}{2}(e^x + e^{-x}).$$

Prove that $\cosh x \geq 1$ for all real numbers x, while $\cosh x = 1$ if and only if $x = 0$.

(vi) Use Figure 1.6 to give another proof of the Arithmetic-Geometric Mean Inequality.

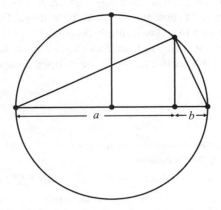

Figure 1.6. Arithmetic–Geometric Mean Inequality.

1.7 * Prove that there is no Pythagorean triple (a, b, c) with $c = 12$.

1.8 * Let (a, b, c) be a Pythagorean triple.

(i) Prove that the legs a and b cannot both be odd.

(ii) Show that the area of $\triangle(a, b, c)$ is an integer.

1.9 * Show that 5 is not the area of a triangle whose side-lengths form a Pythagorean triple.

1.10 * Let (a, b, c) be a Pythagorean triple. If m is a positive integer, prove that (ma, mb, mc) is also a Pythagorean triple.

1.11 (**Converse of Pythagorean Theorem**). * Let $\triangle = \triangle(a, b, c)$ be a triangle with sides of lengths a, b, c (positive real numbers, not necessarily integers). Prove that if $a^2 + b^2 = c^2$, then $\triangle$ is a right triangle.

> **Hint.** Construct a right triangle $\triangle'$ with legs of lengths a, b, and prove that $\triangle'$ is congruent to $\triangle$ by side-side-side.

1.12 * Prove that every Pythagorean triple (a, b, c) arises from a right triangle $\triangle(a, b, c)$ having sides of lengths a, b, c.

1.13 If $P = (a, b, c)$ is a Pythagorean triple, define $r(P) = c/a$. If we label the Pythagorean triples on Plimpton 322 as $P_1, \ldots, P_{15}$, show that $r(P_i)$ is decreasing: $r(P_i) > r(P_{i+1})$ for all $i \leq 14$.

1.14 * If (a, b, c) is a Pythagorean triple, show that $(a/c, b/c)$ is a point on the graph of $x^2 + y^2 = 1$. What is the graph of $x^2 + y^2 = 1$?

1.15 **Preview.** Let L be the line through $(-1, 0)$ with slope t.

- (i) If $t = \frac{1}{2}$, find all the points where L intersects the graph of $x^2 + y^2 = 1$.

    Answer. $(\frac{3}{5}, \frac{4}{5})$.

- (ii) If $t = \frac{3}{2}$, find all the points where L intersects the graph of $x^2 + y^2 = 1$.

    Answer. $(\frac{-5}{13}, \frac{12}{13})$.

- (iii) Pick a rational number t, not $\frac{1}{2}$ or $\frac{3}{2}$, and find all the points where L intersects the graph of $x^2 + y^2 = 1$.

- (iv) Suppose ℓ is a line that contains $(-1, 0)$ with slope r. If r is a rational number, show that ℓ intersects the graph of $x^2 + y^2 = 1$ in two points, each of which has rational number coordinates.

1.16 **Preview.** A *Gaussian integer* is a complex number $a + bi$ where both a and b are integers. Pick six Gaussian integers $r + si$ with $r > s > 0$ and square them. State something interesting that you see in your results.

1.17 **Preview.** Consider a complex number $z = q + ip$, where $q > p$ are positive integers. Prove that

$$(q^2 - p^2, 2qp, q^2 + p^2)$$

is a Pythagorean triple by showing that $|z^2| = |z|^2$.

> If z is a complex number, say $z = a + bi$, then we define $|z| = \sqrt{a^2 + b^2}$.

1.18 **Preview.** Show, for all real numbers m and n, that

$$\left[\tfrac{1}{2}(m + n) + \tfrac{1}{2}(m - n)i\right]^2 = mn + \tfrac{1}{2}(m^2 - n^2)i.$$

1.2 Diophantus

We are going to classify Pythagorean triples using a geometric method of Diophantus that describes *all* Pythagorean triples.

Historical Note. We know very little about the life of Diophantus. He was a mathematician who lived in Alexandria, Egypt, but his precise dates are unknown; most historians believe he lived around 250 CE. His extant work shows systematic algebraic procedures and notation, but his leaps of intuition strongly suggest that he was thinking geometrically; indeed, Newton called Diophantus's discussion of Pythagorean triples the *chord method* (see Figure 1.7). Thus, geometry (the Pythagorean Theorem) and applied problems (teaching) suggested an algebraic problem (find all Pythagorean triples), and we now return to geometry to solve it. Here is evidence that the distinction between algebra and geometry is an artificial one; both are parts of the same subject.

Geometry and Pythagorean Triples

Before we get into the technicalities of Diophantus's classification of Pythagorean triples, let's note that geometry is lurking nearby. Exercise 1.14 above makes a natural observation: if (a, b, c) is a Pythagorean triple, then

$$\left(\frac{a}{c}\right)^2 + \left(\frac{b}{c}\right)^2 = 1,$$

a point on the **unit circle**, the circle having radius 1, center the origin, and equation $x^2 + y^2 = 1$. Dividing through by c^2 is a good idea. For example, $(6, 8, 10)$ is a "duplicate" of $(3, 4, 5)$, and both of these Pythagorean triples determine the same point, $(3/5, 4/5)$, on the unit circle.

Here is the main idea of Diophantus. Even though those points arising from Pythagorean triples are special (for example, they lie in the first quadrant and both their coordinates are rational numbers), let's parametrize *all* the points P on the unit circle. Choose a point on the unit circle "far away" from the first quadrant; the simplest is $(-1, 0)$, and let $\ell = \ell(P)$ be the line joining it to P. We shall see that the slopes of such lines parametrize all the points on the unit circle. In more detail, any line ℓ through $(-1, 0)$ (other than the tangent)

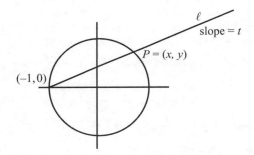

Figure 1.7. Geometric Idea of Diophantus.

intersects the unit circle in a unique second point, $P = (x, y)$; let t be the slope of ℓ. As t varies through all real numbers, $-\infty < t < \infty$, the intersection points P of ℓ and the unit circle trace out the entire circle (except for $(-1, 0)$).

Proposition 1.2. *The points P on the unit circle (other than $(-1, 0)$) are parametrized as*

$$P = \left(\frac{1 - t^2}{1 + t^2}, \frac{2t}{1 + t^2}\right), \qquad where \ -\infty < t < \infty.$$

Proof. The line through points (a, b) and (c, d) has equation $y - b = t(x - a)$, where $t = (d - b)/(c - a)$, so the line ℓ through $(-1, 0)$ and a point $P = (x, y)$ on the unit circle has an equation of the form $y = t(x + 1)$, so that $x = (y - t)/t$. Thus, (x, y) is a solution of the system

$$y = t(x + 1)$$
$$x^2 + y^2 = 1.$$

An obvious solution of this system is $(-1, 0)$, because this point lies on both the line and the circle. Let's find x and y in terms of t. If the slope $t = 0$, then ℓ is the x-axis and the other solution is $(1, 0)$. To find the solutions when $t \neq 0$, eliminate x: the equations

$$\frac{y - t}{t} = x \qquad \text{and} \qquad x^2 + y^2 = 1$$

give

$$\left(\frac{y - t}{t}\right)^2 + y^2 = 1.$$

Expanding and simplifying, we obtain

$$y\left[(1 + t^2)y - 2t\right] = 0.$$

We knew at the outset that $y = 0$ makes this true. If $y \neq 0$, then canceling gives

$$y = \frac{2t}{1 + t^2},$$

and solving for x gives

$$x = \frac{y - t}{t} = \frac{\frac{2t}{1+t^2} - t}{t} = \frac{1 - t^2}{1 + t^2}. \qquad \blacksquare$$

In Exercise 1.12 on page 7, we saw that every Pythagorean triple (a, b, c) arises from a right triangle $\triangle(a, b, c)$ having sides of integral lengths a, b, c. Conversely, the Pythagorean Theorem says that every right triangle $\triangle(a, b, c)$ whose sides have integral length gives the Pythagorean triple (a, b, c). Thus, Pythagorean triples and certain right triangles are merely two ways to view the same idea, one algebraic, one geometric. At any given time, we will adopt that viewpoint which is most convenient.

We have already run across distinct Pythagorean triples that are essentially the same; Exercise 1.10 on page 7 shows that if (a, b, c) is a Pythagorean triple, then so is (ma, mb, mc), where m is a positive integer. The right triangles $\triangle(a, b, c)$ and $\triangle(ma, mb, mc)$ determined by these Pythagorean triples are similar, for their sides are proportional. More generally, the Pythagorean triples $(6, 8, 10)$ and $(9, 12, 15)$ are not really different, for each arises from $(3, 4, 5)$; however, neither $(6, 8, 10)$ nor $(9, 12, 15)$ is obtained from the other by multiplying its terms by some *integer m*.

Definition. Two Pythagorean triples (a, b, c) and (u, v, z) are ***similar*** if their right triangles $\triangle(a, b, c)$ and $\triangle(u, v, z)$ are similar triangles.

The method of Diophantus will give a formula for certain special Pythagorean triples, and it will then show that every Pythagorean triple is similar to a special one.

Definition. A point (x, y) in the plane is a ***rational point*** if both x and y are rational numbers.

A ***Pythagorean point*** is a rational point in the first quadrant, lying on the unit circle, and above the ***diagonal*** line with equation $y = x$.

Remember that we regard Pythagorean triples (a, b, c) and (b, a, c) as the same. Recall some analytic geometry: if $a \neq b$, $P = (a, b)$, and $Q = (b, a)$, then the diagonal is the perpendicular bisector of the segment $\overline{PQ}$. (The line through P and Q has equation $y = -x + a + b$; it is perpendicular to the diagonal for the product of their slopes is -1; the line intersects the diagonal in the point $(\frac{a+b}{2}, \frac{a+b}{2})$, which is equidistant from P and Q. If $a \leq b$, then P is above the diagonal and Q is below.)

Proposition 1.3. *A triple (a, b, c) of integers is a Pythagorean triple if and only if $(a/c, b/c)$ is a Pythagorean point.*

Proof. Let (a, b, c) be a Pythagorean triple. Dividing both sides of the defining equation $a^2 + b^2 = c^2$ by c^2 gives

$$(a/c)^2 + (b/c)^2 = 1,$$

so that the triple gives an ordered pair of positive *rational* numbers $(x, y) = (a/c, b/c)$ with $x^2 + y^2 = 1$. Thus, the rational point $P = (x, y)$ lies in the first quadrant. As both (a, b, c) and (b, a, c) are the same Pythagorean triple, we may assume that

$$x = a/c \leq b/c = y,$$

so that (x, y) lies above the diagonal line with equation $y = x$. Hence, (x, y) is a Pythagorean point.

Conversely, let's now see that a Pythagorean point (x, y) gives rise to a Pythagorean triple. Write the rational numbers $x \leq y$ with the same denominator, say, $x = a/c$ and $y = b/c$, where a, b, and c are positive integers and $a \leq b \leq c$. Now

$$1 = x^2 + y^2 = \frac{a^2}{c^2} + \frac{b^2}{c^2},$$

so that $a^2 + b^2 = c^2$ and hence (a, b, c) is a Pythagorean triple.   ∎

In summary, the problem of finding all Pythagorean triples corresponds to the problem of finding all Pythagorean points. This is exactly what the geometric idea of Diophantus does. In fact, a Pythagorean point (x, y) gives rise to infinitely many Pythagorean triples. Write the coordinates with another denominator, say $x = u/z$ and $y = v/z$. The calculation at the end of the proof of Proposition 1.3 shows that (u, v, z) is another Pythagorean triple arising from (x, y).

Etymology.   Here are sources of some common words of mathematics.

- **Proposition**. From Latin, meaning a statement or something pictured in the mind.

- **Theorem**. From the Greek word meaning "spectacle" or "contemplate." Related words are "theory" and "theater." Theorems are important propositions.

- **Corollary**. From the Latin word meaning "flower." In ancient Rome, it meant a "gratuity;" flowers were left as tips. In mathematics, corollaries follow easily from theorems; they are gifts bequeathed to us.

- **Lemma**. From Greek; it meant something taken for granted. In mathematics nowadays, it is usually a technical result, a minor theorem, which can be used in the course of proving a more important theorem.

- **Proof**. From Medieval French, meaning an argument from evidence establishing the truth. The adage, "The exception proves the rule," uses the word in the sense of testing: it originally meant a kind of indirect proof. We test whether a rule is true by checking whether an exception to it leads to a contradiction. Nowadays, this adage seems to have lost its meaning.

The Method of Diophantus

Proposition 1.2 parametrizes *all* the points P on the unit circle other than $(-1, 0)$. We are now going to see which values of t produce Pythagorean points: rational points on the unit circle lying in the first quadrant above the diagonal line with equation $y = x$.

Theorem 1.4. *Let $P = (x, y) \neq (-1, 0)$ be a point on the unit circle, and let t be the slope of the line ℓ joining $(-1, 0)$ and P.*

(i) *The slope t is a rational number if and only if P is a rational point.*

(ii) *The point P is a Pythagorean point if and only if t is a rational number satisfying $\sqrt{2} - 1 < t < 1$.*

Proof. (i) The parametrization $P = (x, y)$ gives a pair of equations:

$$x = \frac{1 - t^2}{1 + t^2} \quad \text{and} \quad y = \frac{2t}{1 + t^2}.$$

Clearly, if t is rational, then both x and y are rational. Conversely, if $P = (x, y)$ is a rational point, then the slope t of ℓ is $t = \frac{y - 0}{x - (-1)} = \frac{y}{x+1}$, and so t is a rational number.

(ii) Pythagorean points correspond to rational points on the unit circle that lie in the first quadrant above the line $y = x$. Points on the circle lying in the first quadrant arise from lines having slope t with $0 < t < 1$. The point in the first quadrant that is the intersection of the unit circle and the line $y = x$ is $(\frac{\sqrt{2}}{2}, \frac{\sqrt{2}}{2})$, and the slope of the line joining $(-1, 0)$ to $(\frac{\sqrt{2}}{2}, \frac{\sqrt{2}}{2})$ is

The slope of the line joining $(-1, 0)$ to $(0, 1)$ is 1.

$$t = \frac{\frac{\sqrt{2}}{2}}{1 + \frac{\sqrt{2}}{2}} = \sqrt{2} - 1 \approx .414.$$

Therefore, Pythagorean points correspond to the lines ℓ through $(-1, 0)$ having rational slope t satisfying $\sqrt{2} - 1 < t < 1$. ∎

Let's look at this more closely. If $t = p/q$ is a rational number between $\sqrt{2} - 1$ and 1, then the Pythagorean point it gives can be expressed in terms of p and q:

$$\left(\frac{1-t^2}{1+t^2}, \frac{2t}{1+t^2}\right) = \left(\frac{1-\left(\frac{p}{q}\right)^2}{1+\left(\frac{p}{q}\right)^2}, \frac{2\left(\frac{p}{q}\right)}{1+\left(\frac{p}{q}\right)^2}\right)$$

$$= \left(\frac{q^2-p^2}{q^2+p^2}, \frac{2qp}{q^2+p^2}\right). \tag{1.4}$$

Theorem 1.5 (Diophantus). *Every Pythagorean triple (a, b, c) is similar to a Pythagorean triple of the form*

$$(2qp, q^2 - p^2, q^2 + p^2),$$

where p and q are positive integers with $q > p > \sqrt{2} - 1$.

Proof. Since (a, b, c) is a Pythagorean triple, $P = (a/c, b/c)$ is a Pythagorean point. By Eq. (1.4),

$$\left(\frac{a}{c}, \frac{b}{c}\right) = \left(\frac{1-t^2}{1+t^2}, \frac{2t}{1+t^2}\right) = \left(\frac{q^2-p^2}{q^2+p^2}, \frac{2qp}{q^2+p^2}\right).$$

It follows that $\triangle(a, b, c)$ is similar to $\triangle(2pq, q^2 - p^2, q^2 + p^2)$, because their sides are proportional. Therefore, the Pythagorean triple (a, b, c) is similar to $(2qp, q^2 - p^2, q^2 + p^2)$, as claimed.   ■

How to Think About It. The strategy of Diophantus is quite elegant. The problem of determining all Pythagorean triples is reduced from finding three unknowns, a, b, and c, to two unknowns, $x = a/c$ and $y = b/c$, to one unknown, $t = p/q$. In effect, all Pythagorean triples are parametrized by t; that is, as t varies over all rational numbers between $\sqrt{2} - 1$ and 1, the formulas involving t vary over all Pythagorean points and hence over all Pythagorean triples.

We can now show that the Babylonians had, in fact, found all Pythagorean triples.

Corollary 1.6. *Every Pythagorean triple is similar to one arising from the Babylonian method.*

Proof. By Theorem 1.5, every Pythagorean triple is similar to one of the form $(2qp, q^2 - p^2, q^2 + p^2)$, where $q > p$ are positive integers. If both q and p are even, then we can replace $q > p$ by $\frac{1}{2}q > \frac{1}{2}p$, obtaining a Pythagorean triple $(\frac{1}{4}2qp, \frac{1}{4}(q^2 - p^2), \frac{1}{4}(q^2 + p^2))$ similar to the original one. If both parameters of the new triple are still even, replace $\frac{1}{2}q > \frac{1}{2}p$ by $\frac{1}{4}q > \frac{1}{4}p$. Eventually, we arrive at a Pythagorean triple $(2rs, r^2 - s^2, r^2 + s^2)$, similar to the original triple, that arises from parameters $r > s$, at least one of which is odd.

We are tacitly using a technique of proof called **Infinite Descent**. If, for a given positive integer n with certain properties, there always exists a strictly smaller positive integer n_1 having the same properties, then there are infinitely many such integers. But this is impossible; there are only finitely many integers with $n > n_1 > n_2 > \cdots > 0$.

There are two possibilities. If r and s have different parity, define $m = r+s$ and $n = r-s$. Both m and n are odd, and the Pythagorean triple given by the Babylonian method from $m > n$ is

$$B = \left(mn, \tfrac{1}{2}(m^2 - n^2), \tfrac{1}{2}(m^2 + n^2)\right).$$

Substitute:

$$mn = (r+s)(r-s) = r^2 - s^2, \ \tfrac{1}{2}(m^2 - n^2) = 2rs, \text{ and } \tfrac{1}{2}(m^2 - n^2) = r^2 + s^2.$$

Thus, the Pythagorean triple B is similar to $(2rs, r^2 - s^2, r^2 + s^2)$.

If both r and s are odd, then the Pythagorean triple given by the Babylonian method from $r > s$ is $(rs, \tfrac{1}{2}(r^2 - s^2), \tfrac{1}{2}(r^2 + s^2))$ which is similar to $(2rs, r^2 - s^2, r^2 + s^2)$. ∎

Not every Pythagorean triple (a, b, c) is *equal* to $(2qp, q^2 - p^2, q^2 + p^2)$ for some $q > p$, nor does the theorem say that it is; the theorem asserts only that (a, b, c) is *similar* to a Pythagorean triple arising from the formula. For example, let us show that $(9, 12, 15)$ is not of this form. Since the leg 9 is odd, the even leg 12 must be $2qp$, so that $qp = 6$, and the only possible parameters are $3 > 2$ or $6 > 1$. But $3 > 2$ gives $(5, 12, 13)$ and $6 > 1$ gives $(12, 35, 37)$, neither of which is similar to $(9, 12, 15)$. However, $(9, 12, 15)$ is similar to $(3, 4, 5)$, and $(3, 4, 5)$ arises from $2 > 1$.

A Pythagorean triple (a, b, c) is **primitive** if there is no integer $d > 1$ that is a divisor of $a, b,$ and c. Thus, $(3, 4, 5)$ is primitive but $(9, 12, 15)$ is not. In Theorem 1.25, we'll give a rigorous proof that every Pythagorean triple is similar to exactly one primitive Pythagorean triple.

Exercises

1.19 Find q and p in Theorem 1.5 for each of the following Pythagorean triples.

(i) $(7, 24, 25)$.

Answer. $q = 5$ and $p = 3$.

(ii) $(129396, 261547, 291805)$.

Answer. $q = 526$ and $p = 123$.

1.20 * Show that every Pythagorean triple (x, y, z) with x, y, z having no common factor $d > 1$ is of the form

$$(r^2 - s^2, \ 2rs, \ r^2 + s^2)$$

for positive integers $r > s$ having no common factor > 1; that is,

$$x = r^2 - s^2, \quad y = 2rs, \quad z = r^2 + s^2.$$

1.21 A line in the plane with equation $y = mx + c$ is called a **rational line** if m and c are rational numbers. If P and Q are distinct rational points, prove that the line joining them is a rational line.

1.22 A **lattice point** is a point in the plane whose coordinates are integers. Let $P = (x, y)$ be a Pythagorean point and ℓ the line through P and the origin. Prove that if $Q = (a, b)$ is a lattice point on ℓ and c is the distance from Q to the origin, then (a, b, c) is a Pythagorean triple.

1.23 * Let $P = (x_0, y_0)$ be a Pythagorean point and L the line joining P and the origin (so the equation of L is $y = mx$, where $m = y_0/x_0$). Show that if $(a/c, b/c)$ is a rational point on L, then (a, b, c) is a Pythagorean triple.

1.24 Does every rational point in the plane correspond to a Pythagorean point? If so, prove it. If not, characterize the ones that do.

Answer. No. For example, $(\frac{1}{2}, \frac{1}{2})$ does not correspond.

1.25 * Prove the identity $\left(x^2 + y^2\right)^2 = \left(x^2 - y^2\right)^2 + (2xy)^2$.

1.26 *

   (i) Show that the same number can occur as a leg in two nonsimilar Pythagorean triangles.

   (ii) Prove that the area of $\triangle(a, b, c)$, a right triangle with integer side lengths, is an integer.

   (iii) A **Heron triangle** is a triangle with integer side lengths and area. Find a Heron triangle that is not a right triangle.

   Hint. Use parts (i) and (ii).

1.27 Show that every integer $n \geq 3$ occurs as a leg of some Pythagorean triple.

   Hint. The cases n even and n odd should be done separately.

1.28 Distinct Pythagorean triples can have the same hypotenuse: both $(33, 56, 65)$ and $(16, 63, 65)$ are Pythagorean triples. Find another pair of distinct Pythagorean triples having the same hypotenuse.

1.29 * If $(\cos \theta, \sin \theta)$ is a rational point, prove that both $\cos(\theta + 30°)$ and $\sin(\theta + 30°)$ are irrational.

Fermat's Last Theorem

About fourteen centuries after Diophantus, Fermat (1601–1665) proved that there are no positive integers a, b, c with $a^4 + b^4 = c^4$. He was studying his copy of Diophantus's *Arithmetica*, published in 1621, and he wrote in its margin,

> ... it is impossible for a cube to be written as a sum of two cubes or a fourth power to be written as a sum of two fourth powers or, in general, for any number which is a power greater than the second to be written as a sum of two like powers. I have discovered a truly marvelous demonstration of this proposition which this margin is too narrow to contain.

Fermat was not the first mathematician to write a marginal note in a copy of Diophantus. Next to the same problem, the Byzantine mathematician Maximus Planudes wrote, *Thy soul, Diophantus, be with Satan because of the difficulty of your theorems.*

Fermat never returned to this problem (at least, not publicly) except for his proof of the case $n = 4$, which we give below. The statement: *If $n > 2$, there are no positive integers a, b, c with $a^n + b^n = c^n$*, was called **Fermat's Last Theorem**, perhaps in jest. The original text in which Fermat wrote his famous marginal note is lost today. Fermat's son edited the next edition of Diophantus, published in 1670; this version contains Fermat's annotations, including his famous "Last Theorem;" it contained other unproved assertions as well, most true, some not. By the early 1800s, only Fermat's Last Theorem remained undecided. It became a famous problem, resisting the attempts of mathematicians of the highest order for 350 years, until it was finally proved, in 1995, by Wiles. His proof is very sophisticated, and most mathematicians believe that Fermat did not have a correct proof. The quest for a proof of Fermat's Last Theorem generated much beautiful mathematics. In particular, it led to an understanding of complex numbers, factorization, and polynomials. We'll see, in the Epilog, that extending the method of Diophantus from quadratics to cubics involves *elliptic curves*, the study of which is the setting for Wiles' proof of Fermat's Last Theorem.

Fermat proved the next theorem (which implies the case $n = 4$ of Fermat's Last Theorem) because he was interested in the geometric problem of determining which right triangles having all sides of rational length have integer area (we'll soon discuss this problem in more detail).

Theorem 1.7 (Fermat). *There is no triple (x, y, z) of positive integers with*

$$x^4 + y^4 = z^2. \tag{1.5}$$

Proof. The proof will be by infinite descent (Fermat invented infinite descent for this very problem). Given a triple of positive integers (x, y, z) satisfying Eq. (1.5), we'll show there is another triple (u, v, w) of the same sort with $w < z$, and so repeating this process leads to a contradiction.

Let's say that integers x and y are **relatively prime** if there is no integer $d > 1$ dividing both of them; that is, it's not true that $x = da$ and $y = db$.

We can assume that x and y are relatively prime, for otherwise a common factor of x and y would also be a factor of z, and we could divide it out. It follows (and we'll prove it in the next chapter) that x^2 and y^2 are also relatively prime. And note that $x^4 + y^4 = z^2$ implies that

$$\left(x^2\right)^2 + \left(y^2\right)^2 = z^2$$

so that (x^2, y^2, z) form a Pythagorean triple.

We also observe that x^2 and y^2 can't both be odd; if $x^2 = 2k + 1$ and $y^2 = 2j + 1$, then

$$(2k + 1)^2 + (2j + 1)^2 = z^2.$$

Expanding and collecting terms gives $z^2 = 4h + 2$ for some integer h. But you can check that the square of any integer is either of the form $4h$ or $4h + 1$.

We can now assume that (x^2, y^2, z) is a Pythagorean triple in which x and y are relatively prime, x is odd, and y is even. By Exercise 1.20 on page 13, there are relatively prime integers r and s with $r > s > 0$ such that

$$x^2 = r^2 - s^2, \ y^2 = 2rs, \text{ and } z = r^2 + s^2.$$

The first equation says that $x^2 + s^2 = r^2$; that is, (x, s, r) is another Pythagorean triple with x odd. Moreover, x and s have no common factor (why?), so that Exercise 1.20 gives relatively prime integers a and b such that

$$x = a^2 - b^2, \ s = 2ab, \text{ and } r = a^2 + b^2.$$

Now,

$$y^2 = 2rs = 2(a^2 + b^2)(2ab) = 4ab(a^2 + b^2).$$

Since y is even, we have an equation in integers:

$$\left(\frac{y}{2}\right)^2 = ab(a^2 + b^2). \tag{1.6}$$

This proof is not difficult, but it uses several elementary divisibility results we'll prove later. Since we feel that this is the appropriate place for this theorem, we'll just refer to the needed things.

As a and b are relatively prime (no common factor $d > 1$), each pair from the three factors on the right-hand side of Eq. (1.6) is relatively prime. Since the left-hand side $(y/2)^2$ is a square, each factor on the right is a square (Exercise 2.12 on page 59). In other words, there are integers u, v, and w such that

$$a = u^2, \ b = v^2, \text{ and } a^2 + b^2 = w^2.$$

And, since a and b are relatively prime, so, too, are u and v relatively prime. Hence, we have

$$u^4 + v^4 = w^2.$$

This is our "smaller" solution to Eq. (1.5), for

$$0 < w < w^2 = a^2 + b^2 = r < r^2 < r^2 + s^2 = z.$$

We can now repeat this process on (u, v, w). By infinite descent, there is no solution to Eq. (1.5). ∎

Corollary 1.8 (Fermat's Last Theorem for Exponent 4). *There are no positive integers x, y, z with*

$$x^4 + y^4 = z^4.$$

Proof. If such a triple existed, we'd have

$$x^4 + y^4 = \left(z^2\right)^2,$$

and that's impossible, by Theorem 1.7. ∎

Call an integer $n \geq 2$ *good* if there are no positive integers a, b, c with $a^n + b^n = c^n$. If n is good, then so is any multiple nk of it. Otherwise, there are positive integers r, s, t with $r^{nk} + s^{nk} = t^{nk}$, and this gives the contradiction $a^n + b^n = c^n$, where $a = r^k$, $b = s^k$, and $c = t^k$. For example, Corollary 1.8 shows that that any positive integer of the form $4k$ is good. Since every $n \geq 2$ is a product of primes, it follows that Fermat's Last Theorem would be true if every odd prime is good.

Connections: Congruent Numbers

Fermat's motivation for Theorem 1.7 came, not from a desire to prove there are no non-trivial integer solutions to $x^4 + y^4 = z^4$, but from a problem in the intersection of arithmetic and geometry. In more detail, suppose that $\triangle = \triangle(a, b, c)$ is the right triangle arising from a Pythagorean triple (a, b, c). Since $\triangle$ is a right triangle, the leg a is an altitude and the area of $\triangle(a, b, c)$ is $\frac{1}{2}ab$; since (a, b, c) is a Pythagorean triple, the area is an integer (Exercise 1.8 on page 6). Tipping this statement on its head, we ask which integers are areas of right triangles having integer side-lengths. Certainly 6 is, because it's the area of $\triangle(3, 4, 5)$. But 5 is not the area of such a triangle (Exercise 1.9 on page 7).

However, we claim that 5 *is* the area of a right triangle whose side-lengths are *rational* numbers. Consider the Pythagorean triple $(9, 40, 41)$; its right triangle $\triangle = \triangle(9, 40, 41)$ has area $\frac{1}{2}(9 \cdot 40) = 180$. Now $180 = 36 \cdot 5$. Scaling the side-lengths of $\triangle$ by $\frac{1}{6}$ scales the area by $\frac{1}{36}$, so that $\triangle\left(\frac{3}{2}, \frac{20}{3}, \frac{41}{6}\right)$ has area $180/36 = 5$.

So, the question arises: "Is every integer the area of a right triangle with rational side-lengths?" Fermat showed that 1 and 2 are not, and his proof for 2 involved Eq. (1.5).

Theorem 1.9. *There is no right triangle with rational side-lengths and area* 2.

Proof. Suppose, on the contrary, that the rational numbers r, s, t are the lengths of the sides of a right triangle with area 2. Then we have two equations:

$$r^2 + s^2 = t^2$$
$$\tfrac{1}{2}rs = 2.$$

Multiply the first equation by r^2 to obtain

$$r^4 + (rs)^2 = (rt)^2,$$

so that (since $rs = 4$),

$$r^4 + 2^4 = (rt)^2.$$

Write the rational numbers r and t as fractions with the same denominator: $r = a/c$ and $t = b/c$. When we clear denominators, we get $a^4 + z^4c^4 = t^2$, an equation in integers x, y, z of the form

$$x^4 + y^4 = z^2.$$

This is Eq. (1.5), and Theorem 1.7 says that this cannot occur. ∎

So, not every positive integer is the area of a right triangle with rational side-lengths.

Definition. A *congruent number* is a positive integer n that is the area of a right triangle having rational side-lengths.

Theorem 1.9 says that 2 is not a congruent number. Using similar ideas, Fermat showed that 1 is not a congruent number (Exercise 1.31 below).

One way to generate congruent numbers is to scale a Pythagorean triple using the largest perfect square that divides its area. For example, the area of $\triangle(7, 24, 25)$ is $84 = 2^2 \cdot 21$. Since $4 = 2^2$ is the largest perfect square in 84, scaling the sides by 2 will produce a triangle of area 21, so that 21 is the area of $\triangle(\frac{7}{2}, 12, \frac{25}{2})$ and, hence, 21 is a congruent number. More generally, we have

We have already used this method on the Pythagorean triple $(9, 40, 41)$ when we showed that 5 is a congruent number.

Proposition 1.10. *Let* (a, b, c) *be a Pythagorean triple. If its right triangle* $\triangle(a, b, c)$ *has area* m^2n, *where* n *is squarefree, then* n *is a congruent number. Moreover, every squarefree congruent number is obtained in this way.*

Proof. Since (a, b, c) is a Pythagorean triple, $\triangle = \triangle(a, b, c)$ is a right triangle. Now area$(\triangle) = m^2n = \frac{1}{2}ab$, so that

$$\text{area}\left(\triangle(\tfrac{a}{m}, \tfrac{b}{m}, \tfrac{c}{m})\right) = \tfrac{1}{2}\left(\frac{a}{m}\frac{b}{m}\right) = \frac{m^2n}{m^2} = n,$$

and so n is a congruent number.

Conversely, if n is a square-free congruent number, then there are rational numbers r, s, and t so that

$$r^2 + s^2 = t^2$$
$$\tfrac{1}{2}rs = n.$$

Clearing denominators, we find integers a, b, c, and m so that

$$a^2 + b^2 = c^2$$
$$\tfrac{1}{2}ab = m^2 n. \quad \blacksquare$$

The first few congruent numbers are

$$5, \ 6, \ 7, \ 13, \ 14, \ 15, \ 20, \ 21, \ 22, \ 23.$$

In light of Exercise 1.33 on page 20, we now have a method for determining all congruent numbers: generate the areas of all Pythagorean triangles (we know how to do that), and then divide out its largest perfect square factor: case closed.

Not quite. The trouble with this method is that you have no idea how many triangle areas to calculate before (if ever) you get to an area of $m^2 n$ for a particular n. For some congruent numbers, it takes a *long* time. For example, 157 is a congruent number, but the smallest rational right triangle with area 157 has side lengths

$$\frac{224403517704336969924557513090674863160948472041}{8912332268928859588025535178967163570016480830},$$

$$\frac{6803298487826435051217540}{411340519227716149383203}, \quad \frac{411340519227716149383203}{21666555693714761309610}.$$

A method for effectively determining whether or not an integer is a congruent number is an unsolved problem (this problem is at least a thousand years old, for historians have found it in manuscripts dating from the late tenth century). A detailed discussion of the ***Congruent Number Problem*** is in [19].

This triangle was found by Don Zagier, using sophisticated techniques investigating elliptic curves, and using a substantial amount of computer power (see [19] for more details).

A readable account of the congruent problem, with more examples than we provide here, can be found at www.math.uconn.edu/~kconrad/blurbs/

How to Think About It. Proposition 1.10 shows that every squarefree congruent number n is the area of a scaled Pythagorean triangle. But there might be more than one Pythagorean triangle whose area has n as its squarefree part. The search for more than one rational right triangle with the same area leads to some fantastic calculations. For example, we saw that 5 is the area of $\triangle \left(\frac{3}{2}, \frac{20}{3}, \frac{41}{6} \right)$, which comes from the Pythagorean triangle $\triangle(9, 40, 41)$ whose area is $5 \cdot 6^2$. But 5 is also the area of

$$\triangle \left(\frac{1519}{492}, \frac{4920}{1519}, \frac{3344161}{747348} \right),$$

and this comes from the Pythagorean triangle $\triangle(2420640, 2307361, 3344161)$ whose area is $5 \cdot 747348^2$.

As usual, this isn't magic; in Chapter 9, we'll show how to find infinitely many rational right triangles with the same congruent number as area.

There's a surprising connection between congruent numbers and 3-term arithmetic sequences of perfect squares of rational numbers: positive rationals $s^2 < t^2 < u^2$ with $u^2 - t^2 = t^2 - s^2$, like $1, 25, 49$ (common difference 24) and

$$\frac{961}{36}, \quad \frac{1681}{36}, \quad \frac{2401}{36}$$

(common difference 20). Note that $24 = 4 \cdot 6$ and $20 = 4 \cdot 5$. So, for these examples, at least, the common difference is 4 times a congruent number. This suggests that something's going on. One approach is due to Fibonacci.

Our two equations

$$a^2 + b^2 = c^2$$
$$ab = 2n$$

might lead us to think that we could find a and b by finding their sum and product, for this would lead to a quadratic equation whose roots are a and b.

Well, we know ab, and

$$(a + b)^2 = a^2 + b^2 + 2ab = c^2 + 4n.$$

So, $a + b = \sqrt{c^2 + 4n}$ (why can we take the positive square root?), and hence a and b are roots of the quadratic equation

$$x^2 - \sqrt{c^2 + 4n}\, x + 2n.$$

The quadratic formula gives us a and b:

$$a = \frac{\sqrt{c^2 + 4n} + \sqrt{(c^2 + 4n) - 4(2n)}}{2} = \frac{\sqrt{c^2 + 4n} + \sqrt{c^2 - 4n}}{2}$$

and

$$b = \frac{\sqrt{c^2 + 4n} - \sqrt{(c^2 + 4n) - 4(2n)}}{2} = \frac{\sqrt{c^2 + 4n} - \sqrt{c^2 - 4n}}{2}.$$

But we want a and b to be rational, so we want $c^2 \pm 4n$ to be perfect squares. That produces an arithmetic sequence of three perfect squares:

$$c^2 - 4n, \quad c^2, \quad c^2 + 4n.$$

There are details to settle, but that's the gist of the proof of the following theorem.

Theorem 1.11. *An integer n is a congruent number if and only of there is a 3-term arithmetic sequence of perfect squares whose common difference is $4n$.*

rational

Exercises

1.30 * Show that 1 is not a congruent number.

1.31 Show that there are no positive rational numbers x and y so that

$$x^4 \pm 1 = y^2.$$

1.32 Show that if n is a congruent number and m is an integer, then $m^2 n$ is also a congruent number.

1.33 Show that there are no right triangles with rational side-lengths whose area is a perfect square or twice a perfect square.

1.34 Show that 7 and 14 are congruent numbers.

1.35 Take It Further. Show that 13 is a congruent number.

1.36 * Prove Theorem 1.11.

1.3   Euclid

Euclid of Alexandria (ca. 325 BCE–ca.265 BCE) is one of the most prominent mathematicians of antiquity. He is best known for *The Elements*, his treatise consisting of thirteen books: six on plane geometry, four on number theory, and three on solid geometry. *The Elements* has been used for over two thousand years, which must make Euclid the leading mathematics teacher of all time. We do not know much about Euclid himself other than that he taught in Alexandria, Egypt around 270 BCE. We quote from Sir Thomas Heath [16], the great translator and commentator on *The Elements*.

Pappus (ca. 290 CE–ca. 350 CE), was one of the last great classic geometers.

> *It is most probable that Euclid received his mathematical training in Athens from the students of Plato; for most of the geometers who could have taught him were of that school ... Pappus says ... such was (Euclid's) scrupulous fairness and his exemplary kindliness towards all who could advance mathematical science to however small an extent; (he was) in no wise contentious and, though exact, yet no braggart.*

Eight hundred years after Euclid, Proclus (412 CE–485 CE) wrote:

> *Not much younger than these (pupils of Plato) is Euclid, who put together* The Elements, *collecting many of Eudoxus's theorems, perfecting many of Theaetetus's, and also bringing to irrefragable demonstration the things which were only somewhat loosely proved by his predecessors. This man lived in the time of the first Ptolemy (323 BCE − 283 BCE). For Archimedes, who came immediately after the first Ptolemy makes mention of Euclid; and further they say that Ptolemy once asked him if there were a shorter way to study geometry than* The Elements, *to which he replied that there was no royal road to geometry. He is therefore younger than Plato's circle, but older than Eratosthenes and Archimedes; for these were contemporaries, as Eratosthenes somewhere says.*

The Elements is remarkable for the clarity with which its theorems are stated and proved. The standard of rigor was a goal (rarely achieved!) for the inventors of calculus centuries later. As Heath writes in the preface to the second edition of his translation [16] of *The Elements*,

> *... so long as mathematics is studied, mathematicians will find it necessary and worthwhile to come back again and again ... to the twenty-two-centuries-old book which, notwithstanding its imperfections, remains the greatest elementary textbook in mathematics that the world is privileged to possess.*

More than one thousand editions of *The Elements* have been published since it was first printed in 1482. In the *Encyclopedia Britannica*, van der Waerden wrote,

> *Almost from the time of its writing and lasting almost to the present,* The Elements *has exerted a continuous and major influence on human affairs. It was the primary source of geometric reasoning, theorems, and methods at least until the advent of non-Euclidean geometry in the 19th century. It is sometimes said that, next to the Bible,* The Elements *may be the most translated, published, and studied of all the books produced in the Western world.*

Greek Number Theory

In spite of the glowing reviews of *The Elements*, we must deviate a bit from Euclid, for the Greeks, and Euclid in particular, recognized neither negative numbers nor zero.

Notation. The *natural numbers* is the set

$$\mathbb{N} = \{0, 1, 2, 3, \ldots\}.$$

The set of all *integers*, positive, negative, and 0, is denoted by

$$\mathbb{Z} = \{\pm n : n \in \mathbb{N}\}.$$

The set of integers is denoted by $\mathbb{Z}$ because the German word for numbers is Zahlen.

We are going to assume that the set $\mathbb{N}$ of natural numbers satisfies a certain property—a generalized version of Infinite Descent.

Definition. The *Least Integer Axiom* (often called the *Well-Ordering Axiom*) states that every nonempty collection C of natural numbers contains a smallest element; that is, there is a number $c_0 \in C$ with $c_0 \le c$ for all $c \in C$.

Note that the set of positive rationals $\mathbb{Q}^+$ does not satisfy an analogous property: the nonempty subset $\{x \in \mathbb{Q}^+ : x^2 > 2\}$ contains no smallest element.

This axiom is surely plausible. If $0 \in C$, then $c_0 = 0$. If $0 \notin C$ and $1 \in C$, then $c_0 = 1$. If $0, 1 \notin C$ and $2 \in C$, then $c_0 = 2$. Since C is not empty, you will eventually bump into C, and c_0 is the first number you'll meet.

We now define some familiar terms.

Definition. If a and b are integers, then a *divides* b, denoted by

$$a \mid b,$$

if there is an *integer* c with $b = ca$. We also say that a is a *divisor* (or a *factor*) of b, and that b is a *multiple* of a.

$b = ca \Rightarrow \frac{b}{a} = c$

Example 1.12. Consider some special cases. Every number a divides itself, for $a = a \times 1$; similarly, 1 divides every number. Every number a divides 0: taking $c = 0$, we have $0 = a \times 0$. On the other hand, if 0 divides b, then $b = 0$, for $b = 0 \times c = 0$. Note that $3 \mid 6$, because $6 = 3 \times 2$, but $3 \nmid 5$ (that is, 3 does not divide 5): even though $5 = 3 \times \frac{5}{3}$, the fraction $\frac{5}{3}$ is not an integer. ▲

Note that 0 divides itself: $0 \mid 0$ is true. Do not confuse the notation $a \mid b$, which is the relation "a is a divisor of b" with a/b, which is a number. In particular, we are not saying that $0/0$ is a number.

Lemma 1.13. *If a and b are positive integers and $a \mid b$, then $a \le b$.*

Inequalities are discussed in Appendix A. 4.

Proof. There is a positive integer c with $b = ca$; note that $1 \leq c$, for 1 is the smallest positive integer. Multiplying by the positive number a, we have $a \leq ac = b$. ∎

Every integer a has $1, -1, a, -a$ as divisors. A positive integer $a \neq 1$ having only these divisors is called *prime*.

Definition. An integer a is *prime* if $a \geq 2$ and its only divisors are ± 1 and $\pm a$; if $a \geq 2$ has other divisors, then it is called *composite*.

The first few primes are $2, 3, 5, 7, 11, 13, \ldots$. We will soon see that there are infinitely many primes.

The reason we do not consider 1 to be a prime is that theorems about primes would then require special cases treating the behavior of 1. For example, we will prove later that every positive integer $a \geq 2$ has exactly one factorization of the form $a = p_1 p_2 \cdots p_t$, where $p_1 \leq p_2 \leq \cdots \leq p_t$ are primes. This statement would be more complicated if we allowed 1 to be a prime.

We allow products having only one factor; it's okay to say that a single prime is a product of primes.

Proposition 1.14. *Every integer $a \geq 2$ is a product of primes.*

Proof. Let C be the set of all natural numbers $a \geq 2$ that are not products of primes. If the proposition is false, then C is nonempty, and the Least Integer Axiom gives a smallest such integer, say, c_0. Since $c_0 \in C$, it is not prime; hence, it factors, say, $c_0 = ab$, where $a, b \neq 1$. As $a \mid c_0$, we have $a \leq c_0$, by Lemma 1.13; but $a \neq c_0$, lest $b = 1$, so that $a < c_0$. Therefore, $a \notin C$, for c_0 is the smallest number in C, and so a is a product of primes: $a = p_1 \cdots p_m$ for $m \geq 1$. Similarly, b is a product of primes: $b = q_1 \cdots q_n$. Therefore, $c_0 = ab = p_1 \cdots p_m q_1 \cdots q_n$ is a product of primes, a contradiction, and so C is empty. ∎

Division and Remainders

Dividing an integer b by a positive integer a gives

$$b/a = q + r/a,$$

where q is an integer and $0 \leq r/a < 1$. If we clear denominators, we get the statement $b = qa + r$ which involves only integers. For example, $\frac{22}{5} = 4 + \frac{2}{5}$ becomes $22 = 4 \cdot 5 + 2$.

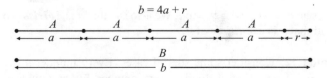

Figure 1.8. Division Algorithm.

Euclid viewed division geometrically, as in Figure 1.8. Suppose B is a line segment of length b, and that A is a shorter segment of length a. Lay off copies of A along B as long as possible. If there's nothing left over, then a is a divisor of b; if some segment of length, say, r, is left over, then r is the remainder.

Theorem 1.15 (Division Algorithm). *If a and b are positive integers, then there are unique (i.e.,exactly one) integers q (the **quotient**) and r (the **remainder**) with*

$$b = qa + r \quad and \quad 0 \le r < a.$$

Proof. We first prove that q and r exist; afterward, we'll prove their uniqueness.

If $b < a$, set $q = 0$ and $r = b$. Now $b = qa + r = 0 \cdot a + b$, while $0 \le b < a$. Hence, we may assume that $b \ge a$; that is, $b - a \ge 0$. Consider the sequence $b > b - a > b - 2a > b - 3a > \cdots$. There must be an integer $q \ge 1$ with $b - qa \ge 0$ and $b - (q + 1)a < 0$ (this is just *Infinite Descent*, described on page 12; in more down-to-earth language, there can be at most b steps before this sequence becomes negative). If we define $r = b - qa$, then $b = qa + r$. We also have the desired inequalities. Clearly, $0 \le r$. If $r = b - qa \ge a$, then $b - qa - a \ge 0$; that is, $b - (q + 1)a \ge 0$, contradicting the definition of q.

Let's prove uniqueness. If there are integers Q and R with $b = Qa + R$ and $0 \le R < a$, then $qa + r = b = Qa + R$ and

$$(Q - q)a = r - R.$$

If $Q \ne q$, there is no loss in generality in assuming that $Q > q$; that is, $0 < Q - q$. By Lemma 1.13, $a \le (Q - q)a = r - R$. But $r < a$ and $R \ge 0$ gives $r - R < r < a$. Therefore, $a \le (Q - q)a = r - R < a$; that is, $a < a$, a contradiction. Hence $Q = q$. It follows that $R = r$, and we are done. ∎

For example, there are only two possible remainders after dividing by 2, namely, 0 and 1. An integer b is even if the remainder is 0; b is odd if the remainder is 1. Thus, either $b = 2q$ or $b = 2q + 1$.

The equation $b = qa + r$ is of no value at all without the restriction on the remainder r. For example, the equations $1000 = 3 \cdot 25 + 925$ and $1000 = 2 \cdot 53 + 894$ are true and useless.

How to Think About It. We have been trained to regard the quotient q as more important than the remainder; r is just the little bit left over. But our viewpoint now is just the reverse. Given a and b, the important question for us is whether a is a divisor of b. The remainder is the **obstruction**: $a \mid b$ if and only if $r = 0$. This will be a common strategy: to see whether $a \mid b$, use the Division Algorithm to get $b = qa + r$, and then try to show that $r = 0$.

The next result shows that there is no largest prime. The proof shows, given any finite set of primes, that there always exists another one.

Corollary 1.16. *There are infinitely many primes.*

Proof. (**Euclid**) Suppose, on the contrary, that there are only finitely many primes. If $p_1, p_2, \ldots, p_k$ is the complete list of all the primes, define

$$M = (p_1 \cdots p_k) + 1.$$

By Proposition 1.14, M is a product of primes. But M has no prime divisor p_i, for dividing M by p_i gives remainder 1 and not 0. For example, dividing

The hypothesis of Theorem 1.15 can be weakened to $a, b \in \mathbb{Z}$ and $a \ne 0$; the inequalities for the remainder now read $0 \le r < |a|$.

If $a \le b$, the quotient q is the largest multiple qa with $qa \le b$. This is very much the way young children are taught to find the integer quotient in division when a and b are small.

M by p_1 gives $M = p_1(p_2 \cdots p_k) + 1$, so that the quotient and remainder are $q = p_2 \cdots p_k$ and $r = 1$; dividing M by p_2 gives $M = p_2(p_1 p_3 \cdots p_k) + 1$, so that $q = p_1 p_3 \cdots p_k$ and $r = 1$; and so forth. The assumption that there are only finitely many primes leads to a contradiction, and so there must be an infinite number of them.   ■

Linear Combinations and Euclid's Lemma

The greatest common divisor of two integers is a fundamental tool in studying factorization.

Definition. A *common divisor* of integers a and b is an integer c with $c \mid a$ and $c \mid b$. The *greatest common divisor* of a and b, denoted by $\gcd(a, b)$ (or, more briefly, by (a, b)), is defined by

$$\gcd(a, b) = \begin{cases} 0 & \text{if } a = 0 = b \\ \text{the largest common divisor of } a \text{ and } b & \text{otherwise.} \end{cases}$$

We saw, in Lemma 1.13, that if a and m are positive integers with $a \mid m$, then $a \leq m$. It follows, if at least one of a, b is not zero, that gcd's exist: there are always common divisors (1 is always a common divisor), and there are only finitely many positive common divisors $\leq \max\{|a|, |b|\}$.

Lemma 1.17. *If p is a prime and b is an integer, then*

$$\gcd(p, b) = \begin{cases} p & \text{if } p \mid b \\ 1 & \text{otherwise.} \end{cases}$$

Proof. A common divisor c of p and b is, in particular, a divisor of p. But the only positive divisors of p are p and 1, and so $\gcd(p, b) = p$ or 1; it is p if $p \mid b$, and it is 1 otherwise.   ■

If $b \geq 0$, then $\gcd(0, b) = b$ (why?).

Definition. A *linear combination* of integers a and b is an integer of the form

$$sa + tb,$$

where $s, t \in \mathbb{Z}$ (the numbers s, t are allowed to be negative).

Example 1.18. The equation $b = qa + r$ in the Division Algorithm displays b as a linear combination of a and r (for $b = qa + 1 \cdot r$). Note that 0 is a linear combination of any pair of integers: $0 = 0 \cdot a + 0 \cdot b$. There are infinitely many linear combinations of 12 and 16, each of which is divisible by 4 (why?). It follows that 5, for example, is not such a linear combination.   ▲

The next result is one of the most useful properties of gcd's.

See Exercise 1.47 on page 30.

Theorem 1.19. *If a and b are integers, then $\gcd(a, b)$ is a linear combination of a and b.*

Proof. We may assume that at least one of a and b is not zero (otherwise, the gcd is 0 and the result is obvious). Consider the set I of all the linear combinations of a and b:

$$I = \{sa + tb : s, t \in \mathbb{Z}\}.$$

Both a and b are in I (take $s = 1$ and $t = 0$ or vice versa). It follows that I contains positive integers (if $a < 0$, then $-a$ is positive, and I contains $-a = (-1)a + 0b$); hence, the set C of all those positive integers lying in I is nonempty. By the Least Integer Axiom, C contains a smallest positive integer, say, d; we claim that d is the gcd.

Since $d \in I$, it is a linear combination of a and b: there are integers s and t with

$$d = sa + tb.$$

We'll show that d is a common divisor by trying to divide each of a and b by d. The Division Algorithm gives integers q and r with $a = qd + r$, where $0 \leq r < d$. If $r > 0$, then

$$r = a - qd = a - q(sa + tb) = (1 - qs)a + (-qt)b \in C,$$

contradicting d being the smallest element of C. Hence $r = 0$ and $d \mid a$; a similar argument shows that $d \mid b$.

Finally, if c is a common divisor of a and b, then Exercise 1.46 on page 29 shows that c divides every linear combination of a and b; in particular, $c \mid d$. By Lemma 1.13, we have $c \leq d$. ∎

If $d = \gcd(a, b)$ and if c is a common divisor of a and b, then $c \leq d$, by Lemma 1.13. The next corollary shows that more is true: c is a divisor of d; that is, $c \mid d$ for every common divisor c.

Corollary 1.20. *Let a and b be integers. A nonnegative common divisor d is their gcd if and only if $c \mid d$ for every common divisor c of a and b.*

Proof. Necessity (the implication $\Rightarrow$). We showed that any common divisor of a and b divides $\gcd(a, b)$ at the end of the proof of Theorem 1.19.

Sufficiency (the implication $\Leftarrow$). Let $d = \gcd(a, b)$, and let $D \geq 0$ be a common divisor of a and b with $c \mid D$ for every common divisor c of a and b. Now D is a common divisor, so that $d \mid D$, by hypothesis; hence, $d \leq D$, by Lemma 1.13. But the definition of gcd (d is the *greatest* common divisor) gives $D \leq d$, and so $D = d$. ∎

The next theorem is of great interest: not only is it very useful, but it also characterizes prime numbers.

Theorem 1.21 (Euclid's Lemma). *If p is a prime and $p \mid ab$ for integers a, b, then $p \mid a$ or $p \mid b$. Conversely, if $m \geq 2$ is an integer such that $m \mid ab$ always implies $m \mid a$ or $m \mid b$, then m is a prime.*

Proof. ($\Rightarrow$): Suppose that $p \mid ab$ and that $p \nmid a$; that is, p does not divide a; we must show that $p \mid b$. Since $\gcd(p, a) = 1$ (by Lemma 1.17), Theorem 1.19 gives integers s and t with $1 = sp + ta$. Hence,

$$b = spb + tab.$$

The proof of Theorem 1.19 contains an idea that will be used again, as in Exercise 1.49 on page 30.

In other words, d is the smallest positive linear combination of a and b.

In some treatments of number theory, Corollary 1.20 is taken as the *definition* of gcd. Later, we will want to define greatest common divisor in other algebraic structures. It often will not make sense to say that one element of such a structure is greater than another, but it will make sense to say that one element divides another. Corollary 1.20 will allow us to extend the notion of gcd.

The *contrapositive* of a statement P implies Q is "not Q" implies "not P". For example, the contrapositive of "If I live in Chicago, then I live in Illinois" is "If I don't live in Illinois, then I don't live in Chicago." A statement and its contrapositive are either both true or both false. Thus, to prove a statement, it suffices to prove its contrapositive.

Now p divides both expressions on the right, for $p \mid ab$, and so $p \mid b$, by Exercise 1.46 on page 29.

($\Leftarrow$): We prove the contrapositive. If m is composite, then $m = ab$, where $a < m$ and $b < m$. Now $m \mid m = ab$, but $m \nmid a$ and $m \nmid b$, by Lemma 1.13. Thus, m divides a product but it divides neither factor. ■

To illustrate: $6 \mid 12$ and $12 = 4 \times 3$, but $6 \nmid 4$ and $6 \nmid 3$. Of course, 6 is not prime.

We will generalize Euclid's Lemma in the next chapter. Theorem 2.8 says that if p is a prime and $p \mid a_1 \cdots a_n$ for integers $a_1, \ldots, a_n$, where $n \geq 2$, then $p \mid a_i$ for some i.

Definition. Call integers a and b **relatively prime** if their gcd is 1.

Thus, a and b are relatively prime if their only common divisors are ± 1. For example, 2 and 3 are relatively prime, as are 8 and 15.

Here is a generalization of Euclid's Lemma having the same proof.

Corollary 1.22. *Let a, b, and c be integers. If c and a are relatively prime and $c \mid ab$, then $c \mid b$.*

Proof. Theorem 1.19 gives integers s and t with $1 = sc + ta$. Hence, $b = scb + tab$. Now c divides both expressions on the right, for $c \mid ab$, and so $c \mid b$, by Exercise 1.46 on page 29. ■

How to Think About It.

We have just seen one reason why it is important to know proofs: Corollary 1.22 does not follow from the statement of Euclid's Lemma, but it does follow from its proof. See Exercise 1.54 on page 34 for another example of this.

Proposition 1.23. *Let a and b be integers.*

(i) $\gcd(a, b) = 1$ *(that is, a and b are relatively prime) if and only if 1 is a linear combination of a and b.*

(ii) *If $d = \gcd(a, b) \neq 0$, then the integers a/d and b/d are relatively prime.*

Proof. (i) By Theorem 1.19, the gcd d is a linear combination; here, $d = 1$. Conversely, if $1 = sa + tb$ and c is a common divisor of a and b, then $c \mid 1$, by Exercise 1.46 on page 29. Hence, $c = \pm 1$.

(ii) There are integers s and t with $d = sa + tb$. Divide both sides by $d = \gcd(a, b)$:

$$1 = s\left(\frac{a}{d}\right) + t\left(\frac{b}{d}\right) = \left(\frac{s}{d}\right)a + \left(\frac{t}{d}\right)b.$$

Since d is a common divisor, both a/d and b/d are integers, and part (i) applies. ■

Definition. An expression a/b for a rational number (where a and b are integers and $b \neq 0$) is in **lowest terms** if a and b are relatively prime.

Proposition 1.24. *Every nonzero rational number a/b has an expression in lowest terms.*

Proof. If $d = \gcd(a, b)$, then $a = a'd$, $b = b'd$, and $\dfrac{a}{b} = \dfrac{a'd}{b'd} = \dfrac{a'}{b'}$. But $a' = \dfrac{a}{d}$ and $b' = \dfrac{b}{d}$, so $\gcd(a', b') = 1$ by Lemma 1.23. ∎

We can now complete our discussion of Pythagorean triples.

Definition. A Pythagorean triple (a, b, c) is ***primitive*** if a, b, c have no common divisor $d \geq 2$; that is, there is no integer $d \geq 2$ which divides each of a, b, and c.

Theorem 1.25 (Diophantus). *Every Pythagorean triple (a, b, c) is similar to a unique primitive Pythagorean triple.*

Proof. We show first that (a, b, c) is similar to a primitive Pythagorean triple. If d is a common divisor of a, b, c, then $a = du$, $b = dv$, and $c = dz$, and (u, v, z) is a Pythagorean triple similar to (a, b, c) (why?). If d is the largest common divisor of a, b, c, we claim that (u, v, z) is primitive. Otherwise, there is an integer $e \geq 2$ with $u = eu'$, $v = ev'$, and $z = ez'$; hence, $a = du = deu'$, $b = dv = dev'$, and $c = dz = dez'$. Thus, $de > d$ is a common divisor of a, b, c, contradicting d being the largest such.

To prove uniqueness, suppose that (a, b, c) is similar to two primitive Pythagorean triples, say (u, v, z) and (r, s, t). It follows that the right triangles $\triangle(u, v, z)$ and $\triangle(r, s, t)$ are similar, and so their sides are proportional, so there is some positive number h with

$$u = hr, \quad v = hs, \text{ and } z = ht.$$

Since the side lengths are integers, h is rational, say $h = m/\ell$, and we may assume that it is in lowest terms; that is, $\gcd(m, \ell) = 1$. Cross multiply:

$$mu = \ell r, \quad mv = \ell s \quad \text{and} \quad mz = \ell t.$$

By Corollary 1.22, ℓ is a common divisor of u, v, and z and m is a common divisor of r, s, and t. Since both (u, v, z) and (r, s, t) are primitive, $\ell = 1 = m$, and so $(u, v, z) = (r, s, t)$. ∎

This next result is significant in the history of mathematics.

Proposition 1.26. *There is no rational number a/b whose square is 2.*

Proof. Suppose, on the contrary, that $(a/b)^2 = 2$. We may assume that a/b is in lowest terms; that is, $\gcd(a, b) = 1$. Since $a^2 = 2b^2$, Euclid's Lemma gives $2 \mid a$, and so $2m = a$. Hence, $4m^2 = a^2 = 2b^2$, and $2m^2 = b^2$. Euclid's Lemma now gives $2 \mid b$, contradicting $\gcd(a, b) = 1$. ∎

An ***indirect proof*** or ***proof by contradiction*** has the following structure. We assume that the desired statement is false and reach a contradiction. We conclude that the original statement must be true.

It follows that the legs of a Pythagorean triple (a, b, c) cannot be equal, for if $a = b$, then $a^2 + a^2 = c^2$, which implies that $2 = (c/a)^2$.

Proposition 1.26 is often stated as "$\sqrt{2}$ is irrational," which is a stronger statement than what we've just proved. We can assert that $\sqrt{2}$ is irrational only if we further assume that there exists a number u with $u^2 = 2$.

Our proof can be made more elementary; we need assume only that at least one of a, b is odd. Also, see Exercise 1.75 on page 41.

Historical Note. The ancient Greeks defined *number* to mean "positive integer." Rationals were not viewed as numbers but, rather, as ways of comparing two lengths. They called two segments of lengths a and b **commensurable** if there is a third segment of length c with $a = mc$ and $b = nc$ for positive integers m and n. That $\sqrt{2}$ is irrational was a shock to the Pythagoreans (ca. 500 BCE); given a square with sides of length 1, its diagonal and side are not commensurable; that is, $\sqrt{2}$ cannot be defined in terms of numbers (positive integers) alone. Thus, there is no numerical solution to the equation $x^2 = 2$, but there is a geometric solution.

By the time of Euclid, around 270 BCE, this problem had been resolved by splitting mathematics into two disciplines: number theory and geometry.

In ancient Greece, algebra as we know it did not really exist. Euclid and the Greek mathematicians did *geometric algebra*. For simple ideas, e.g., $(a + b)^2 = a^2 + 2ab + b^2$ or completing the square, geometry clarifies algebraic formulas (for example, see the right-hand part of Figure 1.2 on page 2 without the dashed lines). For more difficult ideas, say equations of higher degree, the geometric figures involved are very complicated, so that geometry is no longer clarifying. As van der Waerden writes in [34], p. 266,

> *one has to be a mathematician of genius, thoroughly versed in transforming proportions with the aid of geometric figures, to obtain results by this extremely cumbersome method. Anyone can use our algebraic notation, but only a gifted mathematician can deal with the Greek theory of proportions and with geometric algebra.*

The problem of defining *number* has arisen several times since the classical Greek era. Mathematicians had to deal with negative numbers and with complex numbers in the 1500s after the discovery of the Cubic Formula, because that formula often gives real roots of a cubic polynomial, even integer roots, in unrecognizable form (see Chapter 3). The definition of real numbers generally accepted today dates from the late 1800s. But there are echos of ancient Athens in our time. Kronecker (1823–1891) wrote,

> *Die ganzen Zahlen hat der liebe Gott gemacht, alles andere ist Menschenwerk.* (God created the integers; everything else is the work of Man.)

Even today some logicians argue for a new definition of number.

To bridge the gap between numbers and geometric magnitudes, Eudoxus (408 BCE–355 BCE) introduced the sophisticated notion of proportions (this idea, discussed in The Elements, is equivalent to our contemporary definition of real numbers).

Exercises

1.37 True or false, with reasons. Of course, it is important to get the right answer, but most attention should be paid to your reasoning.

(i) 6 | 2.   **Answer.** False.   (ii) 2 | 6.   **Answer.** True.

(iii) 6 | 0.   **Answer.** True.   (iv) 0 | 6.   **Answer.** False.

(v) 0 | 0.   **Answer.** True.

1.38 True or false, with reasons.

(i) $\gcd(n, n+1) = 1$ for every natural number n.   **Answer**. True.

(ii) $\gcd(n, n+2) = 2$ for every natural number n.   **Answer**. False.

(iii) 113 is a sum of distinct powers of 2.   **Answer**. True.

(iv) If a and b are natural numbers, there there are natural numbers s and t with $\gcd(a, b) = sa + tb$.   **Answer**. False.

(v) If an integer m is a divisor of a product of integers ab, then m is a divisor of either a or b (or both).   **Answer**. False.

1.39 Prove, or disprove and salvage if possible.

(i) $\gcd(0, b) = b$

(ii) $\gcd(a^2, b^2) = \gcd(a, b)^2$

(iii) $\gcd(a, b) = \gcd(a, b + ka)$ for all $k \in \mathbb{Z}$

(iv) $\gcd(a, a) = a$

(v) $\gcd(a, b) = \gcd(b, a)$

(vi) $\gcd(a, 1) = 1$

(vii) $\gcd(a, b) = -\gcd(-a, b)$

(viii) $\gcd(a, 2b) = 2\gcd(a, b)$

> "Disprove" here means "give a concrete counterexample." "Salvage" means "add a hypothesis to make it true."

1.40 * If x is a real number, let $\lfloor x \rfloor$ denote the greatest integer n with $n \le x$. (For example, $3 = \lfloor \pi \rfloor$ and $5 = \lfloor 5 \rfloor$.) If q is the quotient in Theorem 1.15, show that $q = \lfloor b/a \rfloor$.

1.41 *

(i) Given integers a and b (possibly negative) with $a \ne 0$, prove that there exist unique integers q and r with $b = qa + r$ and $0 \le r < |a|$.

> Hint: try $a = 5, b = 23$
> $a = 5, b = -23$
> $a = -5, b = 23$
> $a = -5, b = -23$

Hint. Use the portion of the Division Algorithm that has already been proved.

(ii) If b and a are positive integers, do b and $-b$ have the same remainder after dividing by a?   **Answer**. No.

1.42 For each of the following pairs a, b, find the largest nonnegative integer n with $n \le b/a < n + 1$.

(i) $a = 4$ and $b = 5$.   **Answer**. $n = 1$.

(ii) $a = 5$ and $b = 4$.   **Answer**. $n = 0$.

(iii) $a = 16$ and $b = 36$.   **Answer**. $n = 2$.

(iv) $a = 36$ and $b = 124$.   **Answer**. $n = 3$.

(v) $a = 124$ and $b = 1028$.   **Answer**. $n = 7$.

1.43 Let $p_1, p_2, p_3, \ldots$ be the list of the primes in ascending order: $p_1 = 2$, $p_2 = 3$, $p_3 = 5$, and so forth. Define $f_k = 1 + p_1 p_2 \cdots p_k$ for $k \ge 1$. Find the smallest k for which f_k is not a prime.

Hint. $19 \mid f_7$, but 7 is not the smallest k.

1.44 What can you say about two integers a and b with the property that $a \mid b$ and $b \mid a$? What if both a and b are positive?

1.45 * Show that if a is positive and $a \mid b$, then $\gcd(a, b) = a$. Why do we assume that a is positive?

1.46 *(**Two Out of Three**). Suppose that m, n, and q are integers and $m = n + q$. If c is an integer that divides any two of m, n, q, show that c divides the third one as well.

1.47 *

Allow for positive and negative values of s and t.

(i) For each a and b, give the smallest positive integer d that can be written as $sa + tb$ for integers s and t:

- $a = 12$ and $b = 16$.   **Answer.** $d = 4$.
- $a = 12$ and $b = 17$.   **Answer.** $d = 1$.
- $a = 12$ and $b = 36$.   **Answer.** $d = 12$.
- $a = 0$ and $b = 4$.   **Answer.** $d = 4$.
- $a = 4$ and $b = 16$.   **Answer.** $d = 4$.
- $a = 16$ and $b = 36$.   **Answer.** $d = 4$.
- $a = 36$ and $b = 124$.   **Answer.** $d = 4$.
- $a = 124$ and $b = 1028$.   **Answer.** $d = 4$.

(ii) How is "smallest positive integer d expressible as $sa + tb$" related to a and b in each case? Is d a divisor of both a and b?

1.48 * Show that the set of all linear combinations of two integers is precisely the set of all multiples of their gcd.

1.49 * Let I be a subset of $\mathbb{Z}$ such that

(i) $0 \in I$

(ii) if $a, b \in I$, then $a - b \in I$

(iii) if $a \in I$ and $q \in \mathbb{Z}$, then $qa \in I$.

Prove that there is a nonnegative integer $d \in I$ with I consisting precisely of all the multiples of d.

1.50 How might one define the $\gcd(a, b, c)$ of three integers? When applied to a *primitive* Pythagorean triple (a, b, c), your definition should say that $\gcd(a, b, c) = 1$.

Euclidean Algorithm

Our discussion of gcd's is incomplete. What is gcd(12327, 2409)? To ask the question another way, is the expression 2409/12327 in lowest terms? The next result enables us to compute gcd's efficiently. We first prove another lemma from Greek times.

Lemma 1.27. *Let a and b be integers.*

(i) *If $b = qa + r$, then $\gcd(a, b) = \gcd(r, a)$.*

(ii) *If $b \geq a$, then $\gcd(a, b) = \gcd(b - a, a)$.*

Proof. (i) In light of Corollary 1.20, it suffices to show that an integer c is a common divisor of a and b if and only if it is a common divisor of a and r. Since $b = qa + r$, this follows from Exercise 1.46 on page 29.

(ii) This follows from part (i) because $b = 1 \cdot a + (b - a)$.   ∎

The hypothesis $b \geq a$ in part (ii) of Lemma 1.27 is not necessary; it is there only to put you in the mood to accept the next example showing a method the Greeks probably used to compute gcd's. This method of computation is nowadays called the Euclidean Algorithm; it is Theorem 1.29.

Example 1.28. In this example, we will abbreviate $\gcd(b, a)$ to (b, a). Computing (b, a) is simple when a and b are small. If $b \geq a$, then Lemma 1.27

allows us to replace (b, a) by $(b - a, a)$; indeed, we can continue replacing numbers, $(b - 2a, a)$, $(b - 3a, a), \ldots, (b - qa, a)$ as long as $b - qa > 0$. Since the natural numbers $b - a, b - 2a, \ldots, b - qa$ are strictly decreasing, the Least Integer Axiom (or Infinite Descent) says that they must reach a smallest such integer: $r = b - qa$; that is, $0 < r < a$. Now $(b, a) = (r, a)$. (We see the proof of the Division Algorithm in this discussion.) Since $(r, a) = (a, r)$ and $a > r$, they could continue replacing numbers: $(a, r) = (a - r, r) = (a - 2r, r) = \cdots$ (remember that the Greeks did not recognize negative numbers, so it was natural for them to reverse direction). This process eventually ends, computing gcd's; we call it the *Euclidean Algorithm*. The Greek term for this method is **antanairesis**, a free translation of which is "back and forth subtraction." Let us implement this idea before we state and prove the Euclidean Algorithm.

Antanairesis computes $\gcd(326, 78)$ as follows:

$$(326, 78) = (248, 78) = (170, 78) = (92, 78) = (14, 78).$$

So far, we have been subtracting 78 from the other larger numbers. At this point, we now start subtracting 14 (this is the reciprocal, direction-changing, aspect of antanairesis), for $78 > 14$.

$$(78, 14) = (64, 14) = (50, 14) = (36, 14) = (22, 14) = (8, 14).$$

Again we change direction:

$$(14, 8) = (6, 8).$$

Change direction once again to get $(8, 6) = (2, 6)$, and change direction one last time to get

$$(6, 2) = (4, 2) = (2, 2) = (0, 2) = 2.$$

Thus, $\gcd(326, 78) = 2$.

The Division Algorithm and Lemma 1.27(i) give a more efficient way of performing antanairesis. There are four subtractions in the passage from $(326, 78)$ to $(14, 78)$; the Division Algorithm expresses this as

$$326 = 4 \cdot 78 + 14.$$

There are then five subtractions in the passage from $(78, 14)$ to $(8, 14)$; the Division Algorithm expresses this as

$$78 = 5 \cdot 14 + 8.$$

There is one subtraction in the passage from $(14, 8)$ to $(6, 8)$:

$$14 = 1 \cdot 8 + 6.$$

There is one subtraction in the passage from $(8, 6)$ to $(2, 6)$:

$$8 = 1 \cdot 6 + 2,$$

and there are three subtractions from $(6, 2)$ to $(0, 2) = 2$:

$$6 = 3 \cdot 2. \quad \blacktriangle$$

The beginning of the proof
of the theorem gives the
algorithm.

Theorem 1.29 (Euclidean Algorithm I). *If a and b are positive integers, there is an algorithm computing $\gcd(a, b)$.*

Proof. Let us set $b = r_0$ and $a = r_1$, so that the equation $b = qa + r$ reads $r_0 = q_1 a + r_2$. There are integers q_i and positive integers r_i such that

$$
\begin{aligned}
b = r_0 &= q_1 a + r_2, & r_2 &< a \\
a = r_1 &= q_2 r_2 + r_3, & r_3 &< r_2 \\
r_2 &= q_3 r_3 + r_4, & r_4 &< r_3 \\
&\ \ \vdots & &\ \ \vdots \\
r_{n-3} &= q_{n-2} r_{n-2} + r_{n-1}, & r_{n-1} &< r_{n-2} \\
r_{n-2} &= q_{n-1} r_{n-1} + r_n, & r_n &< r_{n-1} \\
r_{n-1} &= q_n r_n
\end{aligned}
$$

Lamé (1795–1870) proved
that the number of steps
in the Euclidean Algorithm
cannot exceed 5 times
the number of digits in the
smaller number (see [26],
p. 49).

(remember that all q_j and r_j are explicitly known from the Division Algorithm). There is a last remainder: the procedure stops (by Infinite Descent!) because the remainders form a strictly decreasing sequence of nonnegative integers (indeed, the number of steps needed is less than a).

We now show that the last remainder r_n is the gcd.

$$
\begin{aligned}
b &= q_1 a + r_2 & &\Rightarrow & \gcd(a, b) &= \gcd(a, r_2) \\
a &= q_2 r_2 + r_3 & &\Rightarrow & \gcd(a, r_2) &= \gcd(r_2, r_3) \\
r_2 &= q_3 r_3 + r_4 & &\Rightarrow & \gcd(r_2, r_3) &= \gcd(r_3, r_4) \\
& & & & &\ \ \vdots \\
r_{n-2} &= q_{n-1} r_{n-1} + r_n & &\Rightarrow & \gcd(r_{n-2}, r_{n-1}) &= \gcd(r_{n-1}, r_n) \\
r_{n-1} &= q_n r_n & &\Rightarrow & \gcd(r_{n-1}, r_n) &= r_n.
\end{aligned}
$$

All the implications except the last follow from Lemma 1.27. The last one follows from Exercise 1.45 on page 29. ■

Let's rewrite the previous example in the notation of the proof of Theorem 1.29. The passage from one line to the line below it involves moving the boldface numbers "southwest."

$$
\begin{aligned}
\mathbf{326} &= 4 \cdot \mathbf{78} + \mathbf{14} & &(1.7) \\
\mathbf{78} &= 5 \cdot \mathbf{14} + \mathbf{8} & &(1.8) \\
\mathbf{14} &= 1 \cdot \mathbf{8} + \mathbf{6} & &(1.9) \\
\mathbf{8} &= 1 \cdot \mathbf{6} + \mathbf{2} & &(1.10) \\
\mathbf{6} &= 3 \cdot \mathbf{2}.
\end{aligned}
$$

The Euclidean Algorithm also allows us to find a pair of integers s and t expressing the gcd as a linear combination.

Theorem 1.30 (Euclidean Algorithm II). *If a and b are positive integers, there is an algorithm computing a pair of integers s and t with $\gcd(a, b) = sa + tb$.*

Proof. It suffices to show, given equations

$$b = qa + r$$
$$a = q'r + r'$$
$$r = q''r' + r'',$$

how to write r'' as a linear combination of b and a (why?). Start at the bottom, and write

$$r'' = r - q''r'.$$

Now rewrite the middle equation as $r' = a - q'r$, and substitute:

$$r'' = r - q''r' = r - q''(a - q'r) = (1 - q''q')r - q''a.$$

Now rewrite the top equation as $r = b - qa$, and substitute:

$$r'' = (1 - q''q')r - q''a = (1 - q''q')(b - qa) - q''a.$$

Thus, r'' is a linear combination of b and a. ∎

We use the equations to find coefficients s and t expressing 2 as a linear combination of 326 and 78. Work from the bottom up.

$$
\begin{aligned}
2 &= \mathbf{8} - 1 \cdot \mathbf{6} && \text{by Eq. (1.10)}\\
&= \mathbf{8} - 1 \cdot (\mathbf{14} - 1 \cdot \mathbf{8}) && \text{by Eq. (1.9)}\\
&= 2 \cdot \mathbf{8} - 1 \cdot \mathbf{14}\\
&= 2 \cdot (\mathbf{78} - 5 \cdot \mathbf{14}) - 1 \cdot 14 && \text{by Eq. (1.8)}\\
&= 2 \cdot \mathbf{78} - 11 \cdot \mathbf{14}\\
&= 2 \cdot \mathbf{78} - 11 \cdot (\mathbf{326} - 4 \cdot \mathbf{78}) && \text{by Eq. (1.7)}\\
&= 46 \cdot \mathbf{78} - 11 \cdot \mathbf{326}.
\end{aligned}
$$

Thus, $s = 46$ and $t = -11$.

How to Think About It. The algorithm produces one pair of coefficients that works. However, it's not the only pair. For example, consider $\gcd(2, 3) = 1$. A moment's thought gives $s = -1$ and $t = 1$; but another moment's thought gives $s = 2$ and $t = -1$ (see Exercise 1.57 on page 35). However, the Euclidean Algorithm always produces a specific pair of coefficients; assuming that no mistakes in arithmetic are made, two people using the algorithm always come up with the same s and t.

Students usually encounter greatest common divisors in elementary school, sometimes as early as the fifth grade, when they learn how to add fractions and put the sum in lowest terms. As we have seen, putting a fraction in lowest terms involves the gcd of numerator and denominator. The preferred method of finding gcd's in early grades involves prime factorization, for if integers a and b are small, then it is easy to factor them into primes: after several cancellations, the expression a/b is in lowest terms. Pedagogically, this may be the right choice, but finding gcd's using prime factorization is practical only when numbers are small; can you put the fraction $167291/223377$ in lowest terms using prime factorization?

Putting a fraction in lowest terms is not always wise. For example,

$$\frac{2}{3} + \frac{1}{5} = \frac{10}{15} + \frac{3}{15}$$
$$= \frac{13}{15}.$$

How to Think About It. In calculating gcd's with the Euclidean Algorithm, many students get confused keeping track of the divisors and remainders. We illustrate one way to organize the steps that has been effective with high school students. Arrange the steps computing gcd(124, 1028) as on the left:

$$\mathbf{4} = 36 - 2 \cdot \mathbf{16}$$

$$= 36 - 2 \cdot (124 - 3 \cdot 36)$$
$$= -2 \cdot 124 + 7 \cdot \mathbf{36}$$

$$= -2 \cdot 124 + 7 \cdot (1028 - 8 \cdot 124)$$
$$= 7 \cdot 1028 - 58 \cdot 124$$

The last nonzero remainder is the gcd, so gcd(124, 1028) = 4. This arrangement can be used to read off coefficients s and t so that $4 = 124s + 128t$. Start at the next to last division and solve for each remainder.

Exercises

1.51 If a and b are positive integers, then $\gcd(a, b) = sa + tb$. Prove that either s or t is negative.

1.52 * Use Infinite Descent to prove that every positive integer a has a factorization $a = 2^k m$, where $k \geq 0$ and m is odd. Now prove that $\sqrt{2}$ is irrational using this fact instead of Euclid's Lemma.

1.53 Prove that if n is **squarefree** (i.e., $n > 1$ and n is not divisible by the square of any prime), then there is no rational number x with $x^2 = n$.

Hint. Adapt the proof of Proposition 1.26.

1.54 * Assuming there is a real number x with $x^3 = 2$, prove that x is irrational.

1.55 (i) Find $d = \gcd(326, 78)$, find integers s and t with $d = 326s + 78t$, and put the expression 326/78 in lowest terms.

Answer. $d = 2$, $s = -11$, $t = 46$, and $\frac{163}{39}$.

(ii) Find $d = \gcd(12327, 2409)$, find integers s and t with $d = 12327s + 2409t$, and put the expression 2409/12327 in lowest terms.

Answer. $d = 3$, $s = 299$, $t = -1530$, and $\frac{803}{4109}$.

(iii) Find $d = \gcd(7563, 526)$, and express d as a linear combination of 7563 and 526.

Answer. $d = 1$, $s = -37$, $t = 532$.

(iv) Find $d = \gcd(73122, 7404621)$ and express d as a linear combination of 73122 and 7404621.

Answer. $d = 21$, $s = 34531$, $t = -7404621$.

1.56 * Prove that if $\gcd(r, m) = 1$ and $\gcd(r', m) = 1$, then $\gcd(rr', m) = 1$. Conclude that if both r and r' are relatively prime to m, then so is their product rr'.

Hint. If $ar + bm = 1$ and $sr' + tm = 1$, consider $(ar + bm)(sr' + tm)$.

1.57 * Let a, b, and d be integers. If $d = sa + tb$, where s and t are integers, find infinitely many pairs of integers (s_k, t_k) with $d = s_k a + t_k b$.

Hint. If $2s + 3t = 1$, then $2(s + 3) + 3(t - 2) = 1$.

1.58 * If a and b are relatively prime and each divides an integer n, prove that their product ab also divides n.

Hint. Use Corollary 1.22.

1.59 If $m > 0$, prove that $m \gcd(b, c) = \gcd(mb, mc)$. (We must assume that $m > 0$ lest $m \gcd(b, c)$ be negative.)

Hint. Show that if k is a common divisor of mb and mc, then $k \mid m \gcd(b, c)$.

1.60 Write $d = \gcd(a, b)$ as a linear combination of a and b.

(i) $a = 4$ and $b = 16$.

Answer. $d = 4 = 5 \cdot 4 + (-1) \cdot 16$ (or, $4 = 1 \cdot 4 + 0 \cdot 16$).

(ii) $a = 16$ and $b = 36$.

Answer. $d = 4 = (-2) \cdot 16 + 1 \cdot 36$.

(iii) $a = 36$ and $b = 124$.

Answer. $d = 4 = 7 \cdot 36 + (-2) \cdot 124$.

(iv) $a = 124$ and $b = 1028$.

Answer. $d = 4 = (-58) \cdot 124 + 7 \cdot 1028$.

1.61 Given integers a, b, and c with $c \mid a$ and $c \mid b$, prove that c divides every linear combination $sa + tb$.

1.62 Is anything wrong with this calculation? Explain your answer.

$$
\begin{array}{r}
4 \\
7{\overline{)}\,37} \\
28 \\
\hline
9
\end{array}
$$

1.63 Given integers b, c, d, and e satisfying $b = 7c + 2$ and $d = 7e + 4$,

(i) What's the remainder when $b + d$ is divided by 7?

Answer. 6.

(ii) What's the remainder when bd is divided by 7?

Answer. 1.

(iii) Explain your answers.

1.64 A *lattice point* is a point (x, y) in the plane with both x and y integers.

(i) Which lattice points are on the line whose equation is $4x + 6y = 24$?

(ii) Which lattice points are on the line whose equation is $3x + 6y = 24$?

(iii) Find a line whose equation has integer coordinates but that never passes through a lattice point.

(iv) Explain how to tell whether the line with equation $y = ax + b$ contains lattice points.

1.65 Consider the calculation of gcd(124, 1028) on page 34. Show that the integer pairs

$$(124, 1028), \quad (36, 124), \quad (16, 36), \quad (4, 16), \quad (0, 4)$$

have the same greatest common divisor.

1.66 Most calculators have functions computing quotients and remainders. Let $r(b, a)$ denote the remainder when b is divided by a, and let $q(b, a)$ denote the quotient. Find $r(b, a)$ and $q(b, a)$ if

(i) $a = 12, b = 16$.   **Answer.** $q(16, 12) = 1, r(16, 12) = 4$.

(ii) $a = 16, b = 12$.   **Answer.** $q(12, 16) = 0, r(12, 16) = 12$.

(iii) $a = 124, b = 1028$.   **Answer.** $q(1028, 124) = 8, r(1028, 24) = 36$.

(iv) $a = 78, b = 326$.   **Answer.** $q(326, 78) = 4, r(326, 78) = 14$.

The functions can be programmed into a calculator.

1.67 Preview. Using the notation in Exercise 1.66, consider the pair of recursively defined functions on $\mathbb{N}$:

$$s(a, b) = \begin{cases} 0 & a = 0 \\ t\left(r(b, a), a\right) - q(b, a) \cdot s\left(r(b, a), a\right) & a > 0 \end{cases}$$

$$t(a, b) = \begin{cases} 1 & a = 0 \\ s\left(r(b, a), a\right) & a > 0. \end{cases}$$

Find $s(a, b)$ and $t(a, b)$ if

(i) $a = 124, b = 1028$.   **Answer.** $s(124, 1028) = -58, t(124, 1028) = 7$.

(ii) $a = 36, b = 124$.   **Answer.** $s(36, 124) = 7, t(36, 124) = -2$.

(iii) $a = 78, b = 326$.   **Answer.** $s(78, 326) = 46, t(78, 326) = -11$.

(iv) $a = 12327, b = 2409$.   **Answer.** $s(1237, 2409) = 1186, t(1237, 2409) = -609$.

(v) $a = 7563, b = 526$.   **Answer.** $s(7563, 526) = -37, t(7563, 526) = 532$.

(vi) $a = 167291, b = 223377$.   **Answer.** $s(167291, 223377) = -4, t(167291, 223377) = 3$.

1.4   Nine Fundamental Properties

We now focus on a small number (nine) of properties of arithmetic, for it turns out that many of the usual rules follow from them. This obviously simplifies things, making explicit what we are allowed to assume. But we have an ulterior motive. The properties will eventually be treated as axioms that will describe addition and multiplication in other systems, such as complex numbers, polynomials, and modular arithmetic; these systems lead naturally to their common generalization, *commutative rings*.

Notation. The set of all rational numbers is denoted by $\mathbb{Q}$, and the set of all real numbers is denoted by $\mathbb{R}$.

We begin by stating some basic properties of real numbers (of course, integers and rationals are special cases). These properties undergird a great deal of high school algebra; they are essential for the rest of this book and, indeed, for abstract algebra.

Functions are discussed in Appendix A.1.

Addition and multiplication are functions $\mathbb{R} \times \mathbb{R} \to \mathbb{R}$, namely, $(a, b) \mapsto a + b$ and $(a, b) \mapsto ab$. The **Laws of Substitution** say that if a, a', b, b' are real numbers with $a = a'$ and $b = b'$, then

$$a + b = a' + b' \quad \text{and} \quad ab = a'b'.$$

The Laws of Substitution are used extensively (usually tacitly) when solving equations or transforming expressions, and they merely say that addition and multiplication are single-valued. For example, since $-5 + 5 = 0$, we have

$$(-5 + 5) \times (-1) = 0 \times (-1) = 0.$$

Here are the properties of addition we are emphasizing.

Addition : For all real numbers a, b, and c,

  (i) **Commutativity**. $a + b = b + a$,

 (ii) $0 + a = a$,

(iii) there is a number $-a$, called the ***negative*** of a (or its ***additive inverse***), with $-a + a = 0$,

(iv) **Associativity**. $a + (b + c) = (a + b) + c$.

Let's say a bit more about associativity. Addition is defined as an operation performed on two numbers at a time, but it's often necessary to add three or more numbers. Associativity says that, when evaluating, say $2 + 5 + 3$, we can first add 2 and 5, giving $7 + 3 = 10$, or we can first add 5 and 3, giving $2 + 8 = 10$. In other words, we don't need parentheses: writing $2 + 5 + 3$ is unambiguous because $(2 + 5) + 3 = 2 + (5 + 3)$. This is not the case with subtraction. What is $8 - 3 - 2$? If we first subtract $8 - 3$, then the answer is $5 - 2 = 3$. However, if we evaluate $8 - (3 - 2) = 8 - 1$, we obtain a different answer. Thus, subtraction $\mathbb{R}^2 \to \mathbb{R}$, defined by $(a, b) \mapsto a - b$, is not associative, and we do need parentheses for it.

> Given associativity for the sum or product of 3 numbers, *generalized associativity* is also true: we don't need parentheses for the sum or product of $n \geq 3$ numbers. A proof is in Appendix A.5.

Here are the properties of multiplication that we are emphasizing; note that they are, formally, the same as those for addition: just replace "plus" by "times" (we usually denote the product of numbers a and b by ab, although we will occasionally write $a \cdot b$ or $a \times b$).

Multiplication : For all real numbers a, b, and c,

  (i) **Commutativity**. $ab = ba$,

 (ii) $1 \cdot a = a$,

(iii) If $a \neq 0$, there is a number a^{-1}, called its (multiplicative) ***inverse*** (or its ***reciprocal***) with $a \cdot a^{-1} = 1$,

(iv) **Associativity**. $a(bc) = (ab)c$.

> Why do we assume that $a \neq 0$? Read on.

Finally, we highlight a property involving both addition and multiplication.

Distributivity. $a(b + c) = ac + ab$.

Reading from left to right, distributivity says that we can "multiply a through;" reading from right to left, distributivity says that we can "factor a out."

Aside from the two Laws of Substitution, one for addition and one for multiplication, we have now listed nine properties of addition and multiplication.

Subtraction and division are defined as follows.

Definition. If a and b are numbers, define *subtraction* by

$$b - a = b + (-a),$$

where $-a$ is the negative of a; that is, $-a$ is the number which, when added to a, gives 0.

Quotient (or division) is defined similarly.

Definition. If a and b are numbers with $b \neq 0$, then the *quotient* of a by b is ab^{-1}, where b^{-1} is the number which, when multiplied by b, gives 1. We often denote ab^{-1} by a/b.

The word *quotient* is used here in a different way than in the Division Algorithm, where it is $\lfloor b/a \rfloor$, the integer part of b/a (see Exercise 1.40 on page 29).

How to Think About It. Almost all the properties just listed for the set $\mathbb{R}$ of real numbers also hold for the set $\mathbb{Z}$ of integers—these properties are "inherited" from $\mathbb{R}$ because integers are real numbers. The only property that $\mathbb{Z}$ doesn't inherit is the existence of multiplicative inverses. While every nonzero integer does have an inverse in $\mathbb{R}$, it may not be an integer; in fact, the only nonzero integers whose inverses also lie in $\mathbb{Z}$ are 1 and -1. There are other familiar algebraic systems that are more like $\mathbb{Z}$ than $\mathbb{R}$ in the sense that multiplicative inverses may not exist in the system. For example, all polynomials in one variable with rational coefficients form such a system, but the multiplicative inverse $1/x$ of x is not a polynomial.

Other familiar "rules" of arithmetic are easy consequences of these fundamental ones. Here are some of them.

Proposition 1.31. *For every number a, we have $0 \times a = 0$.*

Proof. By Addition Rule (ii), we have $0 + 0 = 0$. Therefore,

$$0 \times a = (0 + 0)a = (0 \times a) + (0 \times a).$$

Now subtract $0 \times a$ from both sides to obtain $0 = 0 \times a$. ∎

What is the meaning of division by 5? When we say that $20 \div 5 = 4$, we mean that $20/5$ is a number (namely 4), and that $(20/5) \times 5 = 20$. Dividing is the "opposite" of multiplying: dividing by 5 undoes multiplying by 5. This agrees with our formal definition. The inverse of 5 is 5^{-1}, and $20 \cdot 5^{-1} = 20/5 = 4$.

Multiplication by 5 is a bijection, and we are saying that division by 5 is its inverse function. See Example A.10 in Appendix A.1.

Can we divide by zero? If so, then $1/0$ would be a number with $0 \times (1/0) = 1$. But we have just seen that $0 \times a = 0$ for any number a. In particular, $0 \times (1/0) = 0$, giving the contradiction $1 = 0$. It follows that $1/0 = 0^{-1}$ is not a number; we cannot divide by 0.

Here is another familiar consequence of the nine fundamental properties.

Proposition 1.32. *For a number a, we have*

$$(-a) \cdot (-1) = a.$$

In particular,

$$(-1) \cdot (-1) = 1.$$

Proof. The distributive law gives

$$0 = 0 \cdot (-a) = (-1 + 1)(-a) = (-1) \cdot (-a) + (-a).$$

Now, add a to both sides to get $a = (-1)(-a)$. ∎

The Law of Substitution allows us to replace 0 by $-1 + 1$.

How to Think About It.

Even though its proof is very simple, Proposition 1.32 is often presented to high school students as something mysterious and almost magical. We can only guess at a reason. From Euclid's time until the 1500s, numbers were always positive; either negative numbers were not recognized at all or, if they did appear, they were regarded with suspicion, as not being bona fide (the complex numbers, which came on the scene around the same time, were also suspected of witchcraft). In the proof of Proposition 1.32, we treated negative numbers without prejudice, and we assumed that they obey the same elementary rules as positive numbers do. And we have reaped a reward for clear thinking.

Addition Rule (iii) states that every real number has a negative, an additive inverse. Can a number a have more than one negative? Intuition tells us no, and this can be proved using the nine fundamental properties.

Proposition 1.33. *Negatives in* $\mathbb{R}$ *are unique; that is, for* $a \in \mathbb{R}$, *there is exactly one number* b *in* $\mathbb{R}$ *with* $b + a = 0$.

Multiplicative inverses of nonzero real numbers are unique; that is, for nonzero $c \in \mathbb{R}$, *there is exactly one real number* d *with* $cd = 1$.

Proof. Suppose b is a number with $a + b = 0$. Add $-a$ to both sides:

$$-a + (a + b) = -a.$$

We can now use associativity to calculate, like this.

$$-a + (a + b) = -a$$
$$(-a + a) + b = -a$$
$$0 + b = -a$$
$$b = -a.$$

This argument can be adapted to prove uniqueness of multiplicative inverses; merely replace $+$ by $\times$ and "additive inverse" by "multiplicative inverse." ∎

Uniqueness theorems like Proposition 1.33 are useful because they show that certain objects are characterized by their behavior. For example, to show that a number b is equal to $-a$, add b to a and see if you get 0. This is the strategy in the next proof.

Corollary 1.34. *For every real number* a, *we have* $-a = (-1)a$. *Similarly, if* $b \neq 0$, *then* $(b^{-1})^{-1} = b$.

Proof. We add $(-1)a$ to a and see if we get 0.

$$(-1)a + a = (-1)a + 1 \cdot a = a(-1 + 1) = a \cdot 0 = 0.$$

We do get 0, and so Proposition 1.33 guarantees that $-a = (-1)a$.

To prove the second statement, interpret the equation $bb^{-1} = 1$ as saying that b is an element which, when multiplied by b^{-1}, gives 1. ∎

We can now prove the distributive law for subtraction.

Corollary 1.35. *If a, b, c are real numbers, then $a(b - c) = ab - ac$.*

Proof. By definition, $b - c = b + (-c)$. But $b - c = b + (-1)c$, by Corollary 1.34. Therefore, distributivity gives

$$\begin{aligned}
a(b - c) &= a\big(b + (-1)c\big) \\
&= ab + a(-1)c \\
&= ab + (-1)(ac) \\
&= ab - ac. \quad \blacksquare
\end{aligned}$$

We have just displayed some properties of addition and multiplication of real numbers following from the nine fundamental properties. The proofs follow *only* from the nine properties; we did not use any other properties of $\mathbb{R}$, such as decimal expansions or inequalities. Hence, if we show, for example, that addition and multiplication of complex numbers or of polynomials satisfy the nine properties, then each of these systems satisfy the "other properties," Propositions 1.31, 1.32, and 1.33, as well.

Exercises

1.68 (i) Prove the additive cancellation law using only the nine properties: if a, b, c are real numbers with $a + c = b + c$, then $a = b$.

(ii) Prove the multiplicative cancellation law for real numbers using only the nine properties: if a, b, c are real numbers with $ac = bc$ and $c \neq 0$, then $a = b$.

1.69 Suppose that $b \neq 0$. Show that a/b is the unique real number whose product with b is a.

1.70 (i) Prove that a real number a is a square if and only if $a \geq 0$.

(ii) Prove that every complex number is a square.

1.71 * Let a, b, c be numbers.

(i) Prove that $-ac$, the negative of ac, is equal to $(-a)c$; that is, $ac + (-a)c = 0$.

(ii) In the proof of Corollary 1.35, we stated that

$$ab + a(-1)c = ab - ac.$$

Prove this.

Hint. Evaluate $a(0 + 0)$ in two ways.

1.72 * Suppose that e and f are integers and let $m = \min\{e, f\}$ and $M = \max\{e, f\}$. Show that

$$m + M = e + f.$$

1.73 *

  (i) If a is a positive real number such that $a^n = 1$ for an integer $n \geq 1$, prove that $a = 1$.

  (ii) If a is a real number such that $a^n = 1$ for an integer $n \geq 1$, prove that $a = \pm 1$.

1.74 The Post Office has only 5 and 8 cent stamps today. Which denominations of postage can you buy?

1.75 * Later in this book, we'll prove Theorem 2.10: every integer can be factored into primes in essentially only one way. You may use this theorem here.

  (i) If $a \in \mathbb{Z}$, prove that every prime p that divides a^2 shows up with even exponent; that is, if $p \mid a^2$, then $p^2 \mid a^2$.

  (ii) Show that there are no integers a and b so that $2a^2 = b^2$.

  (iii) Use part(ii) to show that there is no rational number x with $x^2 = 2$.

1.76 Use Euclid's idea of a geometric Division Algorithm (see Figure 1.8 on page 22) to give a geometric version of the Euclidean Algorithm that uses repeated geometric division. Apply your geometric algorithm to

  (i) two segments of length 12 and 90.

  (ii) the diagonal and the side of a square.

1.5 Connections

This section applies the method of Diophantus to trigonometry and to calculus.

Trigonometry

The formulas $x = (1 - t^2)/(1 + t^2)$ and $y = 2t/(1 + t^2)$, where t is a real number, parametrize all the points on the unit circle except $(-1, 0)$. But we know that if $A = (x, y)$ is a point on the unit circle, then $x = \cos\theta$ and $y = \sin\theta$, where $\theta = \angle DOA$ (see Figure 1.9).

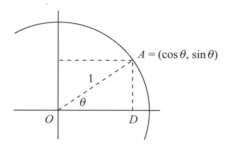

Figure 1.9. Cosine and Sine.

If $\theta = 30°$, then $(\cos\theta, \sin\theta) = (\frac{\sqrt{3}}{2}, \frac{1}{2})$; one coordinate is irrational and one is rational. Are there any acute angles θ with both $\cos\theta$ and $\sin\theta$ rational? If (x, y) is the Pythagorean point arising from $(3, 4, 5)$, then $x = \cos\theta = \frac{3}{5}$ and $y = \sin\theta = \frac{4}{5}$. With a little more work, we can prove that there are infinitely many angles θ with both $\cos\theta$ and $\sin\theta$ rational (is it obvious that Pythagorean triples arising from distinct Pythagorean points are not similar?)

and also infinitely many angles with both cosine and sine irrational (see Exercise 1.29 on page 14).

The parametrization of the unit circle in Proposition 1.2,

$$\cos\theta = \frac{1-t^2}{1+t^2} \quad \text{and} \quad \sin\theta = \frac{2t}{1+t^2}, \qquad -\infty < t < \infty,$$

enables us to prove some trigonometric identities. For example, let's prove the identity

$$\frac{1+\cos\theta+\sin\theta}{1+\cos\theta-\sin\theta} = \sec\theta + \tan\theta.$$

First, rewrite everything in terms of $\sin\theta$ and $\cos\theta$. The left-hand side is fine; the right-hand side is $(1/\cos\theta) + (\sin\theta/\cos\theta)$. Now replace these by their formulas in t. The left-hand side is

$$\frac{1 + \frac{1-t^2}{1+t^2} + \frac{2t}{1+t^2}}{1 + \frac{1-t^2}{1+t^2} - \frac{2t}{1+t^2}},$$

and this simplifies to a rational function of t (that is, a quotient of two polynomials). Similarly, the right-hand side is also a rational function of t, for $\sec\theta = \dfrac{1}{\cos\theta} = \dfrac{1+t^2}{1-t^2}$ and $\tan\theta = \dfrac{2t}{1-t^2}$. Thus, verifying whether the trigonometric identity is true is the same thing as verifying whether one rational expression is equal to another. This problem involves no ingenuity at all. Just cross multiply and check whether the polynomials on either side are equal; that is, check whether the monomials on either side having the same degree have the same coefficients.

Integration

The parametrization of the unit circle is useful for certain integration problems. In Figure 1.10, we see that $\triangle AOB$ is isosceles, for two sides are radii; thus, the

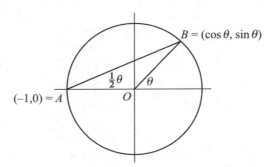

Figure 1.10. Tangent Half-Angle.

We denote the line determined by points A and B by $L(A, B)$, but the notation $\overleftrightarrow{AB}$ is the convention in geometry and precalculus books.

base angles are equal. But the exterior angle θ is their sum, and so $\angle BAO = \theta/2$. Therefore,

$$t = \frac{\sin\theta}{1+\cos\theta} = \text{slope } L(A, B) = \tan(\theta/2);$$

$t = \tan(\theta/2)$ is called the ***tangent half-angle formula***. Now

$$\theta = 2\arctan t \quad \text{and} \quad d\theta = \frac{2\,dt}{1+t^2}.$$

Let's apply this substitution. In most calculus courses, the indefinite integral $\int \sec\theta\, d\theta = \log|\sec\theta + \tan\theta|$ is found by some unmotivated trick, but this integration is quite natural when we use the method of Diophantus.

$$\int \sec\theta\, d\theta = \int \frac{d\theta}{\cos\theta} = \int \frac{1+t^2}{1-t^2}\cdot\frac{2\,dt}{1+t^2} = \int \frac{2\,dt}{1-t^2}.$$

Since

$$\frac{2}{1-t^2} = \frac{1}{1+t} + \frac{1}{1-t},$$

we have

$$\int \frac{2\,dt}{1-t^2} = \int \frac{dt}{1+t} + \int \frac{dt}{1-t} = \log|1+t| - \log|1-t|.$$

The hard part is now done;

$$\log|1+t| - \log|1-t| = \log\left|\frac{1+t}{1-t}\right|,$$

and it is cosmetic to rewrite, using the formula relating t and θ

$$\frac{1+t}{1-t} = \frac{(1+t)^2}{1-t^2} = \frac{1+2t+t^2}{1-t^2} = \frac{1+t^2}{1-t^2} + \frac{2t}{1-t^2} = \sec\theta + \tan\theta.$$

Other integrands can also be integrated using the tangent half-angle formula (see Exercise 1.78 below). Similar parametrizations of other conic sections also lead to integration formulas (see Exercises 1.80–1.82 below and [28, pp. 86–97]).

Exercises

1.77 Verify the following trigonometric identities.

(i) $1 + \csc\theta = \dfrac{\cos\theta\cot\theta}{1-\sin\theta}.$

(ii) $\dfrac{1}{\csc\theta - \cot\theta} - \dfrac{1}{\csc\theta + \cot\theta} = 2\cot\theta.$

(iii) $\cot^4\theta + \cot^2\theta = \csc^4\theta - \csc^2\theta.$

1.78 Integrate the following using the tangent half-angle formula.

(i) $\displaystyle\int \frac{\sin\theta}{2+\cos\theta}\,d\theta.$

Answer. $\ln\left|\dfrac{1+t^2}{3+t^2}\right|$, where $t = \dfrac{\sin\theta}{1+\cos\theta}.$

(ii) $\displaystyle\int \frac{\sin\theta - \cos\theta}{\sin\theta + \cos\theta}\,d\theta.$

Answer. $\ln\left|\dfrac{1+t^2}{1+2t=t^2}\right|$, which leads to $-\ln|\cos\theta + \sin\theta|.$

1.79 * **Preview.**

(i) Sketch the graph of $x^2 - xy + y^2 = 1$.

(ii) Find a "sweeping lines" parametrization for the points on the graph of $x^2 - xy + y^2 = 1$.

(iii) Find a scalene triangle with integer side lengths and a $60°$ angle.

1.80 **Take It Further.**

(i) Find a "sweeping lines" parametrization for the points on the graph of the parabola $x = y^2$, using lines joining $A = (0, 0)$ to points $P = (x, y)$ on the parabola.

(ii) Use this parametrization to evaluate $\displaystyle\int \frac{dx}{1 + \sqrt{x}}$.

1.81 **Take It Further.** Show that a "sweeping lines" parametrization for the points on the ellipse $x^2/a^2 + y^2/b^2 = 1$, using lines joining $A = (-a, 0)$ to points $P = (x, y)$ on the ellipse, is

$$x = \frac{a(b^2 - a^2t^2)}{b^2 + a^2t^2} \quad \text{and} \quad \frac{2ab^2t}{b^2 + a^2t^2}.$$

1.82 **Take It Further.** Show that a "sweeping lines" parametrization for the points on the hyperbola $x^2/a^2 - y^2/b^2 = 1$, using lines joining $A = (-a, 0)$ to points $P = (x, y)$ on the hyperbola, is

$$x = \frac{a(b^2 + a^2t^2)}{b^2 - a^2t^2} \quad \text{and} \quad y = \frac{2ab^2t}{b^2 - a^2t^2}.$$

1.83 * **Take It Further.** Most high school texts derive the quadratic formula by "completing the square," a method we'll discuss and generalize in Chapter 3. Here's another way to derive the formula.

(i) Show that if r and s are the roots of $x^2 + bx + c = 0$, then

$$r + s = -b \quad \text{and}$$
$$rs = c$$

(ii) If $r + s = -b$ and $rs = c$, show that

$$(r - s)^2 = b^2 - 4c,$$

so that $r - s = \pm\sqrt{b^2 - 4c}$.

(iii) Solve the system

$$r + s = -b$$
$$r - s = \pm\sqrt{b^2 - 4c}$$

for r and s.

2 Induction

In Chapter 1, we proved some basic theorems of ordinary arithmetic: Division Algorithm; Euclidean Algorithm; prime factorization. We are now going to prove the ***Fundamental Theorem of Arithmetic***: any two people writing an integer as a product of primes always get the same factors. We need a very useful tool in order to do this, and so we interrupt our historical account to introduce *mathematical induction*, a method of proof that finds application throughout mathematics. We'll go on here to use induction to discuss the Binomial Theorem and some combinatorics.

We shall see later that many interesting number systems do not have unique factorization. Indeed, not recognizing this fact is probably responsible for many false "proofs" of Fermat's Last Theorem.

2.1 Induction and Applications

The term *induction* has two meanings. The most popular one is ***inductive reasoning***: the process of inferring a general law from the observation of particular instances. For example, we say that the Sun will rise tomorrow morning because, from the dawn of time, the Sun has risen every morning. Although this notion of induction is used frequently in everyday life, it is not adequate for mathematical proofs, as we now show.

Consider the assertion: "$f(n) = n^2 - n + 41$ is prime for every positive integer n." Evaluating $f(n)$ for $n = 1, 2, 3, \ldots, 40$ gives the numbers

$$41, 43, 47, 53, 61, 71, 83, 97, 113, 131,$$
$$151, 173, 197, 223, 251, 281, 313, 347, 383, 421,$$
$$461, 503, 547, 593, 641, 691, 743, 797, 853, 911,$$
$$971, 1033, 1097, 1163, 1231, 1301, 1373, 1447, 1523, 1601.$$

It is tedious, but not very difficult (see Exercise 2.2 on page 52), to show that every one of these numbers is prime. Inductive reasoning leads you to expect that *all* numbers of the form $f(n)$ are prime. But the next number, $f(41) = 1681$, is not prime, for $f(41) = 41^2 - 41 + 41 = 41^2$, which is obviously composite.

An even more spectacular example of the failure of inductive reasoning is given by the harmonic series $1 + \frac{1}{2} + \frac{1}{3} + \cdots + \frac{1}{n} + \cdots$, which diverges (first proved by Oresme (ca.1320–1382)), and so its partial sums get arbitrarily large. Given a number N, there is a partial sum

$$\Sigma^m = \sum_{n=1}^{m} \frac{1}{n} = 1 + \frac{1}{2} + \cdots + \frac{1}{m}$$

with $\Sigma^m > N$. A high school student, unaware of this, playing with his calculator and seeing that $\Sigma^{315} \approx 6.33137$, would probably make the reasonable guess that $\Sigma^m < 100$ for all m. But he's wrong; the series diverges! It is known that if $m < 1.5 \times 10^{43}$, then $\Sigma^m < 100$. The most generous estimate of the age of the Earth is ten billion (10,000,000,000) years, or 3.65×10^{12} days, a number insignificant when compared to 1.5×10^{43}. Therefore, starting from the Earth's very first day, if the statement $\Sigma^m < 100$ was verified on the mth day, then there would be today as much evidence of the general truth of these statements as there is that the Sun will rise tomorrow morning. And yet most statements $\Sigma^m < 100$ are false!

Inductive reasoning is valuable in mathematics, as it is in natural science, because seeing patterns in data often helps us guess what may be true in general (see Exercise 2.1 on page 52, for example). However, merely checking whether the first few (or first few trillion) statements are true is not enough. We have just seen that checking the first 1.5×10^{43} statements is inadequate to establish a general rule.

Let's now discuss ***mathematical induction***. Suppose we are given a sequence of statements

$$S(1), \quad S(2), \quad S(3), \quad \ldots, \quad S(n), \quad \ldots.$$

For example, the formula $2^n > n$ for all $n \geq 1$ can be viewed as the sequence of statements

$$2^1 > 1, \quad 2^2 > 2, \quad 2^3 > 3, \quad \ldots, \quad 2^n > n, \quad \ldots.$$

Mathematical induction is a technique for proving that *all* the statements are true.

The key idea is just this. Imagine a stairway to the sky. We claim that if its bottom step is white and the next step above any white step is also white, then all the steps of the stairway are white. Here's our reasoning. If some steps aren't white, walk up to the first non-white step; call it Fido. Now Fido can't be at the bottom, for the bottom step is white, and so there is a step just below Fido. This lower step must be white, because Fido is the *first* non-white one. But Fido, being the next step above a white step, must also be white. This is a contradiction; there is no Fido. All the steps are white.

To sum up, given a list of statements, we are claiming that if

(i) the first statement is true, and

(ii) whenever a statement is true, so is the next one,

then all the statements on the list are true.

Let's apply this idea to the list of inequalities $S(n): 2^n > n$. Now $S(1)$ is true, for $2^1 = 2 > 1$. Suppose we believe, for every $n > 1$, that the implication $2^{n-1} > n - 1 \Rightarrow 2^n > n$ is true. Since $S(1)$ is true and $S(1) \Rightarrow S(2)$ is true, we have $S(2)$ true; that is, if $2^1 > 1$ and $2^1 > 1 \Rightarrow 2^2 > 2$ are both true, then $2^2 > 2$. Since $2^2 > 2$ is true and $2^2 > 2 \Rightarrow 2^3 > 3$ is true, we have $2^3 > 3$; since $2^3 > 3$ is true and $2^3 > 3 \Rightarrow 2^4 > 4$ is true, we have $2^4 > 4$; and so forth. Mathematical induction replaces the phrase *and so forth* with statement (ii), which guarantees, for every n, that there is never an obstruction in the passage from the truth of any statement $S(n - 1)$ to the truth of the next one $S(n)$. We will prove $2^n > n$ for all $n \geq 1$ in Proposition 2.2.

Here is the formal statement of mathematical induction.

The symbol $\Rightarrow$ means *implies*.

Theorem 2.1 (Mathematical Induction). *Let k be an integer. If $S(k)$, $S(k + 1)$, $S(k + 2), \ldots$ is a sequence of statements such that*

(i) **Base Step**: *$S(k)$ is true, and*

(ii) **Inductive Step**: *If, for $n > k$, $S(n - 1)$ being true implies $S(n)$ true,*

then the statements $S(n)$ are true for all $n \geq k$.

We'll prove this in Theorem 2.17 (you'll see then that the proof is our story about Fido), but let's use the theorem now to prove some interesting results. We start by completing our argument that $2^n > n$.

Many people prefer to write the inductive step as $S(n) \Rightarrow S(n + 1)$ instead of $S(n - 1) \Rightarrow S(n)$ as we do. The difference is cosmetic; the important thing is the passage from one statement to the next one.

Proposition 2.2. $2^n > n$ *for all $n \geq 1$.*

Proof. Here $k = 1$, and the statements are

$$S(1) : 2^1 > 1, \quad \ldots, \quad S(n - 1) : 2^{n-1} > n - 1, \quad S(n) : 2^n > n, \quad \ldots$$

Base Step: If $n = 1$, then $2^1 = 2 > 1$, so $S(1)$ is true.

Inductive Step: We need to show that if $n > 1$ and $S(n - 1)$ is true, then $S(n)$ is true. It is always a good idea to write the statements out so that we can see what needs to be proved. Here, we must show that *if* the **inductive hypothesis**

$$S(n - 1) : \quad 2^{n-1} > n - 1$$

is true, *then* so is $S(n)$; that is, $2^{n-1} > n - 1$ implies $2^n > n$. Multiply both sides of the inequality $S(n - 1)$ by 2: if $2^{n-1} > n - 1$, then

$$2^n = 2 \cdot 2^{n-1} > 2(n - 1) = (n - 1) + (n - 1) \geq (n - 1) + 1 = n$$

(the last inequality holds because $n > 1$ implies $n - 1 \geq 1$). Thus, if $2^{n-1} > n - 1$ is true, then $2^n > n$ is also true.

Since both the base step and the inductive step hold, Theorem 2.1 says that all the statements are true: $2^n > n$ for all $n \geq 1$. ■

Etymology. The word *induction* comes from the Latin word meaning to lead into or to influence. It is used here because, as we have just seen, the truth of the nth statement arises from the truth of the previous statement.

Usually the base step in an inductive proof occurs when $k = 1$, although many proofs occur when $k = 0$ (see Exercise 2.4 on page 52). Here is an example of an induction whose base step occurs when $k = 5$. Consider the statements

$$S(n) : \quad 2^n > n^2.$$

Define $S(0) : 2^0 > 0$. Suppose we had taken the base step in Proposition 2.2 at $k = 0$. Can you write out a proof that $S(0) \Rightarrow S(1)$?

This is not true for small values of n: if $n = 2$ or 4, then there is equality, not inequality; if $n = 3$, the left side, 8, is smaller than the right side, 9. However, $S(5)$ is true: $32 > 25$.

Proposition 2.3. $2^n > n^2$ *for all integers $n \geq 5$.*

Proof. We have just checked the base step $S(5)$. Suppose that $n > 5$ and that

$$2^{n-1} > (n-1)^2. \tag{2.1}$$

Can we use this to show that $2^n > n^2$? Multiply both sides of inequality (2.1) by 2 to obtain

$$2^n > 2(n-1)^2.$$

We'll be done if we show, for $n > 5$, that $2(n-1)^2 > n^2$. Now

$$
\begin{aligned}
2(n-1)^2 &= (n-1)^2 + (n-1)(n-1) \\
&> (n-1)^2 + 4(n-1) \qquad \text{since } n-1 > 4 \\
&\geq (n^2 - 2n + 1) + 4(n-1) \\
&= n^2 + 2n - 3.
\end{aligned}
$$

But $2n - 3$ is positive, because $n > 5$, and so $n^2 + 2n - 3 > n^2$.   ∎

We now use induction to prove a geometric result.

Definition. A polygon P in the plane is ***convex*** if, for every pair of distinct points A, B on its perimeter, the line segment $\overline{AB}$ lies inside of P.

For example, every triangle is convex, but there are quadrilaterals that are not convex. For example, the shaded quadrilateral in Figure 2.1 is not convex, for the line segment joining boundary points A and B is not wholly inside it.

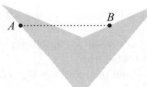

Figure 2.1. Non-convex polygon.

Proposition 2.4. *Let P be a convex polygon with vertices $V_1, \ldots, V_n$. If θ_i is the (interior) angle at V_i, then*

$$\theta_1 + \cdots + \theta_n = (n-2)180°.$$

Proof. The proof is by induction on $n \geq 3$. For the base step $n = 3$, the polygon is a triangle, and it is well known that the sum of the interior angles is $180°$. For the inductive step $n > 3$, let P be a convex polygon with vertices $V_1, \ldots, V_n$. Since P is convex, the segment joining V_1 and V_{n-1} lies wholly inside P; it divides P into the triangle $\Delta = \Delta V_1 V_n V_{n-1}$ and the polygon P' having vertices $V_1, \ldots, V_{n-1}$. Now P' is convex (why?), so that the inductive hypothesis says that the sum of its interior angles is $(n-3)180°$. Figure 2.2 shows that the sum of the interior angles of P is the sum of the angles of Δ and those of P', which is $180° + (n-3)180° = (n-2)180°$.   ∎

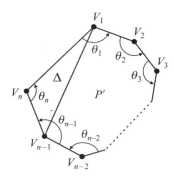

Figure 2.2. Convex polygon.

In any proof by induction, we must verify both the base step and the inductive step; verification of only one of them is insufficient. For example, consider the statements $S(n): n^2 = n$. The base step, $S(1)$, is true, but the inductive step is false; of course, these statements $S(n)$ are false for all $n > 1$. Another example is given by the statements $S(n): n = n + 1$. It is easy to see that the inductive step is true: if $S(n-1)$ is true, i.e., if $n-1 = (n-1)+1$, then adding 1 to both sides gives $n = (n-1)+2 = n+1$, which is the next statement, $S(n)$. But the base step is false; of course, all these statements $S(n)$ are false.

How to Think About It. When first seeing induction, many people suspect that the inductive step is circular reasoning. Why are you allowed to use statement $S(n-1)$, which you don't know is true, to prove that $S(n)$ is true? Isn't the truth of $S(n-1)$ essentially what you are supposed to be proving? A closer analysis shows that this is not at all what is happening. The inductive step, by itself, does not prove that $S(n)$ is true. Rather, it says that *if* $S(n-1)$ is true, *then* $S(n)$ is also true. In other words, the inductive step proves that the *implication* "If $S(n-1)$ is true, then $S(n)$ is true" is correct. The truth of this implication is not the same thing as the truth of its conclusion. For example, consider the two statements: "Your grade on every exam is 100%" and "Your grade for the course is A." The implication "If all your exams are perfect, then you will get the highest grade for the course" is true. Unfortunately, this does not say it is inevitable that your grade for the course will be A. Here is a mathematical example: the implication "If $n-1 = n$, then $n = n+1$" is true, but the conclusion "$n = n + 1$" is false.

From now on, we usually abbreviate *mathematical induction* to **induction**.

Here is the first example of a proof by induction often given in most texts.

Proposition 2.5. *For every integer $n \geq 1$, we have*

$$1 + 2 + \cdots + n = \tfrac{1}{2}n(n+1).$$

Proof. The proof is by induction on $n \geq 1$.

Base step. If $n = 1$, then the left-hand side is 1 and the right-hand side is $\tfrac{1}{2}1(1+1) = 1$, as desired.

Inductive step. The $(n-1)$st statement is

$$S(n-1): 1 + 2 + \cdots + (n-1) = \tfrac{1}{2}(n-1)n,$$

and we must show

$$1 + 2 + \cdots + n = \big[1 + 2 + \cdots + (n-1)\big] + n.$$

By the inductive hypothesis, the right-hand side is

$$\tfrac{1}{2}(n-1)n + n.$$

But $\tfrac{1}{2}(n-1)n + n = \tfrac{1}{2}n(n+1)$. By induction, the formula holds for all $n \geq 1$. ∎

Historical Note. Here is one version of a popular story. As a 7-year old prodigy, Gauss was examined by two mathematicians to evaluate his mathematical ability. When asked to add up all the numbers from 1 to 100, he thought a moment and then said the answer was 5050. Gauss let s denote the sum of all the numbers from 1 to 100: $s = 1 + 2 + \cdots + 99 + 100$. Of course, $s = 100 + 99 + \cdots + 2 + 1$. Arrange these nicely

$$s = 1 + 2 + \cdots + 99 + 100$$
$$s = 100 + 99 + \cdots + 2 + 1$$

and add

$$2s = 101 + 101 + \cdots + 101 + 101,$$

the sum 101 occurring 100 times. We now solve: $s = \tfrac{1}{2}(100 \times 101) = 5050$. This argument is valid for any number n in place of 100 (and there is no obvious use of induction!). Not only does this give a new proof of Proposition 2.5, it shows how the formula could have been discovered.

Example 2.6. Another proof of the formula in Proposition 2.5 comes from an analysis of the square in Figure 2.3.

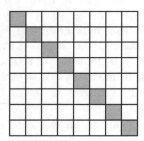

Figure 2.3. $\sum_{k=1}^{n} k$.

We have $n = 7$ in Figure 2.3.

Imagine an $(n+1) \times (n+1)$ square. It contains $(n+1)^2$ small unit squares. Since there are $n+1$ unit squares on the diagonal, there are

$$(n+1)^2 - (n+1) = n^2 + n$$

unit squares off the diagonal. Half of them, $\tfrac{1}{2}(n^2 + n)$, are above the diagonal. But, if you count by rows, there are

$$1 + 2 + \cdots + n$$

unit squares above the diagonal. Hence

$$1 + 2 + \cdots + n = \tfrac{1}{2}(n^2 + n). \quad \blacktriangle$$

How to Think About It. Proposition 2.5 illustrates a common problem students have when learning induction. Induction itself is a technique of proof (involving just two steps), but it is not a method of discovery. However, the two notions of proof and discovery are often intertwined. For example, merely applying mathematical induction, as we did in the proof of Proposition 2.5, is straightforward. But many beginning students get confused because, at the same time as they are following the steps of the proof, they are also wondering where the formula for the sum comes from. In contrast, neither Gauss's proof nor the proof using the $(n + 1) \times (n + 1)$ square is confusing, for the ideas of these proofs and their techniques of proof are separate. In Section 2.3, we'll describe a method for introducing mathematical induction to high school students that usually minimizes this confusion.

Aside from proving statements, induction can also be used to define terms. For example, here is an inductive *definition* of *factorial*.

*Inductive definitions are often called **recursive definitions**.*

Definition. Define $0! = 1$ and, if $n > 0$, define $n! = n \cdot (n - 1)!$. In other words, $n!$ is defined by

$$n! = \begin{cases} 1 & \text{if } n = 0 \\ n \cdot (n - 1)! & \text{if } n > 0. \end{cases}$$

Defining $0! = 1$ is convenient, as we shall see in the next section when we discuss the Binomial Theorem.

Induction allows us to define the *powers* of a number.

Definition. If $a \in \mathbb{R}$, define the *powers* of a, for $n \geq 0$, by induction:

$$a^n = \begin{cases} 1 & \text{if } n = 0 \\ a^{n-1}a & \text{if } n > 0. \end{cases}$$

If $a = 0$, we have defined $0^0 = 1$.

Etymology. The terminology *x square* and *x cube* for x^2 and x^3 is, of course, geometric in origin. Usage of the term *power* in this context arises from a mistranslation of the Greek *dunamis* (from which the word dynamo comes) as used by Euclid. The standard European rendition of *dunamis* was "power;" for example, the first English translation of Euclid's *Elements* by H. Billingsley in 1570, renders a sentence of Euclid as "The power of a line is the square on the same line" (which doesn't make much sense to us). However, contemporaries of Euclid, e.g., Plato and Aristotle, used *dunamis* to mean "amplification." This seems to be a more appropriate translation, for Euclid was probably thinking that a one-dimensional line segment can sweep out a two-dimensional square. We thank Donna Shalev for informing us of the classical usage of *dunamis*.

Proposition 2.7 (Laws of Exponents). *Let $a \in \mathbb{R}$ and $m, n \geq 0$ be integers.*

(i) $a^{m+n} = a^m a^n$.

(ii) $(a^m)^n = a^{mn}$.

Proof. (i) The statement is true for all $n \geq 0$ when $m = 0$. We prove $a^{m+n} = a^m a^n$ is true for all n by induction on $m \geq 1$. The base step says that $aa^n = a^{n+1}$, which is just the definition of powers. For the inductive step,

$$
\begin{aligned}
a^{m+n} &= a^{m+n-1} a && \text{definition of powers} \\
&= a^{m-1+n} a \\
&= a^{m-1} a^n a && \text{inductive hypothesis} \\
&= a^{m-1} a^{n+1} && \text{definition of powers} \\
&= a^{m-1+n+1} && \text{inductive hypothesis} \\
&= a^{m+n}. \quad a^m q^n
\end{aligned}
$$

(ii) The statement is true for all $m \geq 0$ when $n = 0$. We prove $(a^m)^n = a^{mn}$ is true for all m by induction on $n \geq 1$. The base step says that $(a^m)^1 = a^{m1} = a^m$, which is obvious. For the inductive step,

$$
\begin{aligned}
(a^m)^n &= (a^m)^{n-1} a^m && \text{definition of powers} \\
&= a^{m(n-1)} a^m && \text{inductive hypothesis} \\
&= a^{m(n-1)+m} && \text{part (i)} \\
&= a^{mn}. \quad \blacksquare
\end{aligned}
$$

Historical Note. The earliest known occurrence of mathematical induction is in *Sefer ha-Mispar* (also called *Maaseh Hoshev*, whose Hebrew title means practical and theoretical calculating), written by Levi ben Gershon in 1321 (he is also known as Gersonides or as RaLBaG, the acronym for Rabbi Levi ben Gershon). Induction appears later in *Arithmeticorum libri duo*, written by Maurolico in 1557, and also in *Traité du Triangle Arithmetique*, written by Pascal around 1654 (in which Pascal discusses the Binomial Theorem).

Exercises

2.1 * Guess a formula for $1 + \sum_{j=1}^{n} j!\,j$, and use mathematical induction to prove that your formula is correct.

2.2 * Prove that if $m \geq 2$ is an integer not divisible by any prime p with $p \leq \sqrt{m}$, then m is a prime. Use this to prove that the numbers $n^2 - n + 41$ are prime for all $n \leq 40$.

2.3 * Let $m_1, m_2, \ldots, m_n$ be integers such that $\gcd(m_i, m_j) = 1$ for all $i \neq j$. If each m_i divides an integer k, prove that their product $m_1 m_2 \cdots m_n$ also divides k. **Hint**. Use Exercise 1.58 on page 35.

2.4 * If a is positive, give two proofs that

$$
1 + a + a^2 + \cdots + a^{n-1} = \frac{a^n - 1}{a - 1},
$$

by induction on $n \geq 0$ and by multiplying the left-hand expression by $(a - 1)$.

2.5 Let $x > -1$ be a real number. Prove that $(1 + x)^n \geq 1 + nx$ for all $n \geq 1$.

2.6 What is the smallest value of k so that $2^n > n^3$ for all $n \geq k$? Why?

2.7 Assuming the product rule for derivatives, $(fg)' = f'g + fg'$, prove that

$$(x^n)' = nx^{n-1} \quad \text{for all integers } n \geq 1.$$

2.8 In high school, $n!$ is usually defined as $1 \cdot 2 \cdot 3 \cdot \cdots \cdot n$. Show that this agrees with the definition on page 51 for all $n \geq 1$.

2.9 (**Double Induction**) Let k, k' be integers, and let $S(m, n)$ be a doubly indexed family of statements, one for each pair of integers $m \geq k$ and $n \geq k'$. Suppose that

(i) $S(k, k')$ is true,

(ii) if $S(m - 1, k')$ is true, then $S(m, k')$ is true,

(iii) if $S(m, n - 1)$ is true for all $m \geq k$, then $S(m, n)$ is true for all $m \geq k$.

Prove that $S(m, n)$ is true for all $m \geq k$ and $n \geq k'$.

2.10 Prove that $(m + 1)^n > mn$ for all $m, n \geq 1$.

2.11 Prove the Laws of Exponents by Double Induction.

Unique Factorization

Induction is useful in number theory. As a simple example, we generalize Euclid's Lemma to more than two factors.

Theorem 2.8 (Euclid's Lemma). *If p is a prime and $p \mid a_1 a_2 \cdots a_n$, where $n \geq 2$, then $p \mid a_i$ for some i.*

Proof. The proof is by induction on $n \geq 2$. The base step is Theorem 1.21. To prove the inductive step, suppose that $p \mid a_1 a_2 \cdots a_n$. We may group the factors on the right side together so there are only two factors: $(a_1 a_2 \cdots a_{n-1})a_n$. By Theorem 1.21, either $p \mid a_1 a_2 \cdots a_{n-1}$ or $p \mid a_n$. In the first case, the inductive hypothesis gives $p \mid a_i$ for some $i \leq n - 1$ and we are done. In the second case, $p \mid a_n$, and we are also done. ■

"The proof is by induction on $n \geq 2$" not only indicates the base step, it also tells which variable will be changing in the inductive step.

This proof illustrates an empirical fact. It is not always the case, in an inductive proof, that the base step is very simple. In fact, all possibilities can occur: both steps can be easy, both can be difficult, or one can be harder than the other.

Here is an amusing inductive proof (due to Peter Braunfeld) of the existence of the quotient and remainder in the Division Algorithm.

Proposition 2.9. *If a and b are positive integers, then there are integers q and r with $b = qa + r$ and $0 \leq r < a$.*

Proof. We do induction on $b \geq 1$.

The base step: $b = 1$. Now $a \geq 1$, because it is a positive integer. If $a = 1$, choose $q = 1$ and $r = 0$; if $a > 1$, choose $q = 0$ and $r = 1$.

Let's prove the inductive step. The inductive hypothesis is $b - 1 = qa + r$, where $0 \leq r < a$. It follows that $b = qa + r + 1$. Now $r < a$ implies $r + 1 \leq a$. If $r + 1 < a$, we are done. If $r + 1 = a$, then $b = qa + (r + 1) = qa + a = (q + 1)a$, and we are done in this case as well. ■

We often use the word *product* even when there is only one factor. Thus, a prime is a product of primes.

We now use induction to prove unique factorization into primes.

Theorem 2.10 (Fundamental Theorem of Arithmetic). *Every integer $a \geq 2$ is a product of primes. Moreover, if*

$$a = p_1 \cdots p_m \quad and \quad a = q_1 \cdots q_n,$$

where the p's and q's are primes, then $n = m$ and the q's can be re-indexed so that $q_i = p_i$ for all i.

Proof. The existence of a factorization is Theorem 1.14. To prove uniqueness, we may assume that $m \geq n$, and we use induction on $m \geq 1$. The base step is obvious: if $m = 1$, then $n = 1$ and the given equations are $a = p_1 = q_1$. For the inductive step, the equation

$$p_1 \cdots p_m = q_1 \cdots q_n$$

gives $p_m \mid q_1 \cdots q_n$. By Euclid's Lemma, there is some i with $p_m \mid q_i$. But q_i, being a prime, has no positive divisors other than 1 and itself, so that $q_i = p_m$. Re-indexing, we may assume that $q_n = p_m$. Canceling, we have $p_1 \cdots p_{m-1} = q_1 \cdots q_{n-1}$. By the inductive hypothesis, $n - 1 = m - 1$ (so that $n = m$) and the q's may be re-indexed so that $q_i = p_i$ for all $i \leq m$. ∎

Corollary 2.11. *If $a \geq 2$ is an integer, then there are distinct primes p_i and integers $e_i > 0$ with*

$$a = p_1^{e_1} \cdots p_n^{e_n}.$$

Moreover, if there are distinct primes q_j and integers $f_j > 0$ with

$$p_1^{e_1} \cdots p_n^{e_n} = q_1^{f_1} \cdots q_m^{f_m},$$

then $m = n$, $q_i = p_i$ and $f_i = e_i$ for all i (after re-indexing the q's).

Proof. Just collect like terms in a prime factorization. ∎

The Fundamental Theorem of Arithmetic says that the exponents $e_1, \ldots, e_n$ in the prime factorization $a = p_1^{e_1} \cdots p_n^{e_n}$ are well-defined integers determined by a. It would not make sense to speak of the exponent of q dividing a if the Fundamental Theorem were false; if an integer a had two factorizations, say, $a = p^2 q^5 r^6$ and $a = p^2 q^3 s^7$, where p, q, r, s are distinct primes, would its q-exponent be 5 or 3?

It is often convenient to allow factorizations $p_1^{e_1} \cdots p_n^{e_n}$ having some exponents $e_i = 0$, because this allows us to use the same set of primes when factoring two numbers. For example, $168 = 2^3 3^1 7^1$ and $60 = 2^2 3^1 5^1$ may be rewritten as $168 = 2^3 3^1 5^0 7^1$ and $60 = 2^2 3^1 5^1 7^0$.

Lemma 2.12. *Let positive integers a and b have prime factorizations*

$$a = p_1^{e_1} \cdots p_n^{e_n} \quad and \quad b = p_1^{f_1} \cdots p_n^{f_n},$$

where $p_1, \ldots, p_n$ are distinct primes and $e_i, f_i \geq 0$ for all i. Then $a \mid b$ if and only if $e_i \leq f_i$ for all i.

Proof. If $e_i \leq f_i$ for all i, then $b = ac$, where $c = p_1^{f_1-e_1} \cdots p_n^{f_n-e_n}$. Now c is an integer, because $f_i - e_i \geq 0$ for all i, and so $a \mid b$.

Conversely, if $b = ac$, let the prime factorization of c be $c = p_1^{g_1} \cdots p_n^{g_n}$, where $g_i \geq 0$ for all i. It follows from the Fundamental Theorem of Arithmetic that $e_i + g_i = f_i$ for all i, and so $f_i - e_i = g_i \geq 0$ for all i; that is, $e_i \leq f_i$ for all i. ■

Definition. A *common multiple* of integers a and b is an integer m with $a \mid m$ and $b \mid m$. The *least common multiple*, denoted by

$$\mathrm{lcm}(a, b),$$

is the smallest positive common multiple if both $a, b \neq 0$, and it is 0 otherwise.

The following proposition describes gcd's in terms of prime factorizations. This is, in fact, the method usually taught to students in elementary school for putting fractions into lowest terms.

Proposition 2.13. *Let* $a = p_1^{e_1} \cdots p_n^{e_n}$ *and* $b = p_1^{f_1} \cdots p_n^{f_n}$, *where* $p_1, \ldots, p_n$ *are distinct primes and* $e_i \geq 0$, $f_i \geq 0$ *for all* i. *Then*

$$\gcd(a, b) = p_1^{m_1} \cdots p_n^{m_n} \quad and \quad \mathrm{lcm}(a, b) = p_1^{M_1} \cdots p_n^{M_n},$$

where $m_i = \min\{e_i, f_i\}$, *and* $M_i = \max\{e_i, f_i\}$.

Proof. Define $d = p_1^{m_1} \cdots p_n^{m_n}$. Lemma 2.12 shows that d is a (positive) common divisor of a and b; moreover, if c is any (positive) common divisor, then $c = p_1^{g_1} \cdots p_n^{g_n}$, where $0 \leq g_i \leq \min\{e_i, f_i\} = m_i$ for all i. Therefore, $c \mid d$.

A similar argument shows that $D = p_1^{M_1} \cdots p_n^{M_n}$ is a common multiple that divides every other such. ■

Computing the gcd for small numbers a and b using their prime factorizations is more efficient than using Euclidean Algorithm I. For example, since $168 = 2^3 3^1 5^0 7^1$ and $60 = 2^2 3^1 5^1 7^0$, we have $\gcd(168, 60) = 2^2 3^1 5^0 7^0 = 12$ and $\mathrm{lcm}(168, 60) = 2^3 3^1 5^1 7^1 = 840$. However, finding the prime factorization of a large integer is very inefficient, even with today's fanciest computers; it is so inefficient that this empirical fact is one of the main ingredients in *public key cryptography*, the basic reason you can safely submit your credit card number when buying something online.

Corollary 2.14. *If* a *and* b *are positive integers, then*

$$\mathrm{lcm}(a, b) \gcd(a, b) = ab.$$

Proof. The result follows from Proposition 2.13 and Exercise 1.72 on page 40:

$$m_i + M_i = e_i + f_i,$$

where $m_i = \min\{e_i, f_i\}$ and $M_i = \max\{e_i, f_i\}$. ■

Notice how a computational inquiry has given a theorem.

Since gcd's can be computed by Euclidean Algorithm I, this corollary allows us to compute lcm's:

$$\mathrm{lcm}(a, b) = ab/\gcd(a, b).$$

Example 2.15. Sudoku is a popular puzzle. One starts with a 9×9 grid of cells, some filled with numbers. The object is to insert numbers in the blank cells so that every row, every column, and every heavily bordered 3×3 box contains the digits 1 through 9 exactly once.

						6		9
		3	4		3		5	
				1			3	
		8		5	2	3		
4		7	8		3	5		1
		5	7	6		8		
	1			7				
	5		9		6	7		
3		6						

Figure 2.4. Sudoku.

KenKen® is a registered trademark of Nextoy, LLC, 2012, KenKen Puzzle LLC.

KenKen is a variation of Sudoku. As in Sudoku, the object is to fill an $n \times n$ grid with digits: 1 through 4 for a 4×4 grid, 1 through 5 for a 5×5 grid, etc., so that no digit appears more than once in any row or column. That the cells in Sudoku are filled with $1, 2, \ldots, 9$ is not important; one could just as well use the first nine letters a, b, …, i instead. In contrast, KenKen uses arithmetic. KenKen grids are divided into heavily bordered groups of cells, called *cages*, and the numbers in the cells in each cage must produce a target number when combined using a specified mathematical operation—either addition, subtraction, multiplication or division. Here is a 5×5 KenKen puzzle and its solution.

4−	12×		3−	
		12+	1−	
2÷	4−			3
		60×	5+	2÷

Figure 2.5. KenKen® puzzle.

4− 5	12× 3	2	3− 1	4
1	2	12+ 3	1− 4	5
2÷ 2	4− 1	4	5	3 3
4	5	60× 1	5+ 3	2÷ 2
3	4	5	2	1

Figure 2.6. KenKen® solution.

The difficulty in solving a KenKen puzzle arises from there being too many ways to fill in each cage. Sometimes, the Fundamental Theorem of Arithmetic can help. Let's start solving the puzzle in Figure 2.5. We view the grid as a 5×5 matrix, and we'll abbreviate "target-operation" to T-O. Consider the L-shaped cage consisting of 4 cells whose target operation is $60\times$. There are two possibilities: its cells are filled with an arrangement either of 2, 2, 3, 5 or of 1, 3, 4, 5. Assume the first possibility holds. Since we cannot have both 2s in the same row or the same column, one 2 is in position $(4, 3)$; the other 2 is either in position $(5, 1)$ or $(5, 2)$. Suppose 2 sits in the $(5, 1)$ position. There is a cage in the first column with T-O $4-$; its cells must contain 1 and 5. Hence, the other cage, with T-O $2\div$, must contain 3 and 5; it cannot. Thus, 2 sits in position $(5, 2)$. There is a cage in the second column with T-O 4−, and its cells must contain 1 and 5. This says that the top two cells in the second column contain 3 and 4. But the L-shaped cage with T-O 12× must now have 1 in position $(1, 3)$. This forces the column one cage, with T-O 4−, to have 5 in position $(1, 1)$, because it can't be 1. Thus, the last cage in the first row cannot involve 1 or 5. But the only ways to fill in a 2-cell cage with T-O 3− are with 1 and 4 or with 2 and 5. Conclusion: The 4-cell cage with T-O 60× must be an arrangement of 1, 3, 4, 5. The full solution is given in Figure 2.6. ▲

Strong Induction

Certain situations call for a variant of induction, called *Strong Induction* (or the *Second Form of Induction*).

Definition. Given integers k and $n \geq k$, the ***predecessors*** of n are the integers ℓ with $k \leq \ell < n$, namely, $k, k + 1, \ldots, n - 1$ (k has no predecessor).

Theorem 2.16 (Strong Induction). *Let k be an integer. If $S(k)$, $S(k + 1)$, $S(k + 2), \ldots$ is a sequence of statements such that*

(i) ***Base Step***: *$S(k)$ is true, and*

(ii) ***Inductive Step***: *If, for $n > k$, $S(\ell)$ being true for all predecessors ℓ of n implies $S(n)$ true,*

then the statements $S(n)$ are true for all $n \geq k$.

How to Think About It. Let's compare the two forms of induction. Both start by verifying the base step, and both have an inductive step to prove $S(n)$. The inductive hypothesis in the first form is that $S(n - 1)$ is true; the inductive hypothesis in Strong Induction is that all the preceding statements $S(k), \ldots, S(n - 1)$ are true. Thus, Strong Induction has a stronger inductive hypothesis (actually, each of Theorems 2.1 and 2.16 implies the other).

We are going to prove Theorem 2.16 and Theorem 2.1 simultaneously (we haven't yet proved the latter theorem). But first we need an easy technical remark. The Least Integer Axiom says that every nonempty subset C of the natural numbers $\mathbb{N}$ contains a smallest number; that is, there is some $c_0 \in C$ with $c_0 \leq c$ for all $c \in C$. This axiom holds, not only for $\mathbb{N}$, but for any subset

$$\mathbb{N}_k = \{n \in \mathbb{Z} : n \geq k\}$$

as well, where k is a fixed, possibly negative, integer. If $k \geq 0$, then $\mathbb{N}_k \subseteq \mathbb{N}$, and there is nothing to prove; if $k < 0$, then we argue as follows. Let $C \subseteq \mathbb{N}_k$ be a nonempty subset. If C contains no negative integers, then $C \subseteq \mathbb{N}$, and the Least Integer Axiom applies; otherwise, keep asking, in turn, whether $k, k + 1, \ldots, -1$ are in C, and define c_0 to be the first one that lies in C.

We have alreaady seen the basic idea of the next proof, involving Fido, on page 46. Here are the two statements again.

Theorem 2.17 (= Theorems 2.1 and 2.16). *Let k be an integer. If $S(k)$, $S(k + 1)$, $S(k + 2), \ldots$ is a sequence of statements such that*

Base Step: $S(k)$ *is true, and*

Inductive Step: *either*

(i) *if, for* $n > k$, $S(n-1)$ *being true implies* $S(n)$ *true,*

or

(ii) *if, for* $n > k$, $S(\ell)$ *being true for all predecessors* ℓ *of* n *implies* $S(n)$ *true,*

then the statements $S(n)$ *are true for all* $n \geq k$.

Proof. Let $\mathbb{N}_k = \{n \in \mathbb{Z} : n \geq k\}$. We show that there are no integers $n \in \mathbb{N}_k$ for which $S(n)$ is false. Otherwise, the subset $C \subseteq \mathbb{N}_k$, consisting of all n for which $S(n)$ is false, is nonempty, and so C has a smallest element, say, c_0. The truth of the base step says that $k < c_0$, so that $c_0 - 1$ lies in $\mathbb{N}_k$ and hence there is a statement $S(c_0 - 1)$.

Case 1. As $S(c_0)$ is the first false statement, $S(c_0 - 1)$ must be true. Assuming inductive step (i), $S(c_0) = S((c_0 - 1) + 1)$ is true, and this is a contradiction.

Case 2. As $S(c_0)$ is the first false statement, all the statements $S(\ell)$, where ℓ is a predecessor of c_0, are true. Assuming inductive step (ii), the strong version, we again reach the contradiction that $S(c_0)$ is true.

In either case, $C = \varnothing$ (i.e., C is empty), which says that every $S(n)$ is true. ■

Here's a second proof that prime factorizations exist.

Proposition 2.18 (= Proposition 1.14). *Every integer* $n \geq 2$ *is a product of primes.*

Proof. The base step $S(2)$ is true because 2 is a prime. We prove the inductive step. If $n \geq 2$ is a prime, we are done. Otherwise, $n = ab$, where $2 \leq a < n$ and $2 \leq b < n$. As a and b are predecessors of n, each of them is a product of primes:

$$a = pp' \cdots \quad \text{and} \quad b = qq' \cdots .$$

Hence, $n = pp' \cdots qq' \cdots$ is a product of (at least two) primes. ■

The reason why strong induction is more convenient here is that it is more natural to use $S(a)$ and $S(b)$ than to use $S(n-1)$; indeed, it is not at all clear how to use $S(n-1)$.

The next result says that we can always factor out a largest power of 2 from any integer. Of course, this follows easily from the Fundamental Theorem of Arithmetic, but we prove the proposition to illustrate further situations in which strong induction is more appropriate than the first form.

Proposition 2.19. *Every integer* $n \geq 1$ *has a unique factorization* $n = 2^k m$, *where* $k \geq 0$ *and* $m \geq 1$ *is odd.*

Proof. We use strong induction on $n \geq 1$ to prove the existence of k and m. If $n = 1$, take $k = 0$ and $m = 1$. For the inductive step $n \geq 1$, we distinguish two cases. If n is odd, take $k = 0$ and $m = n$. If n is even, then $n = 2b$. Since $b < n$, it is a predecessor of n, and so the inductive hypothesis allows us to assume $b = 2^\ell m$, where $\ell \geq 0$ and m is odd. The desired factorization is $n = 2b = 2^{\ell+1}m$.

Why isn't the first form of induction convenient here?

To prove uniqueness (induction is not needed here), suppose that $2^k m = n = 2^t m'$, where both k and t are nonnegative and both m and m' are odd. We may assume that $k \geq t$. If $k > t$, then canceling 2^t from both sides gives $2^{k-t}m = m'$. Since $k - t > 0$, the left side is even while the right side is odd; this contradiction shows that $k = t$. We may thus cancel 2^k from both sides, leaving $m = m'$. ∎

Exercises

2.12 (i) Prove that an integer $a \geq 2$ is a perfect square if and only if whenever p is prime and $p \mid a$, ~~then $p^2 \mid a$.~~ *occurring in the prime factızn of a are even.*

(ii) Prove that if an integer $z \geq 2$ is a perfect square and $d^4 \mid z^2$, then $d^2 \mid z$.

2.13 Let a and b be relatively prime positive integers. If ab is a perfect square, prove that both a and b are perfect squares.

2.14 * Let a, b, c, n be positive integers with $ab = c^n$. Prove that if a and b are relatively prime, then both a and b are nth powers; that is, there are positive integers k and ℓ with $a = k^n$ and $b = \ell^n$.

2.15 * For any prime p and any positive integer n, denote the highest power of p dividing n by $\mathcal{O}_p(n)$. That is,

$$\mathcal{O}_p(n) = e,$$

Corollary 2.11 guarantees that $\mathcal{O}_p$ is well-defined.

where $p^e \mid n$ but $p^{e+1} \nmid n$. If m and n are positive integers, prove that

(i) $\mathcal{O}_p(mn) = \mathcal{O}_p(m) + \mathcal{O}_p(n)$

(ii) $\mathcal{O}_p(m + n) \geq \min\{\mathcal{O}_p(m), \mathcal{O}_p(n)\}$. When does equality occur?

There is a generalization of Exercise 1.6 on page 6. Using a (tricky) inductive proof (see FCAA [26], p. 11), we can prove the ***Inequality of the Means***: if $n \geq 2$ and $a_1, \ldots, a_n$ are positive numbers, then

$$\sqrt[n]{a_1 \cdots a_n} \leq \tfrac{1}{n}(a_1 + \cdots + a_n).$$

2.16 (i) Using the Inequality of the Means for $n = 3$, prove, for all triangles having a given perimeter, that the equilateral triangle has the largest area.

Hint. Use ***Heron's Formula*** for the area A of a triangle with sides of lengths a, b, c: if the ***semiperimeter*** is $s = \tfrac{1}{2}(a + b + c)$, then

$$A^2 = s(s - a)(s - b)(s - c).$$

(ii) What conditions on a, b, and c ensure that Heron's formula produces 0? Interpret geometrically.

2.17 Let a, b, and c be positive numbers with $a > b > c$, and let $L = \tfrac{1}{3}(a + b + c)$.

(i) Show that either $a > b > L > c$ or $a > L > b > c$.

(ii) Assume that $a > b > L > c$. Show that

$$L^3 - (L - b)^2 c - (L - b)(L - c)c - (L - c)^2 L = abc.$$

(iii) Use part (ii) to prove the Inequality of the Means for three variables.

(iv) Show that a box of dimensions $a \times b \times c$ can be cut up to fit inside a cube of side length L with something left over.

Differential Equations

You may have seen differential equations in other courses. If not, don't worry; the next example is self-contained.

Definition. A *differential equation* is an equation involving a function $y = y(x)$ and its derivatives; a *solution* is a function y that satisfies the equation.

Solving a differential equation generalizes indefinite integration: $\int f(x)\,dx$ is a solution to the differential equation $y' = f$. There may be many solutions: for example, if $y = F(x)$ is an indefinite integral of $\int f(x)\,dx$, then so is $F(x) + c$ for any constant c.

Assume that a differential equation has a solution y that is a power series. Because factorials occur in the coefficients of Taylor series, let's write a solution in the form

$$y(x) = a_0 + a_1 x + \frac{a_2}{2!}x^2 + \cdots + \frac{a_n}{n!}x^n + \cdots . \tag{2.2}$$

> Defining $0! = 1$ allows us to write the coefficient of x^n in Eq. (2.2) as $a_n/n!$ for all $n \geq 0$.

We ignore questions of convergence. Of course, some power series diverge, but we are doing algebra here!

Induction arises here because we can often find y by relating its coefficients a_{n-1} and a_n.

Example 2.20. Consider the differential equation $y' = y$; that is, we seek a function equal to its own derivative (do you know such a function?). Assuming that y is a power series, then y has an expression as in Eq. (2.2). Using term-by-term differentiation, we see that

$$y' = a_1 + a_2 x + \frac{a_3}{2!}x^2 + \cdots + \frac{a_n}{(n-1)!}x^{n-1} + \cdots ,$$

so that

$$\frac{a_n}{(n-1)!}x^{n-1} = \frac{a_{n-1}}{(n-1)!}x^{n-1} \quad \text{for all } n \geq 1;$$

that is, $a_1 = a_0$, $a_2 = a_1$, and, in fact, $a_n = a_{n-1}$ for all $n \geq 1$. Rewrite the equations: there is no restriction on a_0 and, for small n, we see that $a_n = a_0$. If this were true for all n, then

$$y(x) = a_0(1 + x + \tfrac{1}{2}x^2 + \cdots + \tfrac{1}{n!}x^n + \cdots) = a_0 e^x.$$

It is true that $a_n = a_0$ for all n; one proof is by induction (see Exercise 2.18 on page 62). ▲

Differential equations often arise with *initial conditions*: values of $y(0)$, $y'(0)$, $y''(0), \ldots$ are specified.

If y is given by a power series $\sum(a_n/n!)x^n$, then $y(0) = a_0$. Thus, the initial condition $y(0) = 1$ chooses the solution $y = e^x$ in the preceding example.

The next example shows how strong induction can be used in solving differential equations.

Example 2.21. Consider the differential equation

$$y'' = y' + 2y \tag{2.3}$$

with initial conditions

$$y(0) = 2 \quad \text{and} \quad y'(0) = 1.$$

Again, let's see if there is a power series solution

$$y(x) = a_0 + a_1 x + \frac{a_2}{2!} x^2 + \cdots .$$

Substituting y into Eq. (2.3) and equating like powers of x gives

$$a_n = a_{n-1} + 2a_{n-2}.$$

Tabulating a_n for a few values shows a pattern. All the outputs a_n seem to be 1 away from a power of 2, either 1 more or 1 less. The first two entries record the initial conditions.

n	a_n
0	2
1	1
2	$5 = 1 + 2 \cdot 2$
3	$7 = 5 + 2 \cdot 1$
4	$17 = 7 + 2 \cdot 5$
5	$31 = 17 + 2 \cdot 7$
6	65
7	127
8	257
9	511
10	1025

Looking closer, the coefficients seem to satisfy $a_n = 2^n + (-1)^n$. Inductive reasoning suggests the conjecture, and mathematical induction is a natural way to prove it. But there is a problem. The inductive step for a_n involves not only a_{n-1}, but a_{n-2} as well. Strong Induction to the rescue! Before dealing with the details, we show that if the formula can be proved to hold for *all* a_n, then we can complete our discussion of the differential equation. A solution is

$$
\begin{aligned}
y(x) &= \sum \frac{2^n + (-1)^n}{n!} x^n \\
&= \sum \frac{2^n}{n!} x^n + \sum \frac{(-1)^n}{n!} x^n \\
&= \sum \frac{1}{n!} (2x)^n + \sum \frac{1}{n!} (-x)^n \\
&= e^{2x} + e^{-x}.
\end{aligned}
$$

You can check that $e^{2x} + e^{-x}$ works by substituting it into Eq. (2.3); we have solved the differential equation.   ▲

The proof of the equation relating the coefficients is by Strong Induction.

Proposition 2.22. *Suppose, for all $n \geq 0$, that a_n satisfies*

$$a_n = \begin{cases} 2 & n = 0 \\ 1 & n = 1 \\ a_{n-1} + 2a_{n-2} & n > 1. \end{cases}$$

Then $a_n = 2^n + (-1)^n$ for all integers $n \geq 0$.

Proof. Because the definition has two initial values, we need to check two base steps:

$$a_0 = 2 = 2^0 + (-1)^0 \qquad \text{and} \qquad a_1 = 1 = 2^1 + (-1)^1.$$

If $n > 1$ and $a_k = 2^k + (-1)^k$ for all the predecessors of n, $0 \leq k < n$, then

$$\begin{aligned} a_n &= a_{n-1} + 2a_{n-2} \\ &= \left(2^{n-1} + (-1)^{n-1}\right) + 2\left(2^{n-2} + (-1)^{n-2}\right) \\ &= \left(2^{n-1} + (-1)^{n-1}\right) + \left(2 \cdot 2^{n-2} + 2 \cdot (-1)^{n-2}\right) \\ &= \left(2^{n-1} + (-1)^{n-1}\right) + \left(2^{n-1} + 2 \cdot (-1)^{n-2}\right) \\ &= \left(2^{n-1} + 2^{n-1}\right) + \left((-1)^{n-1} + 2 \cdot (-1)^{n-2}\right) \\ &= 2 \cdot 2^{n-1} + (-1)^{n-2}(-1 + 2) \\ &= 2^n + (-1)^{n-2} \\ &= 2^n + (-1)^n. \quad \blacksquare \end{aligned}$$

Exercises

2.18 * Complete the discussion in Example 2.20: show that if

$$y(x) = a_0 + a_1 x + \frac{a_2}{2!}x^2 + \cdots + \frac{a_n}{n!}x^n + \cdots$$

and $y' = y$, then $a_n = a_0$.

2.19 Assume that "term-by-term" differentiation holds for power series: if $f(x) = c_0 + c_1 x + c_2 x^2 + \cdots + c_n x^n + \cdots$, then the power series for the derivative $f'(x)$ is

$$f'(x) = c_1 + 2c_2 x + 3c_3 x^2 + \cdots + nc_n x^{n-1} + \cdots.$$

  (i) Prove that $f(0) = c_0$.
  (ii) Prove, for all $n \geq 0$, that if $f^{(n)}$ is the nth derivative of f, then

$$f^{(n)}(x) = n!c_n + (n+1)!c_{n+1}x + x^2 g_n(x),$$

  where $g_n(x)$ is a power series.

Here is an instance in which it is convenient to write $0! = 1$.

  (iii) Prove that $c_n = f^{(n)}(x)(0)/n!$ for all $n \geq 0$. (This is Taylor's formula.)

This exercise shows why, in Example 2.21, that power series were denoted by $a_0 + a_1 x + (a_2/2!)x^2 + (a_3/3!)x^3 + \cdots$.

2.20 Find the solution to the differential equation

$$2y'' - y' - 3y = 0.$$

subject to the initial conditions $y(0) = y(1) = 1$.

Answer. $y = \frac{1}{5}e^{-x} + \frac{4}{5}e^{3x/2}$.

2.2   Binomial Theorem

We now look at a result, important enough to deserve its own section, which involves both mathematical induction and inductive reasoning. What is the pattern of the coefficients in the formulas for the powers $(1 + x)^n$ of the binomial $1 + x$? Let

$$(1 + x)^n = c_0 + c_1 x + c_2 x^2 + \cdots + c_n x^n.$$

Definition.   The coefficients c_r are called **binomial coefficients**:

$$\binom{n}{r} \text{ is the coefficient } c_r \text{ of } x^r \text{ in } (1 + x)^n.$$

Euler introduced the notation $\left(\frac{n}{r}\right)$, and this symbol evolved into $\binom{n}{r}$, which is generally used today:

The binomial coefficient $\binom{n}{r}$ is pronounced "n choose r" because it also arises in counting problems, as we shall soon see. Thus,

$$(1 + x)^n = \sum_{r=0}^{n} \binom{n}{r} x^r.$$

For example,

$$(1 + x)^0 = 1$$
$$(1 + x)^1 = 1 + 1x$$
$$(1 + x)^2 = 1 + 2x + 1x^2$$
$$(1 + x)^3 = 1 + 3x + 3x^2 + 1x^3$$
$$(1 + x)^4 = 1 + 4x + 6x^2 + 4x^3 + 1x^4.$$

Etymology.   *Binomial* means $a + b$; *trinomial* means $a + b + c$. But *monomial* usually refers to a summand of a polynomial: either ax^e for a polynomial in one variable, or $ax_1^{e_1} \cdots x_n^{e_n}$ for a polynomial in several variables.

The following figure, called **Pascal's triangle**, displays an arrangement of the first few coefficients.

$$
\begin{array}{ccccccccccccccc}
&&&&&&& 1 \\
&&&&&& 1 && 1 \\
&&&&& 1 && 2 && 1 \\
&&&& 1 && 3 && 3 && 1 \\
&&& 1 && 4 && 6 && 4 && 1 \\
&& 1 && 5 && 10 && 10 && 5 && 1 \\
& 1 && 6 && 15 && 20 && 15 && 6 && 1 \\
1 && 7 && 21 && 35 && 35 && 21 && 7 && 1
\end{array}
$$

In Pascal's triangle, an **inside number** (i.e., not a 1 on the border) of the nth row can be computed by going up to the $(n-1)$st row and adding the two neighboring numbers above it. For example, the inside numbers in row 4 can be computed from row 3 as follows:

$$
\begin{array}{ccccccc}
& 1 && 3 && 3 && 1 \\
1 && 4 && 6 && 4 && 1
\end{array}
$$

($4 = 1+3, 6 = 3+3$, and $4 = 3+1$). Let's prove that this observation always holds.

Lemma 2.23. *For all integers $n \geq 1$ and all r with $0 \leq r \leq n$,*

$$
\binom{n}{r} = \begin{cases} 1 & \text{if } r = 0 \text{ or } n = r \\ \binom{n-1}{r-1} + \binom{n-1}{r} & \text{if } 0 < r < n. \end{cases}
$$

Proof. The nth row of Pascal's triangle is the coefficient list for $(1 + x)^n$. The fact that the constant term and the highest degree term have coefficient 1 is Exercise 2.21 on page 67. For the inside terms, we claim that the coefficient of x^r in $(1 + x)^n$ is the sum of two neighboring coefficients in $(1 + x)^{n-1}$. More precisely, we claim that if

$$
(1 + x)^{n-1} = c_0 + c_1 x + c_2 x^2 + \cdots + c_{n-1} x^{n-1},
$$

and $0 < r < n$, then the coefficient of x^r in $(1 + x)^n$ is $c_{r-1} + c_r$. We have

$$
\begin{aligned}
(1 + x)^n &= (1 + x)(1 + x)^{n-1} = (1 + x)^{n-1} + x(1 + x)^{n-1} \\
&= (c_0 + \cdots + c_{n-1} x^{n-1}) + x(c_0 + \cdots + c_{n-1} x^{n-1}) \\
&= (c_0 + \cdots + c_{n-1} x^{n-1}) + (c_0 x + c_1 x^2 + \cdots + c_{n-1} x^n) \\
&= 1 + (c_0 + c_1) x + (c_1 + c_2) x^2 + \cdots.
\end{aligned}
$$

Thus $\binom{n}{r} = c_{r-1} + c_r = \binom{n-1}{r-1} + \binom{n-1}{r}$. ∎

You can also prove Lemma 2.23 by induction. See Exercise 2.22 on page 67.

Pascal's triangle was known centuries before Pascal's birth; Figure 2.7 shows a Chinese scroll from the year 1303 depicting it. Pascal's contribution (around 1650) is a formula for the binomial coefficients.

Proposition 2.24 (Pascal). *For all $n \geq 0$ and all r with $0 \leq r \leq n$,*

$$
\binom{n}{r} = \frac{n!}{r!(n-r)!}.
$$

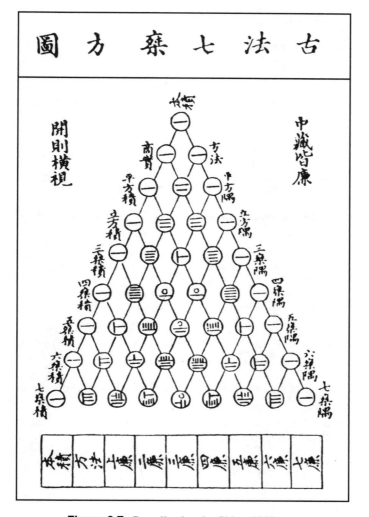

Figure 2.7. Pascal's triangle, China, 1303 CE.

Proof. The proof is by induction on $n \geq 0$. If $n = 0$, then

$$\binom{0}{0} = \frac{0!}{0!0!} = 1.$$

For the inductive step, note first that the formula holds when $r = 0$ and $r = n$:

$$\binom{n}{0} = 1 = \frac{n!}{0!(n-0)!}$$

and

$$\binom{n}{n} = 1 = \frac{n!}{n!\,0!}.$$

Pascal probably discovered this formula by regarding $\binom{n}{r}$ in a different way. We'll look at this in a moment.

Here is another place showing that defining $0! = 1$ is convenient.

If $0 < r < n$, then

$$\binom{n}{r} = \binom{n-1}{r-1} + \binom{n-1}{r} \qquad \text{(Lemma 2.23)}$$

$$= \frac{(n-1)!}{(r-1)!(n-r)!} + \frac{(n-1)!}{r!(n-r-1)!} \qquad \text{(inductive hypothesis)}$$

$$= \frac{(n-1)!}{(r-1)!(n-r-1)!}\left(\frac{1}{n-r} + \frac{1}{r}\right)$$

$$= \frac{(n-1)!}{(r-1)!(n-r-1)!}\left(\frac{n}{r(n-r)}\right) = \frac{n!}{r!(n-r)!}. \qquad \blacksquare$$

Theorem 2.25 (Binomial Theorem).   (i) *For all real numbers x and all integers $n \geq 0$,*

$$(1+x)^n = \sum_{r=0}^{n}\binom{n}{r}x^r = \sum_{r=0}^{n}\frac{n!}{r!(n-r)!}x^r.$$

(ii) *For all real numbers a and b and all integers $n \geq 0$,*

$$(a+b)^n = \sum_{r=0}^{n}\binom{n}{r}a^{n-r}b^r = \sum_{r=0}^{n}\left(\frac{n!}{r!(n-r)!}\right)a^{n-r}b^r.$$

Proof.   (i) This follows from replacing $\binom{n}{r}$ by Pascal's formula in Proposition 2.24.

(ii) The result is trivially true when $a = 0$ (we have agreed that $0^0 = 1$). If $a \neq 0$, set $x = b/a$ in part (i), and observe that

$$\left(1 + \frac{b}{a}\right)^n = \left(\frac{a+b}{a}\right)^n = \frac{(a+b)^n}{a^n}.$$

Hence,

$$(a+b)^n = a^n\left(1 + \frac{b}{a}\right)^n = a^n\sum_{r=0}^{n}\binom{n}{r}\frac{b^r}{a^r} = \sum_{r=0}^{n}\binom{n}{r}a^{n-r}b^r. \qquad \blacksquare$$

There are many beautiful connections between Pascal's triangle and number theory. For example, while it is not generally true that $n \mid \binom{n}{r}$ (for example, $4 \nmid 6 = \binom{4}{2}$), this result is true when n is prime.

Proposition 2.26. *If p is a prime, then $p \mid \binom{p}{r}$ for all r with $0 < r < p$.*

Proof. By Pascal's Theorem,

$$\binom{p}{r} = \frac{p!}{r!(p-r)!} = \frac{p(p-1)\cdots(p-r+1)}{r!},$$

and cross multiplying gives

$$r!\binom{p}{r} = p(p-1)\cdots(p-r+1);$$

that is, $p \mid r!\binom{p}{r}$. But each factor of $r!$ is strictly less than p, because $r < p$, so that p is not a divisor of any of them. Therefore, Euclid's Lemma says that $p \nmid r!$ and, hence, that p must divide $\binom{p}{r}$. ∎

Example 2.27. The Binomial Theorem can be used to express the sum of the nth powers of two variables a and b in terms of the "elementary symmetric functions" $a + b$ and ab. Here are some examples for $n = 2, 3, 4$; from

$$(a + b)^2 = a^2 + 2ab + b^2$$

we have

$$a^2 + b^2 = (a + b)^2 - 2ab.$$

From

$$(a + b)^3 = a^3 + 3a^2b + 3ab^2 + b^3$$

we conclude

$$a^3 + b^3 = (a + b)^3 - 3ab(a + b).$$

For $n = 4$,

$$\begin{aligned}(a + b)^4 &= a^4 + 4a^3b + 6a^2b^2 + 4ab^3 + b^4 \\ &= (a^4 + b^4) + 4ab(a^2 + b^2) - 6(ab)^2.\end{aligned}$$

Hence,

$$a^4 + b^4 = (a + b)^4 - 4ab(a^2 + b^2) + 6(ab)^2$$

We can now replace $a^2 + b^2$ by the already computed expression $(a+b)^2 - 2ab$, collect like terms, and have an expression for $a^4 + b^4$ in terms of $a + b$ and ab.

We could proceed inductively, expressing $a^n + b^n$ in terms of $a + b$ and ab for $n \geq 5$. Try a few more examples; you'll get the sense that there's a general method expressing $a^n + b^n$ in terms of $a + b$, ab, and other terms like $a^k + b^k$ with $k < n$. ▲

Exercises

2.21 * Show, without using the Binomial Theorem, that, if $n \geq 0$ is an integer, then

    (i) the degree of $(1 + x)^n$ is n

    (ii) the leading coefficient of $(1 + x)^n$ is 1

    (iii) the constant term of $(1 + x)^n$ is 1.

2.22 Prove Lemma 2.23 by induction on $n \geq 1$.

2.23 Prove that the binomial coefficients are *symmetric*: for all r with $0 \leq r \leq n$,

$$\binom{n}{r} = \binom{n}{n - r}.$$

2.24 Find a formula for the sum of the entries in the nth row of Pascal's triangle and prove your assertion.

2.25 If $n \geq 1$, find a formula for the alternating sum of the binomial coefficients in the nth row of Pascal's triangle:

$$\binom{n}{0} - \binom{n}{1} + \binom{n}{2} - \cdots + (-1)^n \binom{n}{n}.$$

Prove what you say.

Hint. Consider $f(x) = (1 + x)^n$ when $x = -1$.

2.26 If $n \geq 1$, find a formula for the sum of the squares of the binomial coefficients in the nth row of Pascal's triangle:

$$\binom{n}{0}^2 + \binom{n}{1}^2 + \binom{n}{2}^2 \cdots + \binom{n}{n}^2.$$

Prove what you say.

2.27 Prove, for a given $n \geq 1$, that the sum of all the binomial coefficients $\binom{n}{r}$ with r even is equal to the sum of all those $\binom{n}{r}$ with r odd.

2.28 The *triangular numbers* count the number of squares in a staircase of height n. Figure 2.8 displays the staircases of height n for $1 \leq n \leq 5$.

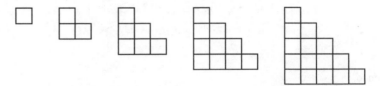

Figure 2.8. Triangular numbers.

(i) Find a formula for the nth triangular number in terms of binomial coefficients, and prove your assertion. Compare this exercise with the discussion of the $(n + 1) \times (n + 1)$ square in Example 2.6.

(ii) Show that the sum of two consecutive triangular numbers is a perfect square.

2.29 **Take It Further.** Using the notation of Example 2.27, use the Binomial Theorem and induction to show that $a^n + b^n$ can be expressed in terms of $a + b$ and ab.

2.30 Pascal's triangle enjoys a sort of *hockey stick* property: if you start at the end of any row and draw a hockey stick along a diagonal, as in Figure 2.9, the sum of the entries on the handle of the stick is the entry at the tip of the blade. Express the hockey stick property as an identity involving binomial coefficients and prove the identity.

2.31 (**Leibniz**) A function $f : \mathbb{R} \to \mathbb{R}$ is called a C^∞*-function* if it has an nth derivative $f^{(n)}(x)$ for every $n \geq 1$. Prove that if f and g are C^∞-functions, then

$$(fg)^{(n)}(x) = \sum_{k=0}^{n} \binom{n}{k} f^{(k)}(x) \cdot g^{(n-k)}(x).$$

In spite of the strong resemblance, there is no routine derivation of the Leibniz formula from the Binomial Theorem (there is a derivation using an idea from hypergeometric series).

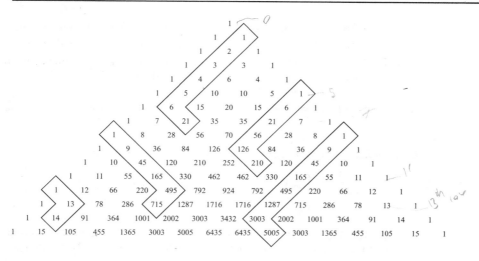

Figure 2.9. Hockey sticks.

2.32 * If p is a prime and a and b are integers, prove that there is an integer c with

$$(a + b)^p = a^p + b^p + pc.$$

Combinatorics

Binomial coefficients have a combinatorial interpretation. Given a set X with n elements, define an ***r-subset*** of X to be a subset having exactly r elements. How many r-subsets of X are there?

Example 2.28. There are ten 3-element subsets of the 5-element set $X = \{A, B, C, D, E\}$. Think of forming a 3-person committee from 5 people. A committee either contains Elvis or doesn't. The committees are

$$\{A, B, C\} \quad \{A, B, D\} \quad \{B, C, D\} \quad \{A, C, D\}$$
$$\{A, B, E\} \quad \{A, C, E\} \quad \{A, D, E\} \quad \{B, C, E\} \quad \{B, D, E\} \quad \{C, D, E\}$$

The first row consists of the 3-subsets that don't contain Elvis (there are four such); the second row displays the 3-subsets that do contain Elvis (there are six of these).   ▲

In general, if X has n elements and $0 \le r \le n$, denote the number of its r-element subsets by

$$[n, r];$$

that is, $[n, r]$ is the number of ways one can choose r things from a box of n things. Note that:

(i) $[n, 0] = 1$ (there's only one 0-subset, the empty set $\varnothing$).

(ii) $[n, n] = 1$ (there's only one n-subset of X, X itself).

When $n = 0$, items (i) and (ii) give the same answer. Why does this make sense?

If $0 < r < n$, you can compute $[n, r]$ using the committee idea in Example 2.28. If $X = \{a_1, a_2, \ldots, a_n\}$ and you want to build an r-subset, first choose a "distinguished" element of X, say a_n, and call him Elvis. Either your subset contains Elvis or it doesn't.

Case 1. If Elvis is in your r-subset, then you must pick $r - 1$ elements from the remaining $n - 1$; by definition, there are $[n - 1, r - 1]$ ways to do this.

Case 2. If Elvis is not in your r-subset, then you must pick all r elements from the remaining $n - 1$; there are $[n - 1, r]$ ways to do this.

It follows that $[n, r] = [n - 1, r - 1] + [n - 1, r]$. We have proved the following result.

Lemma 2.29. *For all integers $n \geq 1$ and all r with $0 \leq r \leq n$,*

$$[n, r] = \begin{cases} 1 & \text{if } n = 0 \text{ or } n = r \\ [n - 1, r] + [n - 1, r] & \text{if } 0 < r < n. \end{cases}$$

The similarity between Lemmas 2.23 and 2.29 inspires the next theorem. It is also the reason why the binomial coefficient $\binom{n}{r}$ is usually pronounced "n choose r."

Theorem 2.30 (Counting Subsets). *If n and r are nonnegative integers with $0 \leq r \leq n$, then*

$$\binom{n}{r} = [n, r].$$

Proof. Use induction on $n \geq 0$. If $n = 0$, the inequality $0 \leq r \leq n$ forces $r = 0$, and

$$\binom{0}{0} = 1 = [0, 0].$$

Suppose the result is true for $n - 1$. If $0 \leq r \leq n$, then

$$\binom{n}{r} = \binom{n - 1}{r - 1} + \binom{n - 1}{r} \qquad \text{Lemma 2.23}$$

$$= [n - 1, r - 1] + [n - 1, r] \qquad \text{inductive hypothesis}$$

$$= [n, r] \qquad \text{Lemma 2.29.} \quad \blacksquare$$

Theorems can often be proved in several ways. The following discussion gives another proof of Theorem 2.30 using Pascal's formula (Proposition 2.24) instead of Lemmas 2.23 and 2.29.

We first compute $[n, r]$ by considering a related question.

Definition. Given an "alphabet" with n (distinct) letters and an integer r with $1 \leq r \leq n$, an *r-anagram* is a sequence of r of these letters with no repetitions.

For example, the 2-anagrams on the alphabet a, b, c are

$$ab, \quad ba, \quad ac, \quad ca, \quad bc, \quad cb$$

(note that aa, bb, cc are not on this list). How many r-anagrams are there on an alphabet with n letters? We count the number of such anagrams in two ways.

(i) There are n choices for the first letter; since no letter is repeated, there are only $n - 1$ choices for the second letter, only $n - 2$ choices for the third letter, and so forth. Thus, the number of r-anagrams is

$$n(n - 1)(n - 2) \cdots (n - (r - 1)) = n(n - 1)(n - 2) \cdots (n - r + 1).$$

In the special case $n = r$, the number of n-anagrams on n letters is $n!$.

(ii) Here is a second way to count the anagrams. First choose an r-subset of the alphabet (consisting of r letters); there are $[n, r]$ ways to do this, for this is exactly what the symbol $[n, r]$ means. For each chosen r-subset, there are $r!$ ways to arrange the r letters in it (this is the special case of our first count when $n = r$). The number of r-anagrams is thus

$$r! \, [n, r].$$

We conclude that

$$r! \, [n, r] = n(n - 1)(n - 2) \cdots (n - r + 1),$$

from which it follows that

$$[n, r] = \frac{n(n - 1)(n - 2) \cdots (n - r + 1)}{r!} = \frac{n!}{(n - r)! \, r!}.$$

Therefore, Pascal's formula gives

$$[n, r] = \binom{n}{r}.$$

If you piece together the results of this section, you'll see that we have shown that the following ways to define binomial coefficients are all equivalent: starting from any one of them, you can derive the others.

Algebraic: $\binom{n}{r}$ is the coefficient of x^r in the expansion of $(1 + x)^n$.

Pascal:

$$\binom{n}{r} = \frac{n!}{r! \, (n - r)!}.$$

Combinatorial: $\binom{n}{r}$ is the number of r-element subsets of an n-element set.

Inductive:

$$\binom{n}{r} = \begin{cases} 1 & \text{if } n = 0 \text{ or } n = r \\ \binom{n-1}{r-1} + \binom{n-1}{r} & \text{if } 0 < r < n. \end{cases}$$

Example 2.31. If you replace the symbols by their definition, Theorem 2.30 says something that is far from obvious: the coefficient of x^r in $(1 + x)^n$ is the same as the number of r-element subsets of an n-element set. The proof by induction of Theorem 2.30 establishes this, but many people are left wondering if there is a more intuitive reason why the expansion of $(1 + x)^n$ contains all the information about subsets of various sizes from an n-element set.

If you were going to multiply out $(1 + x)^5$ by hand, you could view the calculation like this:

$$(1 + x)^5 = (1 + x)(1 + x)(1 + x)(1 + x)(1 + x)$$

The expansion is carried out by taking one term (1 or x) from each binomial factor $1 + x$, multiplying them together, and then collecting like powers of x.

For example, you could take a "1" from each of the first three binomials and an x from the last two. That would produce $1 \cdot 1 \cdot 1 \cdot x \cdot x = x^2$. But that's not the only way to get an x^2. You could have taken an x from the first and third binomials and 1 from the rest. Or an x from the first two binomials and 1 from the last three. Do this in every possible way; the coefficient of x^2 in the expansion will be the number of ways you can pick two binomials from the set of five to be "x terms." And there are precisely $10 = [5, 2]$ ways to do this.

Generalizing, view $(1 + x)^n$ as a product of n binomials:

$$(1 + x)^n = \underbrace{(1 + x)(1 + x)(1 + x) \ldots (1 + x)}_{n \text{ times.}}$$

The coefficient of x^r in this product is the number of ways you can choose r of the binomials to be "x terms" (and the rest to be 1). This number is precisely $[n, r]$. Hence

$$(1 + x)^n = \sum_{r=0}^{n} [n, r] x^r.$$

When combined with the definition of binomial coefficients on page 63, this gives another proof that $\binom{n}{r} = [n, r]$.   ▲

Exercises

2.33 *

(i) For each value of r, $0 \le r \le 4$, how many r-element subsets of the set $\{A, B, C, D\}$ are there?

(ii) For each value of r, $0 \le r \le 5$, how many r-element subsets of the set $\{A, B, C, D, E\}$ are there?

2.34 How many subsets (of any size) are there in an n-element set? Prove your assertion.

2.35 Show that

$$\binom{2n}{r} = \sum_{k=0}^{r} \binom{n}{k} \binom{n}{r - k}.$$

Hint. Split a $2n$-element set into two equal pieces.

2.36 Show that

$$\binom{3n}{r} = \sum_{k=0}^{r} \binom{n}{k} \binom{2n}{r - k}.$$

2.37 If m, n, and r are nonnegative integers, prove **Vandermonde's Identity**:

$$\binom{m+n}{r} = \sum_{k=0}^{r} \binom{m}{k}\binom{n}{r-k}.$$

Hint. $(1+x)^{m+n} = (1+x)^m (1+x)^n$.

2.38 Show that

$$\sum_{k=0}^{n} k\binom{n}{k} = n2^{n-1}.$$

2.39 How many ways can you choose two hats from a closet containing 14 different hats? (One of our friends does not like the phrasing of this exercise. After all, you can choose two hats with your left hand, with your right hand, with your teeth, ..., but we continue the evil tradition.)

2.40 Let D be a collection of ten different dogs, and let C be a collection of ten different cats. Prove that there are the same number of quartets of dogs as there are sextets of cats.

2.41 (i) What is the coefficient of x^{16} in $(1+x)^{20}$?

(ii) How many ways are there to choose 4 colors from a palette containing paints of 20 different colors?

2.42 A weekly lottery asks you to select 5 different numbers between 1 and 45. At the week's end, 5 such numbers are drawn at random, and you win the jackpot if all your numbers match the drawn numbers. What is your chance of winning?

The number of selections of 5 numbers is "45 choose 5", which is $\binom{45}{5} = 1,221,759$. The odds against your winning are more than a million to one.

2.3   Connections

An Approach to Induction

Teaching mathematical induction to high school students is often tough. In particular, many students fall into the trap we described on page 49: in spite of all our explanations to the contrary, they think that the inductive hypothesis assumes what it is they are supposed to be proving. In this section, we look at a well-tested method that avoids this trap.

Suppose you ask a class to come up with a function that agrees with the table

Input	Output
0	4
1	7
2	10
3	13
4	16
5	19

We've found that about half a high school class (beginning algebra, say) comes up with a *closed form definition*, something like $f(n) = 3n+4$, while the other half comes up with an *inductive definition*—something like "start with 4, and each output is 3 more than the previous one." This inductive definition can be written more formally:

$$g(n) = \begin{cases} 4 & \text{if } n = 0 \\ g(n-1) + 3 & \text{if } n > 0. \end{cases}$$

Of course, beginning students don't usually write the definition of g using this case notation. Technology can help get them used to it.

That inductive definitions (or **recursive definitions**, as they are often called) seem to be, in some sense, natural for students, can be exploited to help students understand proof by induction.

Computer algebra systems (CAS) let you model the two definitions. Building such computational models allows students to experiment with both functions, and it also provides an opportunity to launch some important ideas. For example, a teacher can use the models to discuss the domain of each function— f accepts any real number, but g will accept only nonnegative integers.

The part of g's definition involving n, the equation $g(n) = g(n-1) + 3$, is called a *recurrence*.

Here's what's most germane to this section. If you try some values in a spreadsheet or calculator, it seems, for a while, that f and g produce the same output when given the same input. But at some point (exactly where depends on the system), f outputs an integer but g surrenders. Suppose this happens, for example, at 255: both functions return 766 at 254, but $f(255) = 769$ while g returns an error. Is this because f and g really are not equal at 255? Or is it because of the limitations of the technology? Many students will say immediately that the functions *are* equal when $n = 255$; it's just that the computer can't compute the value of g there. Tell the students, "I believe you that $f(254) = g(254)$. Convince me that they are also equal at 255."

After some polishing and a little help, their argument usually goes something like this.

$$\begin{aligned} g(255) &= g(254) + 3 \quad \text{(this is how } g \text{ is defined)} \\ &= f(254) + 3 \quad \text{(the calculator said so—they both output 766)} \\ &= (3 \cdot 254 + 4) + 3 \quad \text{(this is how } f \text{ is defined)} \\ &= (3 \cdot 254 + 3) + 4 \quad \text{(algebra)} \\ &= 3(254 + 1) + 4 \quad \text{(more algebra)} \\ &= 3 \cdot 255 + 4 \quad \text{(arithmetic)} \\ &= f(255) \quad \text{(this is how } f \text{ is defined).} \end{aligned}$$

There's nothing special about 255 here. If you had a more powerful calculator, one that handled inputs to g up to, say, 567 (and then crashed at 568), you could argue that f and g were equal at 568, too.

$$\begin{aligned} g(568) &= g(567) + 3 \quad \text{(this is how } g \text{ is defined)} \\ &= f(567) + 3 \quad \text{(the powerful calculator said so)} \\ &= (3 \cdot 567 + 4) + 3 \quad \text{(this is how } f \text{ is defined)} \\ &= (3 \cdot 567 + 3) + 4 \quad \text{(algebra)} \\ &= 3(567 + 1) + 4 \quad \text{(more algebra)} \\ &= 3 \cdot 568 + 4 \quad \text{(arithmetic)} \\ &= f(568) \quad \text{(this is how } f \text{ is defined).} \end{aligned}$$

Computing $g(568)$ is the same as computing $g(255)$. Now imagine that you had a virtual calculator, one that showed that f and g agreed up to some integer

$n - 1$, but then crashed when you asked for $g(n)$. You could show that f and g are equal at n by the same argument:

$$\begin{aligned} g(n) &= g(n-1) + 3 \quad \text{(this is how g is defined)} \\ &= f(n-1) + 3 \quad \text{(the virtual calculator said so)} \\ &= (3 \cdot (n-1) + 4) + 3 \quad \text{(this is how f is defined)} \\ &= 3n + 4 \quad \text{(algebra)} \\ &= f(n) \quad \text{(this is how f is defined)}. \end{aligned}$$

So, every time f and g are equal at one integer, they are equal at the next one. Since f and g are equal at 0 (in fact, since they are equal at every integer between 0 and 254), they are equal at every nonnegative integer.

This argument is the essence of mathematical induction. In the example, it shows that if two functions f and g are equal at one integer, then they are equal at the next one. Coupled with the fact that they are equal at 0, it makes sense that they are equal for all integers greater than or equal to 0; that is, $f(n) = g(n)$ for all nonnegative integers n.

We have seen that induction applies in much more general situations than this one. But this simple context is quite effective in starting students onto a path that helps them understand induction.

Fibonacci Sequence

Many interesting investigations in high school center around the following sequence, which describes a pattern frequently found in nature and in art.

Definition. The *Fibonacci sequence* is defined by:

$$F_n = \begin{cases} 0 & n = 0 \\ 1 & n = 1 \\ F_{n-1} + F_{n-2} & n > 1. \end{cases}$$

There are two base steps in the definition: $n = 0$ and $n = 1$. The Fibonacci sequence begins: $0, 1, 1, 2, 3, 5, 8, 13, \ldots$.

Historical Note. The Fibonacci sequence is related to the *golden ratio*, a number mentioned in Euclid, Book 6, Proposition 30. It is said that the ancient Greeks thought that a rectangular figure is most pleasing to the eye (such rectangles can be seen in the Parthenon in Athens) if its edges a and b are in the proportion

$$\frac{a}{b} = \frac{b}{a+b}.$$

In this case, $a(a+b) = b^2$, so that $b^2 - ab - a^2 = 0$; that is, $(b/a)^2 - (b/a) - 1 = 0$. The quadratic formula gives $b/a = \frac{1}{2}(1 \pm \sqrt{5})$. Therefore,

$$b/a = \gamma = \tfrac{1}{2}(1 + \sqrt{5}) \quad \text{or} \quad b/a = \delta = \tfrac{1}{2}(1 - \sqrt{5}).$$

But δ is negative, and so we must have

$$b/a = \gamma = \tfrac{1}{2}(1 + \sqrt{5}).$$

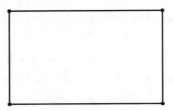

Figure 2.10. Golden rectangle.

The number $\gamma = 1.61803\ldots$, is called the **golden ratio**. Since both γ and δ are roots of $x^2 - x - 1$, we have

$$\gamma^2 = \gamma + 1 \quad \text{and} \quad \delta^2 = \delta + 1. \tag{2.4}$$

So, what's the connection of the golden ratio to the Fibonacci sequence? We discovered the closed form for the sequence c_n in Proposition 2.22 by tabulating the first few terms of the sequence and looking for regularity—it seemed "almost" exponential, off by 1 from a power of 2.

Let's tabulate the first few ratios of consecutive terms F_n/F_{n-1} of the Fibonacci sequence.

$$F_1/F_0 = 1/1 = 1$$
$$F_2/F_1 = 2/1 = 2$$
$$F_3/F_2 = 3/2 = 1.5$$
$$F_4/F_3 = 5/3 = 1.666$$
$$F_5/F_4 = 8/5 = 1.6$$
$$F_6/F_5 = 13/8 = 1.625.$$

If you tabulate a few more ratios (try it), a conjecture emerges—it appears that the ratio of two consecutive terms in the Fibonacci sequence might converge to the golden ratio $\gamma \approx 1.61803$ (if the ratios were actually *constant*, F_n would be a geometric sequence (why?)). This is, in fact, the case, and you'll see, in Exercise 2.50 on page 78, how to refine the conjecture into the statement of the following theorem (the exercise will also help you develop a method that will let you find closed forms for many 2-term recurrences).

Theorem 2.32. *For all $n \geq 0$, the nth term of the Fibonacci sequence satisfies*

$$F_n = \tfrac{1}{\sqrt{5}}(\gamma^n - \delta^n),$$

where $\gamma = \tfrac{1}{2}(1 + \sqrt{5})$ and $\delta = \tfrac{1}{2}(1 - \sqrt{5})$.

Proof. We use strong induction because the inductive step involves the formulas for both F_{n-1} and F_{n-2}. The base steps $S(0)$ and $S(1)$ are true:

$$\tfrac{1}{\sqrt{5}}(\gamma^0 - \delta^0) = 0 = F_0$$
$$\tfrac{1}{\sqrt{5}}(\gamma - \delta) = \tfrac{1}{\sqrt{5}}\left(\tfrac{1}{2}(1 + \sqrt{5}) - \tfrac{1}{2}(1 - \sqrt{5})\right) = 1 = F_1.$$

If $n \geq 2$, then

$$F_n = F_{n-1} + F_{n-2}$$
$$= \tfrac{1}{\sqrt{5}}(\gamma^{n-1} - \delta^{n-1}) + \tfrac{1}{\sqrt{5}}(\gamma^{n-2} - \delta^{n-2})$$
$$= \tfrac{1}{\sqrt{5}}\left[(\gamma^{n-1} + \gamma^{n-2}) - (\delta^{n-1} + \delta^{n-2})\right]$$
$$= \tfrac{1}{\sqrt{5}}\left[\gamma^{n-2}(\gamma + 1) - \delta^{n-2}(\delta + 1)\right]$$
$$= \tfrac{1}{\sqrt{5}}\left[\gamma^{n-2}(\gamma^2) - \delta^{n-2}(\delta^2)\right] \qquad \text{by Eq. (2.4)}$$
$$= \tfrac{1}{\sqrt{5}}(\gamma^n - \delta^n). \quad \blacksquare$$

Isn't it curious that the integers F_n are expressed in terms of the irrational number $\sqrt{5}$?

Corollary 2.33. $F_n > \gamma^{n-2}$ *for all integers* $n \geq 3$, *where* $\gamma = \tfrac{1}{2}(1 + \sqrt{5})$.

If $n = 2$, then $F_2 = 1 = \gamma^0$, and so there is equality, not inequality.

Proof. The proof is by induction on $n \geq 3$. The base step $S(3)$ is true, for $F_3 = 2 > \gamma \approx 1.618$. For the inductive step, we must show that $F_{n+1} > \gamma^{n-1}$. By the inductive hypothesis,

$$F_{n+1} = F_n + F_{n-1} > \gamma^{n-2} + \gamma^{n-3}$$
$$= \gamma^{n-3}(\gamma + 1) = \gamma^{n-3}\gamma^2 = \gamma^{n-1}. \quad \blacksquare$$

Exercises

2.43 Show that the following functions agree on all natural numbers.

$$f(n) = 3n + 5 \quad \text{and} \quad g(n) = \begin{cases} 5 & \text{if } n = 0 \\ g(n-1) + 3 & \text{if } n > 0. \end{cases}$$

2.44 Show that the following two functions agree on all natural numbers.

$$f(n) = 4^n \quad \text{and} \quad g(n) = \begin{cases} 4 & \text{if } n = 0 \\ 4g(n-1) & \text{if } n > 0. \end{cases}$$

2.45 Define the function h inductively:

$$h(n) = \begin{cases} 4 & \text{if } n = 0 \\ h(n-1) + 2n & \text{if } n > 0 \end{cases}$$

Find a polynomial function p that agrees with h on all natural numbers, and prove that your functions are equal on $\mathbb{N}$.

Answer. $n^2 + n + 4$.

2.46 Define the function m inductively:

$$m(n) = \begin{cases} 0 & \text{if } n = 0 \\ m(n-1) + n^2 & \text{if } n > 0 \end{cases}$$

Find a polynomial function s that agrees with m on all natural numbers, and prove that your functions are equal on $\mathbb{N}$.

Answer. $\dfrac{2n^3 + 3n^2 + n}{6} = \dfrac{n(n+1)(2n+1)}{6}$.

2.47 Consider the two functions f and g:

$$f(x) = x^4 - 6x^3 + 14x^2 - 6x + 2$$

and

$$g(x) = \begin{cases} 2 & \text{if } x = 0 \\ s(x-1) + 6x - 3 & \text{if } x > 0. \end{cases}$$

Are f and g equal on $\mathbb{N}$?

2.48 Find a formula for $0^2 + 1^2 + 2^2 + \cdots + (n-1)^2$ as a function of n, and prove your assertion.

2.49 You saw, on page 76, that the ratio of two consecutive terms seems to converge to the golden ratio. Using only the recurrence

$$F_n = F_{n-1} + F_{n-2} \quad \text{for all } n \geq 2,$$

show that

$$\lim_{n \to \infty} \frac{F_n}{F_{n-1}} = \tfrac{1}{2}(1 + \sqrt{5}).$$

2.50 * You saw, on page 76, that the Fibonacci sequence seems to be "almost" exponential.

(i) Suppose the Fibonacci sequence actually was exponential: $F_n = r^n$. Show that r would have to be either

$$\gamma = \frac{1 + \sqrt{5}}{2} \quad \text{or} \quad \delta = \frac{1 - \sqrt{5}}{2}.$$

(ii) Show that the sequences γ^n and δ^n satisfy the recurrence

$$f_n = f_{n-1} + f_{n-2}.$$

(iii) If a and b are any real numbers, show that $a\gamma^n + b\delta^n$ satisfies the recurrence

$$f_n = f_{n-1} + f_{n-2}.$$

(iv) Without using Theorem 2.32, find a and b so that

$$a\gamma^n + b\delta^n = F_n.$$

2.51 Ms. D'Amato likes to take a different route to work every day. She will quit her job the day she has to repeat her route. Her home and work are pictured in the grid of streets in Figure 2.11. If she never backtracks (she only travels north or east), how many days will she work at her job?

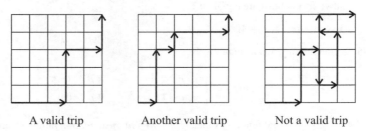

A valid trip　　　　Another valid trip　　　　Not a valid trip

Figure 2.11. Ms. D'Amato.

2.52 Find a closed form for each of the following functions and prove your assertions.

(i) $f(n) = \begin{cases} 4 & n = 0 \\ f(n-1) + 3 & n > 0. \end{cases}$

   Answer. $f(n) = 3n + 4.$

(ii) $f(n) = \begin{cases} 4 & n = 0 \\ 3f(n-1) & n > 0. \end{cases}$

   Answer. $f(n) = 4 \cdot 3^n.$

(iii) $f(n) = \begin{cases} 2 & n = 0 \\ 4 & n = 1 \\ 4f(n-1) - 3f(n-2) & n > 1. \end{cases}$

   Answer. $f(n) = 3^n + 1.$

(iv) $f(n) = \begin{cases} 4 & n = 0 \\ 4 & n = 1 \\ 4f(n-1) - 3f(n-2) & n > 1. \end{cases}$

   Answer. $f(n) = 4.$

2.53 Find a closed form for the following function and prove your assertion.

$$f(n) = \begin{cases} 3 & n = 0 \\ 4 & n = 1 \\ 14 & n = 2 \\ 4f(n-1) - f(n-2) - 6f(n-3) & n > 2. \end{cases}$$

2.54 Take It Further. Find or generate a copy of the first 30 rows of Pascal's triangle. Color the odd numbers red and the even numbers black. Explain any patterns that you see. (Alternatively, you can use a spreadsheet to generate the triangle of 0s and 1s that are the remainders when each entry is divided by 2.) For more on this exercise, see

ecademy.agnesscott.edu/~lriddle/ifs/siertri/Pascalmath.htm

3

Renaissance

For centuries, the Western World believed that the high point of civilization took place from the Greek and Roman eras through the beginning of Christianity. But this worldview began to change dramatically about five hundred years ago. The printing press was invented around 1450, by Johannes Gutenberg, Christopher Columbus landed in North America in 1492, Martin Luther began the Reformation in 1517, and Nicolas Copernicus published *De Revolutionibus* in 1530.

Mathematics was also developing. A formula giving the roots of certain cubic polynomials, similar to the quadratic formula, was discovered by Scipione del Ferro around 1515; by 1545, it was extended to all cubics by Fontana (Tartaglia) and Cardano. The cubic formula contributed to the change in worldview that was the essence of the Renaissance, for it was one of the first mathematical results not known to the ancients. But its impact on contemporary mathematics was much deeper, for it introduced complex numbers. As we shall see, the cubic formula is not as useful for numerical computations as we'd like, because it often gives roots in unrecognizable form. Its importance, however, lies in the ideas it generated. Trying to understand the formula, searching for generalizations of it, and studying questions naturally arising from such endeavors, were driving forces in the development of abstract algebra.

In many high school algebra courses today, the complex numbers, usually denoted by $\mathbb{C}$, are introduced to find the roots of $ax^2 + bx + c$ when $b^2 - 4ac < 0$. That's not how it happened. Square roots of negative real numbers occur in the cubic formula, but not as roots; indeed, in the 16th century, complex roots would have been considered useless. But complex numbers arose in the middle of calculations, eventually producing real numbers (we will see this explicitly in the next section). To understand this phenomenon, mathematicians were forced to investigate the meaning of *number*; are complex numbers bona fide numbers? Are negative numbers bona fide numbers?

Section 3.1 discusses the classical formulas giving the roots of cubic and quartic polynomials. We will look more carefully at the complex numbers themselves in Section 3.2. Although initially used in purely algebraic contexts, $\mathbb{C}$ has a rich geometric and analytic structure that, when taken together with its algebraic properties, can tie together many of the ideas in high school mathematics. Indeed, $\mathbb{C}$ finds applications all over mathematics. Section 3.4 uses $\mathbb{C}$ to solve some problems that are especially useful for teachers (and interesting for all mathematicians). Just as the method of Diophantus was used to create Pythagorean triples, $\mathbb{C}$ can be used to invent problems whose solutions "come out nice."

3.1   Classical Formulas

As Europe emerged from the Dark Ages, a major open problem in mathematics was finding solutions to polynomial equations. The quadratic formula had been known for about four thousand years and, arising from a tradition of public mathematical contests in Pisa and Venice, formulas for the roots of cubics and quartics had been found in the early 1500s. Let's look at these formulas in modern algebraic notation; we will assume for now that the complex numbers obey the usual laws of arithmetic (neither of these simplifying steps was available to mathematicians of the 16th century).

Historical Note.   Modern arithmetic notation was introduced in the late 1500s, but it was not generally agreed upon in Europe until after the influential book of Descartes, *La Géométrie*, was published in 1637 (before then, words and abbreviations were used as well as various competing notations). The symbols $+$, $-$, and $\sqrt{}$, as well as the symbol / for division, as in $4/5$, were introduced by Widman in 1486. The equality sign, $=$, was invented by Recorde in 1557. Designating variables by letters was invented by Viète in 1591, who used consonants to denote constants and vowels to denote variables; the modern notation, using the letters $a, b, c, \ldots$ to denote constants and the letters x, y, z at the end of the alphabet to denote variables, was introduced by Descartes in 1637. The exponential notation $2^2, 2^3, 2^4, \ldots$ was essentially invented by Hume in 1636, who wrote $2^{\mathrm{ii}}, 2^{\mathrm{iii}}, 2^{\mathrm{iv}}, \ldots$. The symbol $\times$ for multiplication was introduced by Oughtred in 1631; the symbol $\div$ for division was introduced by Rahn in 1659. See Cajori [6].

Cubics

The following familiar fact (to be proved in Chapter 6) was known and used by Renaissance mathematicians, and we will use it in this section.

Proposition 6.15.   *If r is a root of a polynomial $f(x)$, then $x - r$ is a factor of $f(x)$; that is, $f(x) = (x - r)g(x)$ for some polynomial $g(x)$.*

One of the simplest cubics is $f(x) = x^3 - 1$. Obviously, 1 is root of f, and so $x^3 - 1 = (x - 1)g(x)$, where

$$g(x) = (x^3 - 1)/(x - 1) = x^2 + x + 1.$$

The roots of g (and, hence, also of f) are

$$\omega = \tfrac{1}{2}(-1 + i\sqrt{3}) \quad \text{and} \quad \overline{\omega} = \tfrac{1}{2}(-1 - i\sqrt{3}),$$

by the quadratic formula. Both ω and $\overline{\omega}$ are called *cube roots of unity*, for $\omega^3 = 1 = \overline{\omega}^3$. Note that $\overline{\omega} = \omega^2 = 1/\omega$.

We know that a positive number a has two square roots. By convention, $\sqrt{a}$ denotes the positive square root, so that the two square roots are $\pm\sqrt{a}$. Any real number a has three cube roots. By convention, $\sqrt[3]{a}$ denotes the real cube root, so that the three cube roots are $\sqrt[3]{a}$, $\omega\sqrt[3]{a}$, $\omega^2\sqrt[3]{a}$. Thus, cube roots of unity generalize $\pm$.

The general cubic equation $aX^3 + bX^2 + cX + d = 0$ can be simplified by dividing both sides by a; this procedure does not affect the roots, and so we may assume that $a = 1$. Thus, we seek the roots of the cubic polynomial $F(X) = X^3 + bX^2 + cX + d$, where $b, c, d \in \mathbb{R}$. The change of variable

$$X = x - \tfrac{1}{3}b$$

yields a simpler polynomial, $f(x) = F(x - \tfrac{1}{3}b) = x^3 + qx + r$, where q and r are expressions in b, c, and d. We call f the **reduced polynomial** arising from F.

Lemma 3.1. *Let $f(x) = x^3 + qx + r$ be the reduced polynomial arising from $F(X) = X^3 + bX^2 + cX + d = 0$. If u is a root of f, then $u - \tfrac{1}{3}b$ is a root of F.*

Proof. Since $f(x) = F(x - \tfrac{1}{3}b)$ for all x, we have $0 = f(u) = F(u - \tfrac{1}{3}b)$; that is, $u - \tfrac{1}{3}b$ is a root of $F(X)$. ∎

We will use the following consequence of the quadratic formula.

Lemma 3.2. *Given a pair of numbers M and N, there are (possibly complex) numbers g and h with $g + h = M$ and $gh = N$. In fact, g and h are roots of $x^2 - Mx + N$.*

Proof. We have

$$(x - g)(x - h) = x^2 - (g + h)x + gh.$$

Thus, the roots g, h of $f(x) = x^2 - Mx + N$ (which exist, thanks to the quadratic formula) satisfy the given equations $g + h = M$ and $gh = N$. ∎

Let's try to find a general method for solving cubic equations—a method that doesn't depend on the specific values of the coefficients—by first solving a numerical equation.

Consider the polynomial $f(x) = x^3 - 18x - 35$. Since the constant term $35 = 5 \cdot 7$, we check whether $\pm 1, \pm 5, \pm 7$ are roots. It turns out that 5 is a root and, dividing by $x - 5$, we can find the other two roots by solving the quadratic $f(x)/(x - 5) = x^2 + 5x + 7$. But we are looking for a general method applicable to other cubics, so let's pretend we don't know that 5 is a root.

It's natural to look for a polynomial identity having the same form as the equation we are trying to solve. Example 2.27 provides one. From

$$a^3 + b^3 = (a + b)^3 - 3ab(a + b),$$

we have the identity

$$(a + b)^3 - 3ab(a + b) - \left(a^3 + b^3\right) = 0.$$

Thinking of $a + b$ as a single "chunk," say, $x = a + b$, the correspondence looks like this:

$$
\begin{array}{ccccccc}
(a+b)^3 & -3ab & \cdot & (a+b) & -(a^3+b^3) & = & 0 \\
\downarrow & \downarrow & & \downarrow & \downarrow & & \\
\square^3 & -3ab & \cdot & \square & -(a^3+b^3) & = & 0 \\
\uparrow & \uparrow & & \uparrow & \uparrow & & \\
x^3 & -18 & \cdot & x & -35 & = & 0.
\end{array}
$$

Exercise 3.5 on page 89 asks you to check that the coefficient of x^2 in $F(x - \tfrac{1}{3}b) = (x - \tfrac{1}{3}b)^3 + b(x - \tfrac{1}{3}b)^2 + c(x - \tfrac{1}{3}b) + d$ is zero.

You can check that the other two roots are complex. Renaissance mathematicians would have dismissed these as meaningless. But stay tuned—we'll soon see that they, too, can be generated by the emerging method.

So, if we can find two numbers a and b such that

$$-3ab = -18 \quad \text{and} \quad a^3 + b^3 = 35,$$

then $a + b$ will be a root of the cubic. Hence we want a and b so that $ab = 6$ and $a^3 + b^3 = 35$. There's an obvious solution here, namely, $a = 3$ and $b = 2$, but we're looking for a general method. Cubing both sides of $ab = 6$, we get

$$a^3 b^3 = 216 \quad \text{and} \quad a^3 + b^3 = 35.$$

By Lemma 3.2, a^3 and b^3 are roots of the quadratic equation

$$x^2 - 35x + 216 = 0.$$

The roots of this are 27 and 8. So we can take $a^3 = 27$, $b^3 = 8$; surprise! $a = 3$ and $b = 2$. Hence, $3 + 2 = 5$ is a root of our original cubic.

The next theorem is usually attributed to Scipione del Ferro; we'll use complex numbers and modern notation in its statement and proof, neither of which was available at the time. In light of Lemma 3.1, we may assume that cubics are reduced.

Theorem 3.3 (Cubic Formula). *The roots of* $f(x) = x^3 + qx + r$ *are*

$$g + h, \quad \omega g + \omega^2 h, \quad \text{and} \quad \omega^2 g + \omega h,$$

where $\omega = \frac{1}{2}\left(-1 + i\sqrt{3}\right)$ *is a cube root of unity,*

$$g^3 = \frac{-r + \sqrt{R}}{2}, \quad h = -\frac{q}{3g}, \quad \text{and} \quad R = r^2 + \frac{4q^3}{27}.$$

Proof. Let u be a root of $f(x) = x^3 + qx + r$ and, as in the discussion above, we try

$$u = g + h.$$

We are led to

$$g^3 + h^3 = -r$$
$$gh = -\tfrac{1}{3}q.$$

Cube $gh = -\tfrac{1}{3}q$, obtaining the pair of equations

$$g^3 + h^3 = -r$$
$$g^3 h^3 = -\tfrac{1}{27}q^3.$$

Lemma 3.2 gives a quadratic equation whose roots are g^3 and h^3:

$$x^2 + rx - \tfrac{1}{27}q^3 = 0. \tag{3.1}$$

The quadratic formula gives

$$g^3 = \tfrac{1}{2}\left(-r + \sqrt{r^2 + \tfrac{4}{27}q^3}\right) = \tfrac{1}{2}\left(-r + \sqrt{R}\right)$$

and

$$h^3 = \tfrac{1}{2}\left(-r - \sqrt{r^2 + \tfrac{4}{27}q^3}\right) = \tfrac{1}{2}\left(-r - \sqrt{R}\right).$$

Now there are three cube roots of g^3, namely, g, ωg, and $\omega^2 g$. Because of the constraint $gh = -\tfrac{1}{3}q$, each has a "mate," namely, $-q/(3g) = h$, $-q/(3\omega g) = \omega^2 h$, and $-q/(3\omega^2 g) = \omega h$. Thus, the roots of f are

$$g + h, \qquad \omega g + \text{its mate}, \qquad \omega^2 g + \text{its mate};$$

that is, the roots of f are $g + h$, $\omega g + \omega^2 h$, and $\omega^2 g + \omega h$. ∎

Example 3.4 (Good Example). If $f(x) = x^3 - 15x - 126$, then $q = -15$, $r = -126$, $R = 15376$, and $\sqrt{R} = 124$. Hence, $g^3 = 125$, so we can take $g = 5$. Thus, $h = -q/3g = 1$. Therefore, the roots of f are

$$6, \quad 5\omega + \omega^2 = -3 + 2i\sqrt{3}, \quad 5\omega^2 + \omega = -3 - 2i\sqrt{3}.$$

For Renaissance mathematicians, this cubic would have only one root—they would have ignored the complex roots. ▲

> Alternatively, having found one root to be 6, the other two roots can be found as the roots of the quadratic $f(x)/(x - 6) = x^2 + 6x + 21$.

But things don't always work out as we expect, as the next surprising example shows.

Example 3.5 (Bad Example). The cubic formula may give the roots in unrecognizable form. Let

$$f(x) = (x - 1)(x - 2)(x + 3) = x^3 - 7x + 6;$$

the roots of f are, obviously, $1, 2$, and -3. But the cubic formula gives

$$g^3 = \tfrac{1}{2}\left(-6 + \sqrt{\tfrac{-400}{27}}\right) \quad \text{and} \quad h^3 = \tfrac{1}{2}\left(-6 - \sqrt{\tfrac{-400}{27}}\right).$$

It is not at all obvious that $g + h$ is a real number, let alone an integer! ▲

Imagine yourself, standing in Piazza San Marco in Venice in 1520, participating in a contest. Your opponent challenges you to find a root of $f(x) = x^3 - 7x + 6$ (he invented the cubic, so he knows that it comes from $(x - 1)(x - 2)(x + 3)$). Still, you are a clever rascal; your mentor taught you the cubic formula. You do as you were taught, and triumphantly announce that a root is $g + h$, where g^3, h^3 are the awful expressions above. Most likely, the judges would agree that your opponent, who says that 1 is a root, has defeated you. After all, $f(1) = 1 - 7 + 6 = 0$, so that 1 is, indeed, a root. The judges even snickered when they asked you to evaluate $f(g + h)$.

With head hung low, you return home. Can you simplify your answer? Why is $g + h$ equal to 1? Let's pretend you have modern notation. Well,

$$g^3 = \tfrac{1}{2}\left(-6 + \sqrt{\tfrac{-400}{27}}\right) = -3 + i\tfrac{10\sqrt{3}}{9}.$$

The first question is how to compute cube roots of "numbers" of the form $a + bi$, where $i^2 = -1$. Specifically, we want $u + iv$ with

$$(u + iv)^3 = -3 + i\tfrac{10\sqrt{3}}{9}.$$

Hmm! Perhaps it's smart to separate terms involving i from honest numbers.

$$(u + iv)^3 = u^3 + 3u^2 iv + 3u(iv)^2 + (iv)^3$$
$$= u^3 + 3u^2 iv - 3uv^2 - iv^3$$
$$= u^3 - 3uv^2 + i(3u^2 v - v^3).$$

But see Example 3.36, or try to solve the system $u^3 - 3uv^2 = -3$ and $3u^2 v - v^3 = \frac{10\sqrt{3}}{9}$ with a computer.

Let's see if the separation pays off. We want numbers u, v with $u^3 - 3uv^2 = -3$ and $3u^2 v - v^3 = \frac{10\sqrt{3}}{9}$. These equations are intractible! Sigh.

Cube roots are tough. Let's simplify things; perhaps solving a simpler problem, say, finding square roots, can give a clue to finding cube roots. And this we can do.

Proposition 3.6. *Every complex number $a + bi$ has a square root.*

Proof. If $b = 0$, then $a + ib = a$. If $a \geq 0$, then $\sqrt{a}$ is well-known; if $a < 0$, then $a = -c$, where $c > 0$, and $\sqrt{a} = i\sqrt{c}$. We can now assume that $b \neq 0$, and our task is to find $u + iv$ with

$$(u + iv)^2 = u^2 + 2iuv - v^2 = a + ib;$$

that is, we seek numbers u, v such that

$$u^2 - v^2 = a \tag{3.2}$$

and

$$2uv = b. \tag{3.3}$$

Since $b \neq 0$, Eq. (3.3) gives $u \neq 0$; define $v = b/2u$. Substituting into Eq. (3.2), we have

$$u^2 - (b/2u)^2 = a;$$

rewriting,

$$4u^4 - 4au^2 - b^2 = 0.$$

This is a quadratic in u^2, and the quadratic formula gives

$$u^2 = \tfrac{1}{8}\left(4a \pm \sqrt{16a^2 + 16b^2}\right)$$
$$= \tfrac{1}{2}\left(a \pm \sqrt{a^2 + b^2}\right).$$

Since $a^2 + b^2 > 0$, it has a real square root. Now $\tfrac{1}{2}\left(a + \sqrt{a^2 + b^2}\right)$ is positive (because $b^2 > 0$ implies $a < \sqrt{a^2 + b^2}$); hence, we can find its (real) square root u as well as $v = b/2u$. ∎

For example, our proof gives a method finding a square root of i. Set $a = 0$ and $b = 1$ to obtain

$$i = \left(\tfrac{1}{\sqrt{2}}(1 + i)\right)^2.$$

We'll see how to find the roots of a complex number in Section 3.3.

Alas, this square root success doesn't lead to a cube root success, although it does give us some confidence that our manipulations may be legitimate.

You can now appreciate the confusion produced by the cubic formula; a cloud enveloped our ancestors. First of all, what are these "numbers" $a + ib$? Sometimes they can help. Can we trust them to always give us the truth? Is it true that we can separate terms involving i from those that don't? When are two complex numbers equal? Does it make sense to do arithmetic with these guys? Do they obey the nine properties of arithmetic on page 40 that familiar numbers do? It took mathematicians about 100 years to become comfortable with complex numbers, and another 100 years until all was set on a firm foundation.

Quartics

A method for solving fourth degree equations was found by Lodovico Ferrari in the 1540s, but we present the version given by Descartes in 1637.

Consider the quartic $F(X) = X^4 + bX^3 + cX^2 + dX + e$. The change of variable $X = x - \frac{1}{4}b$ yields a simpler polynomial, $f(x) = F(x - \frac{1}{4}b) = x^4 + qx^2 + rx + s$, whose roots give the roots of F: if u is a root of f, then $u - \frac{1}{4}b$ is a root of F. Write f as a product of two quadratics:

See Exercise 3.6 on page 89.

$$f(x) = x^4 + qx^2 + rx + s = (x^2 + jx + \ell)(x^2 - jx + m),$$

and determine j, ℓ, and m (note that the coefficients of the linear terms in the quadratic factors are j and $-j$ because f has no cubic term). Expanding and equating like coefficients gives the equations

$$\ell + m - j^2 = q,$$
$$j(m - \ell) = r,$$
$$\ell m = s.$$

The first two equations give

Since $j(m - \ell) = r$, we have $-\ell + m = r/j$.

$$2m = j^2 + q + \frac{r}{j},$$
$$2\ell = j^2 + q - \frac{r}{j}.$$

Substituting these values for m and ℓ into the third equation and simplifying yield a degree 6 polynomial which is a cubic in j^2 (called the **resolvent cubic**):

$$(j^2)^3 + 2q(j^2)^2 + (q^2 - 4s)j^2 - r^2.$$

The cubic formula gives a root j^2, from which we can determine m and ℓ and, hence, the roots of the quartic.

This process is an algorithm that can easily be encoded in a computer algebra system; it is known as the **quartic formula**. The quartic formula has the same disadvantage as the cubic formula: even though it gives correct answers, the values it gives for the roots are usually unrecognizable. But there are some good examples.

Example 3.7. Let's find the roots of

$$f(x) = x^4 - 10x^2 + 1.$$

First, factor f:

$$x^4 - 10x^2 + 1 = (x^2 + jx + \ell)(x^2 - jx + m);$$

in our earlier notation, $q = -10, r = 0$, and $s = 1$. The quartic formula shows us how to find j, ℓ, m. Since $r = 0$, we have $2\ell = j^2 - 10 = 2m$; hence, $\ell = m$. But $\ell m = 1$, so that either $\ell = 1$ and $j^2 = 12$ or $\ell = -1$ and $j^2 = 8$. Taking $\ell = 1$ and $j^2 = 12$ gives

$$f(x) = (x^2 + \sqrt{12}x + 1)(x^2 - \sqrt{12}x + 1),$$

and the quadratic formula gives the four roots of f:

$$\alpha = \sqrt{2} + \sqrt{3}, \quad \beta = -\sqrt{2} + \sqrt{3}, \quad \gamma = \sqrt{2} - \sqrt{3}, \quad \delta = -\sqrt{2} - \sqrt{3}. \quad \blacktriangle$$

The quadratic formula can be derived in a way similar to the derivations of the cubic and quartic formulas (in Chapter 1, we derived the formula by completing the square). The change of variable $X = x - \frac{1}{2}b$ replaces the polynomial $F(X) = X^2 + bX + c$ with the simpler polynomial $f(x) = x^2 + q$, where $q = c - \frac{1}{4}b^2$; the roots $u = \pm\sqrt{-q}$ of $f(x)$ give the roots $u - \frac{1}{2}b$ of F. Since the roots of f are

$$u = \pm\sqrt{-q} = \pm\sqrt{-(c - \frac{1}{4}b^2)} = \pm\frac{1}{2}\sqrt{b^2 - 4c},$$

the roots of F are our old friends

$$\pm\frac{1}{2}\sqrt{b^2 - 4c} - \frac{1}{2}b = \frac{1}{2}\left(-b \pm \sqrt{b^2 - 4c}\right).$$

It is now tempting, as it was for our ancestors, to try to find the roots of the general quintic $F(X) = X^5 + bX^4 + cX^3 + dX^2 + eX + f$ and to express them in a form similar to those for quadratic, cubic, and quartic polynomials; that is, using only extraction of roots, addition, subtraction, multiplication, and division (of course, our ancestors hoped to find roots of polynomials of any degree). They began with the change of variable $X = x - \frac{1}{5}b$ to eliminate the X^4 term. It was natural to expect that some further ingenious substitution together with the formulas for roots of polynomials of lower degree, analogous to the resolvent cubic, would yield the roots of F. For almost 300 years, no such formula was found. But, in 1824, Abel proved that there is no such quintic formula.

How to Think About It. Abel's theorem is often misquoted. It says: there is no formula involving only extraction of roots and the four basic operations of arithmetic that expresses the roots of the general quintic polynomial in terms of its coefficients. Succinctly, the general quintic is not *solvable by radicals*. But there are other kinds of formulas giving roots of polynomials. For example, here is a formula, due to Viète, giving the roots in terms of trigonometric functions. If $f(x) = x^3 + qx + r$ has three real roots, then its roots are $t \cos \theta$, $t \cos(\theta + 120°)$, $t \cos(\theta + 240°)$, where $t = \sqrt{-4q/3}$ and $\cos(3\theta) = -4r/t^3$ (there are variations using cosh and sinh when f has complex roots ([26], p. 445–447)). You may recall Newton's method giving the roots as $\lim_{n \to \infty} x_n$, where $x_{n+1} = x_n - f(x_n)/f'(x_n)$. Now *some* quintic polynomials are solvable by radicals; for example, we'll see in Section 3.3 that $x^5 - 1$ is one such. Another theorem of Abel gives a class of polynomials, of any degree, which are solvable by radicals. Galois, the young wizard who was killed before his 21st birthday, characterized *all* the polynomials which are solvable by radicals, greatly generalizing Abel's theorem. We will look at this more closely in Chapter 9.

Exercises

3.1 For each equation, find all roots in $\mathbb{R}$ and in $\mathbb{C}$

(i) $x^2 - 2x = 15$ (ii) $x^2 - 2x = 16$ (iii) $x^2 - 2x = -16$

(iv) $6x^2 + x = 15$ (v) $6x^2 + x = 16$ (vi) $6x^2 + x = -16$

(vii) $x^2 = 1$ (viii) $x^3 = 1$ (ix) $x^4 = 1$

(x) $x^3 = 8$

3.2 * We know that i satisfies $x^2 + 1 = 0$ in $\mathbb{C}$ (is there another solution?).

(i) Show, for all $n \in \mathbb{Z}$, that the value of i^n is one of $1, i, -1, -i$.

(ii) Use the Division Algorithm to decide which of the four values i^{247} will have.

3.3 Let $\omega = \frac{1}{2}(-1 + i\sqrt{3})$ be a cube root of unity.

(i) Show, for every integer n, the value of ω^n is one of $1, \omega, \omega^2$.

(ii) Use the Division Algorithm to decide, for any fixed n, which of the three values ω^n will have.

3.4 Find two numbers whose

(i) sum is 5 and product is 6.   (ii) sum is 0 and product is -2.

(iii) sum is 3 and product is 3.   (iv) sum is -1 and product is 1.

(v) sum is b and product is c (in terms of b and c).

3.5 * If $F(X) = X^3 + bX^2 + cX + d$, show that the change of variable $X = x - \frac{1}{3}b$ produces a polynomial f with no quadratic term,

$$f(x) = F(x - \tfrac{1}{3}b) = x^3 + qx + r.$$

Express q and r in terms of b, c, and d.

3.6 *

(i) Suppose that $F(X) = X^4 + bX^3 + cX^2 + dX + e$.

  (a) Show that the change of variable $X = x - \frac{1}{4}b$ produces a polynomial f with no cubic term,

$$f(x) = F(x - \tfrac{1}{4}b) = x^4 + qx^2 + rx + s.$$

    Express q, r, and s in terms of b, c, d, and e.

  (b) Show that if u is a root of f, then $u - \frac{1}{4}b$ is a root of F.

(ii) In general, let

$$F(X) = X^n + a_{n-1}X^{n-1} + a_{n-2}X^{n-2} + \cdots + a_0$$

be a polynomial of degree n.

  (a) Show that the change of variable $X = x - \frac{1}{n}a_{n-1}$ produces a polynomial f with no term of degree $n - 1$,

$$f(x) = F(x - \tfrac{1}{n}a_{n-1}) = x^n + q_{n-2}x^{n-2} + \cdots + q_0.$$

  (b) Show that if u is a root of f, then $u - \frac{1}{n}a_{n-1}$ is a root of F.

3.7 Take It Further. Suppose that g and h are complex numbers and

$$\omega = \tfrac{1}{2}\left(-1 + i\sqrt{3}\right).$$

Show that

$$g^3 + h^3 = (g + h)(\omega g + \omega^2 h)(\omega^2 g + \omega h).$$

3.8 In Example 3.7, we found the roots of $x^4 - 10x^2 + 1$ by factoring it into two quadratics (which came from taking, in the notation of page 87, $\ell = 1$ and $j^2 = 12$). Another choice was $\ell = -1$ and $j^2 = 8$.

(i) Using the alternate choice, get a different factorization of the quartic into quadratic factors.

(ii) Show that the two factorizations produce the same linear factors.

3.9 The following problem, from an old Chinese text, was solved by Qin Jiushao (Ch'in Chiu-shao) in 1247. There is a circular castle (see Figure 3.1) whose diameter is unknown; it is provided with four gates, and two lengths out of the north gate there is a large tree, which is visible from a point six lengths east of the south gate. What is the length of the diameter? (The answer is a root of a cubic polynomial.)

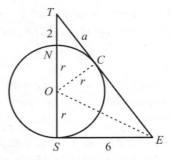

Figure 3.1. Castle problem.

3.10 Show that there is no real number whose square is -1.

3.11 (i) Find the roots of $x^3 - 3x + 1 = 0$.

(ii) Find the roots of $x^4 - 2x^2 + 8x - 3 = 0$.

3.12 Find a complex number s so that $s^3 = 9 - 46i$.

3.13 Find the roots of $x^3 - 21x + 20$.

(i) by finding a root and reducing the cubic to a quadratic.

(ii) by the cubic formula.

(iii) Verify that the answers are the same.

3.14 Suppose that α and β are roots of the quadratic equation $x^2 + bx + c = 0$. Find expressions in terms of b and c for

(i) $\alpha + \beta$

(ii) $\alpha^2 + \beta^2$

(iii) $\alpha^3 + \beta^3$

(iv) $(\alpha - \beta)^2$

(v) Use parts (i) and (iv) to derive the quadratic formula.

3.15 Suppose that α, β, and γ are roots of the cubic equation $x^3 + bx^2 + cx + d = 0$. Show that

(i) $\alpha + \beta + \gamma = -b$

(ii) $\alpha\beta + \alpha\gamma + \beta\gamma = c$

(iii) $\alpha\beta\gamma = -d$.

3.16 (i) Show that a rectangle is determined by its perimeter and area.

(ii) Is a rectangular box determined by its volume and surface area? Explain.

3.17 Suppose that α, β, and γ are roots of the cubic equation $x^3 + bx^2 + cx + d = 0$. Find, in terms of b, c, and d,

Exercise 3.17 can be done without the cubic formula.

(i) $\alpha^2 + \beta^2 + \gamma^2$

(ii) $\alpha^3 + \beta^3 + \gamma^3$

(iii) $\alpha^2\beta^2\gamma + \alpha^2\beta\gamma^2 + \alpha\beta^2\gamma^2$

3.18 Take It Further.

(i) Suppose that α, β, and γ are three numbers whose sum is 0. Show that

$$\left[(\alpha - \beta)(\alpha - \gamma)(\beta - \gamma)\right]^2 + 4(\alpha\beta + \alpha\gamma + \beta\gamma)^3 + 27(\alpha\beta\gamma)^3 = 0.$$

(ii) Suppose that α, β, and γ are roots of $x^3 + qx + r$. Show that

$$-((\alpha - \beta)(\alpha - \gamma)(\beta - \gamma))^2 = 27r^2 + 4q^3.$$

3.19 Take It Further. When finding the roots of $x^3 + qx + r$ with the cubic formula, you are led to Eq. (3.1): $x^2 + rx - \frac{1}{27}q^3$, whose roots are g^3 and h^3.

(i) Show that the discriminant δ of this quadratic is

$$\delta = r^2 + \tfrac{4}{27}q^3$$

(ii) If $\delta > 0$, show that the cubic has one real root and two complex conjugate roots.

The *discriminant* of the cubic $x^3 + qx + r$ is defined to be $\Delta = -4q^3 - 27r^2$.

(iii) If $\delta = 0$, show that the cubic has two real roots, one of them with multiplicity 2.

(iv) If $\delta < 0$, show that the cubic has three distinct real roots.

3.2 Complex Numbers

Before the cubic formula, mathematicians had no difficulty in ignoring negative numbers or square roots of negative numbers. For example, consider the problem of finding the sides x and y of a rectangle having area A and perimeter p. The equations $xy = A$ and $2x + 2y = p$ give the equation $2x^2 - px + 2A = 0$, and the quadratic formula gives

$$x = \tfrac{1}{4}\left(p \pm \sqrt{p^2 - 16A}\right).$$

If $p^2 - 16A \geq 0$, the problem is solved. If $p^2 - 16A < 0$, people didn't invent fantastic rectangles whose sides involve square roots of negative numbers. Instead, they merely said that there is no rectangle whose area and perimeter are so related. But the cubic formula doesn't allow us to avoid "imaginary" numbers, for we have just seen, in Example 3.5, that an "honest" real and positive root can appear in terms of such expressions. Complex numbers arose, not as an attempt to get roots of equations involving square roots of negative real numbers, but as a device to solve cubic equations having real coefficients and real roots.

The cubic formula was revolutionary. For the next 100 years, mathematicians were forced to reconsider the meaning of *number*, calculating with strange objects of the form $a + ib$ (where a and b are real numbers) as if they were

In Chapter 7, using ideas of abstract algebra, we'll see that the naive way of thinking about complex numbers, as polynomials in i obeying the rule $i^2 = -1$, can be made precise.

actual numbers enjoying the simplification rule $i^2 = -1$. It was during this time that the terms *real* and *imaginary* arose. In this section and the next, we'll develop complex numbers in a more careful and formal way, and we'll see that complex numbers are as real as real numbers!

The Complex Plane

When considering expressions of the form $a + bi$, it is natural to separate the two summands. Geometry rears its head.

Definition. A *complex number* is an ordered pair $z = (a, b)$ of real numbers, denoted by $z = a + bi$. We call a the *real part* of z, denoting it by $\Re(z) = a$, and b the *imaginary part* of z, denoting it by $\Im(z) = b$.

Right now, $a + bi$ is just an "alias" for (a, b) but, in your previous experience with $\mathbb{C}$, the summand bi denoted the *product* of b and i. We'll soon recover this notion.

Both the real and the imaginary parts of a complex number are real numbers. Moreover, equality of ordered pairs says that complex numbers $z = a + bi$ and $z' = a' + b'i$ are equal if and only if $\Re(z) = \Re(z')$ and $\Im(z) = \Im(z')$; that is, $a = a'$ and $b = b'$. Thus, one equation of complex numbers is the same as two equations of real numbers.

There is an immediate geometric interpretation of complex numbers: they can be viewed as points in the plane. Real numbers are complex numbers z with $\Im(z) = 0$; that is, they correspond to points $(a, 0)$ on the x-axis (which is called the *real axis* in this context). We usually abbreviate $(a, 0)$ to a; thus, the set of real numbers $\mathbb{R}$ is a subset of $\mathbb{C}$. We denote the complex number $(0, 1)$ by i, so that the *purely imaginary* complex numbers z, those with $\Re(z) = 0$, correspond to points on the y-axis (which is called the *imaginary axis* in this context). When we view points as complex numbers, the plane $\mathbb{R}^2$ is called the *complex plane*, and it is denoted by $\mathbb{C}$.

You hear the following message when you call one of our friends. "The number you have reached is imaginary; please rotate your phone 90 degrees."

Thus, an ordered pair (a, b) of real numbers has two interpretations: algebraic, as the complex number $z = a + bi$, and geometric, as the point P in the plane $\mathbb{R}^2$ having coordinates a and b. We will use both interpretations, algebraic and geometric, depending on which is more convenient for the context in which we are working.

Historical Note. Surprisingly, it took a very long time for people to embrace the idea of representing the elements of $\mathbb{C}$ as points in the plane. It wasn't until Wessel presented a paper in 1797 to the Royal Danish Academy of Sciences, entitled *On the Analytic Representation of Direction: An Attempt*, did this representation crystallize. Wessel's discovery was not adopted immediately but, by 1830, most mathematicians routinely used the bijection $a + bi \leftrightarrow (a, b)$ between complex numbers and points of the plane. The complex plane has gone by other names in its history: for example, *Argand Diagram* and *Gaussian Plane*.

A *bijection* is a one-to-one correspondence. See Appendix A.1, page 416, for the precise definition.

Algebraic Operations

In Section 3.1, you saw that mathematicians were forced to add and multiply complex numbers. However, without precise definitions of the operations or of the complex numbers themselves, they could not trust many of their results. The complex plane allows us to resolve the many doubts our ancestors had about the algebra of complex numbers.

In a linear algebra course, $\mathbb{R}^2$ is often viewed as a vector space with real scalars; we continue using these operations in the complex plane.

Definition. Define *addition* $\mathbb{C} \times \mathbb{C} \to \mathbb{C}$ by

$$(a + bi) + (c + di) = (a + c) + (b + d)i.$$

In terms of ordered pairs, $(a, b) + (c, d) = (a + c, b + d)$.

Define *scalar multiplication* $\mathbb{R} \times \mathbb{C} \to \mathbb{C}$ by

$$r(a + bi) = ra + rbi,$$

where $r \in \mathbb{R}$. In terms of ordered pairs, $r(a, b) = (ra, rb)$.

As in linear algebra, it is useful to look at each point in the plane (and, hence, each complex number) as an ***arrow*** with tail at the origin (sometimes we say *vector* instead of arrow). For example, we'll think of $z = 3 + 2i$ either as the point $P = (3, 2)$ or as the arrow $\overrightarrow{OP}$ [where $O = (0, 0)$]. The context will make it clear which interpretation we are using.

Addition is illustrated by the ***parallelogram law*** (see Figure 3.2). If $P = (a, b)$ and $Q = (c, d)$, then $R = (a + c, b + d)$. Of course, this needs a geometric proof, especially when points don't lie in the first quadrant or they are collinear; see Exercise 3.33 on page 98.

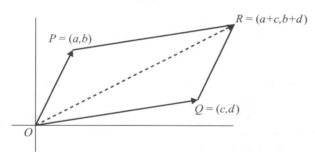

Figure 3.2. Parallelogram Law.

Scalar multiplication of complex numbers has the same geometric interpretation as scalar multiplication of vectors. View a complex number $z = a + ib$ as $\overrightarrow{OP}$, where $P = (a, b)$. If $r \in \mathbb{R}$, then we may view rz as the vector $r\overrightarrow{OP}$; that is, if $r \geq 0$, then it's an arrow in the same direction as $\overrightarrow{OP}$ whose length has been stretched by a factor of r (if $r > 1$) or shrunk by a factor of r (if $r < 1$); if $r < 0$, then $r\overrightarrow{OP}$ is the arrow in the reverse direction whose length has been changed by a factor of $|r|$.

The eight properties listed in the next proposition are precisely the defining properties of a vector space with scalars in $\mathbb{R}$.

Proposition 3.8. *Let $z = a + bi$, $w = c + di$, and $u = e + fi$ be complex numbers, and suppose that $r, s \in \mathbb{R}$.*

(i) $z + w = w + z$

(ii) $z + (w + u) = (z + w) + u$

(iii) $z + 0 = z$

In Exercise 3.21 on page 98, you'll show that negatives are unique.

(iv) *There is a complex number $-z$ such that $z + (-z) = 0$*

(v) $r(sz) = (rs)z$

(vi) $1z = z$

(vii) $r(z + w) = rz + rw$

(viii) $(r + s)z = rz + sz$

Proof. The proofs are routine, just reducing each to a familiar statement about real numbers, and so we'll only prove the longest such: associativity of addition. It is clearer if we use ordered pairs.

$$
\begin{aligned}
z + (w + u) &= (a, b) + [(c, d) + (e, f)] \\
&= (a, b) + (c + e, d + f) \\
&= (a + (c + e), b + (d + f)) \\
&= ((a + c) + e, (b + d) + f) \\
&= (a + c, b + d) + (e, f) \\
&= [(a, b) + (c, d)] + (e, f) \\
&= (z + w) + u. \quad \blacksquare
\end{aligned}
$$

In linear algebra, every vector (a, b) has a decomposition into components $ae_1 + be_2$ with respect to the standard basis $e_1 = (1, 0), e_2 = (0, 1)$:

$$
\begin{aligned}
a + bi &= (a, b) \\
&= (a, 0) + (0, b) \\
&= a(1, 0) + b(0, 1).
\end{aligned}
$$

It follows that the $+$ in the notation $a + bi$ really does mean *add* and that bi is the product of b and i; that is, $bi = b(0, 1) = (0, b)$.

The set $\mathbb{C}$ of complex numbers has more algebraic structure: any two complex numbers can be multiplied, not just when one of them is real. The definition arises from pretending that

$$
\begin{aligned}
(a + bi)(c + di) &= ac + adi + bci + bdi^2 \\
&= (ac - bd) + i(ad + bc),
\end{aligned}
$$

where we have set $i^2 = -1$. This is precisely what our ancestors did, which motivates the formal definition. But our definition involves no pretending.

Definition. Define *multiplication* $\mathbb{C} \times \mathbb{C} \to \mathbb{C}$ by

$$
(a + ib)(c + id) = (ac - bd) + i(ad + bc).
$$

In terms of ordered pairs, $(a, b)(c, d) = (ac - bd, ad + bc)$.

Notice that

$$
i^2 = (0, 1)(0, 1) = (-1, 0) = -1,
$$

for $ac = 0 = ad = bc$.

We are now obliged to prove that the familiar properties of multiplication actually do hold for complex multiplication.

Proposition 3.9. *Let* $z = a + bi$, $w = c + di$, *and* $u = e + fi$ *be complex numbers.*

(i) $zw = wz$

(ii) $z(wu) = (zw)u$

(iii) $1z = z$

(iv) $z(w + u) = zw + zu$.

Proof. Again, the proofs are routine, for each reduces to a familiar statement about real numbers. We'll only prove associativity. As in the proof of Proposition 3.8, it is clearer if we use ordered pairs.

$$\begin{aligned}
z(wu) &= (a,b)\,[(c,d)(e,f)] \\
&= (a,b)(ce - df, de + cf)) \\
&= (a(ce - df) - b(de + cf), b(ce - df) + a(de + cf)) \\
&= (ace - adf - bde - bcf, bce - bdf + ade + acf).
\end{aligned}$$

On the other hand,

$$\begin{aligned}
(zw)u &= [(a,b)(c,d)]\,(e,f) \\
&= (ac - bd, bc + ad)(e,f) \\
&= (ac - bd)e - (bc + ad)f, (bc + ad)e + (ac - bd)f) \\
&= (ace - bde - bcf - adf, bce + ade + acf - bdf).
\end{aligned}$$

Hence, $z(wu) = (zw)u$. ∎

The operations of addition and multiplication in $\mathbb{C}$ extend the definitions in $\mathbb{R}$; for example, if r and s are real numbers, then their sum $r + s$ is the same, whether you think of doing the addition in $\mathbb{R}$ or in $\mathbb{C}$. See Exercise 3.22 on page 98.

In Section 1.4, we displayed nine properties of addition and multiplication in $\mathbb{R}$, and we have just seen that eight of them also hold in $\mathbb{C}$. We could now define subtraction, as we did there, and prove results like $z(w - u) = zw - zu$, $z \cdot 0 = 0$ for all complex numbers z, and the Binomial Theorem. We don't have to repeat all of this. As we said then, once this is established, the proofs of other properties of addition and multiplication, such as $0 \cdot z = 0$ and $(-z)(-w) = zw$, go through verbatim.

The ninth property describes reciprocals: If $z = a + bi \neq 0$, there is a number z^{-1} such that $z \cdot z^{-1} = 1$ (Exercise 3.21 on page 98 shows that such a number z^{-1}, if it exists, is unique.) Here is an explicit formula for $z^{-1} = 1/z$ when $z = a + bi \neq 0$. If $b = 0$, then $z = a$ is a nonzero real number, and we know it has a reciprocal $1/a$. If $b \neq 0$, we can easily find $x + yi$ so that $(x + yi)(a + bi) = 1$. Multiply and equate real and imaginary parts:

$$xa - yb = 1 \quad \text{and} \quad ya + xb = 0.$$

The second equation gives $x = -ay/b$; substitute this into the first equation and obtain

$$y = -\frac{b}{a^2 + b^2} \quad \text{and} \quad x = \frac{a}{a^2 + b^2}.$$

In either case ($b = 0$ or $b \neq 0$),

$$z^{-1} = \frac{a}{a^2 + b^2} - i\frac{b}{a^2 + b^2}.$$

There is a more elegant derivation of this formula. The denominator $a^2 + b^2$ can be factored in $\mathbb{C}$:

Hence, $a^2 + b^2 = z\overline{z}$.

$$a^2 + b^2 = (a + bi)(a - bi).$$

This leads to the useful notion of *complex conjugate*.

Definition. The ***complex conjugate*** $\overline{z}$ of a complex number $z = a + bi$ is defined to be

$$\overline{z} = a - bi.$$

The function $\mathbb{C} \to \mathbb{C}$, given by $z \mapsto \overline{z}$, is called ***complex conjugation***.

If $z = (a, b)$, then $\overline{z} = (a, -b)$, so, geometrically, $\overline{z}$ is obtained from z by reflection in the real axis.

Complex conjugation interacts well with addition and multiplication.

Proposition 3.10. *If $z = a + bi$ and $w = c + di$ are complex numbers, then*

(i) $\overline{z + w} = \overline{z} + \overline{w}$

(ii) $\overline{zw} = \overline{z}\ \overline{w}$

(iii) $z \in \mathbb{R}$ *if and only if* $z = \overline{z}$

(iv) $\overline{\overline{z}} = z$

Proof. We'll prove (i), leaving the rest to Exercise 3.25 on page 98.

$$\begin{aligned}
\overline{z + w} &= \overline{(a + c) + (b + d)i} \\
&= (a + c) - (b + d)i \\
&= (a - bi) + (c - di) \\
&= \overline{z} + \overline{w}. \quad \blacksquare
\end{aligned}$$

Using induction, the first two statements in the proposition can be generalized:

$$\overline{z_1 + \cdots + z_n} = \overline{z}_1 + \cdots + \overline{z}_n$$
$$\overline{z_1 \cdots z_n} = \overline{z}_1 \cdots \overline{z}_n.$$

The formula for the multiplicative inverse of a complex number can be written in terms of conjugates. Informally, cancel $\overline{z}$ to see that $\overline{z}/z\overline{z} = 1/z$.

Proposition 3.11. *Every nonzero complex number $z = a + bi$ has an inverse:*

$$z^{-1} = \frac{a - bi}{a^2 + b^2} = \frac{\overline{z}}{z\,\overline{z}}.$$

Proof. It's enough to see that you get 1 if you multiply z by $(a-bi)/(a^2+b^2)$. And so it is:

$$z\left(\frac{1}{z\,\overline{z}}\overline{z}\right) = \left(\frac{1}{z\,\overline{z}}\right)z\,\overline{z} = 1. \quad \blacksquare$$

It wouldn't be worth introducing the new term *complex conjugation* if our only use of it was to give a neat proof of the formula for reciprocals. The notion has many other uses as well. For example, if $f(x) = ax^2 + bx + c$ has real coefficients, then the quadratic formula implies that whenever z is a complex root of f, then so is $\overline{z}$. In fact, this is true for polynomials of any degree, and the proof depends only on Proposition 3.10.

Is this true if z is real?

Theorem 3.12. *If $f(x)$ is a polynomial with real coefficients and a complex number z is a root of f, then so is $\overline{z}$.*

Proof. Suppose that

$$f(x) = a_0 + a_1 x + \cdots + a_i x^i + \cdots + a_n x^n,$$

where each $a_i \in \mathbb{R}$. Saying that z is a root means that

$$0 = a_0 + a_1 z + \cdots + a_i z^i + \cdots + a_n z^n.$$

Hence,

$$
\begin{aligned}
0 = \overline{0} &= \overline{a_0 + a_1 z + \cdots + a_i z^i + \cdots + a_n z^n} \\
&= \overline{a_0} + \overline{a_1 z} + \cdots + \overline{a_i z^i} + \cdots + \overline{a_n z^n} \\
&= \overline{a_0} + \overline{a_1}\,\overline{z} + \cdots + \overline{a_i}\,\overline{z}^i + \cdots + \overline{a_n}\,\overline{z}^n \\
&= a_0 + a_1\overline{z} + \cdots + a_i\overline{z}^i + \cdots + a_n\overline{z}^n \quad \text{(because all } a_i \text{ are real)} \\
&= f(\overline{z}).
\end{aligned}
$$

Therefore, $\overline{z}$ is a root of f. $\quad \blacksquare$

Exercises

3.20 In Appendix A.4, we considered the subset P of all (strictly) positive real numbers; it satisfies:

If we define $a < b$ to mean $b - a \in P$, then we can prove all the familiar properties of inequality. For example, if $a < b$ and $c < 0$, then $bc < ac$. See page 441.

- if $a, b \in P$, then $a + b \in P$ and $ab \in P$;

- if $r \in \mathbb{R}$, then exactly one of the following is true:

$$r \in P, \quad r = 0, \quad \text{or} \quad -r \in P.$$

(i) Using only the two properties of P, prove that if $a \in \mathbb{R}$, then either $a = 0$ or $a^2 \in P$.

(ii) Prove that there is no subset $Q \subseteq \mathbb{C}$, closed under addition and multiplication, such that if $z \in \mathbb{C}$, then exactly one of the following is true:

$$z \in Q, \quad z = 0, \quad \text{or} \quad -z \in Q.$$

Conclude that it's impossible to order the complex numbers in a way that preserves the basic rules for inequality listed in Proposition A.51.

3.21 * Suppose that z is a complex number. Generalize Propositions 1.33 and 1.32.

(i) Show that $-z$ is unique.

(ii) Show that $-z = (-1)z$.

(iii) If $z \neq 0$, show that z^{-1} is unique.

3.22 *

(i) We may think of real numbers r and s as complex numbers. Show that their sum $r + s$ and their product rs in $\mathbb{C}$ are the same as their sum and product in $\mathbb{R}$.

(ii) If z is a complex number and r is a real number, you can think of the complex number rz in two ways: as the product of scalar multiplication, or as the product of two complex numbers $r + 0i$ and z. Show that the two calculations give the same result.

3.23 If $z \in \mathbb{C}$, show that $z + \overline{z} = 2(\Re z)$ and $z - \overline{z} = 2(\Im z)$.

3.24 Find a complex number z such that $z + \overline{z} = 14$ and $z\overline{z} = 49$.

3.25 * Finish the proof of Proposition 3.10. If z and w are complex numbers, prove

(i) $\overline{zw} = \overline{z}\,\overline{w}$

(ii) $\overline{\overline{z}} = z$

(iii) $\overline{z} = z$ if and only if z is a real number

3.26 If z is a complex number and n is a natural number, show that

$$\overline{z^n} = (\overline{z})^n .$$

Is this equation true if $z \neq 0$ and n is a negative integer?

3.27 * Let z be a complex number and r a real number. Show how to locate rz in the complex plane in terms of z.

Hint. $r(a, b) = (ra, rb)$.

3.28 Solve the following equation for z.

$$(3 + 2i)z = -3 + 11i.$$

3.29 Find real numbers a and b such that

(i) $a + bi = (8 + i)/(3 + 2i)$

(ii) $a + bi = (8 + i)/(3 + i)$

(iii) $(a + bi)^2 = -5 + 12i$

(iv) $(a + bi)^2 = 1 + i$.

3.30 What's wrong with this "proof" that $6 = -6$?

$$6 = \sqrt{36} = \sqrt{(-9)(-4)} = \sqrt{-9}\sqrt{-4} = 3i \cdot 2i = 6i^2 = -6.$$

3.31 Establish the identity

$$(a^2 + b^2)(c^2 + d^2) = (ac - bd)^2 + (bc + ad)^2$$

for all complex numbers a, b, c, d.

3.32 Use Theorem 3.12 to prove that every cubic polynomial with real coefficients has a real root.

3.33 * Let $z = a + bi$ and $w = c + di$. Show that in the complex plane $z + w$ is the fourth vertex of the (possibly degenerate) parallelogram whose other vertices are $0, z$, and w.

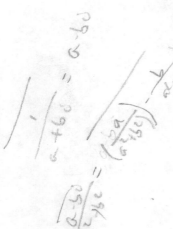

Absolute Value and Direction

We've already seen that addition can be viewed as the parallelogram law, multiplication by a real number can be viewed as scalar multiplication, and conjugation can be viewed as reflection in the real axis. There is a beautiful geometric interpretation of complex multiplication; it is best understood if, first, we consider a different way to describe an arrow in the complex plane using *absolute value* and *direction*.

Definition. The *absolute value* (or *length* or *modulus*) of $z = a + bi$ is

$$|z| = \sqrt{a^2 + b^2}.$$

The absolute value of a real number is its distance to the origin, and so we have just extended the notion of absolute value from $\mathbb{R}$ to $\mathbb{C}$. Thus, if $z = a + bi$, then $|z|$ is the distance from the point $P = (a, b)$ to the origin O; equivalently, it is the length of the arrow $\overrightarrow{OP}$. Because $z\,\overline{z} = a^2 + b^2$, we can write

$$|z| = \sqrt{z\,\overline{z}}.$$

Does this equation hold if z is a real number?

Proposition 3.13. *Let $z = a + bi$ and $w = c + di$.*

(i) $|z| \geq 0$, *and* $|z| = 0$ *if and only if* $z = 0$.

(ii) **(Triangle Inequality).** $|z + w| \leq |z| + |w|$.

(iii) $|zw| = |z|\,|w|$.

Proof.   (i) Both statements follow from the definition $|z| = \sqrt{a^2 + b^2}$, because $a^2 + b^2 = 0$ if and only if $a = 0 = b$.

(ii) If $P = (a, b)$ and $Q = (c, d)$, then z is the arrow $\overrightarrow{OP}$ and w is the arrow $\overrightarrow{OQ}$. As in Figure 3.2, $z + w = \overrightarrow{OR}$. The inequality we want is the usual triangle inequality, which follows from the length of a line segment being the shortest distance between its endpoints.

$$
\begin{aligned}
|zw| &= \sqrt{(zw)\,(\overline{zw})} \\
&= \sqrt{(zw)\,(\overline{z}\,\overline{w})} \\
&= \sqrt{(z\,\overline{z})\,(w\,\overline{w})} \\
&= \sqrt{(z\,\overline{z})}\sqrt{(w\,\overline{w})} \\
&= |z|\,|w|. \quad \blacksquare
\end{aligned}
$$

Make sure you can justify each step in the proof. Would this proof work if either z or w (or both) is real?

What do we mean by *direction*? The most natural way to indicate direction in the plane is to point: "He went thataway!"—an arrow shows the way. The arrow may as well have its tail at the origin and, since the length of the arrow doesn't affect the direction, we may as well assume it is a ***unit vector***; that is, it has length 1. If we denote the tip of the arrow by $P = (a, b)$, then P lies on the unit circle. There are various geometric ways to describe P. One way is to consider the angle θ between the x-axis and $\overrightarrow{OP}$; hence, $P = (a, b) = (\cos\theta, \sin\theta)$. This angle can be described with *degrees*; the ancients divided the circle into 360 equal degrees. The angle can also be described with *radians*; the circumference of the unit circle is 2π, and θ is the length of the arc from

Why did our ancestors divide a circle into 360 parts? We can only guess why. Perhaps it was related to calendars, for a year has about 360 days.

$(1, 0)$ to P. When we view the point $P = (\cos\theta, \sin\theta)$ on the unit circle as a complex number, it is equal to $\cos\theta + i\sin\theta$.

Figure 3.3 shows $z = a + bi$ as the tip of an arrow of length $|z| = r$. The direction of this arrow is the same as the direction of the unit vector $\overrightarrow{OP}$ having the same direction as z. If θ is the angle between the x-axis and $\overrightarrow{OP}$, then the coordinates of P are $|OA| = \cos\theta$ and $|AP| = \sin\theta$.

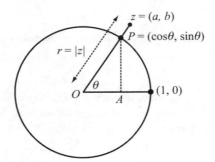

Figure 3.3. Absolute value and argument.

Definition. If z is a nonzero complex number, then its *argument*, denoted by

$$\arg(z),$$

is the counterclockwise angle θ from the positive real axis to $\overrightarrow{OP}$.

Finding $\arg(z)$ requires some way of computing values of inverse trigonometric functions. Nowadays, we use computers; in earlier times, tables of values of cosine and sine were used. Fairly accurate trigonometric tables were known over two thousand years ago.

In Figure 3.3, we see that the coordinates of P are $\cos\theta$ and $\sin\theta$; that is, if $P = (a, b)$, then $a = \cos\theta$ and $b = \sin\theta$. Thus, for any nonzero complex number $z = a + bi$, not necessarily of absolute value 1, the definitions of cosine and sine (in terms of right triangles) give $\arg(z) = \theta$, where $\cos\theta = a/|z|$ and $\sin\theta = b/|z|$. Note that

$$z = |z|\left(\frac{a}{|z|} + i\,\frac{b}{|z|}\right) = |z|\,(\cos\theta + i\sin\theta). \tag{3.4}$$

How to Think About It. Technically, the argument of a complex number is only determined up to a multiple of 360 (if we measure in degrees) or 2π (if we measure in radians). For example, $\arg(1 + i)$ is 45° or $\frac{\pi}{4}$ radians, and this is the same direction as 405° or $\frac{9\pi}{4}$ radians. There is a fussy way to make statements like "$\arg(1 + i) = 45$°" precise (introduce a suitable equivalence relation), but we prefer, as do most people, to be a bit sloppy here; the cure is worse than the disease.

Actually, the trig functions do make angles precise, for $\cos\theta$ and $\sin\theta$ have the same values when θ is replaced by either $\theta + 360n$ degrees or $\theta + 2\pi n$ radians.

The *polar form* of z is

$$z = |z|\,(\cos\theta + i\sin\theta),$$

where $\theta = \arg(z)$. Just as a complex number z is determined by its real and imaginary parts, so too is it determined by its polar form: its absolute value and argument.

Proposition 3.14 (Polar Form). *Every complex number z has a **polar form**:*

$$z = r(\cos\theta + i\sin\theta),$$

where $r \geq 0$ and $0 \leq \theta < 2\pi$. If $z \neq 0$, then this expression is unique.

Proof. Existence is given by Eq. (3.4). Uniqueness is almost obvious. Suppose that $z = r\,(\cos\theta + i\sin\theta)$ for some $r > 0$ and θ. Then

$$|z| = |r\,(\cos\theta + i\sin\theta)| = |r|\,|\cos\theta + i\sin\theta| = |r| = r,$$

since $r \geq 0$ and $|\cos\theta + i\sin\theta| = \cos^2\theta + \sin^2\theta = 1$; thus, $r = |z|$. But then

$$\cos\theta + i\sin\theta = \frac{1}{r}z = \frac{1}{|z|}z = \frac{a}{|z|} + i\frac{b}{|z|},$$

so that

$$\cos\theta = \frac{a}{|z|} \quad \text{and} \quad \sin\theta = \frac{b}{|z|}$$

and $\theta = \arg(z)$. ∎

We may paraphrase uniqueness of polar forms: two vectors are equal if and only if they have the same length and the same direction.

Example 3.15. If $z = 3 + 4i$, the Pythagorean Theorem gives $|z|^2 = 3^2 + 4^2 = 25$, so that $|z| = 5$; your favorite computer gives $\arg(z) = \theta = \cos^{-1}(\frac{3}{5}) \approx 53.13°$. Thus, the polar form of z is

$$z = \cos\theta + i\sin\theta \approx 5\left(\cos 53.13° + i\sin 53.13°\right). \quad \blacktriangle$$

The Geometry Behind Multiplication

We're ready to give a geometric interpretation of complex multiplication. Proposition 3.13(iii) tells part of the story: the absolute value of a product is the product of the absolute values. To finish the geometric analysis of multiplication, we need to know how $\arg(zw)$ is related to $\arg(z)$ and $\arg(w)$. We may as well assume that z and w are unit vectors (why?), so that $z = \cos\alpha + i\sin\alpha$ and $w = \cos\beta + i\sin\beta$. Multiply them together and collect real and imaginary parts.

We know that zw sits on a circle of radius $|z||w|$, centered at the origin. But where?

$$zw = (\cos\alpha + i\sin\alpha)(\cos\beta + i\sin\beta) \tag{3.5}$$
$$= (\cos\alpha\cos\beta - \sin\alpha\sin\beta) + i(\cos\alpha\sin\beta + \sin\alpha\cos\beta). \tag{3.6}$$

Do $\Re(zw)$ and $\Im(zw)$ look familiar? They are the *addition formulas* for sine and cosine:

$$\cos(\alpha + \beta) = \cos\alpha\cos\beta - \sin\alpha\sin\beta$$

and

$$\sin(\alpha + \beta) = \cos\alpha\sin\beta + \sin\alpha\cos\beta.$$

These formulas will give a beautiful characterization of the product of two complex numbers. We now prove them, beginning with a familiar lemma that uses Figure 3.4.

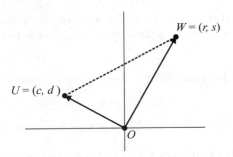

Figure 3.4. Orthogonality and dot product.

We are proving that perpendicularity is equivalent to the dot product being 0.

Lemma 3.16. *If* $W = (r, s)$ *and* $U = (c, d)$, *then the arrows* $\overrightarrow{OW}$ *and* $\overrightarrow{OU}$ *are perpendicular if and only if* $rc + sd = 0$.

Proof. We use the Pythagorean Theorem and its converse, Exercise 1.11 on page 7: $\overrightarrow{OU} \perp \overrightarrow{OW}$ if and only if $|UW|^2 = |OU|^2 + |OW|^2$. Let $h = |UW|$; then

$$h^2 = (r - c)^2 + (s - d)^2$$
$$= r^2 - 2rc + c^2 + s^2 - 2sd + d^2$$
$$= r^2 + s^2 + c^2 + d^2 - 2(rc + sd).$$

But $|OW|^2 = r^2 + s^2$ and $|OU|^2 = c^2 + d^2$. Hence,

$$|OW|^2 + |OU|^2 = (r^2 + s^2) + (c^2 + d^2)$$

and

$$h^2 = r^2 + s^2 + c^2 + d^2 - 2(rc + sd).$$

Therefore, $h^2 = |OU|^2 + |OW|^2$ if and only if $rc + sd = 0$. ∎

Now for the addition formulas.

In Figure 3.4, the coordinates of U are labeled (c, d). This lemma shows that $(c, d) = (-s, r)$.

Theorem 3.17 (Addition Theorem). *Let* α *and* β *be angles.*

(i) $\cos(\alpha + \beta) = \cos \alpha \cos \beta - \sin \alpha \sin \beta$.

(ii) $\sin(\alpha + \beta) = \cos \alpha \sin \beta + \sin \alpha \cos \beta$.

We are looking at points here as elements of $\mathbb{R}^2$, although we'll soon interpret this diagram in the complex plane.

Proof. In Figure 3.5, we have a picture of the unit circle. Let $Z = (a, b) = (\cos \alpha, \sin \alpha)$ and $W = (r, s) = (\cos \beta, \sin \beta)$. Rotate $\triangle OQZ$ counterclockwise through $\angle \beta$ to get $\triangle OQ'Z'$, so that $\triangle OQZ$ and $\triangle OQ'Z'$ are congruent. Thus, $Z' = (\cos(\alpha + \beta), \sin(\alpha + \beta))$. Our task is to find the coordinates of Z' in terms of $r, s, a,$ and b.

Define $U = (-s, r)$. Since $W = (r, s)$ is on the unit circle, we have $r^2 + s^2 = 1$, and so $U = (-s, r)$ is also on the unit circle. Moreover, since $(-s)r + rs = 0$, Lemma 3.16 says that $\overrightarrow{OU}$ is orthogonal to $\overrightarrow{OW}$. Therefore, $OQ'Z'M$ is a rectangle.

Decompose $\overrightarrow{OZ'}$ as the sum of two vectors:

$$\overrightarrow{OZ'} = \overrightarrow{OQ'} + \overrightarrow{OM},$$

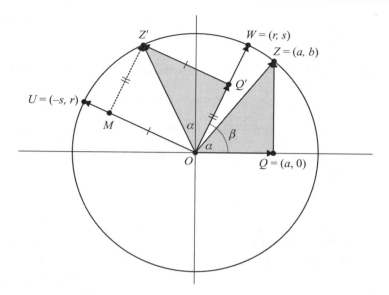

Figure 3.5. Addition Theorem.

where $\overrightarrow{OM}$ is the projection of $\overrightarrow{Q'Z'}$ onto $\overrightarrow{OU}$. We can get explicit expressions for Q' and M. First, $\overrightarrow{OQ'}$ is a scalar multiple of $\overrightarrow{OW}$ and, because $|OQ| = a$, we know the scalar:

$$Q' = a(r, s) = (ar, as).$$

Second, $\overrightarrow{OM}$ is a scalar multiple of $\overrightarrow{OU}$, where $U = (-s, r)$; as $|OM| = |QZ| = b$, we know the scalar:

$$M = b(-s, r) = (-bs, br).$$

Therefore,

$$\overrightarrow{OZ'} = \overrightarrow{OQ'} + \overrightarrow{OM} = (ar, as) + (-bs, br) = (ar - bs, as + br).$$

Making the appropriate substitutions for a, b, r, and s, we have the desired result:

$$\cos(\alpha + \beta) = \cos\alpha\cos\beta - \sin\alpha\sin\beta$$

and

$$\sin(\alpha + \beta) = \cos\alpha\sin\beta + \sin\alpha\cos\beta. \quad \blacksquare$$

Here is the result we have been seeking.

Theorem 3.18 (The Geometry of Multiplication.). *If z and w are complex numbers, then*

(i) $|zw| = |z|\,|w|$, *and*

(ii) $\arg(zw) = \arg(z) + \arg(w)$.

Proof. The first statement is Proposition 3.13, and the second follows from Theorem 3.17 and Eq. (3.5) on page 101. $\quad\blacksquare$

In words, the length of a product is the product of the lengths, and the argument of a product is the sum of the arguments. The equality in Theorem 3.18(ii) holds up to a multiple of 2π.

If we set $\arg(z) = \alpha$ and $\arg(w) = \beta$, then Theorem 3.18 has an especially pleasing restatement in polar form.

Corollary 3.19. *If $z = |z|(\cos\alpha + \sin\alpha)$ and $w = |w|(\cos\beta + \sin\beta)$, then*

$$z \cdot w = |zw|\left(\cos(\alpha + \beta) + i\,\sin(\alpha + \beta)\right).$$

Proof. Both sides equal $|z|(\cos\alpha + i\,\sin\alpha) \cdot |w|(\cos\beta + i\,\sin\beta)$. ∎

It follows easily, by induction on $k \geq 1$, that if z is a complex number and $k \in \mathbb{Z}$, then

$$\left|z^k\right| = |z|^k \quad \text{and} \quad \arg\!\left(z^k\right) = k\,\arg(z).$$

How to Think About It. There's a way to see, without using trigonometry, that angles add in the product of two complex numbers. Essentially, we recast the proof of Theorem 3.17 in terms of complex numbers. Given $z = a + bi$ and $w = r + si$, we want to determine $\arg(zw)$ in terms of $\arg(z) = \alpha$ and $\arg(w) = \beta$. We can assume that z and w are unit vectors; this implies that zw is also a unit vector, by Proposition 3.13. The key insight is that

$$zw = (a + bi)w = aw + (bi)w = aw + b(iw).$$

This approach does not seem to be very well known. It appears in [22] and Kerins, B. "Gauss, Pythagoras, and Heron" (Mathematics Teacher, 96:5, 2003), but we can't find any older sources.

You know the geometric effect of scalar multiplication (Exercise 3.27 on page 98), you know how to add geometrically (parallelogram law), and you know

$$iw = i(r + si) = -s + ri;$$

using Lemma 3.16, it follows that iw is obtained from w by counterclockwise rotation by 90°. Figure 3.6 below is almost the same as Figure 3.5; the difference is that points are now labeled as complex numbers.

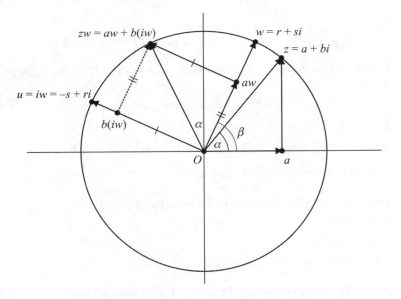

Figure 3.6. Complex multiplication again.

Let's put this all together. Triangle $OZ'Q'$ is congruent to triangle OZQ, so that $\angle Z'OW = \angle ZOQ = \alpha$; hence, $\arg(zw) = \alpha + \beta$. We have shown that Theorem 3.18(ii) follows without any mention of trigonometry.

We have just used plane geometry to derive the geometric interpretation of complex multiplication, avoiding the trigonometric addition formulas. Aside from proving these ideas for students who haven't yet seen the addition formulas, complex numbers can now be used to *derive* these formulas. That's an additional bonus, especially for a precalculus class; it allows us to use complex numbers when trying to establish other trigonometric identities that depend on the addition formulas. For example, to get a formula for $\cos\left(\frac{\pi}{4} + \theta\right)$, calculate like this:

$$\cos\left(\tfrac{\pi}{4} + \theta\right) + i\sin\left(\tfrac{\pi}{4} + \theta\right) = \left(\cos\tfrac{\pi}{4} + i\sin\tfrac{\pi}{4}\right)(\cos\theta + i\sin\theta)$$
$$= \tfrac{1}{\sqrt{2}}(1 + i)(\cos\theta + i\sin\theta)$$
$$= \tfrac{1}{\sqrt{2}}\big((\cos\theta - \sin\theta) + i(\cos\theta + \sin\theta)\big).$$

Hence

$$\cos\left(\tfrac{\pi}{4} + \theta\right) = \tfrac{1}{\sqrt{2}}(\cos\theta - \sin\theta)$$

and, as a bonus,

$$\sin\left(\tfrac{\pi}{4} + \theta\right) = \tfrac{1}{\sqrt{2}}(\cos\theta + \sin\theta).$$

Over the next two centuries, people became comfortable with the fact that polynomial equations with real coefficients can have complex solutions. It was eventually proved that every polynomial $f(x) = x^n + c_{n-1}x^{n-1} + \cdots + c_1 x + c_0$ with real coefficients has a factorization

$$f(x) = (x - \alpha_1) \cdots (x - \alpha_n),$$

where $\alpha_1, \ldots, \alpha_n$ are complex numbers. This amazing result holds for any nonzero polynomial f with *complex* coefficients; it is known as the **Fundamental Theorem of Algebra**. We won't prove this result here because, in spite of its name, it is a theorem of analysis, not of algebra; you can find a readable account in [4], pp. 142–152.

Exercises

3.34 If $z = a + bi$, prove that the arrow corresponding to z, namely $\overrightarrow{OP}$, where $P = (a, b)$, is perpendicular to the arrow corresponding to iz.

3.35 If z and w are complex numbers with $w \neq 0$, show that

$$\arg(z/w) = \arg(z) - \arg(w).$$

3.36 (i) Prove that the quadratic formula holds for polynomials with complex coefficients (use Proposition 3.6).

(ii) Find the roots of $x^2 + 2ix - 1$. Why aren't these roots conjugate?

3.37 If z and w are complex numbers, find a necessary and sufficient condition that $|z + w|$ to equal $|z| + |w|$.

3.38 * If $z = \cos\alpha + i\sin\alpha$, show that

$$\overline{z} = 1/z = \cos(-\alpha) + i\sin(-\alpha).$$

3.39 Let $n \geq 0$ is an integer and $\zeta = \cos(\frac{2\pi}{n}) + i\sin(\frac{2\pi}{n})$. If z is a complex number, give a geometric description of how ζz is located with respect to z on the complex plane.

3.40 Preview. Plot the roots in the complex plane for each of the polynomials $x^2 - 1$, $x^3 - 1$, $x^4 - 1$, $x^6 - 1$, and $x^{12} - 1$.

3.41 Preview. Let $\zeta = \cos(45°) + i\sin(45°)$.

   (i) Show that $\zeta^8 = 1$

   (ii) Show that the distinct roots of $x^8 - 1$ are precisely $1, \zeta, \zeta^2, \ldots, \zeta^7$.

   (iii) Plot these roots in the complex plane.

   (iv) Show that ζ^{147} is a root of $x^8 - 1$.

   (v) Show that ζ^{147} is equal to one of the roots in part (ii). Which one?

3.42 Preview. Let n be a nonnegative integer and $\zeta = \cos(\frac{2\pi}{n}) + i\sin(\frac{2\pi}{n})$.

   (i) Let $\zeta = \cos(\frac{2\pi}{n}) + i\sin(\frac{2\pi}{n})$. Show that $\zeta^n = 1$.

   (ii) If k is any nonnegative integer, show that $\left(\zeta^k\right)^n = 1$.

   (iii) Give a geometric description of the subset $\{\zeta^k : k \geq 0\}$ of the complex plane.

3.43 Take It Further. If $f(x)$ is a polynomial with complex coefficients, define $\overline{f}$ to be the polynomial you get by replacing each coefficient in f by its conjugate. Prove the following statements.

   (i) $\overline{f + g} = \overline{f} + \overline{g}$.

   (ii) $\overline{fg} = \overline{f}\,\overline{g}$.

   (iii) $\overline{f} = f$ if and only if $f(x)$ has real coefficients.

   (iv) $f\,\overline{f}$ has real coefficients.

   (v) $\overline{f(z)} = \overline{f}\,(\overline{z})$.

3.44 Take It Further. Suppose $f(x)$ is a polynomial with coefficients in $\mathbb{C}$. If a complex number z is a root of f, show that $\overline{z}$ is a root of $\overline{f}$.

3.45 Take It Further. Suppose $f(x)$ is a polynomial with coefficients in $\mathbb{C}$. Then by Exercise 3.43 on page 106, if we define $g(x) = f(x)\overline{f}(x)$, then $g(x)$ has coefficients in $\mathbb{R}$. Show that if $g(z) = 0$, either $f(z) = 0$ or $f(\overline{z}) = 0$. Hence conclude that if every polynomial with real coefficients and degree at least 1 has a root in $\mathbb{C}$, then every polynomial with *complex* coefficients and degree at least 1 has a root in $\mathbb{C}$. (The Fundamental Theorem of Algebra says that every polynomial with complex coefficients has all its roots in $\mathbb{C}$. This exercise shows that it's enough to prove this for polynomials with real coefficients.)

3.3   Roots and Powers

We saw in the previous section that every point z on the unit circle can be written as $z = \cos\theta + i\sin\theta$ for some angle θ. Theorem 3.18 tells us that arguments add when complex numbers are multiplied. In particular,

$$
\begin{aligned}
(\cos\theta + i\sin\theta)^2 &= (\cos\theta + i\sin\theta)(\cos\theta + i\sin\theta) \\
&= \cos(\theta + \theta) + i\sin(\theta + \theta) \\
&= \cos(2\theta) + i\sin(2\theta).
\end{aligned}
$$

On the other hand, complex multiplication gives

$$(\cos\theta + i\sin\theta)^2 = \left(\cos^2\theta - \sin^2\theta\right) + i\,2\cos\theta\sin\theta.$$

Equating real parts and imaginary parts gives the *double angle formulas*:

$$\cos(2\theta) = \cos^2\theta - \sin^2\theta$$
$$\sin(2\theta) = 2\cos\theta\sin\theta.$$

We now generalize this to any positive integer power.

Theorem 3.20 (De Moivre). *For every angle θ and all integers $n \geq 0$,*

$$(\cos\theta + i\sin\theta)^n = \cos(n\theta) + i\sin(n\theta).$$

Proof. We prove equality by induction on $n \geq 0$. The theorem is true when $n = 0$, for $\cos 0 = 1$ and $\sin 0 = 0$. Here is the inductive step.

$$
\begin{aligned}
(\cos\theta + i\sin\theta)^n &= (\cos\theta + i\sin\theta)^{n-1}(\cos\theta + i\sin\theta)\\
&= \Big(\cos\big((n-1)\theta\big) + i\sin\big((n-1)\theta\big)\Big)(\cos\theta + i\sin\theta)\\
&= \cos\big((n-1)\theta + \theta\big) + i\sin\big((n-1)\theta + \theta\big)\\
&= \cos(n\theta) + i\sin(n\theta). \quad \blacksquare
\end{aligned}
$$

Example 3.21.   (i)

$$(\cos 3° + i\sin 3°)^{40} = \cos 120° + i\sin 120° = -\tfrac{1}{2} + i\tfrac{\sqrt{3}}{2}.$$

(ii) Let $z = \cos(45°) + i\sin(45°) = \tfrac{1}{\sqrt{2}}(1+i)$. We compute z^6 in two ways:

- With the Binomial Theorem:

$$
\begin{aligned}
z^6 &= \left(\tfrac{1}{\sqrt{2}}(1+i)\right)^6\\
&= \left(\tfrac{1}{\sqrt{2}}\right)^6 (1+i)^6\\
&= \tfrac{1}{8}\left(1 + 6i + 15i^2 + 20i^3 + 15i^4 + 6i^5 + i^6\right)\\
&= \tfrac{1}{8}\left(1 + 6i - 15 - 20i + 15 + 6i - 1\right)\\
&= -i.
\end{aligned}
$$

- With De Moivre's Theorem:

$$
\begin{aligned}
z^6 &= \left[\cos(45°) + i\sin(45°)\right]^6 = \cos(6\cdot 45°) + i\sin(6\cdot 45°)\\
&= \cos(270°) + i\sin(270°) = -i. \quad \blacktriangle
\end{aligned}
$$

Polar Decomposition and De Moivre's Theorem combine to give a nice formula for computing powers of any complex number.

Corollary 3.22. *If $z = r(\cos\alpha + i\sin\alpha)$ is a complex number, then*

$$z^n = r^n\big(\cos(n\alpha) + i\sin(n\alpha)\big).$$

We were unable to find a cube root of $z = a + ib$ earlier, but there's no problem now.

Corollary 3.23. *Let $r(\cos\theta + i\sin\theta)$ be the polar form of a complex number z. If $n \geq 1$ is an integer, then*

$$\left[\sqrt[n]{r}\left(\cos(\theta/n) + i\sin(\theta/n)\right) \right]^n = r(\cos\theta + i\sin\theta) = z.$$

Of course, we must find the polar form of z, which involves finding $\theta = \cos^{-1}(a/|z|)$.

Example 3.24. In Example 3.15, we saw that the polar form of $z = 3 + 4i$ is approximately $5(\cos(53.13°) + i\sin(53.13°))$. Now $(53.13)/3 = 17.71$, and so a cube root of z is approximately $\sqrt[3]{5}(\cos(17.71°) + i\sin(17.71°))$. Our calculator says that

$$\left(\sqrt[3]{5}\left(\cos 17.71° + i\sin 17.71°\right) \right)^3 = 3.000001 + 3.99999i. \quad \blacktriangle$$

We are now going to describe a beautiful formula discovered by Euler. Recall some power series formulas from calculus. For every real number x,

$$e^x = 1 + x + \frac{x^2}{2!} + \cdots + \frac{x^n}{n!} + \cdots,$$

$$\cos x = 1 - \frac{x^2}{2!} + \frac{x^4}{4!} - \cdots + \frac{(-1)^n x^{2n}}{(2n)!} + \cdots,$$

and

$$\sin x = x - \frac{x^3}{3!} + \frac{x^5}{5!} - \cdots + \frac{(-1)^{n-1} x^{2n+1}}{(2n+1)!} + \cdots.$$

We can define convergence of power series $\sum_{n=0}^{\infty} c_n z^n$ for z and c_n complex numbers, and we can then show that the series

$$1 + z + \frac{z^2}{2!} + \cdots + \frac{z^n}{n!} + \cdots$$

converges for every complex number z. The **complex exponential** e^z is defined to be the sum of this series. In particular, the series for e^{ix} converges for all real numbers x, and

$$e^{ix} = 1 + ix + \frac{(ix)^2}{2!} + \cdots + \frac{(ix)^n}{n!} + \cdots.$$

It is said that Euler was delighted by the special case

$$e^{i\pi} + 1 = 0,$$

for it contains five important constants in one equation.

Theorem 3.25 (Euler). *For all real numbers x,*

$$e^{ix} = \cos x + i\sin x.$$

Sketch of proof. We will not discuss necessary arguments involving convergence. As n varies over $0, 1, 2, 3, 4, 5, \ldots$, the powers of i repeat every four steps: that is, the sequence

$$1,\ i,\ i^2,\ i^3,\ i^4,\ i^5,\ i^6,\ i^7,\ i^8,\ i^9,\ i^{10},\ i^{11},\ \ldots$$

is actually

$$1,\ i,\ -1,\ -i,\ 1,\ i,\ -1,\ -i,\ 1,\ i,\ -1,\ -i,\ \ldots;$$

the even powers of i are all real, whereas the odd powers all involve i. It follows, for every real number x, that $(ix)^n = i^n x^n$ takes values

$$1, \; ix, \; -x^2, \; -ix^3, \; x^4, \; ix^5, \; -x^6, \; -ix^7, \; x^8, \; ix^9, \; -x^{10}, \; -ix^{11}, \; \ldots$$

Thus, in the definition of the complex exponential,

$$e^{ix} = 1 + ix + \frac{(ix)^2}{2!} + \cdots + \frac{(ix)^n}{n!} + \cdots,$$

the even powers of ix do not involve i, whereas the odd powers do. Collecting terms, one has e^{ix} = even terms + odd terms. But

$$\begin{aligned}
\text{even terms} &= 1 + \frac{(ix)^2}{2!} + \frac{(ix)^4}{4!} + \cdots \\
&= 1 - \frac{x^2}{2!} + \frac{x^4}{4!} - \cdots
\end{aligned}$$

and

$$\begin{aligned}
\text{odd terms} &= ix + \frac{(ix)^3}{3!} + \frac{(ix)^5}{5!} + \cdots \\
&= i\left(x - \frac{x^3}{3!} + \frac{x^5}{5!} - \cdots\right).
\end{aligned}$$

Therefore, $e^{ix} = \cos x + i \sin x$. ∎

As a consequence of Euler's Theorem, the polar decomposition can be rewritten in exponential form: every complex number z has a factorization

$$z = re^{i\theta},$$

where $r \geq 0$ and $0 \leq \theta < 2\pi$.

We have chosen to denote $\sum_{n=0}^{\infty} \frac{(ix)^n}{n!}$ by e^{ix}, but we cannot assert, merely as a consequence of our notation, that the law of exponents, $e^{ix}e^{iy} = e^{i(x+y)}$, is valid. But this is precisely what Corollary 3.19 says once it is translated into exponential notation.

Theorem 3.26 (Exponential Addition Theorem). *For all real numbers x and y,*

$$e^{ix}e^{iy} = e^{i(x+y)}.$$

Proof. According to Corollary 3.19,

$$\begin{aligned}
e^{ix}e^{iy} &= (\cos x + i \sin x)(\cos y + i \sin y) \\
&= \cos(x + y) + i \sin(x + y) \\
&= e^{i(x+y)}. \quad \blacksquare
\end{aligned}$$

We can also translate De Moivre's Theorem into exponential notation.

Corollary 3.27 (Exponential De Moivre). *For every real number x and all integers $n \geq 1$,*

$$(e^{ix})^n = e^{inx}.$$

Proof. According to De Moivre's Theorem,

$$(e^{ix})^n = (\cos x + i \sin x)^n = \cos(nx) + i \sin(nx) = e^{inx}. \quad \blacksquare$$

It is easier to remember the trigonometric addition formulas in complex form. For example, let's find the **triple angle formulas**. On the one hand, De Moivre's Theorem gives

$$e^{i3x} = \cos(3x) + i \sin(3x).$$

On the other hand,

$$
\begin{aligned}
e^{i3x} &= (e^{ix})^3 \\
&= (\cos x + i \sin x)^3 \\
&= \cos^3 x + 3i \cos^2 x \sin x + 3i^2 \cos x \sin^2 x + i^3 \sin^3 x \\
&= \cos^3 x - 3 \cos x \sin^2 x + i(3 \cos^2 x \sin x - \sin^3 x).
\end{aligned}
$$

Equating real and imaginary parts, we have

$$\cos(3x) = \cos^3 x - 3 \cos x \sin^2 x$$

and

$$\sin(3x) = 3 \cos^2 x \sin x - \sin^3 x.$$

Roots of Unity

De Moivre's Theorem can be used to find the roots of an important family of polynomials: those of the form $x^n - 1$.

Theorem 3.28. *The distinct roots of $x^n - 1$ are*

$$1, \ \zeta, \ \zeta^2, \ \cdots, \ \zeta^{n-1},$$

where $\zeta = \zeta_n = \cos(2\pi/n) + i \sin(2\pi/n)$. These numbers are equally spaced on the unit circle and are the vertices of a regular polygon, called the **unit n-gon.**

Proof. By Corollary 3.23, $\zeta^n = 1$, so that ζ is a root of $x^n - 1$. Furthermore, for any nonnegative integer k, we have $(\zeta^k)^n = (\zeta^n)^k = 1$, so that all ζ^k are also roots of $x^n - 1$. But there are repetitions on the list $1, \ \zeta, \ \zeta^2, \ \cdots$. By the Division Algorithm, for any j, we have $j = qn + r$, where $0 \le r \le n - 1$. Hence,

$$\zeta^j = \zeta^{qn+r} = \zeta^{qn} \zeta^r = \zeta^r,$$

because $\zeta^{qn} = 1$.

On the other hand, all the ζ^k, for $0 \le k \le n - 1$, are distinct. After all, by De Moivre's Theorem,

$$\zeta^k = \cos(2\pi k/n) + i \sin(2\pi k/n),$$

and Proposition 3.14, uniqueness of polar forms, applies, for $0 \le 2\pi k/n < 2\pi$ are n distinct angles. Therefore, we have displayed n distinct roots of $x^n - 1$.

These are *all* the roots of $x^n - 1$, for a polynomial of degree n can have at most n distinct roots. We'll give a proof of this later (see Theorem 6.16) but, since we haven't yet proved this result, we now proceed in a different way.

If $z \in \mathbb{C}$ is a root of $x^n - 1$, then $1 = |z^n| = |z|^n$, so that $|z| = 1$, by Exercise 1.73 on page 41, and $z = \cos\theta + i\sin\theta$ for some θ. By De Moivre's Theorem, $1 = z^n = \cos(n\theta) + i\sin(n\theta)$, so that $n\theta = 2\pi k$ for some integer k; hence, $\theta = 2\pi k/n$. Write $k = qn + r$, where $0 \leq r < n$, and

$$z = \cos(2\pi k/n) + i\sin(2\pi k/n) = \cos(2\pi r/n) + i\sin(2\pi r/n).$$

Thus, z is equal to the root ζ^r already displayed.

Finally, since $\arg(\zeta^k) = k\arg(\zeta) = 2\pi k/n$, the roots ζ^k are equally spaced around the circle and, hence, they are the vertices of a regular n-gon. (See Figure 3.7 for the case $n = 8$.) ∎

Definition. The roots of $x^n - 1$ are called the *nth roots of unity*. An nth root of unity ζ is a *primitive nth root of unity* if n is the smallest positive integer for which $\zeta^n = 1$.

For every $n \geq 1$, we see that $\zeta = e^{2\pi i/n}$ is a primitive nth root of unity, for if $1 \leq m < n$, then $\zeta^m = \cos(2\pi m/n) + i\sin(2\pi m/n) \neq 1$. In particular, $i = \cos(2\pi/4) + i\sin(2\pi/4)$ is a primitive fourth root of unity, and $\omega = \frac{1}{2}(-1 + \sqrt{3}) = \cos(2\pi/3) + i\sin(2\pi/3)$ is a primitive cube root of unity.

Corollary 3.29. *Let $\zeta^k = \cos(2\pi k/n) + i\sin(2\pi k/n)$ be an nth root of unity.*

(i) *ζ^k is a primitive nth root of unity if and only if $\gcd(k, n) = 1$.*

(ii) *If ζ^k is a primitive nth root of unity, then every nth root of unity is a power of ζ^k.*

Proof. (i) Suppose that ζ^k is a primitive nth root of unity. If $d = \gcd(k, n) > 1$, then $n/d < n$, and

$$(\zeta^k)^{n/d} = (\zeta^n)^{k/d} = 1.$$

This contradicts n being the smallest positive integer with $(\zeta^k)^n = 1$.

Suppose that ζ^k is not primitive; that is, $(\zeta^k)^m = 1$ for some $m < n$. Since, by hypothesis, $\gcd(k, n) = 1$, there are integers s and t with $1 = sk + tn$; hence, $m = msk + mtn$. But now

$$\zeta^m = \zeta^{msk+mtn} = \zeta^{msk}\zeta^{mtn} = 1,$$

which contradicts ζ being a primitive nth root of unity.

(ii) Every nth root of unity is equal to ζ^j for some j. If $\gcd(k, n) = 1$, then there are integers s and t with $1 = sk + tn$. Hence,

$$\zeta^j = \zeta^{jsk+jtn} = \zeta^{jsk}\zeta^{jtn} = (\zeta^k)^{js}. \quad ∎$$

Definition. For every integer $n \geq 1$, define the *Euler ϕ-function $\phi(n)$* by

$$\phi(n) = \text{number of } k \text{ with } 1 \leq k \leq n \text{ and } \gcd(k, n) = 1.$$

For example, $\phi(1) = 1$ and, if p is prime, $\phi(p) = p - 1$.

Corollary 3.30. *For every positive integer n, there are exactly $\phi(n)$ primitive nth roots of unity.*

Proof. This follows at once from Corollary 3.29(i). ∎

Example 3.31. The complex number $\cos(\frac{2\pi}{n}) + i\sin(\frac{2\pi}{n})$ is a primitive nth root of unity, by Theorem 3.28. The 8th roots of unity (shown in Figure 3.7) are

$$\cos(\tfrac{2\pi}{8}) + i\sin(\tfrac{2\pi}{8}), \qquad \cos(\tfrac{6\pi}{8}) + i\sin(\tfrac{6\pi}{8}),$$

$$\cos(\tfrac{10\pi}{8}) + i\sin(\tfrac{10\pi}{8}), \qquad \cos(\tfrac{14\pi}{8}) + i\sin(\tfrac{14\pi}{8});$$

that is, the primitive 8th roots of unity are all those $\cos(\frac{2k\pi}{8}) + i\sin(\frac{2k\pi}{8})$ for which $\gcd(k, 8) = 1$. ▲

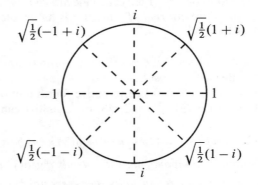

Figure 3.7. 8th roots of unity.

The nth roots of unity enjoy some remarkable properties that we'll use in upcoming chapters; here are some of them. (See Exercise 3.51 on page 115 and Proposition 6.63 for some other interesting properties.)

Theorem 3.32. *Let ζ be an nth root of unity.*

(i) $1 + \zeta^2 + \zeta^3 + \cdots + \zeta^{n-1} = 0$.

(ii) $\overline{\zeta^k} = 1/\zeta^k$ *for every nonnegative integer k.*

(iii) *If $k = qn + r$, then $\zeta^k = \zeta^r$.*

Proof. (i) We have $x^n - 1 = (x - 1)q(x)$, and we find q by long division:

$$(x - 1)\left(1 + x + x^2 + \cdots + x^{n-1}\right) = x^n - 1.$$

Now set $x = \zeta$ to see that

$$(\zeta - 1)(1 + \zeta + \zeta^2 + \cdots + \zeta^{n-1}) = \zeta^n - 1 = 0.$$

But $\zeta - 1 \neq 0$, and the result follows.

(ii) This follows from Exercise 3.38 on page 106.

(iii) This follows from $\zeta^{qn} = 1$. ∎

Theorem 3.28 establishes an intimate connection between the nth roots of unity and the geometry of the unit n-gon. The next examples illustrate this connection for small n.

Example 3.33. The vertices of unit n-gons for small values of n can be calculated with plane geometry.

- The vertices of the unit 3-gon are 1, $\frac{1}{2}(-1 + i\sqrt{3})$, $\frac{1}{2}(-1 - i\sqrt{3})$.
- The vertices of the unit 4-gon are 1, i, -1, $-i$.
- The vertices of the unit 6-gon are

See Exercise 3.49 on page 115 for more examples.

$$1, \quad \tfrac{1}{2}(1+i\sqrt{3}), \quad \tfrac{1}{2}(-1+i\sqrt{3}), \quad -1, \quad \tfrac{1}{2}(-1-i\sqrt{3}), \quad \tfrac{1}{2}(1-i\sqrt{3}). \quad \blacktriangle$$

Example 3.34 (Regular Pentagon). Since a primitive 5th root of unity is $\zeta = \cos(2\pi/5) + i \sin(2\pi/5)$, by Theorem 3.28, the vertices of the unit 5-gon are

$$\zeta = \cos(2\pi/5) + i \sin(2\pi/5)$$
$$\zeta^2 = \cos(4\pi/5) + i \sin(4\pi/5)$$
$$\zeta^3 = \cos(6\pi/5) + i \sin(6\pi/5)$$
$$\zeta^4 = \cos(8\pi/5) + i \sin(8\pi/5)$$
$$\zeta^5 = \cos(10\pi/5) + i \sin(10\pi/5) = 1 = \zeta^0.$$

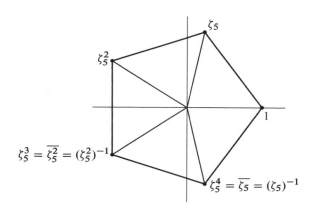

Figure 3.8. Unit 5-gon.

Can we find explicit expressions for these vertices that don't involve trigonometry? We'll obtain such an expression for $\cos(2\pi/5)$, but we'll leave the rest of the details for you (Exercise 3.49(i) on page 115); after all, you can evaluate, say, $\cos(8\pi/5)$.

We have $\bar{\zeta} = \zeta^4$, by Theorem 3.32(ii). Inspired by Lemma 3.2, we define

$$g = \zeta + \zeta^4 \qquad \text{and} \qquad h = \zeta^2 + \zeta^3.$$

Now

$$g = \zeta + \zeta^4 = 2\cos(2\pi/5),$$

a real number that is twice the number we are after. Similarly,

$$h = \zeta^2 + \zeta^3 = 2\cos(4\pi/5),$$

another real number. By Theorem 3.32(i),

$$g + h = \zeta + \zeta^4 + \zeta^2 + \zeta^3$$
$$= -1.$$

Thus, we know that $g + h = -1$. Do we also know gh? Let's see.

$$gh = \left(\zeta + \zeta^4\right)\left(\zeta^2 + \zeta^3\right)$$
$$= \zeta^3 + \zeta^4 + \zeta^6 + \zeta^7$$
$$= \zeta^3 + \zeta^4 + \zeta + \zeta^2 \quad \text{by Theorem 3.32(iii)}$$
$$= -1.$$

Hence, $g + h = -1 = gh$, and so g and h are the roots of

$$x^2 + x - 1.$$

Now $x^2 + x - 1$ has a positive root and a negative one. Since $g > h$ (why?), the positive root is g, and so

$$\cos(2\pi/5) = \tfrac{1}{2}g$$
$$= \tfrac{1}{4}(-1 + \sqrt{5}). \quad \blacktriangle$$

An ancient problem, going back to the Greeks, is to determine which regular n-gons can be constructed with ruler and compass. As we'll see in Chapter 7, the problem comes down to finding an expression for $\cos(2\pi/n)$ that doesn't mention trigonometry, only the operations of arithmetic and iterated square roots (as in Example 3.34). We've essentially shown that the regular pentagon is so constructible. This argument, grouping the ζ^k into convenient subclusters, was greatly generalized and refined by Gauss (when he was only 17 years old!) to show that the vertices of the unit 17-gon can be constructed (Euclid did not know this!). (This and much more is in Gauss's masterpiece, *Disquisitiones Arithmeticae*.) Gauss requested that his tombstone portray a regular 17-gon, but the stonemason was unable to carve it, saying it would look more like a circle than a polygon.

Exercises

3.46 Is De Moivre's Theorem true for negative integer exponents? Explain.

3.47 Let $z = \cos\theta + i\sin\theta$. Show, for all nonnegative integers n, that

$$z^n + (\overline{z})^n = 2\cos n\theta \qquad \text{and} \qquad z^n - (\overline{z})^n = 2\sin n\theta.$$

3.48 This exercise shows that there's something special about a 72° angle: there's only one isosceles triangle (up to similarity) whose base angle is twice the vertex angle, namely, the "72-72-36 triangle." Let the equal sides of such a triangle have length 1, and let q denote the length of the base.

  (i) Bisect one of the base angles of the triangle.

  (ii) Show that the small triangle is similar to the whole triangle.

  (iii) Use (ii) to show that $\frac{1}{q} = \frac{q}{1-q}$, and solve for q.

  (iv) Show that $q/2 = \cos 72°$.

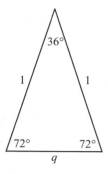

Figure 3.9. 72-72-36 triangle. **Figure 3.10.** Its construction.

3.49 * Find explicit formulas (i.e., without trigonometry) for the vertices of the unit

    (i) pentagon.   (ii) decagon.   (iii) 20-gon.

3.50 * Let n be a positive integer and let $\zeta = e^{2\pi i/n}$.

    (i) Establish the identity

$$x^n - 1 = (x-1)(x-\zeta)\left(x-\zeta^2\right)\cdots\left(x-\zeta^{n-1}\right).$$

    (ii) If x and y are integers, show that

$$x^n - y^n = (x-y)(x-\zeta y)\left(x-\zeta^2 y\right)\cdots\left(x-\zeta^{n-1}y\right).$$

    (iii) If x and y are integers and n is odd , show that

$$x^n + y^n = (x+y)(x+\zeta y)\left(x+\zeta^2 y\right)\cdots\left(x+\zeta^{n-1}y\right).$$

3.51 Take It Further. We saw, on page 111, that $\phi(p) = p-1$, where p is prime and ϕ is the Euler-ϕ function.

    (i) Suppose n is the product of two primes, $n = p_1 p_2$. Show that

$$\phi(n) = (p_1 - 1)(p_2 - 1).$$

Note that if $n = p_1 p_2$, then $(p_1 - 1)(p_2 - 1) = n - \frac{n}{p_1} - \frac{n}{p_2} + \frac{n}{p_1 p_2}$.

    (ii) Suppose n is the product of two primes powers, $n = p_1^{e_1} p_2^{e_2}$. Show that

$$\phi(n) = n - \frac{n}{p_1} - \frac{n}{p_2} + \frac{n}{p_1 p_2} = n\left(1 - \frac{1}{p_1}\right)\left(1 - \frac{1}{p_2}\right).$$

    (iii) Generalize to show that, if $n = p_1^{e_1} p_2^{e_2} \dots p_n^{e_n}$, then

$$\phi(n) = n\prod_{k=1}^{n}\left(1 - \frac{1}{p_k}\right).$$

3.52 Prove or disprove and salvage if possible. If a and b are positive integers,

$$\phi(ab) = \phi(a)\phi(b).$$

3.53 Find explicit formulas (i.e., without trigonometry) for the vertices of the unit n-gon if

    (i) $n = 3$ (ii) $n = 4$

    (iii) $n = 6$ (iv) $n = 8$

    (v) $n = 12$ (vi) $n = 16$

3.54 For all integers n between 3 and 9, find all the primitive nth roots of unity.

3.55 Find a primitive 12th root of unity ζ. Is ζ unique?

3.56 Suppose $\zeta = \cos(\frac{2\pi}{7}) + i \sin(\frac{2\pi}{7})$.

   (i) Plot the roots of $x^7 - 1$ in the complex plane.

   (ii) Show that $\alpha = \zeta + \zeta^6$, $\beta = \zeta^2 + \zeta^5$, and $\gamma = \zeta^3 + \zeta^4$ are real numbers.

   (iii) Find a cubic equation satisfied by $2\cos(\frac{2\pi}{7})$ by finding the values of $\alpha + \beta + \gamma$, $\alpha\beta + \alpha\gamma + \beta\gamma$, and $\alpha\beta\gamma$.

3.57 If $\zeta_n = \cos(\frac{2\pi}{n}) + i \sin(\frac{2\pi}{n})$, evaluate $\sum_{k=0}^{n-1} \zeta_n^k$.

3.58 Show that $\cos(\frac{2\pi}{5}) + \cos(\frac{4\pi}{5}) = -\frac{1}{2}$.

3.59 **Take It Further.** If n is a nonnegative integer, how many irreducible factors over $\mathbb{Z}$ does $x^n - 1$ have? In other words, we're looking for a pattern in the outputs of the function $n \mapsto$ # of factors of $x^n - 1$ over $\mathbb{Z}$. (Use a computer).

n	Number of Factors of $x^n - 1$
1	
2	
3	
4	
5	
6	
7	
8	
9	
10	
11	
12	

3.4 Connections: Designing Good Problems

This section will use complex numbers to help create mathematics problems that "come out nice." When launching a new topic, you want to start with examples which focus on the new idea; there shouldn't be any distractions—for example, numbers should be simple integers or rationals. Indeed, this is why the Babylonians introduced Pythagorean triples.

Norms

We begin by introducing a function $\mathbb{C} \to \mathbb{R}$, called the *norm*, that is closely related to absolute value. It will be an important tool for our applications; it will also be very useful in Chapter 8 when we do some algebraic number theory.

Definition. The *norm* of a complex number $z = a + bi$ is

$$N(z) = z\bar{z} = a^2 + b^2.$$

Here are some basic properties.

Proposition 3.35. *Let $z = a + ib$ and w be complex numbers.*

(i) $N(z)$ *is a nonnegative real number, and* $N(z) = 0$ *if and only if* $z = 0$.

(ii) $N(z) = |z|^2$.

(iii) $N(zw) = N(z)N(w)$.

Proof. (i) This follows at once from $N(a + bi) = a^2 + b^2$.

(ii) This follows at once from $|z| = |a + bi| = \sqrt{a^2 + b^2}$.

(iii) $N(zw) = zw\,\overline{zw} = zw\,\overline{z}\,\overline{w} = z\overline{z}\,w\overline{w} = N(z)N(w)$. ∎

It follows from Proposition 3.35(iii) that

$$N(z^k) = N(z)^k$$

for all z and all $k \geq 0$.

Here is an application of the norm.

Example 3.36. Let's revisit Example 3.5, the "bad example," in which the cubic formula gives the roots of $x^3 - 7x + 6 = (x - 1)(x - 2)(x + 3)$ in unrecognizable form.

Imagine again that you have just left the contest in Piazza San Marco, thinking about how $g + h$ could possibly equal 1, where

$$g^3 = -3 + i\tfrac{10}{9}\sqrt{3} \quad \text{and} \quad h^3 = -3 - i\tfrac{10}{9}\sqrt{3}.$$

Had you known about conjugates, you'd have seen that $\overline{g^3} = h^3$. It would have been natural to guess that the cube roots g and h are also complex conjugates (you'd have guessed right: see Exercise 3.64 on page 127); thus, $g = a + ib$ and $h = a - ib$. Now if $g + h = 1$, as your opponent loudly proclaimed, then $(a + ib) + (a - ib) = 2a = 1$; that is,

$$g = \tfrac{1}{2} + ib \quad \text{and} \quad h = \tfrac{1}{2} - ib.$$

You *really* want to find g and h now—what is b? Using the norm function, you see that

$$N(g)^3 = N(g^3) = (-3)^2 + \left(\tfrac{10\sqrt{3}}{9}\right)^2 = \tfrac{343}{27}.$$

Since norms are always real numbers, you conclude that

$$N(g) = \sqrt[3]{\tfrac{343}{27}} = \tfrac{7}{3}$$

(the other cube roots are complex; they are $\tfrac{7}{3}\omega$ and $\tfrac{7}{3}\omega^2$, where ω is a primitive cube root of unity). But if $g = \tfrac{1}{2} + ib$, then $N(g) = \tfrac{1}{4} + b^2$. Hence, $\tfrac{1}{4} + b^2 = \tfrac{7}{3}$, and $b = \pm\tfrac{5}{2\sqrt{3}}$. Thus,

$$g = \tfrac{1}{2} + i\tfrac{5}{2\sqrt{3}} \quad \text{and} \quad h = \tfrac{1}{2} - i\tfrac{5}{2\sqrt{3}}.$$

Bingo! For these "values" of g and h, we have $g + h = 1$. You were right! Elated, you run back to the square to show off g and h, but everyone has gone home. ▲

To find the other two roots, see Exercise 3.65 on page 127.

Pippins and Cheese

We call this subsection *Pippins and Cheese*, a phrase borrowed from Shakespeare's *Merry Wives of Windsor*, which describes delicious desserts.

Here are five problems. Solve them now; they are not difficult, but the answers may surprise you.

(i) A triangle has vertices $(-18, 49)$, $(15, -7)$, and $(30, -15)$. How long are its sides?

(ii) In Figure 3.11, the side lengths of $\triangle QSU$ are as marked. How big is $\angle Q$?

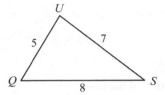

Figure 3.11. Side lengths.

(iii) An open box is formed by cutting out squares from a 7×15 rectangle and folding up the sides (see Figure 3.12). What size cut-out x maximizes the volume of the box?

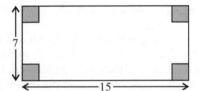

Figure 3.12. Making boxes.

(iv) Find the zeros, extrema, and inflection points of the function

$$f(x) = 140 - 144\,x + 3\,x^2 + x^3.$$

(v) Find the area of the triangle with sides of lengths 13, 14, and 15.

A ***meta-problem*** is a problem that asks how to design "nice" exercises of a particular genre, such as "How do you construct integer-sided scalene triangles having a 60° angle?" As we mentioned earlier, finding Pythagorean triples was one of the first meta-problems; it was invented by teachers who wanted to study and apply side-lengths of right triangles. In Section 1.2, we developed the method of Diophantus for this purpose—rational points on the unit circle correspond to Pythagorean triples. In this section, we'll consider two types of meta-problems: two ways of creating exercises like the five listed above. One meta-problem uses the norm function; the other generalizes Diophantus's chord method of "sweeping lines" by replacing circles with other conic sections. (There are many other kinds of meta-problems, ranging in topic from exponential equations to algebra word problems to trigonometry.)

See Exercises 1.79—1.82 on page 44.

Gaussian Integers: Pythagorean Triples Revisited

In Chapter 1, we saw that the Pythagorean equation $a^2 + b^2 = c^2$ corresponds to a rational point $\left(\frac{a}{c}, \frac{b}{c}\right)$ on the unit circle. But, given what you've just been studying, the Pythagorean equation might conjure up another rewrite in your mind, namely

$$(a + bi)(a - bi) = c^2,$$

or even

$$N(a + bi) = c^2.$$

So, we're looking for complex numbers $z = a + bi$ whose norms are perfect squares of integers. The Pythagorean equation now looks like

$$N(z) = c^2.$$

For example, $N(3 + 4i) = 5^2$, $N(5 + 12i) = 13^2$, and $N(8 + 15i) = 17^2$.

This idea doesn't work for every complex number. What's needed are complex numbers whose real and imaginary parts are *integers* (and, besides, whose norms are perfect squares). We'd like the real and imaginary parts to be *positive* integers, but any integers will do, because changing the sign of the real or imaginary part of a complex number doesn't change its norm (why?).

Definition. The *Gaussian integers* is the set $\mathbb{Z}[i]$ of all complex numbers whose real and imaginary parts are integers. In symbols,

$$\mathbb{Z}[i] = \{a + bi \in \mathbb{C} : a \in \mathbb{Z} \text{ and } b \in \mathbb{Z}\}.$$

Proposition 3.37. (i) *The set $\mathbb{Z}[i]$ of Gaussian integers is closed under addition and multiplication: If $a + bi, c + di \in \mathbb{Z}[i]$, then*

$$(a + bi) + (c + di) = (a + c) + (b + d)i \in \mathbb{Z}[i]$$
$$(a + bi)(c + di) = (ac - bd) + (ad + bc)i \in \mathbb{Z}[i].$$

(ii) *If $z = a + bi$, then*

$$N(z) = a^2 + b^2.$$

Proof. The formula for addition is clear; for multiplication, use the fact that $i^2 = -1$. Of course, part (ii) is just the definition of the norm. ∎

We'll investigate the Gaussian integers in more detail in Chapter 8.

Let's return to the norm equation $N(z) = c^2$ arising from Pythagorean triples, but with z a Gaussian integer. Our question is now "Which Gaussian integers have perfect squares as norms?" The answer comes from Proposition 3.35(iii): if z and w are complex numbers, then $N(zw) = N(z) N(w)$. In particular (letting $z = w$),

$$N\left(z^2\right) = N(z)^2.$$

The left-hand side of this equation is the norm of a Gaussian integer: if $z = a + ib$, then $z^2 = (a^2 - b^2) + i2ab$; moreover, $N(z^2)$ is a sum of two nonzero perfect squares if $a > 0$, $b > 0$, and $a \neq b$. Now the right-hand side is the square of an integer, namely, $N(z)^2$, which produces a Pythagorean triple. For example, if $z = 3 + 2i$, then $N(z) = 13$ and $z^2 = 5 + 12i$, and we get the Pythagorean triple $(5, 12, 13)$, for

$$5^2 + 12^2 = N\left((3 + 2i)^2\right) = N(3 + 2i)^2 = 13^2.$$

We now have a quick way to generate Pythagorean triples (by hand or with a computer; one of our colleagues uses this method to amaze friends at parties). Pick a Gaussian integer $r + si$ (with $r > 0$, $s > 0$, and $r \neq s$), and square it.

The r, s entry in the following table is $[(r + is)^2, N(r + is)]$. For example, the top entry in the first column, arising from $r = 2$ and $s = 1$, is $[(2 + i)^2, N(2 + i)] = [3 + 4i, 5]$; the corresponding Pythagorean triple is $(3, 4, 5)$.

	$s = 1$	$s = 2$	$s = 3$	$s = 4$
$r = 2$	$3 + 4i$, 5			
$r = 3$	$8 + 6i$, 10	$5 + 12i$, 13		
$r = 4$	$15 + 8i$, 17	$12 + 16i$, 20	$7 + 24i$, 25	
$r = 5$	$24 + 10i$, 26	$21 + 20i$, 29	$16 + 30i$, 34	$9 + 40i$, 41
$r = 6$	$35 + 12i$, 37	$32 + 24i$, 40	$27 + 36i$, 45	$20 + 48i$, 52

Eisenstein Integers.

Let's now look at the meta-problem of creating triangles with integer side-lengths and a 60° angle.

Let $\angle C = 60°$ in Figure 3.13, so that $\cos(\angle C) = \frac{1}{2}$. By the Law of Cosines,

$$c^2 = a^2 + b^2 - 2ab \cos \angle C$$
$$= a^2 + b^2 - ab.$$

What's important here is that the right-hand side of the equation, $a^2 - ab + b^2$, is the norm of $a + b\omega$, where $\omega = \frac{1}{2}(-1 + i\sqrt{3})$ is a primitive cube root of unity (Exercise 3.72 on page 128). This leads to the following definition.

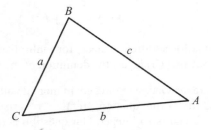

Figure 3.13. $\angle C = 60°$.

Eisenstein did extensive research on complex numbers of the form $a + b\zeta$, where ζ is a primitive nth root of unity. Note that ω is a primitive cube root of unity.

Definition. The *Eisenstein integers* is the set $\mathbb{Z}[\omega]$ of all complex numbers of the form $a + b\omega$, where $\omega = \frac{1}{2}(-1 + i\sqrt{3})$ is a primitive cube root of unity and a, b are integers. In symbols,

$$\mathbb{Z}[\omega] = \{a + b\omega \in \mathbb{C} : a \in \mathbb{Z} \text{ and } b \in \mathbb{Z}\}.$$

Here are some properties of Eisenstein integers.

Proposition 3.38. (i) *The set $\mathbb{Z}[\omega]$ of Eisenstein integers is closed under addition and multiplication: If $(a + b\omega), (c + d\omega) \in \mathbb{Z}[\omega]$, then*

$$(a + b\omega) + (c + d\omega) = (a + c) + (b + d)\omega \in \mathbb{Z}[\omega]$$
$$(a + b\omega)(c + d\omega) = (ac - bd) + (bc + ad - bd)\omega \in \mathbb{Z}[\omega].$$

In $\mathbb{Z}[i]$, $i^2 = -1$. In $\mathbb{Z}[\omega]$, $\omega^2 = -1 - \omega$.

(ii) *If $z = a + b\omega$, then*

$$N(z) = a^2 - ab + b^2.$$

In $\mathbb{Z}[i]$, $i^2 = -1$.
In $\mathbb{Z}[\omega]$, $\omega^2 = -1 - \omega$.

Proof. (i) The formula for addition is clear. For multiplication,

$$(a + b\omega)(c + d\omega) = ac + (bc + ad)\omega + bd\omega^2.$$

Since $\omega^2 + \omega + 1 = 0$, we have $\omega^2 = -1 - \omega$, and

$$(a + b\omega)(c + d\omega) = (ac - bd) + (bc + ad - bd)\omega.$$

(ii) As we said above, this is Exercise 3.72 on page 128. ∎

Definition. An *Eisenstein triple* is a triple of positive integers (a, b, c) such that

$$a^2 - ab + b^2 = c^2.$$

The same idea that produces Pythagorean triples from norms of squares of Gaussian integers applies to produce Eisenstein triples from norms of squares of Eisenstein integers. If z is an Eisenstein integer, then

$$N(z^2) = N(z)^2.$$

The left-hand side of this equation, being the norm of an Eisenstein integer, is of the form $a^2 - ab + b^2$. And the right-hand side is the square of the integer $N(z)$. Hence $a^2 - ab + b^2$ is a perfect square, and we have produced an Eisenstein triple.

Example 3.39. If $z = 3 + 2\omega$, then $N(z) = 3^2 - 3 \cdot 2 + 2^2 = 7$, and we have

$$z^2 = 9 + 12\omega + 4\omega^2$$
$$= 9 + 12\omega + 4(-1 - \omega)$$
$$= 5 + 8\omega.$$

Hence, $5^2 - 5 \cdot 8 + 8^2 = N\left(z^2\right) = N(z)^2 = 7^2$, and $(5, 8, 7)$ is an Eisenstein triple. ▲

In Figure 3.14, we have $\angle Q = 60°$.

We have found a quick way to generate Eisenstein triples (by hand or with a computer). Pick an Eisenstein integer $r + s\omega$ (with $r > 0$, $s > 0$, and $r \neq s$) and square it.

The r, s entry in the following table is $(r + s\omega)^2, N(r + s\omega)$. For example, the top entry in the first column, which arises from $r = 2$ and $s = 1$,

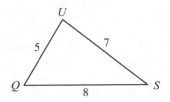

Figure 3.14. All sides have integer length.

is $\big((2 + \omega)^2, N(2 + \omega)\big) = (3 + 3\omega, 3)$; the corresponding Eisenstein triple gives $(3, 3, 3)$, which is an equilateral triangle. One of our friends calls this table a "candy store of patterns." Which entries give equilateral triangles?

	$s = 1$	$s = 2$	$s = 3$	$s = 4$
$r = 2$	$3 + 3\omega$, 3			
$r = 3$	$8 + 5\omega$, 7	$5 + 8\omega$, 7		
$r = 4$	$15 + 7\omega$, 13	$12 + 12\omega$, 12	$7 + 15\omega$, 13	
$r = 5$	$24 + 9\omega$, 21	$21 + 16\omega$, 19	$16 + 21\omega$, 19	$9 + 24\omega$, 21
$r = 6$	$35 + 11\omega$, 31	$32 + 20\omega$, 28	$27 + 27\omega$, 27	$20 + 32\omega$, 28
$r = 7$	$48 + 13\omega$, 43	$45 + 24\omega$, 39	$40 + 33\omega$, 37	$33 + 40\omega$, 37
$r = 8$	$63 + 15\omega$, 57	$60 + 28\omega$, 52	$55 + 39\omega$, 49	$48 + 48\omega$, 48
$r = 9$	$80 + 17\omega$, 73	$77 + 32\omega$, 67	$72 + 45\omega$, 63	$65 + 56\omega$, 61
$r = 10$	$99 + 19\omega$, 91	$96 + 36\omega$, 84	$91 + 51\omega$, 79	$84 + 64\omega$, 76

Eisenstein Triples and Diophantus

See Exercise 1.79 on page 44.

There's another, geometric, way to generate Eisenstein triples, using the same idea as the method of Diophantus in Chapter 1. If (a, b, c) is an Eisenstein triple, so that

$$a^2 - ab + b^2 = c^2,$$

then dividing by c^2 gives

$$(a/c)^2 - (a/c)(b/c) + (b/c)^2 = 1.$$

Thus, $(a/c, b/c)$ is a rational point on the ellipse with equation

$$x^2 - xy + y^2 = 1.$$

See Exercise 3.66 on page 128.

(See Figure 3.15.) As with the unit circle, the graph contains $(-1, 0)$, and we can use the chord method idea of Diophantus.

Proposition 3.40. *Let ℓ be a line through $(-1, 0)$ which intersects the ellipse with equation $x^2 - xy + y^2 = 1$ in a point P. If ℓ has rational slope, then P has rational coordinates, $P = (a/c, b/c)$, and*

$$a^2 - ab + b^2 = c^2.$$

If $P = (a/c, b/c)$ is in the first quadrant, then (a, b, c) is an Eisenstein triple.

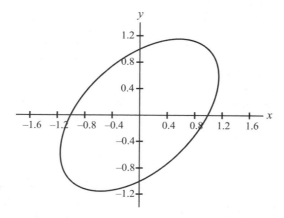

Figure 3.15. The graph of $x^2 - xy + y^2 = 1$.

Proof. The proof is almost identical to the proof of Proposition 1.2. We leave it to you to fill in the details. ∎

For example, if ℓ has slope $\frac{1}{4}$ and equation $y = \frac{1}{4}(x + 1)$, then ℓ intersects the ellipse in $\left(\frac{15}{13}, \frac{7}{13}\right)$, and $(15, 7, 13)$ is an Eisenstein triple. So, the triangle whose side lengths are 15, 7, and 13 has a 60° angle. Which angle is it?

Nice Boxes

Our next application is to a "box problem." In an $a \times b$ rectangle, cut out little squares at the corners, and then fold up the sides to form an open-top box (see Figure 3.16). What size cut-out maximizes the volume of the box? For most rectangles, the best cut-out has irrational side length. The meta-problem:

How can we find a and b to make the optimal cut-out a rational number?

Figure 3.16. Box problem.

As we tell our students, let the size of the cut-out be x. Then the volume of the box is a function of x:

$$V(x) = (a - 2x)(b - 2x)x = 4x^3 - 2(a + b)x^2 + abx,$$

and its derivative is

$$V'(x) = 12x^2 - 4(a + b)x + ab.$$

We want $V'(x)$ to have rational zeros, and so its discriminant

$$16(a + b)^2 - 48ab$$

should be a perfect square. But 16 is a perfect square, and so

$$(a + b)^2 - 3ab = a^2 - ab + b^2$$

should be a perfect square. This will be so if a and b are the legs of an Eisenstein triple (a, b, c).

For example, from the Eisenstein triple $(7, 15, 13)$, we get a 7×15 rectangle that can be used to create a box whose maximum volume occurs at a rational-length cut-out. The volume of the resulting box is

$$V(x) = (7 - 2x)(15 - 2x)x = 4x^3 - 443x^2 + 105x.$$

So, $V'(x) = 12x^2 - 88x + 105$. The roots of $V'(x)$ are $\frac{3}{2}$ and $\frac{35}{6}$. Both are rational, but only $\frac{3}{2}$ fits the context and maximizes V. (Why doesn't $\frac{35}{6}$ fit the context? What significance does it have? Also, see Exercise 3.69 on page 128.)

Nice Functions for Calculus Problems

Our next meta-problem is one that has occupied faculty room discussions about calculus teaching for years.

> How do you find cubic polynomials $f(x)$ with integer coefficients and rational roots, whose extrema and inflection points have rational coordinates?

No cheating: we want the extrema points and inflection points to be distinct. We'll actually create cubics in which all these points have integer coordinates.

Using the notation of Theorem 3.3, we can first assume that the cubic f is reduced; that is, it has form

$$f(x) = x^3 + qx + r.$$

This immediately guarantees that $f''(x) = 6x$ has an integer root, namely 0 (the inflection point of the graph is on the y-axis). Next, if we replace q by $-3p^2$ for some integer p, then $f'(x) = 3x^2 - 3p^2$, and $f'(x)$ also has integer roots. So, our cubic now looks like $f(x) = x^3 - 3p^2x + r$. This will have rational extrema and inflection points (what are they?), so all we have to do is ensure that it has three rational roots.

If f has two rational roots, it has three (why?), and so it's enough to make two roots, say $-\alpha$ and β, rational (we use $-\alpha$ instead of α because we've experimented a bit and found that this makes the calculations come out nicer). But if $f(-\alpha) = f(\beta) = 0$, we have

$$-\alpha^3 + 3p^2\alpha = \beta^3 - 3p^2\beta$$

or

$$\beta^3 + \alpha^3 = 3p^2(\alpha + \beta).$$

We can divide both sides by $\alpha + \beta$, for $\alpha + \beta \neq 0$ (lest $-\alpha = \beta$; remember that we want our roots distinct); we obtain

$$\alpha^2 - \alpha\beta + \beta^2 = 3p^2.$$

Eisenstein integers again. This is the same as

$$N(\alpha + \beta\omega) = 3p^2.$$

This time we want an Eisenstein integer whose norm is 3 times a square.

We're in luck: the equation $a^2 - ab + b^2 = 3$ has several integer solutions, including $(1, 2)$. So, $3 = N(1 - \omega)$. Hence we just need to take $\alpha + \beta\omega$ to be $1 - \omega$ times the square of an Eisenstein integer. Indeed, if

$$\alpha + \beta\omega = (1 - \omega)(r + s\omega)^2,$$

then

$$\begin{aligned}
\alpha^2 - \alpha\beta + \beta^2 = N(\alpha + \beta\omega) &= N\left((1 - \omega)(r + s\omega)^2\right) \\
&= N(1 - \omega)\, N\left((r + s\omega)^2\right) \\
&= 3N(r + s\omega)^2,
\end{aligned}$$

which is 3 times the square of an integer.

Example 3.41. Let's take $s = 1$ and $r = 3$. Then we have

$$\begin{aligned}
\alpha + \beta\omega &= (1 - \omega)(3 + \omega)^2 \\
&= 13 + 2\omega.
\end{aligned}$$

This tells us several things:

(i) Since $N(13 + 2\omega) = 147$, our cubic is

$$f(x) = x^3 - 147x + r.$$

(ii) But because $147 = 3 \cdot 7^2$ (so, $p = 7$), $f'(x) = 3x^3 - 3 \cdot 49$ will have rational roots: ± 7.

(iii) Since two roots of our cubic are $-\alpha$ and β, two roots are -13 and 2.

This lets us find r. Since

$$2^3 - 147 \cdot 2 + r = 0,$$

we have $r = 286$. Hence our cubic is

$$f(x) = x^3 - 147x + 286.$$

You can check that the third root is 11 and that the extrema and inflection points are rational. ▲

Creating examples like this is not hard by hand, but a computer algebra system makes it automatic. The next table was generated by a CAS, and it shows the results of our algorithm for small values of r and s.

	$s = 1$	$s = 2$	$s = 3$
$r = 2$	$54 - 27x + x^3$	$-128 - 48x + x^3$	
$r = 3$	$286 - 147x + x^3$	$286 - 147x + x^3$	$-1458 - 243x + x^3$
$r = 4$	$-506 - 507x + x^3$	$3456 - 432x + x^3$	$-506 - 507x + x^3$
$r = 5$	$-7722 - 1323x + x^3$	$10582 - 1083x + x^3$	$10582 - 1083x + x^3$
$r = 6$	$-35282 - 2883x + x^3$	$18304 - 2352x + x^3$	$39366 - 2187x + x^3$

See Exercise 3.70 on page 128.

All these cubics have coefficient of x^2 equal to 0. If you'd like examples where this is not the case, just replace x by, say, $x + 1$ and simplify. Again, a CAS makes this easy.

Lattice Point Triangles

A lattice point is a point with integer coordinates.

Our last meta-problem arises when illustrating the distance formula.

How can you find three lattice points A, B, and C in the plane so that the distance between any two of them is an integer?

Clearly, solutions are *invariant under translation* by a lattice point; that is, if A, B, and C form a lattice point solution and U is any lattice point, then $A - U$, $B - U$, and $C - U$ form another solution: since $d(A - U, B - U) = d(A, B)$ (where $d(P, Q)$ is the distance between points P and Q), we have $d(A, B) = |A - B|$. Hence, we can assume that one of the points, say C, is at the origin.

Now view the plane as the complex plane, so that lattice points are Gaussian integers. Thus, we want Gaussian integers z and w such that $|z|$, $|w|$, and $|z - w|$ are integers. But if $z = a + bi$, then

$$|z| = \sqrt{a^2 + b^2} = \sqrt{N(z)}.$$

Hence, to make the length an integer, make the norm a perfect square and, to make the norm a perfect square, make the Gaussian integer a perfect square in $\mathbb{Z}[i]$. That is, we want Gaussian integers z and w so that z, w, and $z - w$ are perfect squares in $\mathbb{Z}[i]$. Hence, we choose z and w so that

$$z = \alpha^2 \qquad \text{for some } \alpha \in \mathbb{Z}[i]$$
$$w = \beta^2 \qquad \text{for some } \beta \in \mathbb{Z}[i]$$
$$z - w = \gamma^2 \qquad \text{for some } \gamma \in \mathbb{Z}[i].$$

In other words, we want Gaussian integers α, β, and γ so that

$$\alpha^2 - \beta^2 = \gamma^2$$

or

$$\alpha^2 = \beta^2 + \gamma^2.$$

The punchline is that one of our favorite identities,

$$\left(x^2 + y^2\right)^2 = \left(x^2 - y^2\right)^2 + (2xy)^2,$$

which holds in any commutative ring, holds, in particular, in $\mathbb{Z}[i]$. So, the trick is to pick Gaussian integers x and y, set

See Exercise 1.25 on page 14.

$$\alpha = x^2 + y^2$$
$$\beta = x^2 - y^2,$$

and then let

$$z = \alpha^2$$
$$w = \beta^2.$$

Example 3.42. Pick $x = 2 + i$ and $y = 3 + 2i$. Then

$$\alpha = x^2 + y^2 = 8 + 16i \quad \text{and} \quad \beta = x^2 - y^2 = -2 - 8i.$$

Now put

$$z = \alpha^2 = -192 + 256i \quad \text{and} \quad w = \beta^2 = -60 + 32i.$$

Hence, $(0, 0)$, $(-192, 256)$, and $(-60, 32)$ are vertices of an integer-sided triangle. Moreover, adding a lattice point to each vertex produces another such triangle with no vertex at the origin. Once again, a CAS can be used to generate many more.   ▲

This is just the beginning; many research problems are generalizations of meta-problems. Fermat's Last Theorem started as a search for integer solutions to equations like the Pythagorean equation but with larger exponents.

There are many other meta-problems that yield to these two methods: norms from $\mathbb{Z}[i]$ or $\mathbb{Z}[\omega]$; rational points on the unit circle or on the graph of $x^2 - xy + y^2 = 1$. Still others can be solved with norms from other number systems or from rational points on other curves. In Chapter 9, we will see that congruent numbers lead to rational points on certain cubic curves.

Exercises

3.60 For each integer n between 3 and 9, find a polynomial of smallest degree with integer coefficients whose roots are the primitive nth roots of unity.

3.61 * Let a and b be real numbers, and let z be a complex number.

   (i) Show that $\overline{a + bz} = a + b\overline{z}$.

   (ii) Show that $N(a + bz) = a^2 + 2\Re(z)ab + b^2 N(z)$.

3.62 * If z and w are complex numbers, show that $N(z) < N(w)$ if and only if $|z| < |w|$.

3.63 Let Δ be an isosceles triangle with side lengths 13, 13, and 10.

   (i) Show that the altitude to the base has length 12, and that it divides Δ into two 5, 12, 13 triangles.

   (ii) Show that the altitude to one of the sides of length 13 divides Δ into two right triangles whose side lengths are rational.

   (iii) Each of the side lengths can thus be scaled to get a Pythagorean triple. Show that one triple is similar to $(5, 12, 13)$ and that other comes from $(5 + 12i)^2$.

   (iv) Generalize this result to any isosceles triangle formed by two copies of a Pythagorean triple, joined along a leg.

3.64 * Let g and h be complex numbers such that $\overline{g^3} = h^3$.

   (i) Show that $\overline{g}$ is equal to either h, ωh, or $\omega^2 h$, where $\omega = \frac{1}{2}(-1 + i\sqrt{3})$.

   (ii) If gh is also real, show that $\overline{g} = h$.

3.65 * Suppose that $g = \frac{1}{2} + \frac{5}{2\sqrt{3}}i$, $h = \overline{g}$, and $\omega = \cos(\frac{2\pi}{3}) + i\sin(\frac{2\pi}{3})$ (see Example 3.36). Find the value of

   (i) $g\omega + h\omega^2$

   (ii) $g\omega^2 + h\omega$

3.66 * Sketch the graph of $x^2 - ax + y^2 = 1$ for

   (i) $a = -1$

   (ii) $a = 1$

   (iii) $a = 2$

   (iv) $a = 3$

   (v) $a = \frac{1}{2}$

3.67 In Theorem 1.5, we saw that every Pythagorean triple is similar to one of the form

$$(2xy, x^2 - y^2, x^2 + y^2).$$

Show how this can be obtained via the "norm from $\mathbb{Z}[i]$" method.

3.68 Obtain a formula for Eisenstein triples analgous to the one for Gaussian integers in Theorem 1.5 using norms from $\mathbb{Z}[\omega]$ and rational points on the graph of $x^2 - xy + y^2 = 1$.

3.69 * Assume that the square of the Eisenstein integer $r + s\omega$ is used to generate an Eisenstein triple, and that the triple is used to create a "nice box," as on page 123. Express the volume of the box in terms of r and s.

Replacing x by $x + 1$ just translates the graph by one unit. Which way?

3.70 * Replace x by $x + 1$ in several of the cubics in the table on page 125 to produce nice cubics whose coefficient of x^2 is nonzero. Show that your cubics are indeed nice.

3.71 Describe where Gaussian integers are situated in the complex plane.

3.72 * Suppose that a and b are real numbers and

$$\omega = \cos(\tfrac{2\pi}{3}) + i \sin(\tfrac{2\pi}{3}) = \tfrac{1}{2}(-1 + i\sqrt{3}).$$

Show that

$$N(a + b\omega) = a^2 - ab + b^2.$$

3.73 Describe where Eisenstein integers are situated in the complex plane.

3.74 Find an integer-sided triangle one of whose angles has cosine equal to 3/5.

Hint. Let $\rho = \cos(\tfrac{3}{5}) + i \sin(\tfrac{3}{5})$ and consider norms from $\mathbb{Z}[\rho]$. What conic would help here?

3.75 A **_Heron triangle_** is a triangle with integer side lengths and integer area. In Exercise 1.26 on page 14, you found a Heron triangle by joining two Pythagorean triangles together along a common leg. Show that the following method also produces Heron triangles.

    Pick a rational point $(\cos\theta, \sin\theta)$ on the unit circle, where $0 < \theta < \pi$, and let $\alpha = -\cos\theta + i \sin\theta$. Then pick any number z of the form $r + s\alpha$, where r and s are rational numbers and $r > s > 0$.

   (i) What is the norm of $r + s\alpha$?

   (ii) Show that

$$\alpha^2 + 2\alpha \cos\theta + 1 = 0.$$

   (iii) Show that if $z^2 = a + b\alpha$, then the triangle with side lengths a and b and included angle θ will have a rational number, say c, as its third side length and a rational number as an area. (This triangle can be then scaled to produce a Heron triangle.) Use this method to generate a few Heron triangles.

3.76 Show that a triangle with lattice point vertices and integer side-lengths is a Heron triangle.

3.77 Take It Further.   Here's a typical *current problem*, taken from B. Kerins, Gauss, Pythagoras, and Heron, *Mathematics Teacher*, 2003, 350-357:

> A boat is making a round trip, 135 miles in each direction. Without a current, the boat's speed would be 32 miles per hour. However, there is a constant current that increases the boat's speed in one direction and decreases it in the other. If the round trip takes exactly 9 hours, what is the speed of the current?

(i) Solve the problem.

(ii) Solve the corresponding meta-problem: find a method for generating current problems that come out nice.

4 Modular Arithmetic

Theorems about integers can be generalized to other interesting contexts. For example, an early attack on Fermat's Last Theorem was to factor $x^n + y^n$ (n odd) as in Exercise 3.50 on page 115:

$$x^n + y^n = (x + y)(x + \zeta y)\cdots(x + \zeta^{n-1}y),$$

where $\zeta = e^{2\pi i/n}$ is an nth root of unity. It turns out that the most fruitful way to understand this factorization is within the system $\mathbb{Z}[\zeta]$ of *cyclotomic integers*, the collection of all polynomials in ζ with coefficients in $\mathbb{Z}$ (a common generalization of Gaussian integers $\mathbb{Z}[i]$ and Eisenstein integers $\mathbb{Z}[\omega]$). Numbers in these systems can be added and multiplied, and they satisfy all but one of the nine fundamental properties that ordinary numbers do (reciprocals of cyclotomic integers need not be such); we will call such systems *commutative rings*. But for some roots of unity ζ, the commutative ring $\mathbb{Z}[\zeta]$ does not enjoy the unique factorization property that $\mathbb{Z}$, $\mathbb{Z}[i]$, and $\mathbb{Z}[\omega]$ have, and this caused early "proofs" of Fermat's Last Theorem to be false. Dealing with the lack of unique factorization was one important problem that led naturally to the modern way of studying algebra.

> We'll discuss cyclotomic integers in Chapter 8.

In Section 4.1, we shall see that the distinction between even and odd can be generalized, using *congruences*: studying remainders in the Division Algorithm. It turns out, as we'll see in Section 4.3, that, for any fixed positive integer m, the set of its remainders, $0, 1, ..., m - 1$, can be viewed as a commutative ring, as can cyclotomic integers, and they behave in many, but not all, ways as do ordinary integers. Finally, in Section 4.5, we'll apply these results to an analysis of decimal expansions of rational numbers.

> It turns out that many of the "number systems" studied in high school are commutative rings.

4.1 Congruence

It is often useful to know the **parity** of an integer n; that is, whether n is *even* or *odd* (why else would these words be in the language?). But n being even or odd is equivalent to whether its remainder after dividing by 2 is 0 or 1. Modular arithmetic, introduced by Euler around 1750, studies the generalization of parity arising from considering remainders after dividing by any positive integer. At a low level, it will help us answer questions of the following sort:

- London time is 6 hours ahead of Chicago time; if it is now 9:00 AM in Chicago, what time is it in London?

- If April 12 falls on a Thursday this year, on what day of the week is May 26?

At a more sophisticated level, it will allow us to solve some difficult number theoretic problems.

Note that $a \equiv 0 \bmod m$ if and only if $m \mid a$.

Definition. Let $m \geq 0$ be an integer. If $a, b \in \mathbb{Z}$, then a is ***congruent to*** b ***modulo*** m, denoted by

$$a \equiv b \bmod m,$$

if $m \mid (a - b)$.

Etymology. The number m in the expression $a \equiv b \bmod m$ is called the ***modulus***, the Latin word meaning a standard unit of measurement. The term *modular unit* is used today in architecture: a fixed length m is chosen, say, $m = 1$ foot, and plans are drawn so that the dimensions of every window, door, wall, etc., are integral multiples of m.

We claim that integers a and b have the same parity if and only if $a \equiv b \bmod 2$. Assume that a and b have the same parity. If both are even, then $a = 2a'$ and $b = 2b'$. Hence, $a - b = 2(a' - b')$, $2 \mid (a - b)$, and $a \equiv b \bmod 2$. Similarly, if both are odd, then $a = 2a' + 1$ and $b = 2b' + 1$. Hence, $a - b = (2a' + 1) - (2b' + 1) = 2(a' - b')$, $2 \mid (a - b)$, and $a \equiv b \bmod 2$ in this case as well. Conversely, suppose that $a \equiv b \bmod 2$. If a and b have different parity, then one is even, the other is odd, and so their difference is odd. Hence, $2 \nmid (a - b)$, and $a \not\equiv b \bmod 2$. Having proved the contrapositive, we may now assert that a and b have the same parity.

Modular arithmetic is called "clock arithmetic" in some introductory texts.

Example 4.1. If $a \equiv r \bmod m$, then r is obtained from a by throwing out a multiple of m. For example, let's compute the time of day using a 12-hour clock. When adding 6 hours to 9:00, the answer, 3:00, is obtained by taking $9 + 6 = 15 \equiv 3 \bmod 12$ (i.e., we throw away 12). In more detail, let 0 denote 12:00, 1 denote 1:00, ..., 11 denote 11:00. Three hours after 9:00 is 12:00; that is, $9 + 3 = 12 \equiv 0 \bmod 12$; 4 hours after 9:00 is 1:00; that is, $9 + 4 = 13 \equiv 1 \bmod 12$, and 6 hours after 9:00 is 3:00; that is, $9 + 6 = 15 \equiv 3 \bmod 12$.

The same idea applies to calendars. Let 0 denote Sunday, 1 denote Monday, ..., 6 denote Saturday.

Sun	Mon	Tues	Wed	Thurs	Fri	Sat
0	1	2	3	4	5	6

If today is Tuesday, what day of the week is 90 days from now? Since $2 + 90 = 92 \equiv 1 \bmod 7$, the answer is Monday.

Let's now answer the question: if April 12 falls on Thursday this year, on what day of the week is May 26? There are 18 days to April 30, so there are $18 + 26 = 44$ days until May 26 (for April has only 30 days). Now Thursday corresponds to 4, so that May 26 corresponds to $4 + 44 = 48 \equiv 6 \bmod 7$; therefore, May 26 falls on Saturday. ▲

There are at least two ways to state the solutions of Exercises 3.2 and 3.3 on page 89. We expected you to say then that $i^n = i^m$ if and only if n and m leave the same remainder when divided by 4 and, if ω is a primitive cube root of unity, that $\omega^n = \omega^m$ if and only if n and m leave the same remainder when divided by 3. In light of the next proposition, we can also say that $i^n = i^m$ if and only if $4 \mid (n - m)$; that is, $n \equiv m$ mod 4, and $\omega^n = \omega^m$ if and only if $3 \mid (n - m)$; that is, $n \equiv m$ mod 3.

Proposition 4.2. *Let $m \geq 2$ and $a, b \in \mathbb{Z}$.*

(i) *If $a = qm + r$, then $a \equiv r$ mod m.*

(ii) *$a \equiv b$ mod m if and only if each of a and b have the same remainder after dividing by m.*

Proof. (i) Since $a - r = qm$, we have $m \mid (a - r)$; that is, $a \equiv r$ mod m.

(ii) Assume $a \equiv b$ mod m. Let r, r' be the remainders after dividing a, b, respectively, by m; that is, $a = qm + r$ and $b = q'm + r'$, where $0 \leq r < m$ and $0 \leq r' < m$. We want to show that $r' = r$. If not, suppose that $r' < r$ (the argument is the same if $r < r'$). Then $a - b = m(q - q') + (r - r')$ with $0 < r - r' < m$. Now Exercise 1.46 on page 29 gives $m \mid (r - r')$. Hence, $m \leq r - r'$, by Lemma 1.13, contradicting $r - r' < m$.

Conversely, if $a = qm + r$, $b = q'm + r'$, and $r = r'$, then $a - b = m(q - q')$ and $a \equiv b$ mod m. ■

Notice that Proposition 4.2 generalizes the fact that integers a and b have the same parity if and only if $a \equiv b$ mod 2.

We are now going to see that congruence modulo m behaves very much like ordinary equality; more precisely, it is an *equivalence relation* (see Appendix A.2): it is *reflexive*, *symmetric*, and *transitive*.

Proposition 4.3. *Let $m \geq 0$. For all integers a, b, c, we have*

(i) *$a \equiv a$ mod m;*

(ii) *if $a \equiv b$ mod m, then $b \equiv a$ mod m;*

(iii) *if $a \equiv b$ mod m and $b \equiv c$ mod m, then $a \equiv c$ mod m.*

Proof. All are easy to check. We have $a \equiv a$ mod m, because $a - a = 0$ and $m \mid 0$ is always true (even when $m = 0$). Since $b - a = -(a - b)$, if $m \mid (a - b)$, then $m \mid (b - a)$. Finally, $(a - b) - (b - c) = a - c$, so that if $m \mid (a - b)$ and $m \mid (b - c)$, then $m \mid (a - c)$. ■

How to Think About It. Congruence mod 1 makes sense, but it is not very interesting, for $a \equiv b$ mod 1 if and only if $1 \mid (a - b)$. But this latter condition is always true, for 1 is a divisor of every integer. Thus, every two integers are congruent mod 1. Similarly, congruence mod 0 makes sense, but it, too, is not very interesting, for $0 \mid c$ if and only if $c = 0$. Thus, $a \equiv b$ mod 0 if and only if $0 \mid (a - b)$; that is, $a \equiv b$ mod 0 if and only if $a = b$, and so congruence mod 0 is just ordinary equality. You should not be surprised that we usually assume that $m \geq 2$.

If $0 \mid c$, then there is some k with $c = 0 \cdot k = 0$; that is, $c = 0$.

Corollary 4.4. *If $m \geq 2$, then every integer a is congruent mod m to exactly one integer on the list*

See Exercise 4.5 on page 140 for a generalization.

$$0, 1, \ldots, m - 1.$$

Proof. By the Division Algorithm, we have $a = qm + r$, where $0 \leq r < m$; that is, $a \equiv r \bmod m$.

If a were congruent to two integers on the list, say, $r < r'$, then $r \equiv r' \bmod m$ by transitivity, so that $m \mid (r' - r)$. Since $0 < r' - r < m$, this would contradict Lemma 1.13. ■

Congruence gets along well with addition and multiplication.

Proposition 4.5. *Let $m \geq 0$.*

 (i) *If $a \equiv a' \bmod m$ and $b \equiv b' \bmod m$, then*

$$a + b \equiv a' + b' \bmod m.$$

More generally, if $a_i \equiv a_i' \bmod m$ for $i = 1, \ldots, k$, then

$$a_1 + \cdots + a_k \equiv a_1' + \cdots + a_k' \bmod m.$$

 (ii) *If $a \equiv a' \bmod m$ and $b \equiv b' \bmod m$, then*

$$ab \equiv a'b' \bmod m.$$

More generally, if $a_i \equiv a_i' \bmod m$ for $i = 1, \ldots, k$, then

$$a_1 \cdots a_k \equiv a_1' \cdots a_k' \bmod m.$$

(iii) *If $a \equiv b \bmod m$, then*

$$a^k \equiv b^k \bmod m \ \ \text{for all } k \geq 1.$$

Proof.   (i) If $m \mid (a - a')$ and $m \mid (b - b')$, then $m \mid (a + b) - (a' + b')$, because $(a + b) - (a' + b') = (a - a') + (b - b')$. The generalization to k summands follows by induction on $k \geq 2$.

 (ii) We must show that if $m \mid (a - a')$ and $m \mid (b - b')$, then $m \mid (ab - a'b')$. This follows from the identity

$$ab - a'b' = ab - ab' + ab' - a'b' = a(b - b') + (a - a')b'.$$

The generalization to k factors follows by induction on $k \geq 2$.

(iii) This is the special case of part (ii) in which all $a_i = a$ and all $a_i' = b$. ■

How to Think About It. The key idea in calculating with congruences mod m is that every number can be replaced by its remainder after dividing by m, for this is precisely what Proposition 4.5 permits; it allows you to "reduce as you go" in calculations, as the next example shows.

Example 4.6. The last (units) digit of a positive integer is the remainder when it is divided by 10. What is the last digit of

$$10324^3 + 2348 \cdot 5267?$$

We could do this by brute force: cube 10324, multiply 2348 and 5267, add, and look at the last digit. But, as one of our friends says, why should the calculator have all the fun? You can do this more cleverly using congruence.

- To compute 10324^3, first look at 10324.

$$10324 \equiv 4 \bmod 10, \text{ so that } 10324^3 \equiv 4^3 \bmod 10.$$

Now $4^3 = 64 \equiv 4 \bmod 10$, so that $10324^3 \equiv 4 \bmod 10$, and the last digit of 10324^3 is 4.

- To multiply 2348 and 5267, note that $2348 \equiv 8 \bmod 10$ and $5267 \equiv 7 \bmod 10$. Hence,

$$2348 \cdot 5267 \equiv 8 \cdot 7 = 56 \equiv 6 \bmod 10.$$

- Thus,

$$10324^3 + 2348 \cdot 5267 \equiv 4 + 6 = 10 \bmod 10,$$

and 0 is the last digit; $10324^3 + 2348 \cdot 5267$ is divisible by 10.

Now you try one: what is the last digit of $75284^3 + 10988 \cdot 310767$? ▲

> More simply, think of multiplying 2348 and 5267 by hand. What's the last digit? This is what most middle school students would do. We just want to illustrate the general principle here.

The next example uses congruence to solve more difficult problems.

Example 4.7. (i) *If $a \in \mathbb{Z}$, then $a^2 \equiv 0$, 1, or 4 mod 8.*

If a is an integer, then $a \equiv r \bmod 8$, where $0 \le r \le 7$; moreover, by Proposition 4.5 (iii), $a^2 \equiv r^2 \bmod 8$, and so it suffices to look at the squares of the remainders. We see in Figure 4.1 that only 0, 1, or 4 can be a remainder after dividing a perfect square by 8.

r	0	1	2	3	4	5	6	7
r^2	0	1	4	9	16	25	36	49
$r^2 \bmod 8$	0	1	4	1	0	1	4	1

Figure 4.1. Squares mod 8.

(ii) $n = 1003456789$ *is not a perfect square.*

Since $1000 = 8 \cdot 125$, we have $1000 \equiv 0 \bmod 8$, and so

$$1003456789 = 1003456 \cdot 1000 + 789 \equiv 789 \bmod 8.$$

Dividing 789 by 8 leaves remainder 5; that is, $n \equiv 5 \bmod 8$. But if n were a perfect square, then $n \equiv 0$, 1, or 4 mod 8.

(iii) *There are no perfect squares of the form $3^m + 3^n + 1$, where m and n are positive integers.*

Again, let's look at remainders mod 8. Now $3^2 = 9 \equiv 1 \bmod 8$, and so we can evaluate $3^m \bmod 8$ as follows: if $m = 2k$, then $3^m = 3^{2k} = 9^k \equiv 1 \bmod 8$; if $m = 2k + 1$, then $3^m = 3^{2k+1} = 3^{2k} \cdot 3 \equiv 3 \bmod 8$. Thus,

$$3^m \equiv \begin{cases} 1 \bmod 8 & \text{if } m \text{ is even} \\ 3 \bmod 8 & \text{if } m \text{ is odd.} \end{cases}$$

Replacing numbers by their remainders after dividing by 8, we have the following possibilities for the remainder of $3^m + 3^n + 1$, depending on the parities of m and n:

$$3 + 1 + 1 \equiv 5 \bmod 8$$
$$3 + 3 + 1 \equiv 7 \bmod 8$$
$$1 + 1 + 1 \equiv 3 \bmod 8$$
$$1 + 3 + 1 \equiv 5 \bmod 8.$$

In no case is the remainder 0, 1, or 4, and so no number of the form $3^m + 3^n + 1$ can be a perfect square, by part (i).   ▲

Many beginning algebra students wish that $(a+b)^p = a^p + b^p$ in $\mathbb{Z}$; if only $(a + b)^2 = a^2 + b^2$! The next proposition (which paraphrases Exercise 7.27 on page 293) would delight them. If theorems were movies, Proposition 4.8 would be X-rated: only adults would be allowed to see it.

Proposition 4.8. *If p is a prime and a, b are integers, then*

$$(a + b)^p \equiv a^p + b^p \bmod p.$$

Proof. The Binomial Theorem gives

$$(a + b)^p = \sum_{r=0}^{p} \binom{p}{r} a^{p-r} b^r.$$

But $\binom{p}{r} \equiv 0 \bmod p$ for all r with $0 < r < p$, by Proposition 2.26. The result now follows from Proposition 4.5(i). ■

The next theorem (sometimes called the *Little Fermat Theorem* to distinguish it from Fermat's Last Theorem) turns out to be very useful.

See Corollary 4.67 for another proof.

Theorem 4.9 (Fermat). *Let p be a prime and $a \in \mathbb{Z}$.*

(i) $a^p \equiv a \bmod p$.

(ii) $a^{p^n} \equiv a \bmod p$ *for all $n \geq 1$.*

(iii) *If $p \nmid a$, then $a^{p-1} \equiv 1 \bmod p$.*

Proof. (i) We first prove the statement when $a \geq 0$, by induction on a. The base step $a = 0$ is obviously true. For the inductive step, the inductive hypothesis is $a^p \equiv a \bmod p$. Hence, Proposition 4.8 gives

$$(a + 1)^p \equiv a^p + 1 \equiv a + 1 \bmod p.$$

To complete the proof, consider $-a$, where $a > 0$; now

$$(-a)^p = (-1)^p a^p \equiv (-1)^p a \bmod p.$$

If p is an odd prime (indeed, if p is odd), then $(-1)^p = -1$, and $(-1)^p a = -a$, as desired. If $p = 2$, then $(-a)^2 = a^2 \equiv a \bmod 2$, and we are finished in this case as well.

(ii) The proof is by induction on $n \geq 1$: the base step is part (i), while the inductive step follows from the identity $a^{p^n} = \left(a^{p^{n-1}}\right)^p$.

(iii) By part (i), $p \mid (a^p - a)$; that is, $p \mid a(a^{p-1} - 1)$. Since $p \nmid a$, Euclid's Lemma gives $p \mid (a^{p-1} - 1)$; that is, $a^{p-1} \equiv 1 \bmod p$. ■

Later in this chapter, we will use the next corollary to construct codes that are extremely difficult for spies to decode.

Corollary 4.10. *If p is a prime and $m \equiv 1 \bmod (p-1)$, then $a^m \equiv a \bmod p$ for all $a \in \mathbb{Z}$.*

Proof. If $a \equiv 0 \bmod p$, then $a^m \equiv 0 \bmod p$, and so $a^m \equiv a \bmod p$. Assume now that $a \not\equiv 0 \bmod p$; that is, $p \nmid a$. By hypothesis, $m - 1 = k(p - 1)$ for some integer k, and so $m = 1 + (p - 1)k$. Therefore,

$$a^m = a^{1+(p-1)k} = aa^{(p-1)k} = a\left(a^{p-1}\right)^k \equiv a \bmod p,$$

for $a^{p-1} \equiv 1 \bmod p$, by Theorem 4.9(iii) ∎

We can now explain a well-known divisibility test.

Proposition 4.11. *A positive integer a is divisible by 3 if and only if the sum of its (decimal) digits is divisible by 3.*

Proof. The decimal notation for a is $d_k \ldots d_1 d_0$; that is,

$$a = d_k 10^k + \cdots + d_1 10 + d_0,$$

where $0 \le d_i < 10$ for all i. Now $10 \equiv 1 \bmod 3$, and Proposition 4.5(iii) gives $10^i \equiv 1^i = 1 \bmod 3$ for all i; thus parts (i) and (ii) of Proposition 4.5 give $a \equiv d_k + \cdots + d_1 + d_0 \bmod 3$. Therefore, a is divisible by 3 if and only if $a \equiv 0 \bmod 3$ if and only if $d_k + \cdots + d_1 + d_0 \equiv 0 \bmod 3$. ∎

How to Think About It. The proof of Proposition 4.11 shows more than its statement claims: the sum of the (decimal) digits of any positive integer a is congruent to $a \bmod 3$, whether or not a is divisible by 3. For example,

$$172 \equiv 1 + 7 + 2 \bmod 3;$$

that is, both 172 and 10 (the sum of its digits) are $\equiv 1 \bmod 3$.

Since $10 \equiv 1 \bmod 9$, Proposition 4.11 holds if we replace 3 by 9 (it is often called *casting out 9s*): A positive integer a is divisible by 9 if and only if the sum of its digits, $\Sigma(a)$, is divisible by 9.

Define two operations on the decimal digits of a positive integer a.

> Is the sum of the decimal digits of an integer a congruent mod 9 to a itself?

(i) Delete all 9s (if any) and delete any group of digits whose sum is 9

(ii) Add up all the digits.

It is easy to see that repeated applications of these operations to a positive integer a yields a single digit; call it $r(a)$. For example,

$$5261934 \to 526134 \to 561 \text{ (for } 2 + 3 + 4 = 9) \to 12 \to 3.$$

(It is now clear why this procedure is called *casting out 9s*.) In light of $a \equiv \Sigma(a) \bmod 9$, we have $\Sigma(a) \equiv r(a) \bmod 9$, so that $r(a)$, which seems to depend on a choice of operations (i) and (ii), depends only on a, for the variation of Proposition 4.11 for 9 says that $\Sigma(a)$ is the remainder after dividing a by 9.

Before today's calculators, casting out 9s was used by bookkeepers to detect errors in calculations (alas, it could not detect all errors). For example, suppose the end of a calculation gave the equation

$$(22345 + 5261934)1776 = 9347119504.$$

Casting out 9s from each number gives

$$(7 + 3)3 = 8,$$

for $r(22345) = 7$, $r(5261934) = 3$, $r(1776) = 3$, and $r(9347119504) = 8$. But $(7 + 3)3 = 30 \equiv 3 \bmod 9$, not 8 mod 9, and so there was a mistake in the calculation.

The word "bookkeeper" is unusual in that it has three consecutive double letters: oo, kk, ee. This reminds us of a silly story about a word having six consecutive double letters. A zoo discovered that one of its animals, Ricky the raccoon, was quite remarkable. Ricky was a born showman: he could do somersaults, hang by his tail, and give wonderful soft-shoe dances whenever spectators sang. As his fame spread, the zoo provided him with a special cage containing a private corner where he could unwind after popular performances. Crowds came from far and wide came to see him. Indeed, Ricky became so famous that the zoo was forced to hire attendants to take care of his needs. In particular, someone was sought to maintain Ricky's corner; the job description: raccoonnookkeeper.

The usual decimal notation for the integer 5754 is an abbreviation of

$$5 \cdot 10^3 + 7 \cdot 10^2 + 5 \cdot 10 + 4.$$

But there is nothing special about the number 10.

Example 4.12. Let's write 12345 in "base 7." Repeated use of the Division Algorithm gives

$$12345 = 1763 \cdot 7 + 4$$
$$1763 = 251 \cdot 7 + 6$$
$$251 = 35 \cdot 7 + 6$$
$$35 = 5 \cdot 7 + 0$$
$$5 = 0 \cdot 7 + 5.$$

Back substituting (i.e., working from the bottom up),

$$0 \cdot 7 + 5 = \mathbf{5}$$
$$\mathbf{5} \cdot 7 + 0 = 35$$
$$(0 \cdot 7 + 5) \cdot 7 + 0 = \mathbf{35}$$
$$\mathbf{35} \cdot 7 + 6 = 251$$
$$((0 \cdot 7 + 5) \cdot 7 + 0) \cdot 7 + 6 = \mathbf{251}$$
$$\mathbf{251} \cdot 7 + 6 = 1763$$
$$(((0 \cdot 7 + 5) \cdot 7 + 0) \cdot 7 + 6) \cdot 7 + 6 = \mathbf{1763}$$
$$\mathbf{1763} \cdot 7 + 4 = 12345$$
$$((((0 \cdot 7 + 5) \cdot 7 + 0) \cdot 7 + 6) \cdot 7 + 6) \cdot 7 + 4 = 12345.$$

Expanding and collecting terms gives

$$5 \cdot 7^4 + 0 \cdot 7^3 + 6 \cdot 7^2 + 6 \cdot 7 + 4 = 12005 + 0 + 294 + 42 + 4 = 12345. \quad \blacktriangle$$

This idea works for any integer $b \geq 2$.

Proposition 4.13. *If $b \geq 2$ is an integer, then every positive integer h has an expression in **base** b: there are unique integers d_i with $0 \leq d_i < b$ such that*

$$h = d_k b^k + d_{k-1} b^{k-1} + \cdots + d_0.$$

Proof. We first prove the existence of such an expression, by induction on h. By the Division Algorithm, $h = qb + r$, where $0 \leq r < b$. Since $b \geq 2$, we have $h = qb + r \geq qb \geq 2q$. It follows that $q < h$: otherwise, $q \geq h$, giving the contradiction $h \geq 2q \geq 2h$. By the inductive hypothesis,

$$h = qb + r = (d'_k b^k + \cdots + d'_0)b + r = d'_k b^{k+1} + \cdots + d'_0 b + r.$$

We prove uniqueness by induction on h. Suppose that

$$h = d_k b^k + \cdots + d_1 b + d_0 = e_m b^m + \cdots + e_1 b + e_0,$$

where $0 \leq e_j < b$ for all j. that is, $h = (d_k b^{k-1} + \cdots + d_1)b + d_0$ and $h = (e_m b^{m-1} + \cdots + e_1)b + e_0$. By the uniqueness of quotient and remainder in the Division Algorithm, we have

$$d_k b^{k-1} + \cdots + d_1 = e_m b^{m-1} + \cdots + e_1 \quad \text{and} \quad d_0 = e_0.$$

The inductive hypothesis gives $k = m$ and $d_i = e_i$ for all $i > 0$. $\blacksquare$

Definition. If $h = d_k b^k + d_{k-1} b^{k-1} + \cdots + d_0$, where $0 \leq d_i < b$ for all i, then the numbers $d_k, \ldots, d_0$ are called the b-**adic digits** of h.

> Example 4.12 shows that the 7-adic digits of 12345 are 50664.

That every positive integer h has a unique expression in base 2 says that there is exactly one way to write h as a sum of distinct powers of 2 (for the only binary digits are 0 and 1).

Example 4.14. Let's calculate the 13-adic digits of 441. The only complication here is that we need 13 digits d (for $0 \leq d < 13$), and so we augment 0 through 9 with three new symbols

$$t = 10, \qquad e = 11, \quad \text{and} \quad w = 12.$$

Now

$$441 = 33 \cdot 13 + 12$$
$$33 = 2 \cdot 13 + 7$$
$$2 = 0 \cdot 13 + 2.$$

So, $441 = 2 \cdot 13^2 + 7 \cdot 13 + 12$, and the 13-adic expansion for 441 is

$$27w.$$

Note that the expansion for 33 is just 27. $\blacktriangle$

The most popular bases are $b = 10$ (giving everyday *decimal* digits), $b = 2$ (giving *binary* digits, useful because a computer can interpret 1 as "on" and 0 as "off"), and $b = 16$ (*hexadecimal*, also for computers). The Babylonians preferred base 60 (giving *sexagesimal* digits).

Fermat's Theorem enables us to compute $n^{p^k} \bmod p$ for every prime p and exponent p^k; it says that $n^{p^k} \equiv n \bmod p$. We now generalize this result to compute $n^h \bmod p$ for any exponent h.

Lemma 4.15. *Let p be a prime and let n be a positive integer. If $h \geq 0$, then*

$$n^h \equiv n^{\Sigma(h)} \bmod p,$$

where $\Sigma(h)$ is the sum of the p-adic digits of h.

This lemma generalizes Fermat's Theorem, for if $h = p^k$, then $\Sigma(h) = 1$; see Exercise 4.10 on page 141.

Proof. Let $h = d_k p^k + \cdots + d_1 p + d_0$ be the expression of h in base p. By Fermat's Theorem, $n^{p^i} \equiv n \bmod p$ for all i; thus, $n^{d_i p^i} = (n^{d_i})^{p^i} \equiv n^{d_i} \bmod p$. Therefore,

$$
\begin{aligned}
n^h &= n^{d_k p^k + \cdots + d_1 p + d_0} \\
&= n^{d_k p^k} n^{d_{k-1} p^{k-1}} \cdots n^{d_1 p} n^{d_0} \\
&= \left(n^{p^k}\right)^{d_k} \left(n^{p^{k-1}}\right)^{d_{k-1}} \cdots \left(n^p\right)^{d_1} n^{d_0} \\
&\equiv n^{d_k} n^{d_{k-1}} \cdots n^{d_1} n^{d_0} \bmod p \\
&\equiv n^{d_k + \cdots + d_1 + d_0} \bmod p \\
&\equiv n^{\Sigma(h)} \bmod p. \quad \blacksquare
\end{aligned}
$$

Example 4.16. What is the remainder after dividing 3^{12345} by 7? By Example 4.12, the 7-adic digits of 12345 are 50664. Therefore, $3^{12345} \equiv 3^{21} \bmod 7$ (because $5 + 0 + 6 + 6 + 4 = 21$). The 7-adic digits of 21 are 30 (because $21 = 3 \cdot 7 + 0$), and so $3^{21} \equiv 3^3 \bmod 7$ (because $2 + 1 = 3$). We conclude that $3^{12345} \equiv 3^3 = 27 \equiv 6 \bmod 7$. ▲

Exercises

4.1 Show that if integers a and b are congruent mod m to the same thing, say r, then they are congruent to each other.

4.2 We saw in Exercise 1.41 on page 29 that an integer b and its negative $-b$ can have different remainders, say r and s, after dividing by some nonzero a. Prove that $s \equiv -r \bmod a$.

4.3 Show that if $a \equiv b \bmod n$ and $m \mid n$, then $a \equiv b \bmod m$.

4.4 A *googol* is 10^{100}; that is, 1 followed by 100 zeros. Compute the remainder mod 7 of a googol.

4.5 *

  (i) If $m \geq 2$, show that every integer a is

  (ii) congruent mod m to exactly one integer on the list

$$1, \ 2, \ \ldots, \ m.$$

  (iii) Generalize Corollary 4.4 by showing that if $m \geq 2$, every integer a is congruent mod m to exactly one integer on any list of m consecutive integers.

4.6 (i) Show that every nonnegative integer is congruent mod 6 to the sum of its 7-adic digits.

  (ii) Show that every nonnegative integer is congruent mod 3 to the sum of its 7-adic digits.

  (iii) Suppose b and n are nonnegative integers. If $n \mid (b - 1)$, show that every integer is congruent mod n to the sum of its b-adic digits.

4.7 (i) Show that every nonnegative integer is congruent mod 11 to the alternating sum of its decimal digits.

  (ii) Show that every nonnegative integer is congruent mod $b + 1$ to the alternating sum of its b-adic digits.

4.8 Let a nonnegative integer n have decimal expansion $n = \sum_{i=0}^{k} d_i 10^i$. Define $t(n) = \frac{n - d_0}{10} - 4d_0$.

  (i) Show that n is divisible by 41 if and only if $t(n)$ is.

  (ii) Is $n \equiv t(n) \bmod 41$ for all nonnegative n?

4.9 Find the b-adic digits of 1000 for $b = 2, 3, 4, 5$, and 20.

You will have to invent symbols for some 20-adic digits.

4.10 (i) Find the 11-adic digits of 11^5.

  (ii) What is the b-adic expansion for b^k (k a nonnegative integer)?

4.11 Let a be a positive integer, and let a' be obtained from a by rearranging its (decimal) digits (e.g., $a = 12345$ and $a' = 52314$). Prove that $a - a'$ is a multiple of 9.

4.12 Prove that there are no positive integers a, b, c with

$$a^2 + b^2 + c^2 = 999.$$

4.13 Prove that there is no perfect square whose last two decimal digits are 35.

4.14 Using Fermat's Theorem 4.9, prove that if $a^p + b^p = c^p$, then $a + b \equiv c \bmod p$.

Linear congruences

We are now going to solve *linear congruences*; that is, we'll find all the integers x, if any, satisfying

$$ax \equiv b \bmod m.$$

Later, we will consider several linear congruences in one unknown with distinct moduli (see Theorems 4.21, 4.25, and 4.27). And we'll even consider two linear congruences in more than one unknown (see Theorem 4.44).

Theorem 4.17. *If* $\gcd(a, m) = 1$, *then, for every integer b, the congruence*

$$ax \equiv b \bmod m$$

can be solved for x; in fact, $x = sb$, where $as + mt = 1$. Moreover, any two solutions are congruent mod m.

Proof. Since $\gcd(a, m) = 1$, there are integers s and t with $as + mt = 1$; that is, $as \equiv 1 \bmod m$. Multiplying both sides by b, Proposition 4.5(ii) gives $asb \equiv b \bmod m$, so that $x = sb$ is a solution. If y is another solution, then $ax \equiv ay \bmod m$, and so $m \mid a(x - y)$. Since $\gcd(m, a) = 1$, Corollary 1.22 gives $m \mid (x - y)$; that is, $x \equiv y \bmod m$. ∎

Corollary 4.18. *If p is prime and $p \nmid a$ (i.e., p does not divide a), then the congruence $ax \equiv b$ mod p is always solvable.*

Proof. Since p is a prime, $p \nmid a$ implies $\gcd(a, p) = 1$. $\blacksquare$

Example 4.19. When $\gcd(a, m) = 1$, Theorem 4.17 says that the set of solutions of $ax \equiv b$ mod m is

$$\{sb + km : \text{where } k \in \mathbb{Z} \text{ and } sa \equiv 1 \bmod m\}.$$

Now $sa + tm = 1$ for some integer t, so that s can always be found by Euclidean Algorithm II. When m is small and you are working by hand, it is easier to find such an integer s by trying each of $ra = 2a, 3a, \ldots, (m-1)a$ in turn, at each step checking whether $ra \equiv 1$ mod m.

For example, let's find all the solutions to

$$2x \equiv 9 \bmod 13.$$

Considering each of the products $2 \cdot 2, \ 3 \cdot 2, \ 4 \cdot 2, \ldots$ mod 13 quickly leads to $7 \cdot 2 = 14 \equiv 1$ mod 13; that is, $s = 7$. By Theorem 4.17, $x = 7 \cdot 9 = 63 \equiv 11$ mod 13. Therefore,

$$x \equiv 11 \bmod 13,$$

and the solutions are $\ldots, -15, \ -2, \ 11, \ 24, \ 37, \ldots$. $\blacktriangle$

Example 4.20. *Find all the solutions to $51x \equiv 10$ mod 94.*

Since 94 is large, seeking an integer s with $51s \equiv 1$ mod 94, as in Example 4.19, is tedious. Euclidean Algorithm II gives $1 = -35 \cdot 51 + 19 \cdot 94$, and so $s = -35$. (The formulas in Exercise 1.67 on page 36 implement Euclidean Algorithm II, and they can be programmed on a calculator to produce the value of s. In fact, a CAS can solve specific congruences, but it can't (yet) solve them in general.) Therefore, the set of solutions consists of all integers x with $x \equiv -35 \cdot 10$ mod 94; that is, all numbers of the form $-350 + 94k$.

If you prefer s to be positive, just replace -35 by 59, for $59 \equiv -35$ mod 94. The solutions are now written as all integers x with $x \equiv 59 \cdot 10$ mod 94; that is, numbers of the form $590 + 94k$. $\blacktriangle$

There are problems solved in ancient Chinese manuscripts, arising from studying calendars, that involve simultaneous congruences with relatively prime moduli.

Theorem 4.21 (Chinese Remainder Theorem). *If m and m' are relatively prime, then the two congruences*

$$x \equiv b \bmod m$$
$$x \equiv b' \bmod m'$$

Theorem 4.27 will generalize Theorem 4.21 to any number of moduli.

have a common solution. Moreover, any two solutions are congruent mod mm'.

Proof. Every solution of the first congruence has the form $x = b + km$ for some integer k; hence, we must find k such that $b + km \equiv b' \bmod m'$; that is, $km \equiv b' - b \bmod m'$. Since $\gcd(m, m') = 1$, however, Theorem 4.17 applies at once to show that such an integer k does exist.

If y is another common solution, then both m and m' divide $x - y$; by Exercise 1.58 on page 35, $mm' \mid (x - y)$, and so $x \equiv y \bmod mm'$. ∎

Example 4.22. Let's find all the solutions to the simultaneous congruences

$$x \equiv 5 \bmod 8$$
$$x \equiv 11 \bmod 15.$$

Every solution to the first congruence has the form

$$x = 5 + 8k,$$

for some integer k. Substituting, $x = 5 + 8k \equiv 11 \bmod 15$, so that

$$8k \equiv 6 \bmod 15.$$

But $2 \cdot 8 = 16 \equiv 1 \bmod 15$, so that multiplying by 2 gives

$$16k \equiv k \equiv 12 \bmod 15.$$

We conclude that $x = 5 + 8 \cdot 12 = 101$ is a solution, and the Chinese Remainder Theorem (which applies because 8 and 15 are relatively prime) says that every solution has the form $101 + 120n$ for $n \in \mathbb{Z}$ (because $120 = 8 \cdot 15$). ▲

Example 4.23. We solve the simultaneous congruences

$$x \equiv -6 \bmod 13$$
$$x \equiv 8 \bmod 20.$$

Now $\gcd(13, 20) = 1$, so that we can solve this system as in the proof of the Chinese Remainder Theorem. The first congruence gives

$$x = 13k - 6,$$

for $k \in \mathbb{Z}$, and substituting into the second congruence gives

$$13k - 6 \equiv 8 \bmod 20;$$

that is,

$$13k \equiv 14 \bmod 20.$$

Since $13 \cdot 17 = 221 \equiv 1 \bmod 20$, multiplying by 17 gives $k \equiv 17 \cdot 14 \bmod 20$, that is,

> One finds 17 either by trying each number between 1 and 19 or by using the Euclidean Algorithm.

$$k \equiv 18 \bmod 20.$$

By the Chinese Remainder Theorem, all the simultaneous solutions x have the form

$$x = 13k - 6 \equiv (13 \cdot 18) - 6 \equiv 228 \bmod 260;$$

that is, the solutions are

$$\ldots, -32, 228, 488, \ldots . ▲$$

Remember that 0 denotes
Sunday, ..., 6 denotes
Saturday.

Example 4.24 (A Mayan Calendar). A congruence arises whenever there is cyclic behavior. For example, suppose we choose some particular Sunday as time zero and enumerate all the days according to the time elapsed since then. Every date now corresponds to some integer, which is negative if it occurred before time zero. Given two dates t_1 and t_2, we ask for the number $x = t_2 - t_1$ of days from one to the other. If, for example, t_1 falls on a Thursday and t_2 falls on a Tuesday, then $t_1 \equiv 4 \bmod 7$ and $t_2 \equiv 2 \bmod 7$, and so $x = t_2 - t_1 = -2 \equiv 5 \bmod 7$. Thus, $x = 7k + 5$ for some k and, incidentally, x falls on a Friday.

About 2500 years ago, the Maya of Central America and Mexico developed three calendars (each having a different use). Their religious calendar, called *tzolkin*, consisted of 20 "months," each having 13 days (so that the tzolkin "year" had 260 days). The months were

1. Imix	6. Cimi	11. Chuen	16. Cib
2. Ik	7. Manik	12. Eb	17. Caban
3. Akbal	8. Lamat	13. Ben	18. Etznab
4. Kan	9. Muluc	14. Ix	19. Cauac
5. Chicchan	10. Oc	15. Men	20. Ahau

Let us describe a tzolkin date by an ordered pair

$$[m, d],$$

where $1 \leq m \leq 20$ and $1 \leq d \leq 13$; thus, m denotes the month and d denotes the day. Instead of enumerating as we do (so that Imix 1 is followed by Imix 2, then Imix 3, and so forth), the Maya let both month and day cycle simultaneously; that is, the days proceed as follows:

Imix 1, Ik 2, Akbal 3,..., Ben 13, Ix 1, Men 2,...,
Cauac 6, Ahau 7, Imix 8, Ik 9,....

We now ask how many days have elapsed between Oc 11 and Etznab 5. More generally, let x be the number of days from tzolkin $[m, d]$ to tzolkin $[m', d']$. As we remarked at the beginning of this example, the cyclic behavior of the days gives the congruence

$$x \equiv d' - d \bmod 13,$$

while the cyclic behavior of the months gives the congruence

$$x \equiv m' - m \bmod 20.$$

To answer the original question, Oc 11 corresponds to the ordered pair $[10, 11]$ and Etznab 5 corresponds to $[18, 5]$. Since $5 - 11 = -6$ and $18 - 10 = 8$, the simultaneous congruences are

$$x \equiv -6 \bmod 13$$
$$x \equiv 8 \bmod 20.$$

In the previous example, we found the solutions:

$$x \equiv 228 \bmod 260.$$

It is not clear whether Oc 11 precedes Etznab 5 in a given year (one must look). If it does, then there are 228 days between them; otherwise, there are $260 - 228 = 32$ days between them (the truth is 228). ▲

If we do not assume that the moduli m and m' are relatively prime, then there may be no solutions to a linear system. For example, if $m = m' > 1$, then uniqueness of the remainder in the Division Algorithm shows that there is no solution to

$$x \equiv 0 \bmod m$$
$$x \equiv 1 \bmod m.$$

Theorem 4.25. *Let* $d = \gcd(m, m')$. *The system*

$$x \equiv b \bmod m$$
$$x \equiv b' \bmod m'$$

has a solution if and only if $b \equiv b' \bmod d$.

Exercise 4.19 on page 148 gives a condition guaranteeing uniqueness of solutions.

Proof. If $h \equiv b \bmod m$ and $h \equiv b' \bmod m'$, then $m \mid (h-b)$ and $m' \mid (h-b')$. Since d is a common divisor of m and m', we have $d \mid (h-b)$ and $d \mid (h-b')$. Therefore, $d \mid (b - b')$, because $(h - b') - (h - b) = b - b'$, and so $b \equiv b' \bmod d$.

Conversely, assume that $b \equiv b' \bmod d$, so that there is an integer k with $b' = b + kd$. If $m = dc$ and $m' = dc'$, then $\gcd(c, c') = 1$, by Proposition 1.23. Hence, there are integers s and t with $1 = sc + tc'$. Define $h = b'sc + btc'$. Now

$$\begin{aligned}
h &= b'sc + btc' \\
&= (b + kd)sc + btc' \\
&= b(sc + tc') + kdsc \\
&= b + ksm \\
&\equiv b \bmod m.
\end{aligned}$$

A similar argument, replacing b by $b' - kd$, shows that $h \equiv b' \bmod m'$. ■

Example 4.26. *Solve the linear system*

$$x \equiv 1 \bmod 6$$
$$x \equiv 4 \bmod 15.$$

Here, $b = 1$ and $b' = 4$, while $m = 6, m' = 15$, and $d = 3$; hence, $c = 2$ and $c' = 5$ (for $6 = 3 \cdot 2$ and $15 = 3 \cdot 5$). Now $s = 3$, and $t = -1$ (for $1 = 3 \cdot 1 + (-1) \cdot 4$). Theorem 4.25 applies, for $1 \equiv 4 \bmod 3$. Define

$$h = 4 \cdot 3 \cdot 2 + 1 \cdot (-1) \cdot 5 = 19.$$

We check that $19 \equiv 1 \bmod 6$ and $19 \equiv 4 \bmod 15$. Since $\text{lcm}(6, 15) = 30$, the solutions are $\ldots, -41, -11, 19, 49, 79, \ldots$. ▲

We are now going to generalize the Chinese Remainder Theorem for any number of linear congruences whose moduli are pairwise relatively prime. We shall see in Chapter 6 that this new version, whose solutions are given more explicitly, can be used to reveal a connection with Lagrange Interpolation, a method for finding a polynomial that agrees with a finite set of data.

Consider the following problem, adapted from Qin Jiushao, *Nine Chapters on the Mathematical Art*, 1247 CE.

> *Three farmers equally divide the rice that they have grown. One goes to a market where an 83-pound weight is used, another to a market that uses a 112-pound weight, and the third to a market using a 135-pound weight. Each farmer sells as many full measures as possible, and when the three return home, the first has 32 pounds of rice left, the second 70 pounds, and the third 30 pounds. Find the total amount of rice they took to market.*

We can model the situation in the problem with three congruences:

$$x \equiv 32 \bmod 83$$
$$x \equiv 70 \bmod 112 \tag{4.1}$$
$$x \equiv 30 \bmod 135.$$

Now, you could solve this system using the same method we used in Example 4.22: just write out each congruence in terms of its corresponding divisibility tests, and work from there.

There's another technique for solving Eqs. (4.1) that works in more general settings. The idea is to "localize" a solution x, where "localize" means considering only one modulus at a time, ignoring the other two; that is, making the other two congruent to zero. Suppose we can find integers u, v, w such that

$$u \equiv 32 \bmod 83 \qquad v \equiv 0 \bmod 83 \qquad w \equiv 0 \bmod 83$$
$$u \equiv 0 \bmod 112 \qquad v \equiv 70 \bmod 112 \qquad w \equiv 0 \bmod 112$$
$$u \equiv 0 \bmod 135 \qquad v \equiv 0 \bmod 135 \qquad w \equiv 30 \bmod 135.$$

Now take x to be $u + v + w$. Thanks to Proposition 4.5, we can find the remainder when $u + v + w$ is divided by 83 by first finding the remainders when each of u, v, and w is divided by 83, and then adding the answers:

$$x = u + v + w \equiv 32 + 0 + 0 \bmod 83.$$

Similarly, $x \equiv 70 \bmod 112$ and $x \equiv 30 \bmod 135$.

So, how do we find such u, v, and w? Let's look at what we want u to do:

$$u \equiv 32 \bmod 83$$
$$u \equiv 0 \bmod 112$$
$$u \equiv 0 \bmod 135.$$

It's easy to make u congruent to 0 mod 112 and 0 mod 135: just let it be a multiple of $112 \cdot 135 = 15120$. So, we want u to look like

$$u = k \cdot 112 \cdot 135 = 15120k$$

for some integer k. And we choose k to meet the local condition that u wants to be 32 modulo 83:

$$15120k \equiv 14k \equiv 32 \bmod 83 \tag{4.2}$$

Now comes the important step: since 112 and 135 are relatively prime to 83, so is their product (Exercise 1.56 on page 35). Hence, 14 (which is the same as $112 \cdot 135$ modulo 83) is also relatively prime to 83, and so Theorem 4.17 implies that we can solve Eq. (4.2) for k. There is an integer s with $14s \equiv 1 \bmod 83$, and multiplying both sides of $14k \equiv 32 \bmod 83$ by s gives

$$k = 32s \bmod 83.$$

There are several methods for finding s (since 83 is not so small, the Euclidean Algorithm is probably the most efficient); in fact, $s = 6$, and so k satisfies

$$6 \cdot 32 = 192 \equiv 26 \bmod 83.$$

Hence,

$$u = 26 \cdot 112 \cdot 135 = 393120.$$

To get a feel for this method, it's a good idea to go through it twice more, finding v and w. In fact, that's Exercise 4.22 on page 149.

The method just developed generalizes to a proof of the extended Chinese Remainder Theorem. Let's first introduce some notation.

Notation. Given numbers $m_1, m_2, \ldots, m_r$, define

$$M_i = m_1 m_2 \cdots \widehat{m}_i \cdots m_r = m_1 \cdots m_{i-1} m_{i+1} \cdots m_r;$$

that is, M_i is the product of all m_j other than m_i.

Theorem 4.27 (Chinese Remainder Theorem Redux). *If $m_1, m_2, \ldots, m_r$ are pairwise relatively prime integers, then the simultaneous congruences*

$$x \equiv b_1 \bmod m_1$$
$$x \equiv b_2 \bmod m_2$$
$$\vdots \quad \vdots$$
$$x \equiv b_r \bmod m_r$$

have an explicit solution, namely

$$x = b_1 (s_1 M_1) + b_2 (s_2 M_2) + \cdots + b_r (s_r M_r),$$

where

$$M_i = m_1 m_2 \cdots \widehat{m}_i \cdots m_r \quad and \quad s_i M_i \equiv 1 \bmod m_i \ for \ 1 \le i \le r.$$

Furthermore, any solution to this system is congruent to x mod $m_1 m_2 \cdots m_r$.

Proof. Use our discussion on the previous page as a model for the proof. That the specified x works is a consequence of Proposition 4.5. That all solutions are congruent mod $m_1 m_2 \ldots m_r$ is a consequence of Exercise 1.58 on page 35. ∎

Exercises

4.15 * Complete the proof of Theorem 4.27.

4.16 (i) Solve

$$x \equiv 5 \bmod 7$$
$$x \equiv 2 \bmod 11.$$

(ii) In the year 2000, the remainder after dividing my age by 3 was 2, and the remainder after dividing by 8 was 3. If I was a child when people first walked on the Moon, how old was I in 2000?

(iii) Solve

$$x^7 \equiv 5 \bmod 7$$
$$x^{11} \equiv 2 \bmod 11.$$

4.17 (i) Find a solution v to

$$v \equiv 3 \bmod 17$$
$$v \equiv 0 \bmod 11.$$

Answer. $v \equiv 88 \bmod 187$.

(ii) Find a solution w to

$$w \equiv 0 \bmod 17$$
$$w \equiv 9 \bmod 11.$$

Answer. $w \equiv 119 \bmod 187$.

(iii) Using your v and w from (i) and (ii), show that $v + w$ is a solution to the system

$$x \equiv 3 \bmod 17$$
$$x \equiv 9 \bmod 11.$$

4.18 Solve

$$x \equiv 32 \quad (\bmod\ 83)$$
$$x \equiv 70 \quad (\bmod\ 112)$$
$$x \equiv 30 \quad (\bmod\ 135).$$

4.19 * Theorem 4.25 says that if $d = \gcd(m, m')$, then the system

$$x \equiv b \bmod m$$
$$x \equiv b' \bmod m'$$

has a solution if and only if $b \equiv b' \bmod d$. Prove that any two solutions are congruent mod ℓ, where $\ell = \mathrm{lcm}(m, m')$.

4.20 How many days are there between Akbal 13 and Muluc 8 in the Mayan tzolkin calendar?

4.21 On a desert island, five men and a monkey gather coconuts all day, then sleep. The first man awakens and decides to take his share. He divides the coconuts into five equal shares, with one coconut left over. He gives the extra one to the monkey, hides his share, and goes to sleep. Later, the second man awakens and takes his fifth from the remaining pile; he, too, finds one extra and gives it to the monkey. Each of the remaining three men does likewise in turn. Find the minimum number of coconuts originally present.

Hint. Try -4 coconuts.

4.22 * Finish the calculations solving Qin Jiushao's problem on page 146 by first finding s and t, and then finding the smallest positive solution.

4.23 A band of 17 pirates stole a sack of gold coins. When the coins were divided equally, there were three left over. So, one pirate was made to walk the plank. Again the sack was divided equally; this time there were 10 gold coins left over. So, another unlucky member of the crew took a walk. Now, the gold coins could be distributed evenly with none left over. How many gold coins were in the sack?

4.24 (**Bhaskara I**, **ca. 650** C.E.). If eggs in a basket are taken out 2, 3, 4, 5, and 6 at a time, there are 1, 2, 3, 4, and 5 eggs left over, respectively. If they are taken out 7 at a time, there are no eggs left over. What is the least number of eggs that can be in the basket?

4.2 Public Key Codes

A thief who knows your name and credit card number can use this information to steal your money. So why isn't it risky to buy something online, and pay for it by sending your credit card data? After all, thieves can read the message you are sending. Here's why: the online company's software encodes your information before it is transmitted; the company can decode it, but the thieves cannot. And the reason the thieves cannot decode your message is that codes are constructed in a clever way using number theory.

It is no problem to convert a message in English into a number. Make a list of the 52 English letters (lower case and upper case) together with a space and the 11 punctuation marks

$$. \quad , \quad ; \quad : \quad ! \quad ? \quad - \quad ' \quad " \quad (\quad)$$

In all, there are 64 symbols. Assign a two-digit number to each symbol. For example,

$$a \mapsto 01, \ldots, z \mapsto 26, A \mapsto 27, \ldots, Z \mapsto 52$$
$$\text{space} \mapsto 53, . \mapsto 54, , \mapsto 55, \ldots, (\mapsto 63,) \mapsto 64$$

(we could add more symbols if we wished: say, \$, +, −, =, →, 0, 1, ..., 9). A *cipher* is a code in which distinct letters in the original message are replaced by distinct symbols. It is not difficult to decode a cipher; indeed, many newspapers print daily cryptograms to entertain their readers. In the cipher we have just described, "I love you!" is encoded

$$\text{I love you!} = 3553121522055325152158$$

Notice that any message coded in this cipher has an even number of digits, and so decoding, converting the number into English, is a simple matter. Thus,

$$(35)(53)(12)(15)(22)(05)(53)(25)(15)(21)(58) = \text{I love you!}$$

What makes a good code? If a message is a natural number x (and this is no loss in generality, as we have just seen), we need a way to encode x (in a fairly routine way so as to avoid introducing any errors into the coded message), and we need a (fairly routine) method for the recipient to decode the message. Of utmost importance is security: an unauthorized reader of a coded message should not be able to decode it. An ingenious way to find a code with these properties, now called an **RSA code**, was found in 1978 by Rivest, Shamir, and Adleman; they received the 2002 Turing Award for their discovery.

The following terms describe two basic ingredients of RSA codes.

Why these conditions on e? Read on.

Definition. A *public key* is an ordered pair (N, e), where $N = pq$ is a product of distinct primes p and q, and e is a positive integer with $\gcd(e, p - 1) = 1$ and $\gcd(e, q - 1) = 1$.

The numbers N and e are public—they are published on the web—but the primes p and q are kept secret. In practice, the primes p and q are very large.

If x is a message, encoded by assigning natural numbers to its letters as discussed above, then the encoded message sent is

$$x^e \bmod N.$$

Definition. Given a public key (N, e), a *private key* is a number d such that

$$x^{ed} \equiv x \bmod N \quad \text{for all } x \in \mathbb{Z}.$$

A private key essentially decodes the sent message, for

$$x^{ed} = (x^e)^d \equiv x \bmod N.$$

Only the intended recipients know the private key d. To find d, we'll see that you need to factor N, and that's *very* hard. Indeed, the modulus N being a product of two very big primes—each having hundreds of digits—is what makes factoring N so difficult. Since breaking the code requires knowing p and q, this is the reason RSA codes are secure. Now for the details.

Ease of Encoding and Decoding

Given a public key (N, e) and a private key d, we encode x as x^e, and we send the congruence class $x^e \bmod N$. A recipient who knows the number d can decode this, because

$$(x^e)^d = x^{ed} \equiv x \bmod N.$$

There is a minor problem here, for decoding isn't yet complete: we know the congruence class of the original message x but not x itself; that is, we know $x + kN$ for some $k \in \mathbb{Z}$ but not x. There is a routine way used to get around this; one encodes long *blocks* of text, not just letters (see [18], pp. 88–91).

Given any positive integer m, an efficient computation of $x^m \bmod N$ is based on the fact that computing $x^2 \bmod N$ is an easy task for a computer. Since computing x^{2^i} is just computing i squares, this, too, is an easy task. Now write the exponent m in base 2:

$$m = 2^i + 2^j + \cdots + 2^z.$$

Note that $x^4 = (x^2)^2$, $x^8 = (x^4)^2 = ((x^2)^2)^2$, etc.

Computing 2^m is the same as multiplying several squares:

$$x^m = x^{2^i + 2^j + \cdots + 2^z} = x^{2^i} x^{2^j} \cdots x^{2^z}.$$

In particular, after writing e in base 2, computers can easily encode a message $x \bmod N$ as $x^e \bmod N$ and, after writing ed in base 2, they can easily decode $x^{ed} \bmod N$. Since $x^{ed} \equiv x \bmod N$, this congruence essentially recaptures x.

Finding a Private Key

Let (N, e) be a public key, where $N = pq$. We want to find a private key; that is, a number d so that $x^{ed} \equiv x \bmod N$ for all $x \in \mathbb{Z}$. More generally, let's find conditions on any integer m so that $x^m \equiv x \bmod pq$. By Corollary 4.10, we have $x^m \equiv x \bmod p$ if $m \equiv 1 \bmod (p - 1)$; similarly, $x^m \equiv x \bmod q$ if $m \equiv 1 \bmod (q - 1)$. Now if m satisfies both congruence conditions, then

$$ p \mid (x^m - x) \qquad \text{and} \qquad q \mid (x^m - x). $$

As p and q are distinct primes, they are relatively prime, and so $pq \mid (x^m - x)$, by Exercise 2.20 on page 33. Hence, $x^m \equiv x \bmod pq$ for all x; that is, $x^m \equiv x \bmod N$ for all $x \in \mathbb{Z}$.

Return now to the special case $m = ed$; can we find a private key d so that $ed \equiv 1 \bmod (p-1)$ and $ed \equiv 1 \bmod (q-1)$? By hypothesis, $\gcd(e, p-1) = 1 = \gcd(e, q - 1)$; by Exercise 1.56 on page 35, $\gcd\big(e, (p - 1)(q - 1)\big) = 1$. We can now find d with Proposition 4.17, which shows how to construct an integer d such that

$$ ed \equiv 1 \bmod (p - 1)(q - 1). $$

We have constructed an RSA code.

Example 4.28. Let's create a public key and a private key using $p = 11$ and $q = 13$. (This is just for the sake of illustration; in practice, both p and q need to be extremely large primes.)

The modulus is $N = pq = 11 \cdot 13 = 143$, and so $p - 1 = 10$ and $q - 1 = 12$. Let's choose $e = 7$ (note that $\gcd(7, 10 \cdot 12) = 1$). Hence the public key is

$$ (N, e) = (143, 7). $$

If x is a message in cipher (i.e., a natural number), then the encoded message is the congruence class $x^7 \bmod 143$. To find the private key, we need a number d so that $7d \equiv 1 \bmod 120$. Using Euclidean Algorithm II or a CAS, we find a private key

$$ d = 103, $$

for $7 \cdot 103 = 721 = 6 \cdot 120 + 1$.

Let's encode and decode the word "dog": d = 4; o = 15; g = 7. Thus, the cipher for dog is 041507. In the real world, the encoding is $(41507)^7$, and the message sent out is the congruence class $(41507)^7 \bmod 143$. Decoding involves computing $(41507)^{721} \bmod 143$. As we said earlier, decoding is not finished by finding this congruence class; the numbers in this class are of the form $(41507)^{721} + 143k$, and only one of these must be determined. As we said above, the method used in actual RSA transmissions encodes blocks of letters to get around this ambiguity. For this example, however, we'll use a simpler method—we'll send each letter separately, so that "dog" is sent as as three codes

$$ 04^7, \ 15^7, \ 07^7. $$

This eliminates the ambiguity of recovering a congruence class rather than an integer, because the each letter will correspond to a (unique) integer less than 143.

The encoding is calculated like this:

$$d: \quad 4^7 \equiv 82 \bmod 143$$
$$o: 15^7 \equiv 115 \bmod 143$$
$$g: \quad 7^7 \equiv 6 \bmod 143.$$

To decode these messages, apply the private key:

$$82^{103} \equiv 4 \bmod 143$$
$$115^{103} \equiv 15 \bmod 143$$
$$6^{103} \equiv 7 \bmod 143.$$

We get $4 \leftrightarrow$ d, $15 \leftrightarrow$ o, and $7 \leftrightarrow$ g: "dog," which was the original message.   ▲

How to Think About It.   A CAS can easily tell you that $82^{103} \equiv 4 \bmod 143$, but it's interesting to see how the theorems developed in this chapter can allow you to do the computation by hand. Start with the fact that the reduction of $82^{103} \bmod 143$ is equivalent to two calculations, since 143 is $11 \cdot 13$:

$$82^{103} \bmod 11$$
$$82^{103} \bmod 13.$$

The computations of the remainders when 82^{103} is divided by a prime are made easy via Fermat's Little Theorem and the "reduce as you go" idea:

$$82^{103} \equiv 5^{103} \bmod 11 \quad \text{(because } 82 \equiv 5 \bmod 11\text{)}$$
$$82^{103} \equiv 4^{103} \bmod 13 \quad \text{(because } 82 \equiv 4 \bmod 13\text{)}.$$

Now work on the exponents:

$$5^{103} = 5^{10 \cdot 10 + 3} = \left(5^{10}\right)^{10} 5^3 \equiv 5^3 \bmod 11 \quad \text{(Little Fermat)}$$
$$= 125 \bmod 11 \equiv 4 \bmod 11$$

and

$$4^{103} = 4^{12 \cdot 8 + 7} = \left(4^{12}\right)^{10} 4^7 \equiv 4^7 \bmod 13 \quad \text{(Little Fermat)}$$
$$= 4^3 \cdot 4(4^3) = 64 \cdot 4(64) \equiv (-1) \cdot 4(-1) \bmod 13 \equiv 4 \bmod 13.$$

Constructing Secure RSA Codes

Let's construct a specific type of public key. Choose distinct primes $p \equiv 2 \bmod 3$ and $q \equiv 2 \bmod 3$. Now $p - 1 \equiv 1 \bmod 3$, so that $\gcd(3, p - 1) = 1$, and $q - 1 \equiv 1 \bmod 3$, so that $\gcd(3, q - 1) = 1$. Therefore, $(N, 3)$ is a public key, where $N = pq$. The reason that these RSA codes are so secure is that the factorization of a product $N = pq$ of two very large primes is very difficult. Thieves may know the transmitted message $x^3 \bmod N$, and they may even know N, but without knowing the factorization of $N = pq$, they don't know $p - 1$ and $q - 1$, hence, they don't know d (for $3d \equiv 1 \bmod (p - 1)(q - 1)$), and they can't decode. Indeed, if both p and q have about 200 digits (and, for technical reasons, they are not too close together), then the fastest existing computers need two or three months to factor N. A theorem of Dirichlet ([5], p. 339) says that if $\gcd(a, b) = 1$, then the arithmetic progression $a + bn$, where $n \geq 0$, contains infinitely many primes. In particular, there are infinitely many primes of the form $2 + 3n$; that is, there are infinitely many primes p with $p \equiv 2 \bmod 3$. Hence, we may choose a different pair of primes p and q every month, say, thereby stymying the crooks.

RSA codes have been refined and made even more secure over the years. Some of these refinements make use of *elliptic curves*, which we'll touch on in Chapter 9. The book *In Code* [13] is a readable account of how a high school student contributed to other refinements.

Exercises

4.25 For this exercise, use the primes $p = 5$ and $q = 17$ to create public and private keys.

(i) What will be the modulus N for the public key?

(ii) The exponent e for the public key must have no common factors with $p - 1$ and $q - 1$. List the five smallest numbers relatively prime to $(p - 1)(q - 1)$.

(iii) There are many possibilities for e; for now, use $e = 3$. To encode letters (a computer would do blocks of letters), use the rule $x \mapsto x^3 \bmod 85$.

(iv) Encode the phrase "cell phones" using this method.

(v) The private key d satisfies
$$ed \equiv 1 \bmod (p - 1)(q - 1).$$

Find d, decode your message using the private key, and verify that it is, indeed, what was sent.

> The public key reveals $e = 3$ and $N = pq$, but p and q are not revealed. (Why not?)

4.26 The following message was encoded using the public key $(85, 3)$:

$$01 \ 42 \ 59 \ 10 \ 49 \ 27 \ 56;$$

decode this message. It answers the question, "What do you call a boomerang that doesn't come back when you throw it?"

4.27 Decode the following message encoded using the public key $(91, 5)$:

$$04 \ 31 \ 38 \ 38 \ 23 \ 71 \ 14 \ 31.$$

4.28 Let m and r be nonnegative integers, and p be a prime. If $m \equiv r \bmod (p - 1)$, show that $x^m \equiv x^r$ for all integers x.

4.29 Take It Further. (*Electronic Signatures*) Consider this scenario: Elvis receives an email, encoded with his public key, from his abstract algebra instructor Mr. Jagger, which says that algebra is a waste of time and Elvis should spend all his time watching TV. Elvis suspects that the message didn't really come from Mr. J., but how can he be sure?

Suppose both Elvis and Mr. Jagger have private keys, and each knows the other's public keys. They can communicate in total privacy, with no one able to read their messages. Here's how: if Elvis wants to send a message to Mr. J., he follows these steps:

• Write the message to get x_1.

• Encode the message with his *private* key to get x_2.

• Encode x_2 with Mr. J.'s *public* key to get x_3.

• Send x_3.

When Mr. J. receives the message, he can follow a procedure to get the original message back.

(i) What is the procedure?

(ii) Explain why no one besides Mr. Jagger could read the message from Elvis.

4.30 Take It Further. Elvis is home sick with the flu. He decides to send a message to Mr. Jagger, using the method from Exercise 4.29. Suppose Elvis's public key is $(253, 7)$ and Mr. J.'s public key is $(203, 5)$. Elvis sends the message

FIDO ATE MY HOMEWORK.

What is the encoded message that Mr. J. receives? Show how Elvis encodes it and how Mr. J. decodes it.

> These public keys are not realistic. In reality, public keys use much larger primes.

4.3 Commutative Rings

We begin this section by showing, for an integer $m \geq 2$, that we can add and multiply the remainders $0, 1, \ldots, m-1$ in such a way that the new operations behave very much like ordinary addition and multiplication in $\mathbb{Z}$. Once this is done, we will be able to revisit congruences and understand what "makes them tick."

It is shown in Appendix A.2 that if $\equiv$ is an equivalence relation on a set X, then the *equivalence class* of an element $a \in X$ is

$$[a] = \{x \in X : x \equiv a\}.$$

Now Proposition 4.3 says that congruence mod m is an equivalence relation on $\mathbb{Z}$; the equivalence class of an integer a is called its *congruence class* mod m.

> The congruence class $[a]$ does depend on m, but it is standard practice not to make m part of the notation. In fact, we'll eventually write a instead of $[a]$.

Definition. The *congruence class* mod m of an integer a is

$$[a] = \{k \in \mathbb{Z} : k \equiv a \bmod m\}$$
$$= \{\ldots, a-2m, a-m, a, a+m, a+2m, \ldots\}.$$

The *integers mod m* is the set of all congruences classes:

$$\mathbb{Z}_m = \{[0], [1], \ldots, [m-1]\}.$$

Corollary 4.4 says that the list $[0], [1], \ldots, [m-1]$ is complete; that is, there are no other congruence classes mod m.

For example, $\mathbb{Z}_2$, the integers mod 2, is the set $\{[0], [1]\}$; we may think of $[0]$ as *even* (for $[0] = \{a \in \mathbb{Z} : a \equiv 0 \bmod 2\}$ is the set of all even integers) and $[1]$ as *odd* (for $[1]$ is the set of all odds).

Here is the "theological reason" for introducing congruence classes. We could continue to deal with integers and congruence; this is, after all, what Gauss did. We saw in Proposition 4.5 that $+$ and $\times$ are compatible with congruence: if $a \equiv b \bmod m$ and $a' \equiv b' \bmod m$, then $a + b \equiv a' + b' \bmod m$ and $ab \equiv a'b' \bmod m$. But wouldn't life be simpler if we could replace $\equiv$ by $=$; that is, if we could replace congruence by equality? We state the following special case of Lemma A.16 in Appendix A.2 explicitly:

$$a \equiv b \bmod m \quad \text{if and only if} \quad [a] = [b] \text{ in } \mathbb{Z}_m.$$

We often say "odd + odd = even," which does replace $\equiv$ by = at the cost of replacing integers by their congruence classes. Thus, we should define addition of these congruence classes so that $[1] + [1] = [0]$.

Addition and multiplication of evens and odds leads to the following tables.

+	even	odd
even	even	odd
odd	odd	even

×	even	odd
even	even	even
odd	even	odd

Rewrite these tables using congruence classes mod 2.

+	[0]	[1]
[0]	[0]	[1]
[1]	[1]	[0]

×	[0]	[1]
[0]	[0]	[0]
[1]	[0]	[1]

We saw above that $[1] + [1] = [0]$ says that "odd + odd = even;" note that $[1] \times [1] = [1]$ says "odd × odd = odd." The table above on the left defines addition $\alpha\colon \mathbb{Z}_2 \times \mathbb{Z}_2 \to \mathbb{Z}_2$; the table on the right defines multiplication $\mu\colon \mathbb{Z}_2 \times \mathbb{Z}_2 \to \mathbb{Z}_2$. As usual, we view congruence as generalizing parity, and we now extend the definitions to give addition and multiplication of congruence classes mod m for all $m \geq 2$.

A *binary operation* on a set R is a function $R \times R \to R$ (in particular, R is *closed* under f: if a and b are in R, then $f(a, b)$ is in R). Can you prove associativity of the binary operations α and μ when $R = \mathbb{Z}_2$?

Definition. If $m \geq 2$, **addition** and **multiplication** $\mathbb{Z}_m \times \mathbb{Z}_m \to \mathbb{Z}_m$ are defined by

$$[r] + [s] = [r + s] \quad \text{and} \quad [r][s] = [rs].$$

The definitions are simple and natural. However, we are adding and multiplying congruence classes, not remainders. After all, remainders are integers between 0 and $m - 1$, but the sum and product of remainders can exceed $m - 1$, and hence are not remainders.

Lemma 4.29. *Addition and multiplication $\mathbb{Z}_m \times \mathbb{Z}_m \to \mathbb{Z}_m$ are well-defined functions.*

Proof. To see that addition is well-defined, we must show that if $[r] = [r']$ and $[s] = [s']$, then $[r + s] = [r' + s']$. But this is precisely what was proved in Proposition 4.5. A similar argument shows that multiplication is well-defined. ∎

Binary operations $f\colon R \times R \to R$, being functions, are single-valued. This is usually called the *Law of Substitution* in this context: If $(r, s) = (r', s')$, then $f(r, s) = f(r', s')$. In particular, if $f\colon \mathbb{Z}_m \times \mathbb{Z}_m \to \mathbb{Z}_m$ is addition or multiplication, then $[r] = [r']$ and $[s] = [s']$ imply $[r] + [s] = [r'] + [s']$ and $[r][s] = [r'][s']$.

We are now going to show that these binary operations on $\mathbb{Z}_m$ enjoy eight of the nine fundamental properties of ordinary arithmetic on page 37. We have already seen several number systems in which addition and multiplication satisfy these familiar properties, so let's make these properties into a definition.

Definition. A *commutative ring* is a nonempty set R having two binary operations: *addition* $R \times R \to R$, denoted by $(r, s) \mapsto r + s$, and *multiplication* $R \times R \to R$, denoted by $(r, s) \mapsto rs$, which satisfy the following axioms for all $a, b, c \in R$:

(i) $a + b = b + a$;

(ii) there is $0 \in R$ with $a + 0 = a$ for all $a \in R$;

Negatives are often called *additive inverses*.

(iii) for each $a \in R$, there is $-a \in R$, called its **negative**, such that $-a + a = 0$;

(iv) (**Associativity of Addition**) $a + (b + c) = (a + b) + c$;

(v) (**Commutativity of Multiplication**) $ab = ba$;

(vi) there is $1 \in R$, called its **identity**, with $1 \cdot a = a$ for all $a \in R$;

(vii) (**Associativity of Multiplication**) $a(bc) = (ab)c$;

(viii) (**Distributivity**) $a(b + c) = ab + ac$.

How to Think About It. There are more general (non-commutative) rings in which (v), commutativity of multiplication, is not assumed, while (vi) is modified to say that $1 \cdot a = a = a \cdot 1$ and (viii) is modified to say $a(b + c) = ab + ac$ and $(b + c)a = ba + bc$. A good example is the ring of all 2×2 matrices with entries in $\mathbb{R}$, with identity element $\left[\begin{smallmatrix} 1 & 0 \\ 0 & 1 \end{smallmatrix}\right]$, and binary operations ordinary matrix addition and multiplication:

$$\begin{bmatrix} a & b \\ c & d \end{bmatrix} + \begin{bmatrix} a' & b' \\ c' & d' \end{bmatrix} = \begin{bmatrix} a + a' & b + b' \\ c + c' & d + d' \end{bmatrix}$$

and

$$\begin{bmatrix} a & b \\ c & d \end{bmatrix} \begin{bmatrix} a' & b' \\ c' & d' \end{bmatrix} = \begin{bmatrix} aa' + bc' & ab' + bd' \\ ca' + dc' & cb' + dd' \end{bmatrix}.$$

Since all rings in this book are commutative, we will often abuse language and abbreviate "commutative ring" to "ring."

The ninth fundamental property of real numbers is: If $a \neq 0$, there is a real number a^{-1}, called its (multiplicative) **inverse**, such that $a \cdot a^{-1} = 1$. We will soon consider commutative rings, called *fields*, which enjoy this property as well.

How to Think About It. The notion of *commutative ring* wasn't conceived in a vacuum. Mathematicians noticed that several useful systems shared the basic algebraic properties listed in the definition. Definitions usually emerge in this way, distilling common features of different interesting examples.

Precise definitions are valuable; we couldn't prove anything without them. For example, political discourse is often vapid because terms are not defined: what is a liberal; what is a conservative? A mathematician who asserts that there are infinitely many primes can be believed. But can you believe a politician who says his opponent is a fool because he's a liberal (or she's a conservative)?

Example 4.30.   (i) $\mathbb{Z}$, $\mathbb{Q}$, and $\mathbb{R}$ are commutative rings. The ninth fundamental property, reciprocals, does not hold in $\mathbb{Z}$; for example, $2^{-1} = \frac{1}{2}$ does not lie in $\mathbb{Z}$.

 (ii) Propositions 3.8 and 3.9 show that $\mathbb{C}$ is a commutative ring, while Proposition 3.11 shows that every nonzero complex number has an inverse.

(iii) The set of even integers does not form a commutative ring, for it has no identity.

(iv) The *Gaussian integers* $\mathbb{Z}[i]$ form a commutative ring (see Exercise 4.64 on page 168).

 (v) The *Eisenstein integers* $\mathbb{Z}[\omega]$, where ω is a primitive cube root of unity, form a commutative ring (see Exercise 4.64).

(vi) More generally, the *cyclotomic integers* $\mathbb{Z}[\zeta]$, where ζ is any primitive root of unity, form a commutative ring (see Exercise 4.65 on page 168).

(vii) The next theorem shows that $\mathbb{Z}_m$ is a commutative ring for every integer $m \geq 2$.

(viii) We'll see, in the next chapter, that all polynomials whose coefficients lie in a commutative ring (e.g., all polynomials with coefficients in $\mathbb{Z}$) is itself a commutative ring with the usual addition and multiplication.   ▲

Example 4.31.   (i) If R is a commutative ring and S is a set, let R^S be the set of all functions $S \to R$. Define $u : S \to R$ to be the constant function with value 1, where 1 is the identity element of R: that is, $u(s) = 1$ for all $s \in S$. Define the sum and product of $f, g \in R^S$, for all $s \in S$, by

$$f + g : s \mapsto f(s) + g(s)$$

and

$$fg : s \mapsto f(s)g(s);$$

these operations are called *pointwise addition* and *pointwise multiplication*. We leave the straightforward checking that R^S is a commutative ring as Exercise 4.34. An important special case of this example is $\text{Fun}(R) = R^R$, the ring of all functions from a commutative ring R to itself.

If $R = \mathbb{R}$, then $\text{Fun}(\mathbb{R}) = \mathbb{R}^{\mathbb{R}}$ arises in calculus. After all, what are the functions $x + \cos x$ and $x \cos x$?

(ii) If $X = [a, b]$ is an interval on the line, then

$$C(X) = \{f : X \to \mathbb{R} : f \text{ is continuous}\}$$

is a commutative ring under pointwise operations. If both $f, g \in C(X)$ are continuous, then it is shown in calculus that both $f + g$ and fg are also continuous. The constant function e with $e(t) = 1$ for all $t \in X$ is continuous; we let the reader prove that the other axioms in the definition of commutative ring hold.   ▲

Etymology. The word *ring* was probably coined by Hilbert in 1897 when he wrote *Zahlring*. One of the meanings of the word ring, in German as in English, is "collection," as in the phrase "a ring of thieves." It has also been suggested that Hilbert used this term because, for a commutative ring such as the Gaussian integers $\mathbb{Z}[i]$, powers of some elements "cycle back" to being a linear combination of smaller powers (for example, $i, i^2, i^3, i^4 = 1, i^5 = i$).

Theorem 4.32. $\mathbb{Z}_m$ *is a commutative ring for every integer* $m \geq 2$.

Proof. The proof of each of the eight statements is routine; in essence, they are inherited from the analogous statement in $\mathbb{Z}$ (the inheritance is made possible by Proposition 4.5). We prove only statements (i), (vii), and (viii) in the definition of commutative ring; the other proofs are left to Exercise 4.31 below.

Convince yourself that each step in these proofs is legitimate by supplying a reason.

(i) $[a] + [b] = [a + b] = [b + a] = [b] + [a]$.

(vii) (**Associativity of Multiplication**):

$$[a]([b][c]) = [a][bc] = [a(bc)] = [(ab)c] = [ab][c] = ([a][b])[c].$$

(viii) (**Distributivity**):

$$\begin{aligned} [a]([b] + [c]) = [a][b + c] &= [a(b + c)] \\ &= [ab + ac] = [ab] + [ac] \\ &= [a][b] + [a][c]. \quad \blacksquare \end{aligned}$$

A commutative ring is an algebraic system we view as a generalization of ordinary arithmetic. One remarkable feature of the integers mod m is that an integer a is divisible by m if and only if $[a] = [0]$ in $\mathbb{Z}_m$ (for $m \mid a$ if and only if $a \equiv 0 \bmod m$); that is, we have converted a statement about divisibility into an equation.

Exercises

4.31 * Prove the remaining parts of Theorem 4.32

4.32 Prove that every commutative ring R has a unique identity 1.

4.33 (i) Prove that subtraction in $\mathbb{Z}$ is not an associative operation.

(ii) Give an example of a commutative ring in which subtraction is associative.

4.34 * If R is a commutative ring and S is a set, verify that R^S is a commutative ring under pointwise operations. (See Example 4.31.)

4.35 * Define the **weird integers** W as the integers with the usual addition, but with multiplication $*$ defined by

$$a * b = \begin{cases} ab & \text{if } a \text{ or } b \text{ is odd} \\ -ab & \text{if both } a \text{ and } b \text{ are even.} \end{cases}$$

Prove that W is a commutative ring.
Hint. It is clear that 1 is the identity and that $*$ is commutative; only associativity of $*$ and distributivity must be checked.

4.36 For each integer a between 1 and 11, find all solutions to $[a]x = [9]$ in $\mathbb{Z}_{12}$. (There may be no solutions for some a.)

4.37 In $\mathbb{Z}_8$, find all values of x so that $(x - 1)(x + 1) = 0$.

4.38 Solve the equation $x^2 + 3x - 3 = 0$ in $\mathbb{Z}_5$.

4.39 How many roots does the polynomial $x^2 + 1 = 0$ have in each of the following commutative rings?

    (i) $\mathbb{Z}_5$ (ii) $\mathbb{Z}_7$ (iii) $\mathbb{Z}_{11}$

    (iv) $\mathbb{Z}_{101}$ (v) $\mathbb{Z}_{13}$

Properties of Commutative Rings

One advantage of precise definitions is that they are economical: proving a theorem for general commutative rings automatically proves it for each particular commutative ring. For example, we need not prove that $(-1)(-1) = 1$ holds in the Gaussian integers $\mathbb{Z}[i]$ because we prove below that it holds in all commutative rings. The nice thing here is that some general proofs can be copied verbatim from those in Chapter 1. Alas, this is not always so. For example, the generalization of the Chinese Remainder Theorem does hold in $\mathbb{Z}[i]$, but its proof requires more than merely copying, mutatis mutandis, its proof in $\mathbb{Z}$.

Proposition 4.33. *For every a in a commutative ring R, we have $a \times 0 = 0$.*

Proof. Identical to the proof of Proposition 1.31.   ∎

Can $1 = 0$ in a commutative ring R? The answer is "yes," but not really. If $1 = 0$ in R, then $a = 1a = 0a = 0$ for all $a \in R$, by Proposition 4.33; that is, R consists of only one element, namely, 0. So, $1 \neq 0$ in any commutative ring having more than one element. Commutative rings with only one element are called **zero rings**; they are not very interesting, although they do arise every once in a while. For example, Theorem 4.32 says that $\mathbb{Z}_m$ is a commutative ring for every integer $m \geq 2$. Actually, $\mathbb{Z}_m$ is a commutative ring for $m \geq 0$: we have $\mathbb{Z}_0 = \mathbb{Z}$, and $\mathbb{Z}_1$ the zero ring. Since zero rings arise rarely, we declare that $1 \neq 0$ for all commutative rings in this book unless we say otherwise.

Proposition 4.34. *For any a in a commutative ring R, we have*

$$(-a)(-1) = a.$$

In particular,

$$(-1)(-1) = 1.$$

Proof. Identical to the proof of Proposition 1.32.   ∎

Can an element a in a commutative ring R have more than one negative?

Proposition 4.35. *Let R be a commutative ring. Negatives in R are unique; that is, for each $a \in R$, there is exactly one $a' \in R$ with $a + a' = 0$.*

Multiplicative inverses, when they exist, are unique; that is, for each $b \in R$, there is at most one $b' \in R$ with $bb' = 1$.

Proof. Identical to the proof of Proposition 1.33. As usual, the negative of a is denoted by $-a$, and the inverse of b, when it exists, is denoted by b^{-1}.   ∎

Corollary 4.36. *For every a in a commutative ring R, we have $-a = (-1)a$. Moreover, if an element b has an inverse, then $(b^{-1})^{-1} = b$.*

Proof. Identical to the proof of Corollary 1.34.   ∎

The distributive law for subtraction holds, where $b - c$ is defined as $b + (-1)c$.

Corollary 4.37. *If a, b, c lie in a commutative ring R, then $a(b-c) = ab - ac$.*

Proof. Identical to the proof of Corollary 1.35.   ∎

Definition. Let R be a commutative ring. If $a \in R$, define its **powers** by induction on $n \geq 0$. Set $a^0 = 1$ and, if $n \geq 0$, then $a^{n+1} = aa^n$.

We have defined $a^0 = 1$ for *all* $a \in R$; in particular, $0^0 = 1$.

The notation a^n is a hybrid: a is an element of R while n is an integer. Here is the additive version of this notation.

Definition. If R is a commutative ring and $k > 0$ is an integer, define $ka = a + \cdots + a$, the sum of a with itself k times. If $k = 0$, define $ka = 0a = 0$, where the 0 on the right is the zero element of R. If $k < 0$, then $-k = |k| > 0$, and we define $ka = (-k)(-a)$; that is, ka is the sum of $-a$ with itself $|k|$ times.

The hybrid ka can be viewed as the product of two elements in the commutative ring R. If $e = 1$ (the identity element in R), then $ke \in R$ and $ka = (ke)a$. For example, if $k > 0$, then

$$ka = a + a + \cdots + a = (e + e + \cdots + e)a = (ke)a.$$

We note that we could have defined ka, for $k \geq 0$, by induction. Set $0a = 0$ and, if $k \geq 0$, then $(k + 1)a = a + ka$.

The Binomial Theorem holds in every commutative ring R. Since we have defined ka whenever k is an integer and $a \in R$, the notation $\binom{n}{j}a$ makes sense.

Theorem 4.38 (Binomial Theorem). *Let R be a commutative ring.*

(i) *For all $x \in R$ and all integers $n \geq 0$,*

$$(1 + x)^n = \sum_{j=0}^{n} \binom{n}{j} x^j = \sum_{j=0}^{n} \frac{n!}{j!(n-j)!} x^j.$$

(ii) *For all $a, b \in R$ and all integers $n \geq 0$,*

$$(a + b)^n = \sum_{j=0}^{n} \binom{n}{j} a^{n-j} b^j = \sum_{j=0}^{n} \left(\frac{n!}{j!(n-j)!} \right) a^{n-j} b^j.$$

Proof. Identical to the proof of Theorem 2.25. ∎

Units and Fields

Let's return to the ninth fundamental property of ordinary arithmetic. A nonzero element in a commutative ring may not have an inverse. For example, $[2] \neq [0]$ in $\mathbb{Z}_4$, but there is no $[a] \in \mathbb{Z}_4$ with $[2][a] = [1]$: the products $[2][a]$ are

$$[2][0] = [0], \quad [2][1] = [2], \quad [2][2] = [4] = [0], \quad [2][3] = [6] = [2];$$

none of these is $[1]$.

If $m \geq 2$, which nonzero elements in $\mathbb{Z}_m$ have multiplicative inverses?

Proposition 4.39. *Let $m \geq 2$. An element $[a] \in \mathbb{Z}_m$ has an inverse if and only if $\gcd(a, m) = 1$.*

Proof. Since $\gcd(a, m) = 1$, Theorem 4.17 says that there is an integer s so that $sa \equiv 1 \bmod m$. Translating this congruence to $\mathbb{Z}_m$ (using the definition of multiplication in $\mathbb{Z}_m$), we have

$$[s][a] = [sa] = [1];$$

thus, $[s]$ is the inverse of $[a]$.

Conversely, if $[s][a] = [1]$ in $\mathbb{Z}_m$, then $[sa] = [1]$ and $sa \equiv 1 \bmod m$. Therefore, $m \mid (sa - 1)$, so that $sa - 1 = tm$ for some integer t, and $\gcd(a, m) = 1$, by Proposition 1.23. ■

If a and m are relatively prime, then the coefficients s and t displaying 1 as a linear combination are not unique (see Exercise 1.57 on page 35). However, Proposition 4.35 shows that the congruence class of $s \bmod m$ is unique: if also $1 = s'a + t'm$, then $[s'] = [s]$, for both equal $[a]^{-1}$ in $\mathbb{Z}_m$; that is, $s' \equiv s \bmod m$, for inverses are unique when they exist.

Dividing by an element $a \in R$ means multiplying by a^{-1}. Thus, dividing by zero requires an element $0^{-1} \in R$ with $0^{-1} \times 0 = 1$. But we saw, in Proposition 4.33, that $a \times 0 = 0$ for all $a \in R$; in particular, $0^{-1} \times 0 = 0$. It follows that if $1 \neq 0$ in R; that is, if R has more than one element, then 0^{-1} does not exist; therefore, we cannot divide by 0.

How to Think About It. There is a strong analogy between the method for solving linear equations in elementary algebra and the proof of Theorem 4.17. When solving an equation like $3x = 4$ in first-year algebra, you multiply both sides by the number u with $u3 = 1$, namely, $u = \frac{1}{3}$:

$$3x = 4$$
$$\tfrac{1}{3}(3x) = \tfrac{1}{3}\,4$$
$$\left(\tfrac{1}{3}\,3\right)x = \tfrac{4}{3}$$
$$x = \tfrac{4}{3}.$$

Now look at a congruence like $3x \equiv 4 \bmod 7$ as an equation in $\mathbb{Z}_7$,

$$[3]x = [4],$$

and go through the same steps as above, using the fact that $[5] \cdot [3] = [1]$:

$$[3]x = [4]$$
$$[5]([3]x) = [5] \cdot [4]$$
$$([5] \cdot [3])x = [6]$$
$$x = [6].$$

As we remarked on page 158, the notion of commutative ring allows us to turn congruences into equations that obey the usual rules of elementary algebra.

Notation. The various $\mathbb{Z}_m$ are important examples of commutative rings. It is getting cumbersome, as in the above calculation, to decorate elements of $\mathbb{Z}_m$ with brackets. From now on, we will usually drop the brackets, letting the context make things clear. For example, the calculation in $\mathbb{Z}_7$ above will usually be written

$$3x = 4$$
$$5(3x) = 5 \cdot 4$$
$$(5 \cdot 3)\, x = 6$$
$$x = 6.$$

Definition. An element u in a commutative ring R is a ***unit*** if it has a multiplicative inverse in R; that is, there is $v \in R$ with $uv = 1$.

Note that v must be in R in order that u be a unit in R. For example, 2 is not a unit in $\mathbb{Z}$ because $\frac{1}{2}$ is not in $\mathbb{Z}$; of course, 2 is a unit in $\mathbb{Q}$.

Knowledge of the units in a commutative ring R tells us a great deal about how much elementary algebra carries over to R. For example, knowing whether or not a is a unit in R tells us whether or not we can solve the equation $ax = b$ in R by dividing both sides by a.

Example 4.40.   (i) The only units in $\mathbb{Z}$ are ± 1.

 (ii) Proposition 4.39 describes all the units in $\mathbb{Z}_m$. It says that $[a]$ is a unit in $\mathbb{Z}_m$ if and only if $\gcd(a, m) = 1$.

(iii) Every nonzero element of $\mathbb{Q}$, $\mathbb{R}$, and $\mathbb{C}$ is a unit.   ▲

What are the units in $\mathbb{Z}[i]$? Our work in Chapter 3 lets us find the answer. Every nonzero Gaussian integer z has an inverse in $\mathbb{C}$, but that inverse may not be in $\mathbb{Z}[i]$. Proposition 3.11 shows, in $\mathbb{C}$, that

$$z^{-1} = \frac{\overline{z}}{z\,\overline{z}}.$$

The denominator on the right-hand side is none other than $N(z)$, the norm of z, and this suggests the following proposition.

Proposition 4.41. *A Gaussian integer z is a unit in $\mathbb{Z}[i]$ if and only if $N(z) = 1$.*

Proof. If $N(z) = 1$, the formula $z^{-1} = \overline{z}/(z\,\overline{z}) = \overline{z}/N(z)$ shows that $z^{-1} = \overline{z}$, a Gaussian integer, and so z is a unit in $\mathbb{Z}[i]$.

Conversely, if z is a unit in $\mathbb{Z}[i]$, then there is a Gaussian integer w with $zw = 1$. Take the norm of both sides; Proposition 3.35(iii) gives

$$N(z)N(w) = 1.$$

This is an equation in $\mathbb{Z}$ saying that a product of two integers is 1. The only way this can happen is for each factor to be ± 1. But norms are always nonnegative, by Proposition 3.35, and so $N(z)$ (and also $N(w)$) is equal to 1.   ∎

Proposition 4.41 leads to the question "Which Gaussian integers have norm 1?" If $z = a + bi$ is a Gaussian integer and $N(z) = 1$, then a and b are integers with $a^2 + b^2 = 1$. Using the fact that (a, b) is a lattice point on the unit circle, its distance to the origin is 1, and we see that the only (a, b) satisfying the equation are

$$(1, 0),\ (0, 1),\ (-1, 0),\ (0, -1).$$

Hence, we have

Proposition 4.42. *There are exactly four units in $\mathbb{Z}[i]$, namely*

$$1,\ i\ ,-1,\ -i.$$

We know that 0 is never a unit in a nonzero commutative ring R; what if every nonzero element in R is a unit?

Definition. A *field* is a nonzero commutative ring F in which every nonzero $a \in F$ is a unit; that is, there is $b \in F$ with $ab = 1$.

Familiar examples of fields are $\mathbb{Q}$, $\mathbb{R}$, and $\mathbb{C}$; here is a new example.

Theorem 4.43. *If $m \geq 2$, then $\mathbb{Z}_m$ is a field if and only if m is a prime.*

Proof. If m is prime and $0 < a < m$, then $\gcd(a, m) = 1$, and Proposition 4.39 says that a is a unit in $\mathbb{Z}_m$. Hence, $\mathbb{Z}_m$ is a field.

Conversely, suppose that m is not prime; that is, $m = ab$, where $0 < a$, $b < m$. In $\mathbb{Z}_m$, both a and b are nonzero, and $ab = 0$. If a has an inverse in $\mathbb{Z}_m$, say, s, then $sa = 1$, which gives the contradiction:

$$0 = s0 = s(ab) = (sa)b = 1b = b. \quad \blacksquare$$

We have removed the brackets from the notation for elements of $\mathbb{Z}_m$.

Who would have thought that a field could have a finite number of elements? When one of us was a graduate student, a fellow student was tutoring a 10-year old prodigy. To illustrate the boy's talent, he described teaching him how to multiply 2×2 matrices. As soon as he was shown that the 2×2 identity matrix I satisfies $IA = A$ for all matrices A, the boy immediately began writing; after a few minutes he smiled, for he had just discovered that $A = \begin{bmatrix} a & b \\ c & d \end{bmatrix}$ has an inverse if and only if $ad - bc \neq 0$! Later, when this boy was told the definition of a field, he smiled as the usual examples of $\mathbb{Q}$, $\mathbb{R}$, and $\mathbb{C}$ were trotted out. But when he was shown $\mathbb{Z}_2$, he threw a temper tantrum and ended the lesson.

In Theorem 4.17, we considered linear congruences in one variable. We now consider linear systems in two variables.

Theorem 4.44. *If p is a prime, then the system*

$$ax + by \equiv u \bmod p$$
$$cx + dy \equiv v \bmod p$$

has a solution (x, y) if and only if the determinant $ad - bc \not\equiv 0 \bmod p$.

Proof. Since p is a prime, we know that $\mathbb{Z}_p$ is a field. Now the system of congruences can be considered as a system of equations in $\mathbb{Z}_p$.

$$ax + by = u$$
$$cx + dy = v.$$

You can now complete the proof just as in linear algebra. $\blacksquare$

Example 4.45. *Find the solution in $\mathbb{Z}_7$ of the system*

$$4x - 5y = -2$$
$$2x + 3y = 5.$$

We proceed as in linear algebra. The determinant is $4 \cdot 3 - (-5) \cdot 2 = 22 \neq 0$ in $\mathbb{Z}_7$, and so there is a solution. Now $4^{-1} = 2$ in $\mathbb{Z}_7$ (for $4 \cdot 2 = 8 \equiv 1 \bmod 7$), so the top congruence can be rewritten as $x - 10y \equiv -4$. Since $-10 = -3$, we have

$$x = 3y - 4.$$

Substituting into the bottom equation gives $2(3y - 4) + 3y = 5$; that is, $9y = 13$; rewrite this as $2y = 6$. Multiply by $4 = 2^{-1}$ to obtain $y = 24 = 3$. Finally, $x = 3y - 4 = 9 - 4 = 5$. Therefore, the solution is $(5, 3)$. Let's check this. If $x = 5$ and $y = 3$, then

$$4 \cdot 5 - 5 \cdot 3 = 4 \equiv -2 \bmod 7$$
$$2 \cdot 5 + 3 \cdot 3 = 19 \equiv 5 \bmod 7. \quad \blacktriangle$$

How to Think About It. Had you mimicked the method in the example when proving Theorem 4.44, you would have found *Cramer's Rule*, a generic formula for the solution to the system

$$ax + by = u$$
$$cx + dy = v.$$

The solution is (x, y), where

$$x = \frac{\det \begin{bmatrix} u & b \\ v & d \end{bmatrix}}{\det \begin{bmatrix} a & b \\ c & d \end{bmatrix}} \quad \text{and} \quad y = \frac{\det \begin{bmatrix} a & u \\ c & v \end{bmatrix}}{\det \begin{bmatrix} a & b \\ c & d \end{bmatrix}}.$$

Thus, Cramer's Rule holds, giving us an easily remembered formula for solving 2×2 systems of equations in any field. Most linear algebra courses present a more general Cramer's Rule for $n \times n$ systems.

Exercises

4.40 Give an example of a commutative ring R containing an element a with $a \neq 0$, $a \neq 1$, and $a^2 = a$.

4.41 * The notation in this exercise is that of Example 4.31.

  (i) Find all the units in $\text{Fun}(\mathbb{R}) = \mathbb{R}^{\mathbb{R}}$.

  (ii) Prove that a continuous function $u \colon X \to \mathbb{R}$ is a unit in $C(X)$ if and only if $u(t) \neq 0$ for all $t \in X$.

4.42 Let $R = \mathbb{Z}[\sqrt{3}] = \{a + b\sqrt{3} : a, b \in \mathbb{Z}\}$.

  (i) Show, with the usual addition and multiplication of real numbers, that R is a commutative ring.

  (ii) Show that $u = 2 + \sqrt{3}$ is a unit in R.

  (iii) Show that R has infinitely many units.

4.43 If p is a prime, show that a quadratic polynomial with coefficients in $\mathbb{Z}_p$ has at most two roots in $\mathbb{Z}_p$.

4.44 * Prove or give a counterexample. Let R be a commutative ring.

(i) The product of two units in R is a unit.

(ii) The sum of two units in R is a unit.

4.45 * Describe all the units in the Eisenstein integers $\mathbb{Z}[\omega]$.

4.46 * Just as in $\mathbb{C}$, a *root of unity* in a ring R is an element $a \in R$ with $a^n = 1$ for some positive integer n.

Find all roots of unity in $\mathbb{Z}_m$ for all integers m between 5 and 12.

4.47 * Show that $\mathbb{Z}_m$ contains exactly $\phi(m)$ units, where ϕ is the Euler ϕ-function.

4.48 * Show that an element $u \in \mathbb{Z}_m$ is a unit if and only if u is a root of unity.

4.49 * If u is a unit in $\mathbb{Z}_m$, then Exercise 4.48 says there is some positive integer n with $u^n = 1$; the smallest such n is called the *order* of u in $\mathbb{Z}_m$.

For each integer m between 5 and 12, make a table that shows the units and their orders. Any conjectures about which integers can be orders of units?

4.50 State and prove Cramer's Rule for a 3×3 system of linear equations in a field.

4.51 Solve the system of congruences

$$3x - 2y + z \equiv 1 \bmod 7$$
$$x + y - 2z \equiv 0 \bmod 7$$
$$-x + 2y + z \equiv 4 \bmod 7.$$

4.52 For what values of m will the system

$$2x + 5y = 7$$
$$x + 4y = 9$$

have a unique solution in $\mathbb{Z}_m$?

4.53 Find a system of two linear equations in two unknowns that has a unique solution in $\mathbb{Z}_m$ for all $m \geq 2$.

4.54 (i) Show that

$$M_2 = \left\{ \begin{bmatrix} a & b \\ -b & a \end{bmatrix} : a, b \in \mathbb{Z} \right\}$$

is a commutative ring under matrix addition and multiplication.

(ii) What are the units in M_2?

4.55 *

(i) Show that

$$\mathbb{F}_4 = \left\{ \begin{bmatrix} a & b \\ b & a + b \end{bmatrix} : a, b \in \mathbb{Z}_2 \right\},$$

with binary operations matrix addition and multiplication, is a field having exactly four elements.

(ii) Write out addition and multiplication tables for $\mathbb{F}_4$.

Subrings and Subfields

Sometimes, as with $\mathbb{Z}$ and $\mathbb{Q}$, one ring sits inside another ring.

More precisely, if α is the addition on R, then its restriction $\alpha|(S \times S)$ has image in S, and it is the addition on S. Similarly for multiplication.

Definition. A *subring* of a commutative ring R is a commutative ring S contained in R that has the same 1, the same addition, and the same multiplication as R; that is, $1 \in S$ and if $a, b \in S$, then $a + b \in S$ and $ab \in S$.

Query: Is $\mathbb{Z}_m$ a subring of $\mathbb{Z}$?

Each commutative ring on the list $\mathbb{Z} \subseteq \mathbb{Q} \subseteq \mathbb{R} \subseteq \mathbb{C}$ is a subring of the next one. Example A.20 in Appendix A.3 says that if R is a commutative ring and $k \subseteq R$ is a subring that is a field, then R is a vector space over k. Thus, $\mathbb{C}$ is a vector space over $\mathbb{R}$ (and also over $\mathbb{Q}$), and $\mathbb{R}$ is a vector space over $\mathbb{Q}$.

Proposition 4.46. *A subset S of a commutative ring R is a subring of R if and only if*

(i) $1 \in S$;

(ii) *if $a, b \in S$, then $a + b \in S$;*

(iii) *if $a, b \in S$, then $ab \in S$.*

Proof. If S is a subring of R, then the three properties clearly hold.

Conversely, if S satisfies the three properties, then S contains 1, and so it only remains to show that S is a commutative ring. Items (ii) and (iii) (closure under addition and multiplication) show that the (restrictions of) addition and multiplication are binary operations on S. All the other items in the definition of commutative ring are inherited from R. For example, the distributive law holds: since $a(b+c) = ab+ac$ holds for all $a, b, c \in R$, it holds, in particular, for all $a, b, c \in S \subseteq R$. ∎

Proposition 4.46 is more powerful than it looks. A subset S of a commutative ring R is a subring if, using the same operations as those in R, it satisfies all the conditions in the definition of commutative ring. But there's no need to check all the properties; you need check only three of them. For example, Exercise 4.64 on page 168 asks you to prove that the Gaussian integers $\mathbb{Z}[i]$ and the Eisenstein integers $\mathbb{Z}[\omega]$ are commutative rings. This could be tedious: there are ten things in the definition of commutative ring to check: addition and multiplication are binary operations and the eight axioms. However, if we know that $\mathbb{C}$ is a commutative ring and $\mathbb{Z}[i]$ and $\mathbb{Z}[\omega]$ are subrings of $\mathbb{C}$ (facts that can be established via Proposition 4.46), then both $\mathbb{Z}[i]$ and $\mathbb{Z}[\omega]$ are commutative rings in their own right.

$A + B$ is *exclusive or*; that is, all $x \in X$ lying in either A or B but not in both. In terms of Venn diagrams, this pictures the statement: Take it or leave it!

Example 4.47. Here is an example of a commutative ring arising from set theory. If A and B are subsets of a set X, then their *symmetric difference* is

$$A + B = (A \cup B) - (A \cap B)$$

(see Figure 4.2). If U and V are subsets of a set X, then

$$U - V = \{x \in X : x \in U \text{ and } x \notin V\}.$$

Recall that B^A is the family of all functions from a set A to a set B. Why is this ring denoted by 2^X? We'll see why in Example 5.16.

Let X be a set, let 2^X denote the set of all the subsets of X, define addition on 2^X to be symmetric difference, and define multiplication on 2^X to be intersection. Exercises 4.68 through 4.74 on page 169 essentially show that 2^X is a

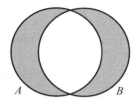

Figure 4.2. Symmetric difference.

commutative ring. The empty set $\varnothing$ is the zero element, for $A + \varnothing = A$, while each subset A is its own negative, for $A + A = \varnothing$. These exercises also show that symmetric difference is associative and that the distributive law holds. Finally, X itself is the identity element, for $X \cap A = A$ for every subset A. We call 2^X a **Boolean ring**.

Suppose now that $Y \subsetneq X$ is a proper subset of X; is 2^Y a subring of 2^X? If A and B are subsets of Y, then $A + B$ and $A \cap B$ are also subsets of Y; that is, 2^Y is closed under the addition and multiplication on 2^X. However, the identity element in 2^Y is Y, not X, and so 2^Y is *not* a subring of 2^X. ▲

The example of 2^X may have surprised you. It was natural for us to introduce the notion of commutative ring, for we had already seen many examples of numbers or of functions in which addition and multiplication make sense and obey the usual rules. But the elements of 2^X are neither numbers nor functions. And even though we call their binary operations addition and multiplication, they are operations from set theory. This is a happy circumstance, which we will exploit in the next chapter. It's not really important what we call addition and multiplication; what is important is that the operations satisfy eight fundamental properties; that is, the axioms in the definition of commutative ring.

Just as the notion of a subring of a commutative ring is useful, so too is the notion of a *subfield* of a field.

Definition. If F is a field, then a **subfield** of F is a subring $k \subseteq F$ that is also a field.

For example, $\mathbb{Q}$ is a subfield of $\mathbb{R}$, and both $\mathbb{Q}$ and $\mathbb{R}$ are subfields of $\mathbb{C}$. There is a shortcut for showing that a subset is a subfield.

Proposition 4.48. *A subring k of a field F is a subfield of F if and only if $a^{-1} \in k$ for all nonzero $a \in k$.*

Proof. This is Exercise 4.57 below. ∎

Exercises

4.56 Give an example of a subring of a field that is not a field.

4.57 * Prove Proposition 4.48.

4.58 (i) Show that $\{0, 2\} \subseteq \mathbb{Z}_4$ has the same addition and multiplication tables as $\mathbb{Z}_2$.

    (ii) Is $\mathbb{Z}_2$ a subring of $\mathbb{Z}_4$?

    (iii) Is $\{0, 2, 4, 6\}$ a subring of $\mathbb{Z}_8$?

4.59 Let $R = \mathbb{Z}[\sqrt{-3}] = \{a + b\sqrt{-3} : a, b \in \mathbb{Z}\}$.

(i) Show that R is a subring of the Eisenstein integers.

(ii) What are the units in R?

4.60 (i) If S and T are subrings of a ring R, show that $S \cap T$ is also a subring of R.

(ii) Show that the intersection of the Gaussian and Eisenstein integers is $\mathbb{Z}$.

4.61 *

(i) If $(S_i)_{i \in I}$ is a family of subrings of a commutative ring R, prove that their intersection $\bigcap_{i \in I} S_i$ is also a subring of R.

(ii) If X is a subset of a commutative ring R, define $G(X)$, the subring **generated by** X, to be the intersection of all the subrings of R that contain X.

Prove that $G(X)$ is the *smallest subring* containing X in the following sense: if S is any subring of R containing X, then $G(X) \subseteq S$.

(iii) Let $(S_i)_{i \in I}$ be a family of subrings of a commutative ring R, each of which is a field. Prove that the subring $\bigcap_{i \in I} S_i$ is a field. Conclude that the intersection of a family of subfields of a field is a subfield.

4.62 Let p be a prime and let A_p be the set of all fractions with denominator a power of p.

(i) Show, with the usual operations of addition and multiplication, that A_p is a subring of $\mathbb{Q}$.

(ii) Describe the smallest subring of $\mathbb{Q}$ that contains both A_2 and A_5.

4.63 Let p be a prime and let $\mathbb{Q}_p$ be the set of rational numbers whose denominator (when written in lowest terms) is not divisible by p.

(i) Show, with the usual operations of addition and multiplication, that $\mathbb{Q}_p$ is a subring of $\mathbb{Q}$.

(ii) Show that $\mathbb{Q}_2 \cap \mathbb{Q}_5$ is a subring of $\mathbb{Q}$.

(iii) Is $\mathbb{Q}_p$ a field? Explain.

(iv) What is $\mathbb{Q}_p \cap A_p$, where A_p is defined in Exercise 4.62?

4.64 *

(i) Prove that $\mathbb{Z}[i] = \{a + bi : i^2 = -1 \text{ and } a, b \in \mathbb{Z}\}$, the Gaussian integers, is a commutative ring.

(ii) Prove that $\mathbb{Z}[\omega] = \{a + b\omega : \omega^3 = 1 \text{ and } a, b \in \mathbb{Z}\}$, the Eisenstein integers, is a commutative ring.

4.65 * Prove that $\mathbb{Z}[\zeta] = \{a + b\zeta^i : 0 \le i < n \text{ and } a, b \in \mathbb{Z}\}$ is a commutative ring, where ζ is a primitive nth root of unity.

4.66 * It may seem more natural to define addition in 2^X as union rather than symmetric difference. Is 2^X a commutative ring if addition $A \oplus B$ is defined as $A \cup B$ and AB is defined as $A \cap B$?

4.67 If X is a finite set with exactly n elements, how many elements are in 2^X?

4.68 * If A and B are subsets of a set X, prove that $A \subseteq B$ if and only if $A = A \cap B$.

4.69 * Recall that if A is a subset of a set X, then its **complement** is

$$A^c = \{x \in X : x \notin A\}.$$

Prove, in the commutative ring 2^X, that $A^c = X + A$.

4.70 * Let A be a subset of a set X. If $S \subseteq X$, prove that $A^c = S$ if and only if $A \cup S = X$ and $A \cap S = \varnothing$.

4.71 Let A, B, C be subsets of a set X.

(i) Prove that $A \cup (B \cap C) = (A \cup B) \cap (A \cup C)$.

(ii) Prove that $A \cap (B \cup C) = (A \cap B) \cup (A \cap C)$.

4.72 If A and B are subsets of a set X, then $A - B = \{x \in A : x \notin B\}$. Prove that $A - B = A \cap B^c$. In particular, $X - B = B^c$, the complement of B.

4.73 * Let A and B be subsets of a set X. Prove the *De Morgan laws*:

$$(A \cup B)^c = A^c \cap B^c \quad \text{and} \quad (A \cap B)^c = A^c \cup B^c,$$

where A^c denotes the complement of A.

4.74 * If A and B are subsets of a set X, define their *symmetric difference* by $A + B = (A - B) \cup (B - A)$ (see Figure 4.2).

(i) Prove that $A + B = (A \cup B) - (A \cap B)$.

(ii) Prove that $(A + B) \cup (A \cap B) = A \cup B$.

(iii) Prove that $A + A = \varnothing$.

(iv) Prove that $A + \varnothing = A$.

(v) Prove that $A + (B + C) = (A + B) + C$.

Hint. Show that each of $A + (B + C)$ and $(A + B) + C$ is described by Figure 4.3.

(vi) Prove that the Boolean ring 2^X is not a field if X has at least two elements.

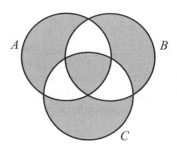

Figure 4.3. Associativity.

4.75 Prove that $A \cap (B + C) = (A \cap B) + (A \cap C)$.

4.4 Connections: Julius and Gregory

On what day of the week was July 4, 1776? We'll use congruence to answer this question. In fact, we'll answer in two ways: with an exact formula computing the day, and with a faster refinement, due to Conway.

Let's begin by seeing why our calendar is complicated. A *year* is the amount of time it takes the Earth to make one complete orbit around the Sun; a *day* is the amount of time it takes the Earth to make a complete rotation about the axis through its north and south poles. There is no reason why the number of days in a year should be an integer, and it isn't; a year is approximately 365.2422 days long. In 46 BCE, Julius Caesar (and his scientific advisors) changed the old Roman calendar, creating the *Julian calendar* containing a *leap year* every four years; that is, every fourth year has an extra day, namely, February 29, and so it contains 366 days (a *common year* is a year that is not a leap year). This would be fine if the year were exactly 365.25 days long, but it has the effect of making the year $365.25 - 365.2422 = .0078$ days (about 11 minutes and 14 seconds) too long. After 128 years, a full day was added to the calendar; that is, the Julian calendar overcounted the number of days. In the year 1582, the vernal equinox (the Spring day on which there are exactly 12 hours of daylight and 12 hours of night) occurred on March 11 instead of on March 21. Pope Gregory XIII (and his scientific advisors) then installed the *Gregorian calendar* by erasing 10 days that year; the day after October 4, 1582 was October 15, 1582. This caused confusion and fear among the people; they thought their lives had been shortened by ten days.

The Gregorian calendar modified the Julian calendar as follows. Call a year y ending in 00 a *century year*. If a year y is not a century year, then it is a leap year if it is divisible by 4; if y is a century year, it is a leap year only if it is divisible by 400. For example, 1900 is not a leap year, but 2000 is a leap year. The Gregorian calendar is the one in common use today, but it was not uniformly adopted throughout Europe. For example, the British empire didn't accept it until 1752, when 11 days were erased, and the Russians didn't accept it until 1918, when 13 days were erased (thus, Trotsky called the Russian revolution, which occurred in 1917, the October Revolution, even though it occurred in November of the Gregorian calendar).

The true number of days in 400 years is about

$$400 \times 365.2422 = 146096.88 \text{ days.}$$

In this period, the Julian calendar has

$$400 \times 365 + 100 = 146,100 \text{ days,}$$

while the Gregorian calendar, which eliminates three leap years from this time period, has 146,097 days. Thus, the Julian calendar gains about 3.12 days every 400 years, while the Gregorian calendar gains only 0.12 days (about 2 hours and 53 minutes.

Historical Note. There are 1628 years from 46 BCE to 1582 CE. The Julian calendar overcounts one day every 128 years, and so it overcounted 12 days in this period (for $12 \times 128 = 1536$). Why didn't Gregory have to erase 12 days? The Council of Nicaea, meeting in the year 325 CE, defined Easter as the first Sunday strictly after the Paschal full moon, which is the first full moon on or after the vernal equinox (now you know why Pope Gregory was interested in the calendar). The vernal equinox in 325 CE fell on March 21, and the Synod of Whitby, in 664 CE, officially defined the vernal equinox to be March 21. The discrepancy observed in 1582 was thus the result of only $1257 = 1582 - 325$ years of the Julian calendar: approximately 10 days.

We now seek a calendar formula. For easier calculation, choose 0000 as our reference year, even though there was no year zero! Assign a number to each day of the week, according to the scheme

Sun	Mon	Tues	Wed	Thurs	Fri	Sat
0	1	2	3	4	5	6

In particular, March 1, 0000, has some number a_0, where $0 \leq a_0 \leq 6$. In the next year 0001, March 1 has number $a_0 + 1$ (mod 7), for 365 days have elapsed from March 1, 0000, to March 1, 0001, and

$$365 = 52 \times 7 + 1 \equiv 1 \bmod 7.$$

Similarly, March 1, 0002, has number $a_0 + 2$, and March 1, 0003, has number $a_0 + 3$. However, March 1, 0004, has number $a_0 + 5$, for February 29, 0004, fell between March 1, 0003, and March 1, 0004, and so $366 \equiv 2 \bmod 7$ days had elapsed since the previous March 1. We see, therefore, that every common year adds 1 to the previous number for March 1, while each leap year adds 2. Thus, if March 1, 0000, has number a_0, then the number a' of March 1, year y, is

$$a' \equiv a_0 + y + L \bmod 7,$$

where L is the number of leap years from year 0001 to year y. To compute L, count all those years divisible by 4, then throw away all the century years, and then put back those century years that are leap years. Thus,

$$L = \lfloor y/4 \rfloor - \lfloor y/100 \rfloor + \lfloor y/400 \rfloor,$$

where $\lfloor x \rfloor$ denotes the greatest integer in x. Therefore, we have

$$a' \equiv a_0 + y + L$$
$$\equiv a_0 + y + \lfloor y/4 \rfloor - \lfloor y/100 \rfloor + \lfloor y/400 \rfloor \bmod 7.$$

We can actually find a_0 by looking at a calendar. Since March 1, 2012, fell on a Thursday,

$$4 \equiv a_0 + 2012 + \lfloor 2012/4 \rfloor - \lfloor 2012/100 \rfloor + \lfloor 2012/400 \rfloor$$
$$\equiv a_0 + 2012 + 503 - 20 + 5 \bmod 7,$$

and so

$$a_0 \equiv -2496 \equiv -4 \equiv 3 \bmod 7$$

(that is, March 1, 0000 fell on Wednesday). We can now determine the day of the week a' on which March 1 will fall in any year $y > 0$, for

$$a' \equiv 3 + y + \lfloor y/4 \rfloor - \lfloor y/100 \rfloor + \lfloor y/400 \rfloor \bmod 7.$$

Historical Note. There is a reason we have been discussing March 1, for it was the first day of the year in the old Roman calendar (753 BCE). There were only ten months: Martius, ..., Iunius, Quintilis, Sextilis, Septembris, ..., Decembris (which explains why September is so named; originally, it was month 7). In 713 BCE, Numa added January and February, and the Julian calendar changed the names of Quintilis and Sextilis to July and August.

Let us now analyze February 28. For example, suppose that February 28, 1600, has number b. As 1600 is a leap year, February 29, 1600, occurs between February 28, 1600, and February 28, 1601; hence, 366 days have elapsed between these two February 28s, so that February 28, 1601, has number $b + 2$. February 28, 1602, has number $b + 3$, February 28, 1603, has number $b + 4$,

February 28, 1604, has number $b + 5$, but February 28, 1605, has number $b + 7$ (for there was a February 29 in 1604).

Let us compare the pattern of behavior of February 28, 1600, namely, b, $b+2, b+3, b+4, b+5, b+7, \ldots$, with that of some date in 1599. If May 26, 1599, has number c, then May 26, 1600, has number $c + 2$, for February 29, 1600, comes between these two May 26s, and so there are $366 \equiv 2 \bmod 7$ intervening days. The numbers of the next few May 26s, beginning with May 26, 1601, are $c + 3, c + 4, c + 5, c + 7$. We see that the pattern of the days for February 28, starting in 1600, is exactly the same as the pattern of the days for May 26, starting in 1599; indeed, the same is true for any date in January or February. Thus, the pattern of the days for any date in January or February of a year y is the same as the pattern for a date occurring in the preceding year $y - 1$: a year preceding a leap year adds 2 to the number for such a date, whereas all other years add 1. Therefore, we pretend we have reverted to the ancient calendar by making New Year's Day fall on March 1, so that any date in January or February is treated as if it had occurred in the previous year.

Historical Note. George Washington's birthday, in the Gregorian calendar, is February 22, 1732. But the Gregorian calendar was not introduced in the British colonies until 1752. Thus, his original birthday was February 11. But New Year's Day was also changed; before 1752, England and its colonies celebrated New Year's Day on March 25; hence, February, which had been in 1731, was regarded, after the calendar change, as being in 1732. George Washington used to joke that not only did his birthday change, but so did his birth year. See Exercise 4.80 on page 176.

How do we find the day corresponding to a date other than March 1? Since March 1, 0000, has number 3 (as we have seen above), April 1, 0000, has number 6, for March has 31 days and $3 + 31 \equiv 6 \bmod 7$. Since April has 30 days, May 1, 0000, has number $6 + 30 \equiv 1 \bmod 7$. Figure 4.4 is the table giving the number of the first day of each month in year 0000.

Remember that we are pretending that March is month 1, April is month 2, and so on. Let us denote these numbers by $1 + j(m)$, where $j(m)$, for $m = 1, 2, \ldots, 12$, is defined by

$$j(m) : \ 2, 5, 0, 3, 5, 1, 4, 6, 2, 4, 0, 3.$$

It follows that month m, day 1, year y, has number

$$1 + j(m) + g(y) \bmod 7,$$

where

$$g(y) = y + \lfloor y/4 \rfloor - \lfloor y/100 \rfloor + \lfloor y/400 \rfloor.$$

Note that $a_0 = 1 + j(1)$, so that the values of $j(m)$ depend on our knowing a_0.
Here's a formula for $j(m)$:

$$j(m) = \lfloor 2.6m - 0.2 \rfloor, \quad \text{where } 1 \leq m \leq 12;$$

the values are displayed in Figure 4.4. This formula is not quite accurate. For example, this number for December, that is, for $m = 10$, is $\lfloor 2.6m - 0.2 \rfloor = 25$; but $j(10) = 4$. However, $25 \equiv 4 \bmod 7$, and so the formula for $j(m)$ really gives the congruence class mod 7.

Date	Number	Date	Number	Date	Number
March 1	2	July 1	5	November 1	2
April 1	5	August 1	1	December 1	4
May 1	0	September 1	4	January 1	0
June 1	3	October 1	6	February 1	3

Figure 4.4. Values of $j(m)$.

Theorem 4.49 (Calendar Formula). *The date with month m, day d, year y has number*

$$d + j(m) + g(y) \bmod 7,$$

where $j(m)$ is given in Figure 4.4,

$$g(y) = y + \lfloor y/4 \rfloor - \lfloor y/100 \rfloor + \lfloor y/400 \rfloor,$$

and dates in January and February are treated as having occurred in the previous year.

Proof. The number mod 7 corresponding to month m, day 1, year y, is

$$1 + j(m) + g(y).$$

It follows that $2 + j(m) + g(y)$ corresponds to month m, day 2, year y, and, more generally, $d + j(m) + g(y)$ corresponds to month m, day d, year y. ■

Let's find the day of the week on which July 4, 1776 fell; here $m = 5$, $d = 4$, and $y = 1776$. Substituting in the formula, we obtain the number

$$4 + 5 + 1776 + 444 - 17 + 4 = 2216 \equiv 4 \bmod 7;$$

therefore, July 4, 1776, fell on a Thursday.

Example 4.50. Does every year y contain a Friday 13? We have

$$5 \equiv 13 + j(m) + g(y) \bmod 7.$$

The question is answered positively if the numbers $j(m)$, as m varies from 1 through 12, give all the remainders 0 through 6 mod 7. And this is what happens. The sequence of remainders mod 7 is

$$2, \ 5, \ [0, \ 3, \ 5, \ 1, \ 4, \ 6, \ 2], \ 4, \ 0, \ 3.$$

Indeed, we see that there must be a Friday 13 occurring between May and November. No number occurs three times on the list, but it is possible that there are three Friday 13s in a year because January and February are viewed as having occurred in the previous year; for example, there were three Friday 13s in 1987 (see Exercise 4.79 on page 176). Of course, we may replace Friday by any other day of the week, and we may replace 13 by any number between 1 and 28. ▲

The word *calendar* comes from the Greek "to call," which evolved into the Latin word for the first day of a month (when accounts were due).

Most of us need paper and pencil (or a calculator) to use the calendar formula in the theorem, but here's a way to simplify the formula so you can do the calculation in your head and amaze your friends. A mnemonic for $j(m)$ is the sentence

My Uncle Charles has eaten a cold supper; he eats nothing hot.

$$2 \quad 5 \quad (7 \equiv 0) \quad 3 \quad 5 \quad 1 \quad 4 \quad\quad 6 \quad 2 \quad 4 \quad (7 \equiv 0) \quad 3$$

Corollary 4.51. *The date with month m, day d, year $y = 100C + N$, where $0 \le N \le 99$, has number*

$$d + j(m) + N + \lfloor N/4 \rfloor + \lfloor C/4 \rfloor - 2C \bmod 7,$$

provided that dates in January and February are treated as having occurred in the previous year.

Proof. If we write a year $y = 100C + N$, where $0 \le N \le 99$, then

$$y = 100C + N \equiv 2C + N \bmod 7,$$
$$\lfloor y/4 \rfloor = 25C + \lfloor N/4 \rfloor \equiv 4C + \lfloor N/4 \rfloor \bmod 7,$$
$$\lfloor y/100 \rfloor = C, \text{ and } \lfloor y/400 \rfloor = \lfloor C/4 \rfloor.$$

Therefore,

$$y + \lfloor y/4 \rfloor - \lfloor y/100 \rfloor + \lfloor y/400 \rfloor \equiv N + 5C + \lfloor N/4 \rfloor + \lfloor C/4 \rfloor \bmod 7$$
$$\equiv N + \lfloor N/4 \rfloor + \lfloor C/4 \rfloor - 2C \bmod 7. \quad \blacksquare$$

This formula is simpler than the first one. For example, the number corresponding to July 4, 1776 is now obtained as

$$4 + 5 + 76 + 19 + 4 - 34 = 74 \equiv 4 \bmod 7,$$

agreeing with our calculation above. The reader may now compute the day of his or her birth.

Example 4.52. The birthday of Rose, the grandmother of Danny and Ella, was January 1, 1909; on what day of the week was she born?

January is counted as belonging to the previous year 1908.

We use Corollary 4.51. If A is the number of the day, then $j(m) = 0$ (for January corresponds to month 11), and

$$A \equiv 1 + 0 + 8 + \lfloor 8/4 \rfloor + \lfloor 19/4 \rfloor - 38$$
$$\equiv -23 \bmod 7$$
$$\equiv 5 \bmod 7.$$

Rose was born on a Friday. ▲

J. H. Conway found an even simpler calendar formula. The day of the week on which the last day of February occurs is called the ***doomsday*** of the year. We can compute doomsdays using Corollary 4.51.

Knowing the doomsday D of a century year $100C$ finds the doomsday D' of any other year $y = 100C + N$ in that century. Since $100C$ is a century year, the number of leap years from $100C$ to y does not involve the Gregorian alteration. Thus,

$$D' \equiv D + N + \lfloor N/4 \rfloor \bmod 7.$$

For example, since doomsday 1900 is Wednesday = 3, we see that doomsday 1994 is Monday = 1, for

$$3 + 94 + 23 = 120 \equiv 1 \bmod 7.$$

February 29, 1600	2	Tuesday
February 28, 1700	0	Sunday
February 28, 1800	5	Friday
February 28, 1900	3	Wednesday
February 29, 2000	2	Tuesday

Figure 4.5. Recent doomsdays.

Proposition 4.53 (Conway). *Let D be doomsday $100C$, and let $0 \leq N \leq 99$. If $N = 12q + r$, where $0 \leq r < 12$, then D', doomsday $100C + N$, is given by*

$$D + q + r + \lfloor r/4 \rfloor \bmod 7.$$

Proof.

$$\begin{aligned}
D' &\equiv D + N + \lfloor N/4 \rfloor \\
&\equiv D + 12q + r + \lfloor (12q + r)/4 \rfloor \\
&\equiv D + 15q + r + \lfloor r/4 \rfloor \\
&\equiv D + q + r + \lfloor r/4 \rfloor \bmod 7. \quad \blacksquare
\end{aligned}$$

For example, what is $D' = $ doomsday 1994? Now $N = 94 = 12 \times 7 + 10$, so that $q = 7$ and $r = 10$. Thus, $D' = 3 + 7 + 10 + 2 \equiv 1 \bmod 7$; that is, doomsday 1994 is Monday, as we saw above.

Once we know doomsday of a particular year, we can use various tricks (e.g., Uncle Charles) to pass from doomsday to any other day in the year. Conway observed that some other dates falling on the same day of the week as the doomsday are

April 4, June 6, August 8, October 10, December 12,

May 9, July 11, September 5, and November 7.

If we return to the everyday listing beginning with January as the first month, then it is easier to remember these dates using the notation month/day:

4/4,   6/6,   8/8,   10/10,   12/12,

5/9,   7/11,   9/5,   11/7.

Since doomsday corresponds to the last day of February, we are now within a few weeks of any date in the year, and we can easily interpolate to find the desired day. For example, let's use this method for July 4, 1776. Notice that July 4 occurs on the same day of the week as July 11, and so we need only find doomsday 1776. By Proposition 4.53,

$$D' \equiv 0 + 76 + \lfloor 76/4 \rfloor = 95 \equiv 4 \bmod 7.$$

We see again that July 4, 1776 fell on a Thursday.

Example 4.54. Let's use Conway's method to compute Rose's birthday again (recall Example 4.52: Rose was born on January 1, 1909). Since Conway's method applies within a given century, there is no need to pretend that January and February live in the preceding year; we can work within 1909. Now doomsday 1900 is 3, so that Proposition 4.53 gives doomsday 1909 = 0; that is, Sunday. By definition, doomsday is the number corresponding to the last date in February, which is here February 28 (for 1909 is not a leap year). Thus, we interpolate that 3 is the number for 1/31, 1/24, 1/3; that is, January 3 fell on Sunday, and so January 1 fell on Friday (which agrees with what we saw in Example 4.52).   ▲

Exercises

4.76 A suspect said that he had spent the Easter holiday April 21, 1893, with his ailing mother; Sherlock Holmes challenged his veracity at once. How could the great detective have been so certain?

Hint. Easter always falls on Sunday. (There is a Jewish variation of this problem, for Yom Kippur must fall on either Monday, Wednesday, Thursday, or Saturday; secular variants can involve Thanksgiving Day, which always falls on a Thursday, or Election Day in the US, which always falls on a Tuesday.)

4.77 How many times in 1900 did the first day of a month fall on a Tuesday?

Hint. The year $y = 1900$ was not a leap year.

4.78 On what day of the week did February 29, 1896 fall?

Hint. On what day did March 1, 1896, fall? Conclude from your method of solution that no extra fuss is needed to find leap days.

4.79 *

(i) Show that 1987 had three Friday 13s.

Hint. See Example 4.50.

(ii) Show, for any year $y > 0$, that $g(y) - g(y - 1) = 1$ or 2, where $g(y) = y + \lfloor y/4 \rfloor - \lfloor y/100 \rfloor + \lfloor y/400 \rfloor$.

(iii) Can there be a year with exactly one Friday 13?

Hint. Either use congruences or scan the 14 possible calendars: there are 7 possible common years and 7 possible leap years, for January 1 can fall on any of the 7 days of the week.

4.80 * JJR's Uncle Ben was born in Pogrebishte, a village near Kiev, and he claimed that his birthday was February 29, 1900. JJR told him that this could not be, for 1900 was not a leap year. Why was JJR wrong?

Hint. When did Russia adopt the Gregorian calendar?

4.5   Connections: Patterns in Decimal Expansions

One of the most beautiful applications of modular arithmetic is to the classification of decimal expansions of rational numbers, a circle of ideas that runs
throughout precollege mathematics.

We now ask what we can infer from knowing the decimal expansion of a
real number x. You probably know Proposition 4.58: x is rational if and only
if its decimal expansion either terminates or eventually repeats. Is there any
nice way to tell ahead of time which fractions terminate? Can you be *sure*
that the ones that don't terminate really do repeat? And, for fractions whose
decimals repeat, can you predict the *period* (the number of digits in its *block*,
the repeating part) as well as the actual sequence of digits in it?

Many conjectures about fractions and decimals come from a careful analysis of numerical calculations. In this section, we (and you) will perform a
great many calculations, looking at patterns you'll observe, with the goal of
analyzing them, and seeing how they are explained by "how the calculation
goes."

Real Numbers

We assume that every real number x has a decimal expansion; for example,
$-\pi = -3.14159\ldots$. This follows from identifying each real number x with
a "point on a number line" having signed distance from a fixed origin on a
coordinatized line. In particular, rational numbers have decimal expansions,
which you can find by long division.

The term *expansion* will be used in a nonstandard way: we restrict the terminology so that, from now on, the ***decimal expansion*** of a real number is the
sequence of digits after the decimal point. With this usage, for example, the
decimal expansion of $-\pi$ is $.14159\ldots$.

We are going to see that decimal expansions of real numbers are unique,
with one possible exception: if there is an infinite string of all 9s. For example,

$$.328 = .327999\ldots.$$

This is explained using the geometric series.

Lemma 4.55. *If r is a real number with $|r| < 1$, then*

$$\sum_{n=0}^{\infty} r^n = 1 + r + r^2 + \cdots = \frac{1}{1-r}.$$

Proof. For every positive integer n, the identity

$$1 - r^n = (1-r)\left(1 + r + r^2 + \cdots + r^{n-1}\right)$$

gives the equation

$$1 + r + r^2 + \cdots + r^{n-1} = \frac{1-r^n}{1-r} = \frac{1}{1-r} - \frac{r^n}{1-r}$$

for every real number $r \neq 1$. Since $|r| < 1$, we have $\lim_{n\to\infty} r^n/(1-r) = 0$. ∎

For example, taking $r = 1/10$, we have

$$.999\ldots = \frac{9}{10} + \frac{9}{10^2} + \frac{9}{10^3} + \cdots$$

$$= \frac{9}{10}\left(1 + \frac{1}{10} + \frac{1}{10^2} + \frac{1}{10^3} + \cdots\right)$$

$$= \frac{9}{10} \cdot \frac{1}{1 - 1/10}$$

$$= \frac{9}{10} \cdot \frac{10}{9} = 1.$$

Hence,

$$.327999\ldots = .327 + .000999\ldots = .327 + \frac{1}{10^3}.999\ldots$$

$$= .327 + \frac{1}{10^3} = .327 + .001 = .328.$$

We'll resolve this ambiguity by choosing, once for all, to avoid infinite strings of 9s. Indeed, we'll soon see that the choice .328 comes from long division.

If we disregard "all nines from some point on," then we can show that every real number has a unique decimal expansion. For this, we need the following corollary to Lemma 4.55.

Corollary 4.56. *If $x = .d_1 d_2 \ldots$ and $d_j < 9$ for some $j > 1$, then*

$$x < \frac{d_1 + 1}{10}.$$

Proof. Each digit d_i is at most 9, and there is some $j > 1$ with d_j strictly less than 9. So, writing x as a series, we have

$$x = \frac{d_1}{10} + \frac{d_2}{10^2} + \frac{d_3}{10^3} + \cdots + \frac{d_j}{10^j} + \cdots$$

$$< \frac{d_1}{10} + \frac{d_2}{10^2} + \frac{d_3}{10^3} + \cdots + \frac{9}{10^j} + \cdots$$

$$\leq \frac{d_1}{10} + \frac{9}{10^2} + \frac{9}{10^3} + \cdots + \frac{9}{10^j} + \cdots$$

$$= \frac{d_1}{10} + \frac{9}{10^2}\left(1 + \frac{1}{10} + \cdots + \frac{1}{10^{j-2}} + \cdots\right)$$

$$= \frac{d_1}{10} + \left(\frac{9}{10^2}\right)\left(\frac{10}{9}\right)$$

$$= \frac{d_1}{10} + \frac{1}{10} = \frac{d_1 + 1}{10}. \quad \blacksquare$$

Proposition 4.57. *Every real number x has a unique decimal expansion that does not end with infinitely many consecutive 9s.*

Proof. Suppose that

$$.d_1 \ldots d_k \ldots = x = .e_1 \ldots e_k \ldots,$$

where $d_1 = e_1, \ldots, d_{k-1} = e_{k-1}$, but that $d_k \neq e_k$. We may assume that $d_k < e_k$, so that $d_k + 1 \leq e_k$.

Multiplying by a power of 10, we see that

$$.d_k d_{k+1} \ldots = .e_k e_{k+1} \ldots .$$

Because there's not an infinite string of 9s in our expansions, we can apply Corollary 4.56 to find that

$$.d_k d_{k+1} \ldots < \frac{d_k + 1}{10} \leq \frac{e_k}{10} \leq .e_k e_{k+1} \ldots ,$$

contradicting the fact that the extreme left-hand and right-hand expressions are equal. ∎

Decimal Expansions of Rationals

Let's now focus on rational numbers. Some decimal expansions of rationals terminate; for example,

$$\frac{1}{10} = .1, \quad \frac{1}{4} = .25, \quad \frac{3}{8} = .375, \quad \frac{1}{40} = .025.$$

And there are some fractions whose decimal expansions repeat (after a possible initial string of digits):

$$\frac{1}{3} = .333\ldots, \quad \frac{1}{7} = .142857142857\ldots, \quad \frac{9}{28} = .32142857142857\ldots$$

Definition. Let a real number x have decimal expansion

$$r = .d_1 d_2 d_3 \ldots ;$$

that is, $x = k.d_1 d_2 d_3 \cdots = k + r$ for some $k \in \mathbb{Z}$.

(i) We say that x **terminates** if there exists an integer N so that $d_i = 0$ for all $i > N$.

(ii) We say that x **repeats** with **period** $m \geq 1$ if

   (a) it doesn't terminate

   (b) there exist positive integers N and m so that $d_i = d_{i+m}$ for all $i > N$

   (c) m is the smallest such integer.

If x repeats, then its **block** is the first occurrence of its repeating part

$$d_i d_{i+1} \ldots d_{i+m-1}.$$

> You can think of .32142857142857...as the 10-adic expansion of 9/28, using negative powers of 10.

We could say that "terminating" and "repeating" decimals are not really different, for terminating rationals have decimal expansions that repeat with period 1 and with block having the single digit 0, but it's convenient and natural to distinguish such rationals from those having infinitely many nonzero digits, as you'll see in Proposition 4.59.

The way to get the decimal expansion for $1/7$ is to divide 7 into 1 via long division, as in Figure 4.6. Each of the remainders 1 through 6 shows up exactly once in this calculation, in the order $3, 2, 6, 4, 5, 1$. Once you get a remainder of 1, the process will start over again, and the digits in the quotient, namely, $1, 4, 2, 8, 5, 7$, will repeat. The block of $1/7$ is 142857. However, even though

$$
\begin{array}{r}
0.142857\ldots \\
7\overline{)1.000000\ldots} \\
7 \\
\hline
30 \\
28 \\
\hline
20 \\
14 \\
\hline
60 \\
56 \\
\hline
40 \\
35 \\
\hline
50 \\
49 \\
\hline
1
\end{array}
$$

$$
\begin{array}{r}
0.153846\ldots \\
13\overline{)2.000000\ldots} \\
1\,3 \\
\hline
70 \\
65 \\
\hline
50 \\
39 \\
\hline
110 \\
104 \\
\hline
60 \\
52 \\
\hline
80 \\
78 \\
\hline
2
\end{array}
$$

Figure 4.6. $1/7 = .142857142857142857\ldots.$ **Figure 4.7.** $2/13 = .153846\ldots.$

142857142857 also repeats, it is not a block because it is too long: $1/7$ has period 6, not 12.

Consider a second example: the calculation of $2/13$ in Figure 4.7. It too has period 6.

Next, we'll see that every rational number terminates or repeats; that is, the two types in the definition are the only possibilities.

Proposition 4.58. *A real number x is rational if and only if it either terminates or repeats. Moreover, if $x = a/b$ is rational, then it has period at most b.*

Proof. The arguments for $1/7$ and $2/13$ generalize. Imagine expressing a fraction a/b (with $a, b > 0$) as a decimal by dividing b into a via long division. There are at most b possible remainders in this process (integers between 0 and $b-1$), so after at most b steps a remainder appears that has shown up before. After that, the process repeats.

Conversely, let's see that if a real number x terminates or repeats, then x is rational. A terminating decimal is just a fraction whose denominator is a power of 10, while a repeating decimal is made up of such a fraction plus the sum of a convergent geometric series. An example is sufficient to see what's going on.

Middle school students practice another method for doing this (for days on end). See Exercise 4.84 on page 181.

$$
\begin{aligned}
.1323232\ldots &= .1 + .0323232\ldots \\
&= .1 + \frac{32}{10^3} + \frac{32}{10^5} + \frac{32}{10^7} + \cdots \\
&= .1 + \frac{32}{10^3}\left(1 + \frac{1}{10^2} + \frac{1}{10^4} + \cdots\right) \\
&= \frac{1}{10} + \frac{4}{25}\left(\frac{1}{1 - \frac{1}{10^2}}\right). \quad \text{(by Lemma 4.55)}
\end{aligned}
$$

The last expression is clearly a rational number. The general proof is a generic version of this idea; it is left as Exercise 4.81 below. ∎

Which rationals terminate? Certainly, any rational a/b whose denominator b is a power of 10 terminates. But some other rationals can also be put in this form; for example,

$$\frac{5}{8} = \frac{5 \cdot 125}{8 \cdot 125} = \frac{625}{1000} = .625.$$

The basic idea is to take a denominator of the form $2^u 5^v$, and multiply top and bottom of the fraction to produce a denominator that's a power of 10.

Theorem 4.61 below gives a necessary and sufficient condition for a/b to repeat.

Proposition 4.59. *Let $x = a/b$ be rational, written in lowest terms. Then x terminates if and only if the only prime factors of b are 2 and 5.*

Proof. If x terminates, say, $x = k.d_1 d_2 \ldots d_m$, then $x = k + D/10^m$ for some $k \in \mathbb{Z}$, where D is the integer with digits $d_1 d_2 \ldots d_m$; thus, x is a fraction whose denominator is divisible only by 2 and 5. Conversely, if $x = k + r = k + a/2^u 5^v$, then

$$r = \frac{a}{2^u 5^v} = \left(\frac{2^v 5^u}{2^v 5^u}\right)\left(\frac{a}{2^u 5^v}\right) = \frac{2^v 5^u a}{10^{u+v}}$$

a fraction whose denominator is a power of 10. Hence, x terminates. ∎

How to Think About It. Exercise 4.83 shows that if r is a rational number and $5^m r$ or $2^\ell r$ terminates, then r also terminates. However, if kr terminates (for some integer k), then r need not terminate; for example, $r = .271333\ldots$ does not terminate, but $3r = .814$ does terminate.

Exercises

4.81 * Complete the proof of Proposition 4.58 that a decimal that eventually repeats is the decimal expansion of a rational number.

4.82 * Let $r = a/b$ be rational.

(i) If r terminates, then kr terminates for every integer k.

(ii) If $\gcd(a, b) = 1$, prove that a/b terminates if and only if $1/b$ terminates.
Hint. $1/b = (sa + tb)/b = sa/b + t$.

4.83 * If $\ell \geq 0$, $m \geq 0$, and $2^\ell 5^m r$ terminates, prove that r terminates.

4.84 * Here's a method used by many precollege texts for converting repeating decimals to fractions. Suppose that you want to convert $.324324\ldots$ to a fraction. Calculate like this: If $x = .324324\ldots$, then $1000x = 324.324324\ldots$, and

$$1000x - x = 999x = 324.$$

Hence, $x = 324/999$.

(i) There is a hidden assumption about geometric series in this method. Where is it?

(ii) Try this method with the following decimal expansions:

(a) $.356356\ldots$ (b) $.5353\ldots$

(c) $.2222\ldots$ (d) $.07593\ldots$

(e) $.0123563563\ldots$

4.85 What's wrong with the following calculation? Let $x = 1 + 2 + 2^2 + \ldots$. Then

$$x = 1 + 2 + 2^2 + \ldots$$
$$2x = 2 + 2^2 + 2^3 + \ldots.$$

Subtract the top equation from the bottom to obtain $x = -1$.

4.86 Calculate decimal expansions for the followings fractions using long division. For each one, what other fractions-to-decimal expansions (if any) do you get for free?

(i) $\dfrac{1}{3}$ (ii) $\dfrac{1}{6}$ (iii) $\dfrac{1}{9}$

(iv) $\dfrac{1}{15}$ (v) $\dfrac{4}{15}$ (vi) $\dfrac{7}{15}$

(vii) $\dfrac{1}{8}$ (viii) $\dfrac{1}{13}$ (ix) $\dfrac{1}{20}$

(x) $\dfrac{1}{19}$ (xi) $\dfrac{1}{31}$ (xii) $\dfrac{1}{37}$

Periods and Blocks

> Corollary 4.62 says that the period of a/n is equal to the period of $1/n$ if $\gcd(a, n) = 1$.

What is the period of a "unit fraction" $1/n$? Our result will come from taking a closer look at how decimal expansions are calculated; the analysis generalizes to the decimal expansion of any rational number.

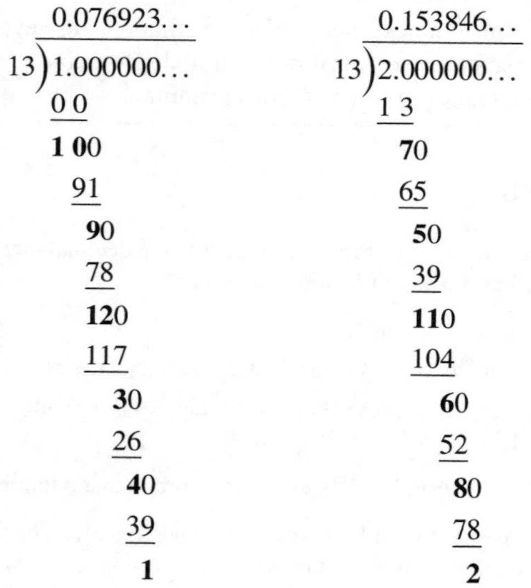

$$
\begin{array}{r}
0.076923\ldots \\
13\overline{)1.000000\ldots} \\
00 \\
\mathbf{1}\,00 \\
91 \\
90 \\
78 \\
120 \\
117 \\
30 \\
26 \\
40 \\
39 \\
1
\end{array}
\qquad
\begin{array}{r}
0.153846\ldots \\
13\overline{)2.000000\ldots} \\
13 \\
70 \\
65 \\
50 \\
39 \\
110 \\
104 \\
60 \\
52 \\
80 \\
78 \\
2
\end{array}
$$

Figure 4.8. Decimal expansions of $1/13$ and $2/13$.

An analysis of the calculation for $1/13$ yields another insight (see Figure 4.8). Pretend that the decimal point isn't there, so we are dividing $1{,}000{,}000 = 10^6$ by 13. Since 1 appears as a remainder, the initial sequence of remainders will repeat, and the period of $1/13$ is 6. Thus, the period of $1/13$ is the smallest power of 10 congruent to 1 mod 13. In other words, the period of $1/13$ is the order of 10 in $\mathbb{Z}_{13}$ (see Exercise 4.49 on page 165).

We will generalize this observation in Theorem 4.61: the period of any fraction $1/n$ is the order of 10 in $\mathbb{Z}_n$ as long as there is some positive integer e with

$10^e \equiv 1 \bmod n$. But, by Exercise 4.48 on page 165, some power of 10 is congruent to 1 mod n if and only if 10 is a unit in $\mathbb{Z}_n$. Now this condition is equivalent to $\gcd(10, n) = 1$; that is, if and only if n is not of the form $2^u 5^v$. Thus, Proposition 4.59 shows why the dichotomy of terminating and repeating rationals is so natural.

To prove the general result for $1/n$, we just need to make sure that the first remainder that shows up twice is, in fact, 1. That's the content of the next lemma.

Lemma 4.60. *If* $\gcd(10, n) = 1$, *then* 1 *occurs as a remainder in the long division of* 1 *by* n; *moreover, there cannot be two identical remainders occurring before* 1 *occurs.*

Proof. As we saw above, because $\gcd(10, n) = 1$, a remainder of 1 will first appear in the long division after e steps where e is the order of 10 in $\mathbb{Z}_n$. We must prove that there is no repeat of some other remainder before that remainder of 1 shows up. First of all, there can't be an earlier 1 (why?). Next, suppose you see the same remainder, say c, occurring earlier, say, at steps $e_1 < e_2 < e$. Then we'd have

$$10^{e_1} \equiv 10^{e_2} \equiv c \bmod n.$$

Since 10 is a unit in $\mathbb{Z}_n$, this would imply that

$$1 \equiv 10^{e_2 - e_1} \bmod n.$$

And since $e_2 - e_1 < e$, this would contradict the fact that e is the order of 10 in $\mathbb{Z}_m$. ∎

Putting it all together, we have a refinement of Propositions 4.58 and 4.59:

Theorem 4.61. *If* $n > 0$ *is an integer, then* $1/n$ *either terminates or repeats.*

(i) $1/n$ *terminates if and only if* $n = 2^u 5^v$ *for nonnegative integers* u *and* v.

(ii) *If* $\gcd(n, 10) = 1$, *then* $1/n$ *repeats with period* m, *where* m *is the order of* 10 *in* $\mathbb{Z}_n$.

Proof. Part (i) was proved in Proposition 4.59. The essence of the proof of part (ii) lies in the discussion on page 182 about the decimal expansion of $1/13$: the expansion for $1/n$ repeats after e steps, where e is the order of 10 in $\mathbb{Z}_n$; that is, the first occurrence of remainder 1 occurs at the eth step of the long division. And Lemma 4.60 shows that there can be no earlier occurrences. ∎

So, if $\gcd(n, 10) = 1$, then $1/n$ repeats, and we know that its period is the order of 10 in $\mathbb{Z}_n$. What about fractions of the form a/n? The next corollary shows that the same thing is true, as long as the fraction is in lowest terms.

Corollary 4.62. *If* $\gcd(a, n) = 1$ *and* $a < n$, *then the period of* a/n *is the same as that of* $1/n$, *namely the order of* 10 *in* $\mathbb{Z}_n$.

Proof. Suppose the period of a/n is ℓ. Then, arguing as in Lemma 4.60, the expansion will repeat only after the remainder a occurs in the long division

p	period for $\frac{1}{p}$		p	period for $\frac{1}{p}$
3	1		97	96
7	6		101	4
11	2		103	34
13	6		107	53
17	16		109	108
19	18		113	112
23	22		127	42
29	28		131	130
31	15		137	8
37	3		139	46
41	5		149	148
43	21		151	75
47	46		157	78
53	13		163	81
59	58		167	166
61	60		173	43
67	33		179	178
71	35		181	180
73	8		191	95
79	13		193	192
83	41		197	98
89	44		199	99

Figure 4.9. Periods of $1/p$ for small primes p.

of a by n (see Exercise 4.88 on page 190). But this implies that ℓ is the smallest positive integer such that

$$a\,10^\ell \equiv a \bmod n.$$

Since a is a unit in $\mathbb{Z}_n$, multiplying by a^{-1} gives

$$10^\ell \equiv 1 \bmod n.$$

It follows that $\ell = m$, the order of 10 in $\mathbb{Z}_n$.   ■

Theorem 4.61 doesn't answer every question about the periods of $1/p$, where p is a prime other than 2 or 5. Sometimes the period is $p - 1$, as when $p = 7$, but this not always so, for $1/13$ has period 6, not 12. In all the entries in Figure 4.9, we see that periods of $1/p$ are *divisors* of $p - 1$. This turns out to be always true, and you'll prove it soon. What about non-prime denominators? Perhaps the length of the period of the expansion of $1/n$ is a factor of $n - 1$? No such luck: $1/21 = .047619047619\ldots$ has period 6 which is not a divisor of 20.

But stay tuned—we'll return to the period of $1/n$ shortly.

Historical Note. In *Disquisitiones Arithmeticae* [14], Gauss conjectured that there are infinitely many primes p that have the property that the decimal expansion for $1/p$ has period $p - 1$. Gauss's conjecture can be restated as follows: there are infinitely many primes p for which the order of 10 in $\mathbb{Z}_p$ is $p - 1$. E. Artin generalized Gauss's conjecture. He claimed that if b is a positive integer that is not a perfect square, then there are infinitely many primes p for which the b-adic expansion of $1/p$ has period $p - 1$. These are still conjectures (as Gauss's conjecture above), and very celebrated ones at that. Many seemingly simple questions in arithmetic are extremely hard to answer.

See Exercises 4.92 and 4.93 on page 190.

But some things *are* known. For example, Gauss proved in *Disquisitiones* that for any prime p, there is always at least one number (not necessarily 10) whose order in $\mathbb{Z}_p$ is $p - 1$. Such a number is called a *primitive root* mod p.

We now know that the period of $1/n$, where $\gcd(10, n) = 1$, is the order of 10 in $\mathbb{Z}_n$. In Exercise 4.49 on page 165, you did some calculations of orders of units. We can now say a little more.

Theorem 4.63. *If u is a unit in $\mathbb{Z}_n$, then*

$$u^{\phi(n)} = 1$$

where ϕ is the Euler ϕ-function.

Proof. By Proposition 4.39, there are $\phi(n)$ units in $\mathbb{Z}_n$. Suppose we list them all:

$$u_1, u_2, \ldots, u_{\phi(n)}.$$

One of these units is u. Now multiply all these units by u; you get

$$uu_1, \ uu_2 \ \ldots, \ uu_{\phi(n)}.$$

All these elements are units (Exercise 4.44 on page 165), and they are distinct (Exercise 4.89 on page 190). This means that the second list contains *all* the units, perhaps in a different order (they are distinct units, and there are $\phi(n)$ of them). Now multiply all the units together, first using the original order, and then using the permuted order:

$$\prod_{i=1}^{\phi(n)} u_i = \prod_{i=1}^{\phi(n)} u\, u_i = u^{\phi(n)} \prod_{i=1}^{\phi(n)} u_i.$$

But $\prod_{i=1}^{\phi(n)} u_i$ is a unit (Exercise 4.44 again), so you can divide both sides by it, and the result follows.   ∎

Corollary 4.64. *The order of a unit in $\mathbb{Z}_n$ is a factor of $\phi(n)$.*

Proof. Suppose that u is a unit in $\mathbb{Z}_n$ with order e.

Divide $\phi(n)$ by e to get a quotient and remainder:

$$\phi(n) = qe + r \quad 0 \le r < e.$$

Then

$$u^{\phi(n)} = u^{qe+r} = (u^e)^q \, u^r.$$

Now use Theorem 4.63 and the fact that e is the minimal positive exponent such that $u^e = 1$ to conclude that $r = 0$.   ∎

Specializing to $u = 10$, we have

Theorem 4.65. *If n is relatively prime to 10 (that is, if $1/n$ repeats), then the period of $1/n$ is a divisor of $\phi(n)$.*

This greatly reduces the number of possibilities. For example, all we could say about the length of the period of $1/231$ before is that it is at most 230. Now we can say it is a factor of $\phi(231) = 120$. Which one is it?

Example 4.66. We saw earlier, on page 184, that $1/21 = .047619047619\ldots$, so that the period of $1/21$ is 6. Now $\phi(21) = 12$ and, of course, 6 is a divisor of 12. ▲

Proposition 4.63 gives us an added bonus: another proof of Fermat's Little Theorem.

Corollary 4.67. *If p is a prime, then $a^p = a$ in $\mathbb{Z}_p$ for all integers a.*

Proof. As in the proof of Theorem 4.9, we have two cases. If $p \mid a$, then

$$a^p \equiv a \equiv 0 \bmod p,$$

and $a^p = a$ in $\mathbb{Z}_p$.

If $\gcd(a, p) = 1$, then a is a unit in $\mathbb{Z}_p$, and Proposition 4.63 gives

$$a^{\phi(p)} = 1$$

in $\mathbb{Z}_p$. But $\phi(p) = p - 1$, because p is prime. Hence,

$$a^{p-1} = 1.$$

Multiplying both sides by a gives $a^p = a$ in $\mathbb{Z}_p$. ■

As another application, we know that the period of $1/n$ is at most $n - 1$. When is it as large as possible?

Corollary 4.68. *If the period of $1/n$ is $n - 1$, then n is prime.*

Proof. If n is not prime, then $\phi(n) < n - 1$, and the period of $1/n$ is not $n - 1$. ■

How to Think About It. The converse of Corollary 4.68 is not true, as the example of $1/13$ shows. As we said on page 184, it's still an open question about which primes p have the property that the decimal expansion for $1/p$ has maximal period. All we can say is that the decimal expansion is a divisor of $\phi(p) = p - 1$, providing an explanation for the evidence gathered in Figure 4.9.

We have discovered information about periods of repeating rationals; let's now look a bit at their blocks. Before continuing, it's worth working out some other decimal expansions to look for interesting patterns. For example, calculate the decimal expansions of

$$\frac{1}{8}, \quad \frac{2}{3}, \quad \frac{1}{15}, \quad \frac{1}{19}, \quad \frac{2}{19}, \quad \frac{1}{13}, \quad \frac{1}{20}$$

to see whether you can come up with some conjectures for connections between the integers a and b and the blocks in the decimal expansion of a/b. Figure 4.10 displays the digits in the blocks of $k/7$ for $1 \leq k \leq 6$. Is there a way to explain where each rearrangement starts?

$$\frac{1}{7} = .142857142857\ldots$$

$$\frac{2}{7} = .285714285714\ldots$$

$$\frac{3}{7} = .428571428571\ldots$$

$$\frac{4}{7} = .571428571428\ldots$$

$$\frac{5}{7} = .714285714285\ldots$$

$$\frac{6}{7} = .857142857142\ldots$$

Figure 4.10. The expansions of $k/7$ for $1 \leq k < 7$.

There are quite a few patterns here. For example, each block consists of six repeating digits—some "cyclic" permutation of 142857:

142857, 285714, 428571, 571428, 714285, 857142

It's the sequence of *remainders* that explains the various decimal expansions of $k/7$—what they are and why they are in a particular order. For example, in calculating $6/7$, you look down the remainder list and see where you get a 6. The process for $6/7$ will start there, as in Figure 4.11.

The point of Figure 4.11 is that you can "pick up" the calculation at any step in the process—in a way, the calculation of $6/7$ is embedded in the calculation of $1/7$. So are the calculations for all the other $k/7$ for $2 \leq k \leq 5$.

So, the sequence of remainders in a long division provides the key to which decimal expansions can be obtained from the same long division. For the rationals $k/7$, there were six remainders before things started to repeat, so we get all the expansions $1/7, 2/7, \ldots, 6/7$ from one calculation. But it isn't always the case that you get all the expansions for k/n (where $1 \leq k < n$) from the calculation of $1/n$. That only happens when the period for the decimal expansion of $1/n$ has the maximal length $n - 1$ (implying that n is prime). For example, for the various $k/13$, you need two calculations, because the period of the expansion for $1/13$ is 6, not 12.

Earlier, on page 187, we listed the blocks for the various $k/7$, noting that there seemed to be no apparent pattern to where each block starts. In fact, a closer analysis of the long division gives us a way to calculate the digits in each block. Consider again the calculation of the expansion for $6/7$. As before, if

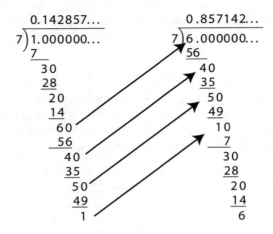

Figure 4.11. The expansion of $6/7$ from that of $1/7$.

we "forget" the decimal point, each new remainder gives the remainder when 6 times a power of 10 is divided by 7. Referring to Figure 4.11, we have

$$6 \equiv 6 \cdot 1 \bmod 7$$
$$4 \equiv 6 \cdot 10 \bmod 7$$
$$5 \equiv 6 \cdot 100 \bmod 7$$
$$1 \equiv 6 \cdot 1000 \bmod 7$$
$$3 \equiv 6 \cdot 10000 \bmod 7$$
$$2 \equiv 6 \cdot 100000 \bmod 7$$
$$6 \equiv 6 \cdot 1000000 \bmod 7.$$

Now, these are the remainders, not the digits in the block. Still, we have an interesting preliminary result.

Lemma 4.69. *Let* $1 \le a < n$, *and suppose that* $\gcd(10, n) = \gcd(a, n) = 1$. *If* e *is the order of* 10 *in* $\mathbb{Z}_n$, *then the* jth *remainder in the long division calculation of* a/n, *where* $0 \le j \le e$, *is the solution* c_j *of the congruence*

$$c_j \equiv a \cdot 10^j \bmod n$$

with $0 \le c_j < n$.

Proof. Imagine dividing n into a with long division, up to j places. Suppose that the remainder is c_j:

$$\frac{.q_1 q_2 q_3 q_4 \ldots q_j}{n \,\big)\, a.\ 0\ 0\ 0\ 0 \ldots\ 0}$$

$$\overline{ c_j.}$$

This says that

$$a = (n \times .q_1\, q_2\, q_3\, q_4 \ldots q_j) + 10^{-j} c_j.$$

Look at the example of $2/13$:
$$2 = (13 \times .1) + 10^{-1} \cdot 7$$
$$2 = (13 \times .15) + 10^{-2} \cdot 5$$
$$2 = (13 \times .13) + 10^{-3} \cdot 11$$
and so on.

Multiply both sides of the equation by 10^j to find

$$a10^j = (n \times q_1 q_2 q_3 q_4 \ldots q_j) + c_j.$$

This says that

$$c_j \equiv a \cdot 10^j \bmod n. \quad \blacksquare$$

What about the digits in the blocks? As in Lemma 4.69, let $\gcd(10, n) = \gcd(a, n) = 1$ and e be the order of 10 in $\mathbb{Z}_n$. Then we know that the eth remainder is a, where

$$a \cdot 10^e \equiv a \bmod n. \tag{4.3}$$

What is the block? Our old friend the Division Algorithm gives the answer:

Theorem 4.70. *Let* $\gcd(10, n) = \gcd(a, n) = 1$. *If* e *is the order of* 10 *in* $\mathbb{Z}_n$, *then the sequence of digits in the block of the decimal expansion for* a/n *is*

$$\frac{a(10^e - 1)}{n}.$$

Proof. The above discussion shows that the block is the partial quotient up to a remainder of a in the division. Rewrite Eq. (4.3) as:

$$a \cdot 10^e = qn + a$$

Solving for q, we have the desired result. $\quad \blacksquare$

Example 4.71. For the various $k/7$,

$$\frac{1}{7} = .142857\ldots \quad \text{and} \quad 1(10^6 - 1)/7 = 142857$$

$$\frac{2}{7} = .285714\ldots \quad \text{and} \quad 2(10^6 - 1)/7 = 285714$$

$$\frac{3}{7} = .428571\ldots \quad \text{and} \quad 3(10^6 - 1)/7 = 428571$$

$$\frac{4}{7} = .571428\ldots \quad \text{and} \quad 4(10^6 - 1)/7 = 571428$$

$$\frac{5}{7} = .714285\ldots \quad \text{and} \quad 5(10^6 - 1)/7 = 714285$$

$$\frac{6}{7} = .857142\ldots \quad \text{and} \quad 6(10^6 - 1)/7 = 857142.$$

Figure 4.12. The blocks of $k/7$ for $1 \le k < 7$.

It's an interesting calculation to go through the same process for the various $k/13$. $\quad \blacktriangle$

Exercises

4.87 Find the order of 10 modulo n (if it exists) for each value of n, and verify that the decimal expansion of $1/n$ has period equal to the order.

(i) 7	(ii) 9	(iii) 3	(iv) 6	(v) 8
(vi) 11	(vii) 13	(viii) 39	(ix) 22	(x) 41
(xi) 73	(xii) 79	(xiii) 123	(xiv) 71	(xv) 61

4.88 * Finish the proof of Corollary 4.62 by showing that the decimal expansion of $1/n$, where $1 \le a < n$ and $\gcd(a, n) = 1$, will repeat only after a remainder of a occurs in the long division of a by n.

4.89 If

$$L = \{u_1, u_2, \ldots, u_{\phi(n)}\}$$

is the list of units in $\mathbb{Z}_n$ and u is any unit, show that the elements of

$$uL = \{uu_1, \ uu_2, \ \ldots, uu_{\phi(n)}\}$$

are all distinct.

4.90 Theorem 4.70 says that if $\gcd(10, n) = \gcd(a, n) = 1$ and e is the order of 10 in $\mathbb{Z}_n$, then the block in the decimal expansion of a/n is $a(10^e - 1)/n$. Why is this latter fraction an integer?

Let $a_0 = a$.

4.91 Suppose that $\gcd(10, n) = \gcd(a, n) = 1$, e is the order of 10 in $\mathbb{Z}_n$, and c_j is the remainder when $a \cdot 10^j$ is divided by n. If the block of a/n is $.a_1 a_2 \ldots a_e$, show, for $1 \le j \le e$, that

$$a_j = \frac{10c_{j-1} - c_j}{n}.$$

4.92 Just as there are b-adic expansions of integers, there are also such expansions for rational numbers. For example, if we are working in base 5, then $1/5 = .1$, $1/5^2 = .01$, $1/5^3 = .001$, and so on. Find rational numbers (written as a/b in the usual way) that are equal to each 5-adic expansion.

(i) .2	(ii) .03	(iii) .1111...
(iv) .171717...	(v) .001001001...	

4.93 Find the 5-adic expansion of each rational number

(i) $\frac{2}{5}$	(ii) $\frac{4}{25}$	(iii) $\frac{1}{4}$
(iv) $\frac{1}{24}$	(v) $\frac{17}{24}$	(vi) $\frac{20}{24}$

4.94 Show that a positive rational number has a terminating b-adic expansion for some positive base b.

4.95 (i) What is the decimal expansion of $1/9801$?

(ii) What is the period of this expansion?

Hint.

$$\frac{10000}{9801} = \frac{1}{\left(1 - \frac{1}{100}\right)^2}$$

4.96 (i) What is the decimal expansion of $1/9899$?

(ii) What is the period of this expansion?

5 Abstract Algebra

Why do mathematics? The answer is simple: we want to understand a corner of our universe. But we are surrounded by so many different things that it makes sense to organize and classify, thereby imposing some order. Naturally, we draw on our experience, so we can decide what we think is important and what is less interesting.

Numbers and calculations have been very useful for thousands of years, and we have chosen to study them. In particular, we have seen that certain arithmetic and geometric ideas help us understand how numbers behave. Sometimes the connections are quite surprising: for example, the relation between Pythagorean triples and the method of Diophantus. We have also developed several tools to facilitate our work: an efficient notation and mathematical induction; the complex numbers and congruences have also enhanced our view.

There are unexpected consequences. As we investigate, we find that even when we find a satisfying answer, new, interesting questions arise. Even though the method of Diophantus explains almost every question we might have about Pythagorean triples, it also suggests that we replace the unit circle by other conic sections, thereby giving insight into some calculus.

It is now time to organize the number theory we have studied. The main idea is to abstract common features of integers, rational numbers, complex numbers, and congruences, as we did when we introduced the definition of commutative ring. This will further our understanding of number theory itself as well as other important topics, such as polynomials.

This chapter continues this adventure. In Section 5.1, we study *domains*, an important class of commutative rings. In Section 5.2, we study polynomials, one of the most important examples of commutative rings. We will show, in particular, that any commutative ring can serve as coefficients in a ring of polynomials. Section 5.3 introduces *homomorphisms*, which allows us to compare and contrast commutative rings, as well as to make precise the idea that two rings have structural similarities.

The rest of the chapter is devoted to the structure of rings of polynomials. Using the results developed in Section 5.3, we'll see how the two main rings in high school mathematics—$\mathbb{Z}$ and polynomials in one variable with coefficients in a field—share many structural similarities. For example, every polynomial has a unique factorization as a product of primes (primes here are called *irreducible polynomials*). And we'll also revisit many of the theorems from advanced high school algebra, like the factor theorem and the fact that polynomials of degree n have at most n roots, putting these results in a more general setting.

5.1 Domains and Fraction Fields

We now introduce a class of commutative rings that satisfy a property enjoyed by our favorite rings: any product of nonzero integers is nonzero. On the other hand, there are commutative rings in which a product of nonzero elements is 0. For example, $2 \times 3 = 0$ in $\mathbb{Z}_6$, even though both 2 and 3 are nonzero. We now promote this property to a definition, for there are interesting examples (e.g., polynomials) where it occurs.

Many texts say *integral domain* instead of *domain*.

Definition. A *domain* D is a nonzero commutative ring in which every product of nonzero elements is nonzero.

A nonzero element a in a commutative ring R is called a *zero divisor* if there is a nonzero $b \in R$ with $ab = 0$. Using this language, we can describe a domain as a commutative ring without zero divisors.

The commutative ring of integers $\mathbb{Z}$ is a domain, but $\mathbb{Z}_m$ is not a domain when m is composite: if $m = ab$ for $0 < a \le b < m$, then $a \ne 0$ and $b \ne 0$, but $ab = m = 0$. Recall the Boolean ring 2^X in Example 4.47: its elements are all the subsets of a set X, and its operations are symmetric difference and intersection. If X has at least two elements, then there are nonempty disjoint subsets A and B; that is, $A \cap B = \varnothing$. Thus, A and B are nonzero elements of 2^X whose product $AB = 0$, and so 2^X is not a domain.

How to Think About It. Everyone believes that $\mathbb{Z}$ is a domain—the product of two nonzero integers is nonzero—but a proof from first principles is surprisingly involved. If you grant that $\mathbb{Z}$ sits inside $\mathbb{R}$, a fact that is a cornerstone of elementary school arithmetic using the "number line representation" of $\mathbb{R}$, and if you grant the fact that $\mathbb{R}$ is a field, then there is a simple proof (see Proposition 5.3). But that's a fair amount of "granting." We'll simply assume that $\mathbb{Z}$ is a domain.

Proposition 5.1. *A nonzero commutative ring D is a domain if and only if it satisfies the* **cancellation law**: *If $ab = ac$ and $a \ne 0$, then $b = c$.*

Proof. Assume that D is a domain. If $ab = ac$ and $a \ne 0$, then $0 = ab - ac = a(b - c)$. Since $a \ne 0$, we must have $b - c = 0$. Hence, $b = c$ and the cancellation law holds.

Conversely, suppose that $ab = 0$, where both a and b are nonzero. Rewrite this as $ab = a0$. Since $a \ne 0$ and the cancellation law holds, we have $b = 0$, a contradiction. Hence, D is a domain. ■

Corollary 5.2. *Every field F is a domain.*

Proof. The cancellation law holds: if $a \in F$ is nonzero and $ab = ac$, then $a^{-1}ab = a^{-1}ac$ and $b = c$. ■

Proposition 5.3. *Every subring S of a field F is a domain.*

Proof. By Corollary 5.2, F is a domain. If $a, b \in S$ are nonzero, then their product (in F, and hence in S) is also nonzero. Hence, S is a domain. ■

For example, if we assume that $\mathbb{R}$ is a field and $\mathbb{Z}$ is a subring of $\mathbb{R}$, then $\mathbb{Z}$ is a domain. The proof of Proposition 5.3 shows more: every subring of a *domain* is a domain.

Fraction Fields

The converse of Proposition 5.3—every domain is a subring of a field—is much more interesting than the proposition. Just as the domain $\mathbb{Z}$ is a subring of the field $\mathbb{Q}$, so, too, is any domain a subring of its *fraction field*. We'll construct such a field containing a given domain using the construction of $\mathbb{Q}$ from $\mathbb{Z}$ as inspiration. This is not mere generalization for generalization's sake; we shall see, for example, that it will show that certain polynomial rings are subrings of fields of rational functions.

How to Think About It. Warning! Over the years, school curricula have tried using the coming discussion to teach fractions to precollege students, even to fourth graders. This is a very bad idea. Experience should precede formalism and, in this particular case, introducing rational numbers as ordered pairs of integers was a pedagogical disaster.

Elementary school teachers often say that $\frac{2}{4} = \frac{3}{6}$ because $2 \times 6 = 4 \times 3$. Sure enough, both products are 12, but isn't this a *non sequitur*? Does it make sense? Why should cross multiplication give equality? Teachers usually continue: suppose you have two pizzas of the same diameter, the first cut into four pieces of the same size, the second into six pieces of the same size; eating two slices of the first pizza is just as filling as eating three slices of the second. This makes more sense, and it tastes better, too. But wouldn't it have been best had the teacher said that if $a/b = c/d$, then multiplying both sides by bd gives $ad = bc$; and, conversely, if $ad = bc$, multiplying both sides by $d^{-1}b^{-1}$ gives $a/b = c/d$?

What is $\frac{1}{2}$? What is a fraction? A fraction is determined by a pair of integers–its numerator and denominator—and so we start with ordered pairs. Let X be the set of all ordered pairs (a, b) of integers with $b \neq 0$ (informally, we are thinking of a/b when we write (a, b)). Define **cross multiplication** to be the relation on X

$$(a, b) \equiv (c, d) \quad \text{if} \quad ad = bc.$$

This is an equivalence relation. It is reflexive: $(a, b) \equiv (a, b)$ because $ab = ba$. It is symmetric: if $(a, b) \equiv (c, d)$, then $(c, d) \equiv (a, b)$ because $ad = bc$ implies $cb = da$. We claim it is transitive: if $(a, b) \equiv (c, d)$ and $(c, d) \equiv (e, f)$, then $(a, b) \equiv (e, f)$. Since $(a, b) \equiv (c, d)$. we have $ad = bc$, so that $adf = bcf$; similarly, $(c, d) \equiv (e, f)$ gives $cf = de$, so that $bcf = bde$. Thus,

$$adf = bcf = bde.$$

Hence, $adf = bde$ and, canceling d (which is not 0), gives $af = be$; that is, $(a, b) \equiv (e, f)$.

How to Think About It. One reason cross multiplication is important is that it converts many problems about fractions into problems about integers.

Lemma 5.4. *If D is a domain and X is the set of all $(a, b) \in D \times D$ with $b \neq 0$, then cross multiplication is an equivalence relation on X.*

Proof. The argument given above for $\mathbb{Z}$ is valid for D. The assumption that D is a domain is present so that we can use the cancellation law to prove transitivity. ∎

Notation. If D is a domain, the equivalence class of $(a, b) \in X \subseteq D \times D$ is denoted by

$$[a, b].$$

Specialize Lemma A.16 in Appendix A.2 to the relation $\equiv$ on X: $[a, b] = [c, d]$ if and only if $(a, b) \equiv (c, d)$; that is, $[a, b] = [c, d]$ if and only if $ad = bc$.

Let's finish the story in the context of arbitrary domains.

Fraction *field*? Stay tuned.

Definition. The ***fraction field*** of a domain D is

$$\operatorname{Frac}(D) = \{[a, b] : a, b \in D \text{ and } b \neq 0\}.$$

How to Think About It. In the back of our minds, we think of $[a, b]$ as the fraction a/b. But, in everyday experience, fractions (especially rational numbers) are used in calculations—they can be added, multiplied, subtracted, and divided. The next theorem equips $\operatorname{Frac}(D)$ with binary operations that will look familiar to you if you keep thinking that $[a, b]$ stands for a/b.

Theorem 5.5. *Let D be a domain.*

(i) $\operatorname{Frac}(D)$ *is a field if we define*

$$[a, b] + [c, d] = [ad + bc, bd] \quad and \quad [a, b][c, d] = [ac, bd].$$

(ii) *The subset D' of $\operatorname{Frac}(D)$, defined by*

$$D' = \{[a, 1] : a \in D\},$$

is a subring of $\operatorname{Frac}(D)$.

(iii) *Every $h \in \operatorname{Frac}(D)$ has the form uv^{-1}, where $u, v \in D'$.*

Proof. (i) Define addition and multiplication on $F = \operatorname{Frac}(D)$ as in the statement. The symbols $[ad + bc, bd]$ and $[ac, bd]$ in the definitions make sense, for $b \neq 0$ and $d \neq 0$ imply $bd \neq 0$, because D is a domain. The proof that F is a field is now a series of routine steps.

We show that addition $F \times F \to F$ is well-defined (i.e., single-valued): if $[a, b] = [a', b']$ and $[c, d] = [c', d']$, then $[ad + bc, bd] = [a'd' + b'c', b'd']$. Now $ab' = a'b$ and $cd' = c'd$. Hence,

$$(ad + bc)b'd' = adb'd' + bcb'd' = (ab')dd' + bb'(cd')$$
$$= a'bdd' + bb'c'd = (a'd' + b'c')bd;$$

that is, $(ad + bc, bd) \equiv (a'd' + b'c', b'd')$, as desired. A similar computation shows that multiplication $F \times F \to F$ is well-defined.

The verification that F is a commutative ring is also routine, and it is left as Exercise 5.5 below, with the hints that the zero element is $[0, 1]$, the identity is $[1, 1]$, and the negative of $[a, b]$ is $[-a, b]$.

To see that F is a field, observe first that if $[a, b] \neq 0$, then $a \neq 0$ (for the zero element of F is $[0, 1] = [0, b]$). We claim that the inverse of $[a, b]$ is $[b, a]$, for $[a, b][b, a] = [ab, ab] = [1, 1]$. Therefore, every nonzero element of F has an inverse in F.

(ii) We show that D' is a subring of F:

$$[1, 1] \in D'$$
$$[a, 1] + [c, 1] = [a + c, 1] \in D'$$
$$[a, 1][c, 1] = [ac, 1] \in D'.$$

(iii) If $h = [a, b]$, where $b \neq 0$, then

$$h = [a, 1][1, b] = [a, 1][b, 1]^{-1}. \quad \blacksquare$$

Notation. From now on, we use standard notation: If D is a domain, then the element $[a, b]$ in $\mathrm{Frac}(D)$ will be denoted by

$$a/b.$$

But be careful: for arbitrary fraction fields, the notation a/b is just an alias for $[a, b]$. For $\mathbb{Q}$, the notation is loaded with all kinds of extra meanings that don't carry over to the general setting (for example, as a number having a decimal expansion obtained by dividing a by b).

Of course, $\mathbb{Q} = \mathrm{Frac}(\mathbb{Z})$. Not surprisingly, elementary school teachers are correct: it is, indeed, true that $a/b = c/d$ if and only if $ad = bc$.

We started this section with two goals: to show that every domain is a subring of a field, and to make precise the notion of "fraction." We've done the second, but we didn't quite show that a domain D is a subring of $\mathrm{Frac}(D)$; instead, we showed that D' is a subring of $\mathrm{Frac}(D)$, where D' consists of all $[a, 1]$ for $a \in D$. Now D and D' do bear a strong resemblance to each other. If we identify each a in D with $[a, 1]$ in D' (which is reminiscent of identifying an integer m with the fraction $m/1$), then not only do elements correspond nicely, but so, too, do the operations: $a + b$ corresponds to $[a + b, 1]$:

$$[a, 1] + [b, 1] = [a \cdot 1 + 1 \cdot b, 1 \cdot 1] = [a + b, 1];$$

similarly, ab corresponds to $[ab, 1] = [a, 1][b, 1]$. In Section 5.3, we will discuss the important idea of *isomorphism* which will make our identification here precise. For the moment, you may regard D and D' as algebraically the same.

Exercises

5.1 Let R be a domain. If $a \in R$ and $a^2 = a$, prove that $a = 0$ or $a = 1$. Compare with Exercise 4.40 on page 164.

5.2 Prove that the Gaussian integers $\mathbb{Z}[i]$ and the Eisenstein integers $\mathbb{Z}[\omega]$ are domains.

5.3 * Prove that $\mathbb{Z}_m$ is a domain if and only if $\mathbb{Z}_m$ is a field. Conclude, using Theorem 4.43, that $\mathbb{Z}_m$ is a domain if and only if m is prime.

5.4 Prove that every finite domain D (i.e., $|D| < \infty$) is a field.

Hint. Use the Pigeonhole Principle, Exercise A.11 on page 419.

5.5 * Complete the proof of Theorem 5.5.

5.6 Let $\mathbb{Q}(i) = \{r + si : r, s \in \mathbb{Q}\}$ be the set of complex numbers whose real and imaginary parts are rational.

(i) Show that $\mathbb{Q}(i)$ is a field.

(ii) True or false? Frac $(\mathbb{Z}[i]) = \{[r + si, 1] : r + si \in \mathbb{Q}(i)\}.$

5.7 *

(i) Show that $\mathbb{Q}(\omega) = \{r + s\omega : r, s \in \mathbb{Q}\}$ is a field, where $\omega = e^{2\pi i/3}$ is a cube root of unity.

(ii) True or false? Frac $(\mathbb{Z}[\omega]) = \{[r + s\omega, 1] : r + s\omega \in \mathbb{Q}(\omega)\}.$ Why?

5.2 Polynomials

You are surely familiar with polynomials; since they can be added and multiplied, it is not surprising that they form commutative rings. However, there are some basic questions about them whose answers may be less familiar. Is a polynomial a function? Is x a variable? If not, just what is x? After all, we first encounter polynomials as real-valued functions having simple formulas; for example, $f(x) = x^3 - 2x^2 + 7$ is viewed as the function $f^{\#}: \mathbb{R} \to \mathbb{R}$ defined by $f^{\#}(a) = a^3 - 2a^2 + 7$ for every $a \in \mathbb{R}$. But some polynomials have complex coefficients. Is it legitimate to consider polynomials whose coefficients lie in any commutative ring R? When are two polynomials equal? Every high school algebra student would say that the functions defined by $f(x) = x^7 + 2x - 1$ and $g(x) = 3x + 6$ are not the same, because they are defined by different polynomials. But these two functions are, in fact, equal when viewed as functions $\mathbb{Z}_7 \to \mathbb{Z}_7$, a fact that you can check by direct calculation. Here's another example. Is it legitimate to treat $2x + 1$ as a polynomial whose coefficients lie in $\mathbb{Z}_4$? If so, then $(2x + 1)^2 = 4x^2 + 4x + 1 = 1$ (for $4 = 0$ in $\mathbb{Z}_4$); that is, the square of this linear polynomial is a constant! Sometimes polynomials are treated as formal expressions in which x is just a symbol, as, for example, when you factor $x^6 - 1$ or expand $(x + 1)^5$. And sometimes polynomials are treated as functions that can be graphed or composed. Both of these perspectives are important and useful, but they are clearly different.

We now introduce polynomials rigorously, for this will enable us to answer these questions. In this section, we'll first study polynomials from the formal viewpoint, after which we'll consider polynomial functions. In the next section, we will see that the notion of *homomorphism* will link the formal and the function viewpoints, revealing their intimate connection.

How to Think About It. As we said on page 193 in the context of fractions, rigorous developments should not be points of entry. One goal of this section is to put polynomials on a firm footing. This will prepare you for any future work you do with beginning algebra students, but it is in no way meant to take the place of all of the informal experience that's necessary before the formalities can be appreciated and understood.

We investigate polynomials in a very formal way, beginning with the allied notion of power series. A key observation is that one should pay attention to where the coefficients of polynomials live.

Definition. If R is a commutative ring, then a *formal power series* over R is a sequence

$$\sigma = (s_0, s_1, s_2, \ldots, s_i, \ldots);$$

the entries $s_i \in R$ are called the *coefficients* of σ.

Be patient. The reason for this terminology will be apparent in a few pages. In the meantime, pretend that $(s_0, s_1, s_2, \ldots, s_i, \ldots)$ is really $s_0 + s_1 x + s_2 x^2 + \cdots + s_i x^i + \cdots$.

A formal power series σ over R is a sequence, but a sequence is just a function $\sigma \colon \mathbb{N} \to R$ (where $\mathbb{N}$ is the set of natural numbers) with $\sigma(i) = s_i$ for all $i \geq 0$. By Proposition A.2 in Appendix A.1, two sequences σ and τ are equal if and only if $\sigma(i) = \tau(i)$ for all $i \in \mathbb{N}$. So, formal power series are equal if and only if they are equal "coefficient by coefficient."

> In linear algebra, you may have seen the example of the vector space V of all polynomials of degree, say, 3 or less, with coefficients in $\mathbb{R}$. As a vector space, V can be thought of as $\mathbb{R}^4$, where the 4-tuple $(5, 6, 8, 9)$ corresponds to the polynomial $5 + 6x + 8x^2 + 9x^3$.

Proposition 5.6. *Formal power series* $\sigma = (s_0, s_1, s_2, \ldots, s_i, \ldots)$ *and* $\tau = (t_0, t_1, t_2, \ldots, t_i, \ldots)$ *over a commutative ring R are equal if and only if* $s_i = t_i$ *for all* $i \geq 0$.

How to Think About It. Discussions of power series in calculus usually involve questions asking about those values of x for which $s_0 + s_1 x + s_2 x^2 + \cdots$ converges. In most commutative rings, however, limits are not defined, and so, in general, convergence of formal power series does not even make sense. Now the definition of formal power series is not very complicated, while limits are a genuinely new and subtle idea (it took mathematicians around 200 years to agree on a proper definition). Since power series are usually introduced at the same time as limits, however, most calculus students (and ex-calculus students!) are not comfortable with them; the simple notion of power series is entangled with the sophisticated notion of limit.

Today's calculus classes do not follow the historical development. Calculus was invented to answer a practical need; in fact, the word *calculus* arose because it described a branch of mathematics involving or leading to calculations. In the 1600s, navigation on the high seas was a matter of life and death, and practical tools were necessary for the safety of boats crossing the oceans. One such tool was calculus, which is needed in astronomical calculations. Newton realized that his definition of integral was complicated; telling a navigator that the integral of a function is some fancy limit of approximations and fluxions would be foolish. To make calculus useful, he introduced power series (Newton discovered the usual power series for $\sin x$ and $\cos x$), he assumed that most integrands occurring in applications have a power series expansion, and he further assumed that term-by-term integration was valid for them. Thus, power series were actually introduced as "long polynomials" in order to simplify using calculus in applications.

Polynomials are special power series.

Definition. A *polynomial* over a commutative ring R is a formal power series $\sigma = (s_0, s_1, \ldots, s_i, \ldots)$ over R for which there exists some integer $n \geq 0$ with $s_i = 0$ for all $i > n$; that is,

$$\sigma = (s_0, s_1, \ldots, s_n, 0, 0, \ldots).$$

The *zero polynomial*, denoted by $\sigma = 0$, is the sequence $\sigma = (0, 0, 0, \dots)$.

A polynomial has only finitely many nonzero coefficients; that is, it is a "short power series."

Some authors define
the degree of the zero
polynomial 0 to be $-\infty$,
where $-\infty + n = -\infty$ for
every integer $n \in \mathbb{N}$ (this
is sometimes convenient).
We choose not to assign
a degree to 0 because,
in proofs, it often must
be treated differently than
other polynomials.

Definition. If $\sigma = (s_0, s_1, \dots, s_n, 0, 0, \dots)$ is a nonzero polynomial, then there is $n \geq 0$ with $s_n \neq 0$ and $s_i = 0$ for all $i > n$. We call s_n the **leading coefficient** of σ, we call n the **degree** of σ, and we denote the degree by $n = \deg(\sigma)$.

The zero polynomial 0 does not have a degree because it has no nonzero coefficients.

Etymology. The word *degree* comes from the Latin word meaning "step." Each term $s_i x^i$ (in the usual notation $s_0 + s_1 x + s_2 x^2 + \cdots + s_i x^i + \cdots$) has degree i, and so the degrees suggest a staircase.

The word *coefficient* means "acting together to some single end." Here, coefficients collectively give one formal power series or one polynomial.

Notation. If R is a commutative ring, then

$$R[[x]]$$

denotes the set of all formal power series over R, and

$$R[x] \subseteq R[[x]]$$

denotes the set of all polynomials over R.

We want to make $R[[x]]$ into a commutative ring, and so we define addition and multiplication of formal power series. Suppose that

$$\sigma = (s_0, s_1, \dots, s_i, \dots) \quad \text{and} \quad \tau = (t_0, t_1, \dots, t_i, \dots).$$

Define their sum by adding term by term:

$$\sigma + \tau = (s_0 + t_0, s_1 + t_1, \dots, s_i + t_i, \dots).$$

What about multiplication? The product of two power series is also computed term by term; multiply formally and collect like powers of x:

$$(s_0 + s_1 x + s_2 x^2 + \cdots + s_i x^i + \cdots)(t_0 + t_1 x + t_2 x^2 + \cdots + t_j x^j + \cdots)$$
$$= s_0(t_0 + t_1 x + t_2 x^2 + \cdots) + s_1 x(t_0 + t_1 x + t_2 x^2 + \cdots) + \cdots$$
$$= (s_0 t_0 + s_0 t_1 x + s_0 t_2 x^2 + \cdots) + (s_1 t_0 x + s_1 t_1 x^2 + s_1 t_2 x^3 + \cdots) + \cdots$$
$$= s_0 t_0 + (s_1 t_0 + s_0 t_1)x + (s_0 t_2 + s_1 t_1 + s_2 t_0)x^2 + \cdots.$$

Motivated by this, we define multiplication of formal power series by

$$\sigma\tau = (s_0 t_0, \; s_0 t_1 + s_1 t_0, \; s_0 t_2 + s_1 t_1 + s_2 t_0, \; \dots);$$

more precisely,

$$\sigma\tau = (c_0, c_1, \dots, c_k, \dots),$$

where $c_k = \sum_{i+j=k} s_i t_j = \sum_{i=0}^{k} s_i t_{k-i}$.

Proposition 5.7. *If R is a commutative ring, then $R[[x]]$, together with the operations of addition and multiplication defined above, is a commutative ring.*

Proof. Addition and multiplication are operations on $R[[x]]$: the sum and product of two formal power series are also formal power series. Define *zero* to be the zero polynomial, define the identity to be the polynomial $(1, 0, 0, \dots)$, and define the negative of $(s_0, s_1, \dots, s_i, \dots)$ to be $(-s_0, -s_1, \dots, -s_i, \dots)$. Verifications of the axioms of a commutative ring are routine, and we leave them as Exercise 5.8 on page 202. The only difficulty that might arise is proving the associativity of multiplication. Hint: if $\rho = (r_0, r_1, \dots, r_i, \dots)$, then the ℓth coordinate of the polynomial $\rho(\sigma\tau)$ turns out to be $\sum_{i+j+k=\ell} r_i(s_j t_k)$, while the ℓth coordinate of the power series $(\rho\sigma)\tau$ turns out to be $\sum_{i+j+k=\ell}(r_i s_j)t_k$; these are equal because associativity of multiplication in R gives $r_i(s_j t_k) = (r_i s_j)t_k$ for all i, j, k. ∎

We'll see in a moment that the subset $R[x]$ of polynomials is a subring of the commutative ring of formal power series $R[[x]]$.

Lemma 5.8. *Let R be a commutative ring and $\sigma, \tau \in R[x]$ be nonzero polynomials.*

(i) *Either $\sigma\tau = 0$ or $\deg(\sigma\tau) \le \deg(\sigma) + \deg(\tau)$.*

(ii) *If R is a domain, then $\sigma\tau \ne 0$ and*

$$\deg(\sigma\tau) = \deg(\sigma) + \deg(\tau).$$

Proof. Let $\sigma = (s_0, s_1, \dots)$ have degree m, let $\tau = (t_0, t_1, \dots)$ have degree n, and let $\sigma\tau = (c_0, c_1, \dots)$.

(i) It suffices to prove that $c_k = 0$ for all $k > m + n$. By definition,

$$c_k = s_0 t_k + \cdots + s_m t_{k-m} + s_{m+1} t_{k-m-1} + \cdots s_k t_0.$$

All terms to the right of $s_m t_{k-m}$ are 0, because $\deg(\sigma) = m$, and so $s_i = 0$ for all $i \ge m + 1$. Now $s_m t_{k-m}$, as well as all the terms to its left, are 0, because $\deg(\tau) = n$, and so $t_j = 0$ for all $j \ge k - m > n$.

(ii) We claim that $c_{m+n} = s_m t_n$, the product of the leading coefficients of σ and τ. Now

$$c_{m+n} = \sum_{i+j=m+n} s_i t_j$$

$$= s_0 t_{m+n} + \cdots + s_{m-1} t_{n+1} + s_m t_n + s_{m+1} t_{n-1} + \cdots.$$

We show that every term $s_i t_j$ in c_{m+n}, other than $s_m t_n$, is 0. If $i < m$, then $m - i > 0$; hence, $j = m - i + n > n$, and so $t_j = 0$; that is, each term to the left of $s_m t_n$ is 0. If $i > m$, then $s_i = 0$, and each term to the right of $s_m t_n$ is 0. Therefore,

$$c_{m+n} = s_m t_n.$$

If R is a domain, then $s_m \ne 0$ and $t_n \ne 0$ imply $s_m t_n \ne 0$; hence, $c_{m+n} = s_m t_n \ne 0$, $\sigma\tau \ne 0$, and $\deg(\sigma\tau) = m + n$. ∎

Exercise 5.22 on page 203 shows that if R is a domain, then $R[[x]]$ is a domain.

Corollary 5.9.   (i) *If R is a commutative ring, then $R[x]$ and R are subrings of $R[[x]]$.*

(ii) *If R is a domain, then $R[x]$ is a domain.*

See Exercise 5.9 on page 202.

Proof.   (i) Let $\sigma, \tau \in R[x]$. Now $\sigma + \tau$ is a polynomial, for either $\sigma + \tau = 0$ or $\deg(\sigma + \tau) \leq \max\{\deg(\sigma), \deg(\tau)\}$. By Lemma 5.8(i), the product of two polynomials is also a polynomial. Finally, $1 = (1, 0, 0, \ldots)$ is a polynomial, and so $R[x]$ is a subring of $R[[x]]$.

It is easy to check that $R' = \{(r, 0, 0, \ldots) : r \in R\}$ is a subring of $R[x]$, and we may view R' as R by identifying $r \in R$ with $(r, 0, 0, \ldots)$.

(ii) If σ and τ are nonzero polynomials, then Lemma 5.8(ii) shows that $\sigma\tau \neq 0$. Therefore, $R[x]$ is a domain.   ∎

We remark that R can't be a subring of $R[x]$ or of $R[[x]]$ because it's not even a subset of these rings. This is why we have introduced the subring R'. A similar thing happened when we couldn't view a domain D as a subring of its fraction field $\text{Frac}(D)$. We shall return to this point when we discuss isomorphisms.

From now on, we view $R[x]$ and $R[[x]]$ as rings, not merely as sets.

Definition.   If R is a commutative ring, then $R[x]$ is called the ***ring of polynomials over*** R, and $R[[x]]$ is called the ***ring of formal power series over*** R.

Here is the link between this discussion and the usual notation.

Definition.   The ***indeterminate*** x is the element

$$x = (0, 1, 0, 0, \ldots) \in R[x].$$

How to Think About It.   Thus, x is neither "the unknown" nor a variable; it is a specific element in the commutative ring $R[x]$, namely, the polynomial $(a_0, a_1, a_2, \ldots)$ with $a_1 = 1$ and all other $a_i = 0$; it is a polynomial of degree 1.

Note that we need the unit 1 in a commutative ring R in order to define the indeterminate in $R[x]$.

Lemma 5.10.   *Let R be a commutative ring.*

(i) *If $\sigma = (s_0, s_1, \ldots, s_j, \ldots) \in R[[x]]$, then*

$$x\sigma = (0, s_0, s_1, \ldots, s_j, \ldots);$$

that is, multiplying by x shifts each coefficient one step to the right.

(ii) *If $n \geq 0$, then x^n is the polynomial having 0 everywhere except for 1 in the nth coordinate.*

(iii) *If $r \in R$ and $(s_0, s_1, \ldots, s_j, \ldots) \in R[[x]]$, then*

$$(r, 0, 0, \ldots)(s_0, s_1, \ldots, s_j, \ldots) = (rs_0, rs_1, \ldots, rs_j, \ldots).$$

Proof. (i) Write $x = (a_0, a_1, \ldots, a_i, \ldots)$, where $a_1 = 1$ and all other $a_i = 0$, and let $x\sigma = (c_0, c_1, \ldots, c_k, \ldots)$. Now $c_0 = a_0 s_0 = 0$, because $a_0 = 0$. If $k \geq 1$, then the only nonzero term in the sum $c_k = \sum_{i+j=k} a_i s_j$ is $a_1 s_{k-1} = s_{k-1}$, because $a_i = 0$ for $i \neq 1$; thus, for $k \geq 1$, the kth coordinate c_k of $x\sigma$ is s_{k-1}, and $x\sigma = (0, s_0, s_1, \ldots, s_i, \ldots)$.

(ii) Use induction and part (i).

(iii) This follows from the definition of multiplication. ■

If we identify $(r, 0, 0, \ldots)$ with r, as in the proof of Corollary 5.9, then Lemma 5.10(iii) reads

See Exercise 5.11 on page 202.

$$r(s_0, s_1, \ldots, s_i, \ldots) = (rs_0, rs_1, \ldots, rs_i, \ldots).$$

We can now recapture the usual polynomial notation.

Proposition 5.11. *Let R be a commutative ring. If $\sigma = (s_0, s_1, \ldots, s_n, 0, 0, \ldots) \in R[x]$ has degree n, then*

$$\sigma = s_0 + s_1 x + s_2 x^2 + \cdots + s_n x^n,$$

where each element $s \in R$ is identified with the polynomial $(s, 0, 0, \ldots)$. Moreover, if $\tau = t_0 + t_1 x + t_2 x^2 + \cdots + t_m x^m$, then $\sigma = \tau$ if and only if $n = m$ and $s_i = t_i$ for all $i \geq 0$.

So, two polynomials are equal in $R[x]$ if and only if they are equal "term by term."

Proof.

$$\begin{aligned}
\sigma &= (s_0, s_1, \ldots, s_n, 0, 0, \ldots) \\
&= (s_0, 0, 0, \ldots) + (0, s_1, 0, \ldots) + \cdots + (0, 0, \ldots, 0, s_n, 0, \ldots) \\
&= s_0(1, 0, 0, \ldots) + s_1(0, 1, 0, \ldots) + \cdots + s_n(0, 0, \ldots, 0, 1, 0, \ldots) \\
&= s_0 + s_1 x + s_2 x^2 + \cdots + s_n x^n.
\end{aligned}$$

The second statement merely rephrases Proposition 5.6, equality of polynomials, in terms of the usual notation. ■

We shall use this familiar (and standard) notation from now on. As is customary, we shall write

$$f(x) = s_0 + s_1 x + s_2 x^2 + \cdots + s_n x^n$$

instead of $\sigma = (s_0, s_1, \ldots, s_n, 0, 0, \ldots)$.

Corollary 5.12. *If R is a commutative ring, then the polynomial ring $R[x]$ is infinite.*

Proof. By Proposition 5.11, $x^i \neq x^j$ if $i \neq j$. ■

If $f(x) = s_0 + s_1 x + s_2 x^2 + \cdots + s_n x^n$, where $s_n \neq 0$, then s_0 is called its **constant term** and, as we have already said, s_n is called its **leading coefficient**. If its leading coefficient $s_n = 1$, then $f(x)$ is called **monic**. Every polynomial other than the zero polynomial 0 (having all coefficients 0) has a degree. A **constant polynomial** is either the zero polynomial or a polynomial of degree 0. Polynomials of degree 1, namely $a + bx$ with $b \neq 0$, are called **linear**, polynomials of degree 2 are **quadratic**, degree 3s are **cubic**, then **quartics**, **quintics**, and so on.

Etymology. Quadratic polynomials are so called because the particular quadratic x^2 gives the area of a square (*quadratic* comes from the Latin word meaning *four*, which reminds us of the 4-sided figure); similarly, cubic polynomials are so called because x^3 gives the volume of a cube. Linear polynomials are so called because the graph of a linear polynomial $ax + b$ in $\mathbb{R}[x]$ is a line.

Exercises

5.8 * Fill in the details and complete the proof of Proposition 5.7.

5.9 * Suppose that R is a commutative ring. In the proof of Corollary 5.9(i), we defined R' as the set of all power series of the form $(r, 0, 0, 0, \dots)$ where $r \in R$, and we said "we may view R' as R by identifying $r \in R$ with $(r, 0, 0, \dots)$." Show, if $r, s \in \mathbb{R}$, that

(i) $r + s$ is identified with $(r, 0, 0, 0, \dots) + (s, 0, 0, 0, \dots)$

(ii) rs is identified with $(r, 0, 0, 0, \dots)(s, 0, 0, 0, \dots)$.

5.10 If $(t_0, t_1, t_2, \dots)$ is a power series over R and $r \in R$, show that

$$(r, 0, 0, 0, \dots)(t_0, t_1, t_2, \dots) = (rt_0, rt_1, rt_2, \dots).$$

5.11 * Suppose that F is a field. Show that $F[[x]]$ is a vector space over F where addition is defined as addition of power series and scalar multiplication is defined by

$$r(s_0, s_1, s_2, \dots) = (rs_0, rs_1, rs_2, \dots).$$

Vector spaces over arbitrary fields are discussed in Appendix A.3.

5.12 If R is the zero ring, what are $R[x]$ and $R[[x]]$? Why?

5.13 Prove that if R is a commutative ring, then $R[x]$ is never a field.

Hint. If x^{-1} exists, what is its degree?

5.14 (i) Let R be a domain. Prove that if a polynomial in $R[x]$ is a unit, then it is a nonzero constant (the converse is true if R is a field).

Hint. Compute degrees.

(ii) Show that $(2x + 1)^2 = 1$ in $\mathbb{Z}_4[x]$. Conclude that $2x + 1$ is a unit in $\mathbb{Z}_4[x]$, and that the hypothesis in part (i) that R be a domain is necessary.

5.15 * If R is a commutative ring and

$$f(x) = s_0 + s_1 x + s_2 x^2 + \cdots + s_n x^n \in R[x]$$

has degree $n \geq 1$, define its formal **derivative** $f'(x) \in R[x]$ by

$$f'(x) = s_1 + 2s_2 x + 3s_3 x^2 + \cdots + ns_n x^{n-1};$$

if f is a constant polynomial, define its derivative to be the zero polynomial. Prove that the usual rules of calculus hold for derivatives in $R[x]$:

$$(f + g)' = f' + g'$$
$$(rf)' = r(f') \quad \text{if } r \in R$$
$$(fg)' = fg' + f'g$$
$$[1pt](f^n)' = nf^{n-1}f' \quad \text{for all } n \geq 1.$$

5.16 Take It Further. Define $\int \colon \mathbb{Q}[x] \to \mathbb{Q}[x]$ by

$$\int f = a_0 x + \tfrac{1}{2} a_1 x^2 + \cdots + \tfrac{1}{n+1} a_n x^{n+1} \in \mathbb{Q}[x],$$

where $f(x) = a_0 + a_1 x + \cdots + a_n x^n \in \mathbb{Q}[x]$.

(i) Prove that $\int (f + g) = \int f + \int g$.

(ii) If D is the derivative, prove that $D \int = 1_{\mathbb{Q}[x]}$, but that $\int D \neq 1_{\mathbb{Q}[x]}$.

$1_{\mathbb{Q}[x]}$ denotes the identity function on the set $\mathbb{Q}[x]$. Why didn't we define $\int \colon R[x] \to R[x]$ for any commutative ring R?

5.17 *Preview. Let R be a commutative ring, let $f(x) \in R[x]$, and let $f'(x)$ be its derivative.

(i) Prove that if $(x - a)^2$ is a divisor of f in $R[x]$, then $x - a$ is a divisor of f' in $R[x]$.

(ii) Prove that if $x - a$ is a divisor of both f and f', then $(x - a)^2$ is a divisor of f.

5.18 (i) If $f(x) = a x^{2p} + b x^p + c \in \mathbb{Z}_p[x]$, prove that $f'(x) = 0$.

(ii) Prove that a polynomial $f(x) \in \mathbb{Z}_p[x]$ has $f'(x) = 0$ if and only if there is a polynomial $g(x) = \sum a_n x^n$ with $f(x) = g(x^p)$; that is, $f(x) = \sum a_n x^{np}$.

5.19 If p is a prime, show, in $\mathbb{Z}_p[x]$, that

$$(x + 1)^p = x^p + 1.$$

5.20 *

(i) If R is a domain and $\sigma = 1 + x + x^2 + \cdots + x^n + \cdots \in R[[x]]$, prove that σ is a unit in $R[[x]]$; in fact, $(1 - x)\sigma = 1$.

(ii) Show that $(1 - x)^2$ is a unit in $\mathbb{Q}[[x]]$, and express $1/(1 - x)^2$ as a power series.

Hint. See Exercise 5.22 below.

5.21 Show that $1 - x - x^2$ is a unit in $\mathbb{Q}[[x]]$, and express $1/(1 - x - x^2)$ as a power series.

5.22 *

(i) Prove that if R is a domain, then $R[[x]]$ is a domain.

Hint. If $\sigma = (s_0, s_1, \dots) \in R[[x]]$ is nonzero, define the ***order*** of σ, denoted by $\operatorname{ord}(\sigma)$, to be the smallest $n \geq 0$ for which $s_n \neq 0$. If R is a domain and $\sigma, \tau \in R[[x]]$ are nonzero, prove that $\operatorname{ord}(\sigma\tau) \geq \operatorname{ord}(\sigma) + \operatorname{ord}(\tau)$, and use this to conclude that $\sigma\tau \neq 0$.

(ii) Let k be a field. Prove that a formal power series $\sigma \in k[[x]]$ is a unit if and only if its constant term is nonzero; that is, $\operatorname{ord}(\sigma) = 0$.

(iii) Prove that if $\sigma \in k[[x]]$ and $\operatorname{ord}(\sigma) = n$, then $\sigma = x^n u$, where u is a unit in $k[[x]]$.

5.23 *

(i) Prove that $\operatorname{Frac}(\mathbb{Z}[x]) = \mathbb{Q}(x)$.

(ii) Let D be a domain with $K = \operatorname{Frac}(D)$. Prove that $\operatorname{Frac}(D[x]) = K(x)$.

5.24 (i) Expand $(C^2 + S^2 - 1)(S^2 + 2CS - C^2)$, where C and S are elements in some commutative ring.

(ii) Establish the trigonometric identity

$$\cos^2 x + 2 \cos^3 x \, \sin x + 2 \cos x \, \sin^3 x + \sin^4 x =$$
$$\cos^4 x + 2 \cos x \, \sin x + \sin^2 x.$$

5.25 Preview. Suppose p is a prime and

$$f_p(x) = \frac{x^p - 1}{x - 1}.$$

(i) Show that $f_p(x) = x^{p-1} + x^{p-2} + \cdots + 1$.

(ii) Show that $f_p(x + 1) = x^p$ in $\mathbb{Z}_p[x]$.

Polynomial Functions

Let's now pass to viewing polynomials as functions. Each polynomial $f(x) = s_0 + s_1 x + s_2 x^2 + \cdots + s_n x^n \in R[x]$ defines its *associated polynomial function* $f^{\#} \colon R \to R$ by *evaluation*:

$$f^{\#}(a) = s_0 + s_1 a + s_2 a^2 + \cdots + s_n a^n \in R,$$

where $a \in R$ (in this way, we can view the indeterminate x as a variable). But polynomials and polynomial functions are different things. For example, Corollary 5.12 says, for every commutative ring R, that there are infinitely many polynomials in $R[x]$. On the other hand, if R is finite (e.g., $R = \mathbb{Z}_m$), then there are only finitely many functions from R to itself, and so there are only finitely many polynomial functions. Fermat's Theorem ($a^p \equiv a \bmod p$ for every prime p) gives a concrete example of distinct polynomials defining the same polynomial function; $f(x) = x^p - x$ is a nonzero polynomial, yet its associated polynomial function $f^{\#} \colon \mathbb{Z}_p \to \mathbb{Z}_p$ is the constant function zero.

In Proposition 6.18, we will see that there's a bijection between polynomials and their associated polynomial functions if R is an infinite field.

Recall Example 4.31: if R is a commutative ring, then $\mathrm{Fun}(R) = R^R$, the set of all functions from R to itself, is a commutative ring under pointwise operations. We have seen that every polynomial $f(x) \in R[x]$ has an associated polynomial function $f^{\#} \in \mathrm{Fun}(R)$, and we claim that

$$\mathrm{Poly}(R) = \{f^{\#} : f(x) \in R[x]\}$$

is a subring of $\mathrm{Fun}(R)$ (we admit that we are being very pedantic, but you will see in the next section that there's a good reason for this fussiness). The identity u of R^R is the constant function with value 1, where 1 is the identity element of R; that is, $u = 1^{\#}$, where 1 is the constant polynomial. We claim that if $f(x), g(x) \in R[x]$, then

$$f^{\#} + g^{\#} = (f + g)^{\#} \quad \text{and} \quad f^{\#} g^{\#} = (fg)^{\#}.$$

(In the equation $f^{\#} + g^{\#} = (f + g)^{\#}$, the plus sign on the left means addition of functions, while the plus sign on the right means the usual addition of polynomials in $R[x]$; a similar remark holds for multiplication.) The proof of these equations is left as Exercise 5.27 on page 206.

Etymology. In spite of the difference between polynomials and polynomial functions, $R[x]$ is often called the ring of all **polynomials in one variable over** R.

Since $k[x]$ is a domain when k is a field, by Corollary 5.9(ii), it has a fraction field.

Definition. If k is a field, then the fraction field $\text{Frac}(k[x])$ of $k[x]$, denoted by

$$k(x),$$

is called the *field of rational functions* over k.

We can define $R(x)$ for arbitrary domains R. See Exercise 5.23 on page 203.

How to Think About It. By convention, the elements of $k(x)$ are called rational "functions" but they are simply elements of the fraction field for $k[x]$. Of course, a rational function can be viewed as an actual function via evaluation at elements of k, in the same way that a polynomial in $k[x]$ gives rise to its associated polynomial function defined on k. But the domain of such a rational function may not be all of k (why?).

We'll use the standard notation for elements in fraction fields (introduced on page 195) for rational functions over a field: $[f, g]$ will be denoted by f/g.

Proposition 5.13. *If p is prime, then the field of rational functions $\mathbb{Z}_p(x)$ is an infinite field containing $\mathbb{Z}_p$ as a subfield.*

Proof. By Corollary 5.12, $\mathbb{Z}_p[x]$ is infinite, because the powers x^n, for $n \in \mathbb{N}$, are distinct. Thus, its fraction field, $\mathbb{Z}_p(x)$, is an infinite field containing $\mathbb{Z}_p[x]$ as a subring. But $\mathbb{Z}_p[x]$ contains $\mathbb{Z}_p$ as a subring, by Corollary 5.9. ∎

Well, $\mathbb{Z}_p(x)$ contains a domain with a "strong resemblance" to $\mathbb{Z}_p[x]$ (see page 195). We'll make this precise in Section 5.3.

Notation. We've been using $\mathbb{Z}_p$ to stand for the integers mod p, and we know that it is a field. There are other finite fields that do not have a prime number of elements; you met one, $\mathbb{F}_4$, in Exercise 4.55 on page 165. It's customary to denote a field with q elements by

$$\mathbb{F}_q.$$

It turns out that $q = p^n$ for some prime p and some $n \geq 1$. Moreover, there exists essentially only one field with q elements, a fact we'll prove in Chapter 7. In particular, there is only one field with exactly p elements and so, from now on, we'll use the notations

$$\mathbb{Z}_p = \mathbb{F}_p$$

interchangeably (we'll use $\mathbb{F}_p$ when we're viewing it as a field).

Let's now consider polynomials over R in two variables x and y. A quadratic polynomial $ax^2 + bxy + cy^2 + dx + ey + f$ can be rewritten as

$$ax^2 + (by + d)x + (cy^2 + ey + f);$$

that is, it is a polynomial in x with coefficients in $R[y]$. If we write $A = R[y]$, then it is clear that $A[x]$ is a commutative ring.

Definition. If R is a commutative ring, then $R[x, y] = A[x]$, where $A = R[y]$, is the ring of all *polynomials over R in two variables*.

By induction, we can form the commutative ring $R[x_1, x_2, \ldots, x_n]$ of all *polynomials in n variables* over R:

$$R[x_1, x_2, \ldots, x_{n+1}] = \big(R[x_1, x_2, \ldots, x_n]\big)[x_{n+1}].$$

Corollary 5.9 can now be generalized, by induction on n, to say that if D is a domain, then so is $D[x_1, x_2, \ldots, x_n]$; we call $\text{Frac}(D[x_1, x_2, \ldots, x_n])$ the ring of *rational functions in n variables*. Exercise 5.23 on page 203 can be generalized to several variables: if $K = \text{Frac}(D)$, then

$$\text{Frac}(D[x_1, x_2, \ldots, x_n]) = K(x_1, x_2, \ldots, x_n);$$

its elements have the form f/g, where $f, g \in K[x_1, x_2, \ldots, x_n]$ and $g \neq 0$.

Exercises

5.26 Let R be a commutative ring. Show that if two polynomials $f(x), g(x) \in R[x]$ are equal, then their associated polynomial functions are equal; that is, $f^{\#} = g^{\#}$.

5.27 * If R is a commutative ring, prove that $\text{Poly}(R)$ is a subring of $\text{Fun}(R) = R^R$.

5.28 True or false, with reasons:

(i) $(x^2 - 9)/(x^2 - 2x - 3) = (x + 3)/(x + 1)$ in $\mathbb{Q}(x)$.

(ii) What are the domains of the functions $x \mapsto (x^2 - 9)/(x^2 - 2x - 3)$ and $x \mapsto (x + 3)/(x + 1)$? Are the functions equal?

5.3 Homomorphisms

The question whether two given commutative rings R and S are somehow the same has already arisen, at least twice.

(i) On page 195 we said

For the moment, you may regard D and D' as algebraically the same.

(ii) And on page 201 we said

If we identify $(r, 0, 0, \ldots)$ with r, then Lemma 5.10(iii) *reads*

$$r(s_0, s_1, \ldots, s_i, \ldots) = (r s_0, r s_1, \ldots, r s_i, \ldots).$$

What does "the same" mean in statement (i)? What does "identify" mean in statement (ii)? More important, if R is a commutative ring, we wish to compare the (formal) polynomial ring $R[x]$ with the ring $\text{Poly}(R)$ of all polynomial functions on R.

We begin our discussion by considering the ring $\mathbb{Z}_2$; it has two elements, the congruence classes 0, 1, and the following addition and multiplication tables.

+	0	1
0	0	1
1	1	0

×	0	1
0	0	0
1	0	1

The two words *even*, *odd* also form a commutative ring, call it $\mathcal{P}$; its addition and multiplication are pictured in the following tables.

+	even	odd
even	even	odd
odd	odd	even

×	even	odd
even	even	even
odd	even	odd

Thus, odd + odd = even and odd × odd = odd. It is clear that the commutative rings $\mathbb{Z}_2$ and $\mathcal{P}$ are distinct; on the other hand, it is equally clear that there is no significant difference between them. The elements of $\mathbb{Z}_2$ are given in terms of numbers; those of $\mathcal{P}$ in terms of words. We may think of $\mathcal{P}$ as a translation of $\mathbb{Z}_2$ into another language. And more than just a correspondence of elements, the operations of addition and multiplication (that is, the two tables) get translated, too.

A reasonable way'to compare two systems is to set up a function between them that preserves certain essential structural properties (we hinted at this idea earlier when we noted that a ring R is essentially a subring of $R[x]$). The notions of homomorphism and isomorphism will make this intuitive idea precise. Here are the definitions; we will discuss what they mean afterward.

Definition. Let R and S be commutative rings. A **homomorphism** is a function $\varphi : R \to S$ such that, for all $a, b \in R$,

(i) $\varphi(a + b) = \varphi(a) + \varphi(b)$,

(ii) $\varphi(ab) = \varphi(a)\varphi(b)$,

(iii) $\varphi(1) = 1$ (the 1 on the left-hand side is the identity of R; the 1 on the right-hand side is the identity of S).

If φ is also a bijection, then φ is called an **isomorphism**. Two commutative rings R and S are called **isomorphic**, denoted by $R \cong S$, if there exists an isomorphism $\varphi : R \to S$ between them.

In the definition of a homomorphism $\varphi : R \to S$, the + on the left-hand side is addition in R, while the + on the right-hand side is addition in S; similarly for products. A more complete notation for a commutative ring would display its addition, multiplication, and unit: instead of R, we could write $(R, +, \times, 1)$. Similarly, a more complete notation for S is $(S, \oplus, \otimes, e)$. The definition of homomorphism can now be stated more precisely

(i) $\varphi(a + b) = \varphi(a) \oplus \varphi(b)$,

(ii) $\varphi(a \times b) = \varphi(a) \otimes \varphi(b)$,

(iii) $\varphi(1) = e$.

Etymology. The word *homomorphism* comes from the Greek *homo*, meaning "same," and *morph*, meaning "shape" or "form." Thus, a homomorphism carries a commutative ring to another commutative ring of similar form. The word *isomorphism* involves the Greek *iso*, meaning "equal," and isomorphic rings have identical form.

Consider the two simple examples above of addition tables arising from the rings $\mathbb{Z}_2$ and $\mathcal{P}$ (the symbol $\mathcal{P}$ stands for "parity."). The rings $\mathbb{Z}_2$ and $\mathcal{P}$

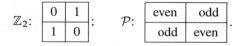

Figure 5.1. Addition tables.

are isomorphic, for the function $\varphi: \mathbb{Z}_2 \to \mathcal{P}$, defined by $\varphi(0) =$ even and $\varphi(1) =$ odd, is an isomorphism, as the reader can quickly check (of course, you must also check the multiplication tables).

Let $a_1, a_2, \ldots, a_j, \ldots$ be a list with no repetitions of all the elements of a ring R. An addition table for R is a matrix whose ij entry is $a_i + a_j$.

$+$	a_1	$\cdots$	a_j	$\cdots$
a_1	$a_1 + a_1$	$\cdots$	$a_1 + a_j$	$\cdots$
a_i	$a_i + a_1$	$\cdots$	$a_i + a_j$	$\cdots$

A multiplication table for R is defined similarly.

The addition and multiplication tables for a ring R depend on the listing of its elements, so that a ring has many tables. Let $a_1, a_2, \ldots, a_j, \ldots$ be a list of all the elements of a ring R with no repetitions. If S is a ring and $\varphi: R \to S$ is a bijection, then $\varphi(a_1), \varphi(a_2), \ldots, \varphi(a_j), \ldots$ is a list of all the elements of S with no repetitions, and so this latter list determines addition and multiplication tables for S. That φ is an isomorphism says that if we superimpose the tables for R (determined by $a_1, a_2, \ldots, a_j, \ldots$) upon the tables for S (determined by $\varphi(a_1), \varphi(a_2), \ldots, \varphi(a_j), \ldots$), then the tables match. In more detail, if $a_i + a_j$ is the ij entry in the given addition table of R, then $\varphi(a_i) + \varphi(a_j)$ is the ij entry of the addition table of S. But $\varphi(a_i) + \varphi(a_j) = \varphi(a_i + a_j)$, because φ is an isomorphism. In this sense, isomorphic rings have the same addition tables and the same multiplication tables (see Figure 5.1). Informally, we say that a homomorphism **preserves** addition, multiplication, and 1. Thus, isomorphic rings are essentially the same, differing only in the notation for their elements and their operations.

Here are two interesting examples of homomorphisms: the first will be used often in this book; the second compares the two different ways we view polynomials.

> Does a homomorphism "preserve 0"? See Lemma 5.17.

Example 5.14. (i) **Reduction** mod m.

We didn't have the language to say it at the time, but Proposition 4.5 sets up a homomorphism $r_m: \mathbb{Z} \to \mathbb{Z}_m$ for any nonnegative integer m, namely, $r_m: n \mapsto [n]$. It's not an isomorphism because $\mathbb{Z}$ is infinite and $\mathbb{Z}_m$ is finite (so there can't be any bijection between them). Another reason is that $r_m(m) = 0 = r_m(2m)$, so r_m can't be injective. Is it surjective?

(ii) **Form to Function**.

In Example 4.31, we saw that every $f(x) \in R[x]$, where R is a commutative ring, determines its associated polynomial function $f^{\#}: R \to R$. The function $\varphi: f \mapsto f^{\#}$ is a homomorphism $R[x] \to \text{Fun}(R) = R^R$: as we saw on page 204, addition of polynomials corresponds to pointwise addition of polynomial functions $(f + g)^{\#} = f^{\#} + g^{\#}$, and multiplication of polynomials corresponds to pointwise multiplication of polynomial functions $(fg)^{\#} = f^{\#}g^{\#}$. Is φ an isomorphism? No, because it's not surjective—not every function on R is a polynomial function. Is φ an isomorphism between $R[x]$ and the subring $\text{im}\,\varphi = \text{Poly}(R)$ of R^R consisting of all polynomial functions? That depends on R. We've seen, for example, that $f(x) = x^7 + 2x - 1$ and $g(x) = 3x + 6$ give the same function ($f^\# = g^\#$) when $R = \mathbb{Z}_7$. But we'll see, in Theorem 6.20, that if R is an infinite field (as it almost always is in high school), then $\varphi: R[x] \to \text{Poly}(R)$ is an isomorphism. ▲

As with all important definitions in mathematics, the idea of homomorphism existed long before the name. Here are some examples that you've encountered so far.

Example 5.15. (i) Complex conjugation $z = a + ib \mapsto \bar{z} = a - ib$ is a homomorphism $\mathbb{C} \to \mathbb{C}$, because $\bar{1} = 1, \overline{z + w} = \bar{z} + \bar{w}$, and $\overline{zw} = \bar{z}\,\bar{w}$; it is a bijection because $\bar{\bar{z}} = z$ (so that it is its own inverse) and, therefore, complex conjugation is an isomorphism.

(ii) Let D be a domain with fraction field $F = \text{Frac}(D)$. In Theorem 5.5, we proved that $D' = \{[a, 1] : a \in D\}$ is a subring of F. We can now identify D' with D, for the function $\varphi: D \to D'$, given by $\varphi(a) = [a, 1] = a/1$, is an isomorphism.

(iii) In the proof of Corollary 5.9, we "identified" an element r in a commutative ring R with the constant polynomial $(r, 0, 0, \dots)$. We said that R is a subring of $R[x]$, but that is not the truth. The function $\varphi: R \to R[x]$, defined by $\varphi(r) = (r, 0, 0, \dots)$, is a homomorphism, and $R' = \{(r, 0, 0, \dots) : r \in R\}$ is a subring of $R[x]$ isomorphic to R, and $\varphi: R \to R'$ is an isomorphism.

(iv) If S is a subring of a commutative ring R, then the inclusion $i: S \to R$ is a ring homomorphism (this is one reason why we insist that the identity of R lie in S).

(v) Recall Example 4.47: if X is a set, then 2^X is the Boolean ring of all the subsets of X, where addition is symmetric difference, multiplication is intersection, and the identity is X. If Y is a proper subset of X, then 2^Y is *not* a subring of 2^X, for the identity of 2^Y is Y, not X. Thus, the inclusion $i: 2^Y \to 2^X$ is not a homomorphism, even though $i(a + b) = i(a) + i(b)$ and $i(ab) = i(a)i(b)$. Therefore, the part of the definition of homomorphism requiring identity elements be preserved is not redundant. ▲

Example 5.16. Example 4.31 on page 157 shows, for a commutative ring R and a set X, that the family R^X of all functions from X to R, equipped with pointwise addition and multiplication, is a commutative ring. We've also used the notation 2^X in Example 4.47 on page 166 to stand for the Boolean ring of all subsets of a set X. The goal of this example is to prove that 2^X and $(\mathbb{Z}_2)^X$ are isomorphic rings.

The basic idea is to associate every subset $A \subseteq X$ with its ***characteristic function*** $f_A \in (\mathbb{Z}_2)^X$, defined by

$$f_A(x) = \begin{cases} 1 & \text{if } x \in A \\ 0 & \text{if } x \notin A. \end{cases}$$

We claim that $\varphi : 2^X \to (\mathbb{Z}_2)^X$, defined by

$$\varphi(A) = f_A,$$

is an isomorphism.

This example is rather dense. It's a good idea to pick a concrete set, say $X = \{1, 2, 3\}$, and work out the characteristic function for each of the 8 subsets of X.

The characteristic function f_A is sometimes called the ***indicator function***, for it tells you whether an element $x \in X$ is or is not in A.

First, φ is a bijection:

(i) *φ is injective*: if $A, B \subseteq X$ and $\varphi(A) = \varphi(B)$, then $f_A = f_B$: for any $x \in X$, we have $f_A(x) = 1$ if and only if $f_B(x) = 1$. Thus, $x \in A$ if and only if $x \in B$; that is, $A = B$.

(ii) *φ is surjective*: given a function $g : X \to (\mathbb{Z}_2)^X$, define $A[g] \subseteq X$ by

$$A[g] = \{x \in X : g(x) = 1\}.$$

It is easy to check that $\varphi(A[g]) = f_{A[g]} = g$.

Finally, φ is a homomorphism:

(i) *φ maps the identity to the identity*: the (multiplicative) identity of 2^X is X, and $\varphi(X) = f_X$. Now $f_X(x) = 1$ for all $x \in X$, because every element of X lies in X! Hence, f_X is the (multiplicative) identity in $(\mathbb{Z}_2)^X$.

(ii) *φ preserves addition*: we must show that

$$\varphi(A + B) = \varphi(A) + \varphi(B)$$

for all $A, B \subseteq X$. Consider the following table:

Recall the different meanings of $+$ in this equation: $A + B$ is symmetric difference, and $\varphi(A) + \varphi(B)$ is pointwise addition.

$f_A(x)$	$f_B(x)$	$f_A(x) + f_B(x)$
1	1	0
1	0	1
0	1	1
0	0	0

It follows that $f_A + f_B = f_{A+B}$, for each of the functions $f_A + f_B$ and f_{A+B} has value 1 if $x \in (A \cup B) - (A \cap B)$ and value 0 otherwise. Therefore,

$$\begin{aligned} \varphi(A + B) &= f_{A+B} \\ &= f_A + f_B \\ &= \varphi(A) + \varphi(B). \end{aligned}$$

(iii) *φ preserves multiplication*: we must show that $\varphi(AB) = \varphi(A)\varphi(B)$ for all $A, B \subseteq X$. The proof is similar to that in part (ii), using a table for $f_A f_B$ to prove that $f_A f_B = f_{AB}$; you will supply the details in Exercise 5.39 on page 212.

We conclude that 2^X and $(\mathbb{Z}_2)^X$ are isomorphic. In Exercise 5.40 on page 212, we will see that if $|X| = n$, then $(\mathbb{Z}_2)^X \cong (\mathbb{Z}_2)^n$, the ring of all n-tuples having coordinates in $\mathbb{Z}_2$ with pointwise operations. ▲

How to Think About It. There are two strategies in trying to show that a homomorphism $\varphi: R \to S$ is an isomorphism. One way is to show that φ is a bijection; that is, it is injective and surjective. A second way is to show that the inverse function $\varphi^{-1}: S \to R$ exists (see Exercise 5.30 on page 211 and Exercise 5.39(ii) on page 212).

Here are some properties of homomorphisms.

Lemma 5.17. *Let R and S be commutative rings, let $\varphi\colon R \to S$ be a homomorphism, and let $a \in R$.*

(i) $\varphi(0) = 0$.

(ii) $\varphi(-a) = -\varphi(a)$.

(iii) $\varphi(na) = n\varphi(a)$ *for all $n \in \mathbb{Z}$.*

(iv) $\varphi(a^n) = \varphi(a)^n$ *for all $n \in \mathbb{N}$.*

(v) *If a is a unit in R, then $\varphi(a)$ is a unit in S, and $\varphi(a^{-1}) = \varphi(a)^{-1}$.*

Proof. (i) Since $0 + 0 = 0$, we have $\varphi(0 + 0) = \varphi(0) + \varphi(0) = \varphi(0)$. Now subtract $\varphi(0)$ from both sides.

(ii) Since $0 = -a + a$, we have $0 = \varphi(0) = \varphi(-a) + \varphi(a)$. But Proposition 4.35 says that negatives are unique: there is exactly one $s \in S$, namely, $s = -\varphi(a)$, with $s + \varphi(a) = 0$. Hence, $\varphi(-a) = -\varphi(a)$.

(iii) If $n \geq 0$, use induction to prove that $\varphi(na) = n\varphi(a)$. Now use (ii) : $\varphi(-na) = -\varphi(na) = -n\varphi(a)$.

(iv) Use induction to show that $\varphi(a^n) = \varphi(a)^n$ for all $n \geq 0$.

(v) By Proposition 4.35, there is exactly one $b \in R$, namely, $b = a^{-1}$, with $ab = 1$. Similarly in S; since $\varphi(a)\varphi(b) = \varphi(ab) = \varphi(1) = 1$, we have $\varphi(a^{-1}) = \varphi(a)^{-1}$. ∎

Example 5.18. If $\varphi\colon A \to B$ is a bijection between finite sets A and B, then they have the same number of elements. In particular, two finite isomorphic commutative rings have the same number of elements. We now show that the converse is false: there are finite commutative rings with the same number of elements that are not isomorphic.

Recall Exercise 4.55 on page 165: there is a field, $\mathbb{F}_4$, having exactly four elements. If $a \in \mathbb{F}_4$ and $a \neq 0$, then $a^2 \neq 0$, for $\mathbb{F}_4$ is a domain (even a field), and so the product of nonzero elements is nonzero. Suppose there were an isomorphism $\varphi\colon \mathbb{F}_4 \to \mathbb{Z}_4$. Since φ is surjective, there is $a \in \mathbb{F}_4$ with $\varphi(a) = 2$. Hence, $\varphi(a^2) = \varphi(a)^2 = 2^2 = 0$. This contradicts φ being injective, for $a^2 \neq 0$ and $\varphi(a^2) = 0 = \varphi(0)$. ▲

Exercises

5.29 Let R and S be commutative rings, and let $\varphi\colon R \to S$ be an isomorphism.

(i) If R is a field, prove that S is a field.

(ii) If R is a domain, prove that S is a domain.

5.30 *

(i) If φ is an isomorphism, prove that its inverse function $\varphi^{-1}\colon S \to R$ is also an isomorphism.

(ii) Show that φ is an isomorphism if and only if φ has an inverse function φ^{-1}.

5.31 (i) Show that the composite of two homomorphisms (isomorphisms) is again a homomorphism (an isomorphism).

(ii) Show that $R \cong S$ defines an equivalence relation on the class of all commutative rings.

5.32 Prove that the weird integers W (see Exercise 4.35 on page 158) is not isomorphic to $\mathbb{Z}$.

5.33 Recall that $\mathbb{Z}[\omega] = \{a + b\omega : a, b \in \mathbb{Z}\}$, where $\omega = -\frac{1}{2} + i\frac{\sqrt{3}}{2}$. Show that $\varphi : \mathbb{Z}[\omega] \to \mathbb{Z}[\omega]$, defined by

$$\varphi : a + b\omega \mapsto a + b\omega^2,$$

is a homomorphism. Is φ an isomorphism?

5.34 If R is a commutative ring and $a \in R$, is the function $\varphi : R \to R$, defined by $\varphi : r \mapsto ar$, a homomorphism? Why?

5.35 Prove that two fields having exactly four elements are isomorphic.

Hint. First prove that $1 + 1 = 0$.

5.36 * Let k be a field that contains $\mathbb{Z}_p$ as a subfield (e.g., $k = \mathbb{Z}_p(x)$). For every integer $n > 0$, show that the function $\varphi_n : k \to k$, given by $\varphi_n(a) = a^{p^n}$, is an injective homomorphism. If k is finite, show that φ_n is an isomorphism.

5.37 * If R is a field, show that $R \cong \mathrm{Frac}(R)$. More precisely, show that the homomorphism $\varphi : R \to \mathrm{Frac}(R)$, given by $\varphi : r \mapsto [r, 1]$, is an isomorphism.

5.38 * Recall, when we constructed the field $\mathrm{Frac}(D)$ of a domain D, that $[a, b]$ denoted the equivalence class of (a, b), and that we then reverted to the usual notation: $[a, b] = a/b$.

(i) If R and S are domains and $\varphi : R \to S$ is an isomorphism, prove that

$$[a, b] \mapsto [\varphi(a), \varphi(b)]$$

is an isomorphism $\mathrm{Frac}(R) \to \mathrm{Frac}(S)$.

(ii) Prove that a field k containing an isomorphic copy of $\mathbb{Z}$ as a subring must contain an isomorphic copy of $\mathbb{Q}$.

(iii) Let R be a domain and let $\varphi : R \to k$ be an injective homomorphism, where k is a field. Prove that there exists a unique homomorphism $\Phi : \mathrm{Frac}(R) \to k$ extending φ; that is, $\Phi | R = \varphi$.

5.39 * In Example 5.16, we proved that if X is a set, then the function $\varphi : 2^X \to (\mathbb{Z}_2)^X$, given by $\varphi(A) = f_A$, the characteristic function of A, is an isomorphism.

(i) Complete the proof in Example 5.16 by showing that $\varphi(AB) = \varphi(A)\varphi(B)$ for all $A, B \in 2^X$.

(ii) Give another proof that $2^X \cong (\mathbb{Z}_2)^X$ by showing that φ^{-1} exists.

5.40 * If n is a positive integer, define $(\mathbb{Z}_2)^n$ to be the set of all n-tuples $(a_1, \ldots, a_n)$ with $a_i \in \mathbb{Z}_2$ for all i (such n-tuples are called **bitstrings**).

(i) Prove that $(\mathbb{Z}_2)^n$ is a commutative ring with pointwise operations

$$(a_1, \ldots, a_n) + (b_1, \ldots, b_n) = (a_1 + b_1, \ldots, a_n + b_n)$$

and

$$(a_1, \ldots, a_n)(b_1, \ldots, b_n) = (a_1 b_1, \ldots, a_n b_n).$$

(ii) If X is a finite set with $|X| = n$, prove that $(\mathbb{Z}_2)^X \cong (\mathbb{Z}_2)^n$. Conclude, in this case, that $2^X \cong (\mathbb{Z}_2)^n$ (see Example 5.16).

Extensions of Homomorphisms

Suppose that a ring R is a subring of a commutative ring E with inclusion $i: R \to E$. Given a homomorphism $\varphi: R \to S$, an ***extension*** Φ of φ is a homomorphism $\Phi: E \to S$ with restriction $\Phi|R = \Phi i = \varphi$.

If $\Phi: U \to Y$ is any function, then its restriction $\Phi|X$ to a subset $X \subseteq U$ is equal to the composite Φi, where $i: X \to U$ is the inclusion.

Some obvious questions about extensions are

(i) Can we extend $\varphi: R \to S$ to a homomorphism $\Phi: E \to S$?

(ii) Can we extend $\varphi: R \to S[x]$ to a homomorphism $\Phi: E \to S[x]$?

Theorems 5.19 and 5.20 below answer the first question when $E = R[x]$ and $E = R[x_1, \ldots, x_n]$; Corollary 5.22 answers the second question when $E = R[x]$. The basic idea is to let φ handle the elements of R, then specify what happens to x, and then use the definition of homomorphism to make sure that the extension preserves addition and multiplication.

Even though the coming proof is routine, we give full details because of the importance of the result.

Theorem 5.19. *Let R and S be commutative rings, and let $\varphi: R \to S$ be a homomorphism. If $s \in S$, then there exists a unique homomorphism*

$$\Phi: R[x] \to S$$

with $\Phi(x) = s$ and $\Phi(r) = \varphi(r)$ for all $r \in R$.

Proof. If $f(x) = \sum_i r_i x^i = r_0 + r_1 x + \cdots + r_n x^n$, define $\Phi: R[x] \to S$ by

$$\Phi(f) = \varphi(r_0) + \varphi(r_1)s + \cdots + \varphi(r_n)s^n.$$

Proposition 5.6, uniqueness of coefficients, shows that Φ is a well-defined function, and the formula shows that $\Phi(x) = s$ and $\Phi(r) = \varphi(r)$ for all $r \in R$.

We now prove that Φ is a homomorphism. First, $\Phi(1) = \varphi(1) = 1$, because φ is a homomorphism.

Second, if $g(x) = a_0 + a_1 x + \cdots + a_m x^m$, then

$$\begin{aligned}
\Phi(f + g) &= \Phi\left(\sum_i (r_i + a_i)x^i\right) \\
&= \sum_i \varphi(r_i + a_i)s^i \\
&= \sum_i \left(\varphi(r_i) + \varphi(a_i)\right)s^i \\
&= \sum_i \varphi(r_i)s^i + \sum_i \varphi(a_i)s^i \\
&= \Phi(f) + \Phi(g).
\end{aligned}$$

Third, let $f(x)g(x) = \sum_k c_k x^k$, where $c_k = \sum_{i+j=k} r_i a_j$. Then

$$\Phi(fg) = \Phi\left(\sum_k c_k x^k\right)$$

$$= \sum_k \varphi(c_k)s^k$$

$$= \sum_k \varphi\left(\sum_{i+j=k} r_i a_j\right) s^k$$

$$= \sum_k \left(\sum_{i+j=k} \varphi(r_i)\varphi(a_j)\right) s^k.$$

On the other hand,

$$\Phi(f)\Phi(g) = \left(\sum_i \varphi(r_i)s^i\right)\left(\sum_j \varphi(a_j)s^j\right) = \sum_k \left(\sum_{i+j=k} \varphi(r_i)\varphi(a_j)\right) s^k.$$

Uniqueness of Φ is easy: if $\Theta\colon R[x] \to S$ is a homomorphism with $\Theta(x) = s$ and $\Theta(r) = \varphi(r)$ for all $r \in R$, then

$$\Theta(r_0 + r_1 x + \cdots + r_d x^d) = \varphi(r_0) + \varphi(r_1)s + \cdots + \varphi(r_d)s^d$$

$$= \Phi(r_0 + r_1 x + \cdots + r_d x^d). \quad \blacksquare$$

This theorem generalizes to polynomial rings in several variables.

Theorem 5.20. *Let R and S be commutative rings and $\varphi\colon R \to S$ a homomorphism. If $s_1, \ldots, s_n \in S$, then there exists a unique homomorphism*

$$\Phi\colon R[x_1, \ldots, x_n] \to S$$

with $\Phi(x_i) = s_i$ for all i and $\Phi(r) = \varphi(r)$ for all $r \in R$.

Proof. The proof is by induction on $n \geq 1$. The base step is Theorem 5.19. For the inductive step, let $n > 1$ and define $A = R[x_1, \ldots, x_{n-1}]$. The inductive hypothesis gives a homomorphism $\psi\colon A \to S$ with $\psi(x_i) = s_i$ for all $i \leq n-1$ and $\psi(r) = \varphi(r)$ for all $r \in R$. The base step gives a homomorphism $\Psi\colon A[x_n] \to S$ with $\Psi(x_n) = s_n$ and $\Psi(a) = \psi(a)$ for all $a \in A$. The result follows, because $R[x_1, \ldots, x_n] = A[x_n]$, $\Psi(x_i) = \psi(x_i) = s_i$ for all $i \leq n-1$, $\Psi(x_n) = \psi(x_n) = s_n$, and $\Psi(r) = \psi(r) = \varphi(r)$ for all $r \in R$. $\quad \blacksquare$

How to Think About It. There is an analogy between Theorem 5.20 and an important theorem of linear algebra, Theorem A.43 in Appendix A.4: Let V and W be vector spaces over a field k; if $v_1, \ldots, v_n$ is a basis of V and $w_1, \ldots, w_n \in W$, then there exists a unique linear transformation $T\colon V \to W$ with $T(v_i) = w_i$ for all i (linear transformations are homomorphisms of vector spaces). The theorem is actually the reason why matrices can describe linear transformations.

Here is a familiar special case of Theorem 5.19.

Definition. If R is a commutative ring and $a \in R$, then ***evaluation at*** a is the function $e_a \colon R[x] \to R$ given by $e_a(f) = f(a)$; that is,

$$ e_a\left(\sum_i r_i x^i\right) = \sum_i r_i a^i. $$

If $f^\#$ is the polynomial function determined by f, then $e_a(f) = f^\#(a)$.

So, in this language, we have:

Corollary 5.21. *If R is a commutative ring and $a \in R$, then $e_a \colon R[x] \to R$, evaluation at a, is a homomorphism.*

Proof. In the notation of Theorem 5.19, set $R = S$, $\varphi = 1_R$ (the identity function $R \to R$), and $s = a \in R$. The homomorphism $\Phi \colon R[x] \to R$ is e_a, which sends $\sum_i r_i x^i$ into $\sum_i r_i a^i$. ∎

As an illustration of Corollary 5.21, if $f, g \in R[x]$ and $h = fg$, then

$$ h(a) = e_a(h) = e_a(f)e_a(g) = f(a)g(a). $$

In other words, we get the same element of R if we first multiply polynomials in $R[x]$ and then substitute a for x, or if we first substitute a for x and then multiply the elements $f(a)$ and $g(a)$. For example, if R is a commutative ring and $a \in R$, then $f(x) = q(x)g(x) + r(x)$ in $R[x]$ implies $f(a) = q(a)g(a) + r(a)$ in R.

Let's return to question (ii) on page 213. Given a homomorphism $\varphi \colon R \to S$, can we extend it to a homomorphism $R[x] \to S[x]$? The basic idea? Let φ handle the coefficients and send x to x.

Corollary 5.22. *If R and S are commutative rings and $\varphi \colon R \to S$ is a homomorphism, then there is a unique homomorphism $\varphi^* \colon R[x] \to S[x]$ given by*

$$ \varphi^* \colon r_0 + r_1 x + r_2 x^2 + \cdots \mapsto \varphi(r_0) + \varphi(r_1)x + \varphi(r_2)x^2 + \cdots . $$

Moreover, φ^* is an isomorphism if φ is.

Proof. The existence of the homomorphism φ^* is a special case of Theorem 5.19. More precisely, consider the following diagram in which $\iota \colon R \to R[x]$ and $\lambda \colon S \to S[x]$ are the usual inclusions viewing elements of R and of S as constant polynomials. The role of $\varphi \colon R \to S$ is now played by the composite $\lambda\varphi \colon R \to S[x]$, namely, $r \mapsto (\varphi(r), 0, 0. \ldots)$.

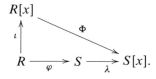

If φ is an isomorphism, then Φ^{-1} is the inverse of the extension of φ^{-1}. ∎

Example 5.23. If $r_m: \mathbb{Z} \to \mathbb{Z}_m$ is reduction mod m, that is, $r_m(a) = [a]$, then the homomorphism $r_m^*: \mathbb{Z}[x] \to \mathbb{Z}_m[x]$ reduces each coefficient of a polynomial mod m:

$$r_m^*: a_0 + a_1 x + a_2 x^2 + \cdots \mapsto [a_0] + [a_1]x + [a_2]x^2 + \cdots .$$

We will usually write a instead of $[a]$ when using r_m^*. ▲

Example 5.24. Complex conjugation extends to an isomorphism $\mathbb{C}[x] \to \mathbb{C}[x]$ in which every polynomial is mapped to the polynomial obtained by taking the complex conjugate of each coefficient. (We have already used this construction in Theorem 3.12 and in Exercises 3.43–3.45 on page 106.) ▲

Exercises

5.41 If R is a commutative ring, prove that $R[x, y] \cong R[y, x]$. In fact, prove that there is an isomorphism Φ with $\Phi(x) = y$, $\Phi(y) = x$, and $\Phi(r) = r$ for all $r \in R$.

Hint. Use Theorem 5.20.

If you look very carefully at the definitions, you'll see that $R[x, y]$ and $R[y, x]$ are different rings. Recall that elements a in a ring A correspond to $(a, 0, \ldots)$ in $A[x]$. In particular, the element $x \in R[x]$ corresponds to $(x, 0, 0, \ldots)$ in $R[x][y]$; that is, we have $x = ((0, 1, 0, \ldots), 0, 0, \ldots)$ so that in $R[x][y]$ the element x has $(0, 1, 0, \ldots)$ in coordinate 1. This is not the same element as x in $R[y][x]$, which has 1 sitting in coordinate 1. However, this exercise allows you to relax and regard these polynomials rings as the same.

5.42 *

(i) If R is a commutative ring and $c \in R$, prove that there is a homomorphism $\varphi: R[x] \to R[x]$ with $\varphi(x) = x + c$ and $\varphi(r) = r$ for all $r \in R$; that is, $\varphi(\sum_i r_i x^i) = \sum_i r_i (x + c)^i$. Is φ an isomorphism?

No calculus is needed for this exercise.

(ii) If $\deg(f) = n$, show that

$$\varphi(f) = f(c) + f'(c)(x + c) + \frac{f''(c)}{2!}(x + c)^2 + \cdots + \frac{f^{(n)}(c)}{n!}(x + c)^n,$$

where $f'(x)$ is the formal derivative of f defined in Exercise 5.15 on page 202.

Kernel, Image, and Ideals

There's a great deal of talk about "modeling" in high school mathematics: we wonder whether a given statement is true, and the idea is to see whether it holds in some model. A homomorphism $R \to S$ is a good illustration of this idea; it transports the structure of R to the structure of S, so that we may test whether a statement in R is true by asking whether its analog in the "model" S is true. For example, is -1 a square in $\mathbb{Z}$; is there $k \in \mathbb{Z}$ with $-1 = k^2$? Now we can list all the squares in $\mathbb{Z}_3$: $0^2 = 0$; $1^2 = 1$; $2^2 = 4 = 1$, and we see that $-1 = 2$ is not a square. But if $-1 = k^2$ in $\mathbb{Z}$, then reduction mod 3, the homomorphism $r_3: \mathbb{Z} \to \mathbb{Z}_3$ taking $a \mapsto [a]$ (see Example 5.14(i)), would give

$$r_3(-1) = r_3(k^2) = r_3(k)^2,$$

contradicting -1 not being a square in $\mathbb{Z}_3$.

We should be cautious when viewing a homomorphism $\varphi: R \to S$ as modeling a ring R. First, φ may not give a *faithful* model of R, thereby losing some information; for example, φ might take different elements of R to the same element of S. Also, some information might get missed: there may be elements of S that don't get "hit" by an element of R. The information that's lost is called the *kernel* of φ; the information that's hit is called its *image*.

Definition. If $\varphi: R \to S$ is a homomorphism, then its **kernel** is

$$\ker \varphi = \{a \in R \text{ with } \varphi(a) = 0\} \subseteq R,$$

and its **image** is

$$\operatorname{im} \varphi = \{s \in S : s = \varphi(a) \text{ for some } a \in R\} \subseteq S.$$

Here are the first properties of these subsets. Note that Lemma 5.17 says, for every homomorphism $\varphi: R \to S$, that $0 \in \ker \varphi$ and $0 \in \operatorname{im} \varphi$. In general, $\operatorname{im} \varphi$ is a subset of S but, as for any function, φ is surjective if and only if $\operatorname{im} \varphi = S$.

Proposition 5.25. *Let R and S be rings and $\varphi: R \to S$ a homomorphism.*

(i) $\operatorname{im} \varphi = \{\varphi(r) : r \in R\}$ *is a subring of S.*

(ii) *If $a, b \in \ker \varphi$, then $a + b \in \ker \varphi$.*

(iii) *If $a \in \ker \varphi$ and $r \in R$, then $ra \in \ker \varphi$.*

Proof. (i) To see that $\operatorname{im} \varphi$ is a subring of S, note first that $1 \in \operatorname{im} \varphi$, because $\varphi(1) = 1$. If $s, t \in \operatorname{im} \varphi$, then there are $a, b \in R$ with $s = \varphi(a)$ and $t = \varphi(b)$. Hence, $s + t = \varphi(a) + \varphi(b) = \varphi(a + b) \in \operatorname{im} \varphi$, and $st = \varphi(a)\varphi(b) = \varphi(ab) \in \operatorname{im} \varphi$. Therefore, $\operatorname{im} \varphi$ is a subring of S.

(ii) If $a, b \in \ker \varphi$, then $\varphi(a) = 0 = \varphi(b)$. Hence, $\varphi(a+b) = \varphi(a)+\varphi(b) = 0 + 0 = 0$, and $a + b \in \ker \varphi$.

(iii) If $a \in \ker \varphi$, then $\varphi(a) = 0$. Hence, $\varphi(ra) = \varphi(r)\varphi(a) = \varphi(r) \cdot 0 = 0$, and so $ra \in \ker \varphi$. ∎

Here are some examples of kernels and images.

Example 5.26. (i) If $\varphi: R \to S$ is an isomorphism, then $\ker \varphi = \{0\}$ and $\operatorname{im} \varphi = S$.

(ii) If φ is injective, then $\ker \varphi = \{0\}$, for if $r \neq 0$, then $\varphi(r) \neq \varphi(0) = 0$. We will soon see that the converse is true, so that φ is injective if and only $\ker \varphi = \{0\}$.

(iii) If $r_m: \mathbb{Z} \to \mathbb{Z}_m$ is reduction mod m, then $\ker r_m$ consists of all the multiples of m.

(iv) Let k be a commutative ring, let $a \in k$, and let $e_a: k[x] \to k$ be the evaluation homomorphism $f(x) \mapsto f(a)$. Now e_a is always surjective: if $b \in k$, then $b = e_a(f)$, where $f(x) = x - a + b$. By definition, $\ker e_a$ consists of all those polynomials g for which $g(a) = 0$.

In particular, let $\varphi: \mathbb{R}[x] \to \mathbb{C}$ be defined by $\varphi(x) = i$ and $\varphi(a) = a$ for all $a \in \mathbb{R}$. Then $\ker \varphi$ is the set of all polynomials $f(x) \in \mathbb{R}[x]$ having i as a root. For example, $x^2 + 1 \in \ker \varphi$. ▲

Proposition 5.25 suggests that $\ker \varphi$ is a subring of R but, in fact, it almost never is because it usually doesn't contain 1. The definition of homomorphism says that $\varphi(1) = 1$. If $1 \in \ker \varphi$, then $\varphi(1) = 0$, and so $1 = 0$ in S; that is, S is the zero ring. We conclude that if S has more than one element, then $\ker \varphi$ is not a subring of R. However, kernels are always *ideals*.

Definition. An *ideal* in a commutative ring R is a subset I of R such that

(i) $0 \in I$

(ii) if $a, b \in I$, then $a + b \in I$

(iii) if $a \in I$ and $r \in R$, then $ra \in I$.

An ideal $I \neq R$ is called a ***proper ideal***.

The ring R itself and $\{0\}$, the subset of R consisting of 0 alone, are always ideals in a commutative ring R. Proposition 5.25 says that the kernel of a homomorphism $\varphi \colon R \to S$ is always an ideal in R; it is a proper ideal if S is not the zero ring because $1 \notin \ker \varphi$.

We have seen ideals in a completely different context. Theorem 1.19, which says that $\gcd(a, b)$ is a linear combination of a, b, involved showing that the set of all linear combinations is an ideal in $\mathbb{Z}$. Indeed, Exercise 1.49 on page 30 makes this explicit (of course, we had not introduced the term *ideal* at that time).

Etymology. As we said on page 131, a natural attempt to prove Fermat's Last Theorem involves factoring $x^p + y^p$ in the ring $\mathbb{Z}[\zeta_p]$ of cyclotomic integers, where ζ_p is a pth root of unity. In Chapter 8, we shall sketch the ideas that show that if this ring has unique factorization into primes, that is, if the analog of the Fundamental Theorem of Arithmetic holds in $\mathbb{Z}[\zeta_p]$, then there are no positive integers a, b, c with $a^p + b^p = c^p$. For some primes p, such an analog is true but, alas, there are primes for which it is false. In his investigation of Fermat's Last Theorem, Kummer invented *ideal numbers* in order to restore unique factorization. His definition was later recast by Dedekind as the ideals we have just defined, and this is why ideals are so called.

Here is a construction of ideals that generalizes that which arose when we studied gcd's. Recall that a ***linear combination*** of elements $b_1, b_2, \ldots, b_n$ in a commutative ring R is an element of R of the form

$$r_1 b_1 + r_2 b_2 + \cdots + r_n b_n,$$

where $r_i \in R$ for all i.

It is very easy to check that $(b_1, b_2, \ldots, b_n)$ is an ideal.

Definition. If $b_1, b_2, \ldots, b_n$ lie in a commutative ring R, then the set of all linear combinations, denoted by

$$(b_1, b_2, \ldots, b_n),$$

is an ideal in R, called the ***ideal generated by*** $b_1, b_2, \ldots, b_n$. In particular, if $n = 1$, then

$$(b) = \{rb : r \in R\}$$

The principal ideal (b) is sometimes denoted by Rb.

consists of all the multiples of b; it is called the ***principal ideal*** generated by b.

Both R and $\{0\}$ are ideals; indeed, both are principal ideals, for $R = (1)$ and, obviously, $\{0\} = (0)$ is generated by 0. Henceforth, we will denote the *zero ideal* $\{0\}$ by (0).

Example 5.27. (i) The even integers comprise an ideal in $\mathbb{Z}$, namely, (2).

(ii) Proposition 5.25 says that if $\varphi \colon R \to S$ is a homomorphism, then $\ker \varphi$ is an ideal in R. In particular, we can generalize part (i): if $r_m \colon \mathbb{Z} \to \mathbb{Z}_m$ is reduction mod m, then $\ker r_m = (m)$.

(iii) If I and J are ideals in a commutative ring R, then it is routine to check that $I \cap J$ is also an ideal in R. More generally, if $(I_j)_{j \in J}$ is a family of ideals in a commutative ring R, then $\bigcap_{j \in J} I_j$ is an ideal in R (see Exercise 5.53 below).

(iv) By Example 5.26(iv), the set I, consisting of all polynomials $f(x)$ in $\mathbb{R}[x]$ having i as a root, is an ideal in $\mathbb{R}[x]$ containing $x^2 + 1$ (it is the kernel of the evaluation e_i). We shall see, in Corollary 6.26, that $I = (x^2 + 1)$. ▲

Example 5.28. Let R be a commutative ring. For a subset A of R, define

$$I = I(A) = \{f(x) \in R[x] : f(a) = 0 \text{ for all } a \in A\}.$$

It is easy to check that I is an ideal in $R[x]$. Clearly, $0 \in I$. If $f \in I$ and $r \in R$, then $(rf)^{\#} = rf^{\#}$, and so $(rf)(a) = r(f(a)) = 0$ for all $a \in A$. Finally, if $f, g \in I$, then $(f + g)^{\#} = f^{\#} + g^{\#}$, so that $(f + g)^{\#} \colon a \mapsto f(a) + g(a) = 0$ for all $a \in A$, and $f + g \in I$. Therefore, I is an ideal. (Alternatively, show that $I(A) = \bigcap_{a \in A} \ker e_a$, where e_a is evaluation at a, and use Exercise 5.53 below that says the intersection is an ideal.)

In the special case when R is a field, then $I(A)$ is a principal ideal. If A is finite, can you find a monic $d(x)$ with $I(A) = (d)$? What if A is infinite? ▲

Theorem 5.29. *Every ideal I in $\mathbb{Z}$ is a principal ideal.*

Proof. If $I = (0)$, then I is the principal ideal with generator 0. If $I \neq (0)$, then there are nonzero integers in I; since $a \in I$ implies $-a \in I$, there are positive integers in I; let $d \in I$ be the smallest such. Clearly, $(d) \subseteq I$. For the reverse inclusion, let $b \in I$. The Division Algorithm gives $q, r \in \mathbb{Z}$ with $b = qd + r$, where $0 \le r < d$. But $r = b - qd \in I$. If $r \neq 0$, then its existence contradicts d being the smallest positive integer in I. Hence, $r = 0$, $d \mid b$, $b \in I$, and $I \subseteq (d)$. Therefore, $I = (d)$. ∎

We'll see in the next chapter that there are commutative rings having ideals that are not principal ideals.

Example 5.30. (i) If an ideal I in a commutative ring R contains 1, then $I = R$, for now I contains $r = r1$ for every $r \in R$. Indeed, if I contains a unit u, then $I = R$, for then I contains $u^{-1}u = 1$.

(ii) It follows from (i) that if R is a field, then the only ideals I in R are (0) and R itself: if $I \neq (0)$, it contains some nonzero element, and every nonzero element in a field is a unit.

Conversely, assume that R is a nonzero commutative ring whose only ideals are R itself and (0). If $a \in R$ and $a \neq 0$, then $(a) = \{ra : r \in R\}$

is a nonzero ideal, and so $(a) = R$; but $1 \in R = (a)$. Thus, there is $r \in R$ with $1 = ra$; that is, a has an inverse in R, and so R is a field.   ▲

Proposition 5.31. *A homomorphism* $\varphi \colon R \to S$ *is an injection if and only if* $\ker \varphi = (0)$.

Proof. If φ is an injection, then $a \neq 0$ implies $\varphi(a) \neq \varphi(0) = 0$. Hence, $\ker \varphi = (0)$. Conversely, assume that $\ker \varphi = (0)$. If $\varphi(a) = \varphi(b)$, then $\varphi(a - b) = \varphi(a) - \varphi(b) = 0$; that is, $a - b \in \ker \varphi = (0)$. Therefore, $a = b$ and φ is an injection.   ■

Corollary 5.32. *If k is a field and* $\varphi \colon k \to S$ *is a homomorphism, where S is a nonzero commutative ring, then φ is an injection.*

Proof. The only proper ideal in k is (0), by Example 5.30; now apply Proposition 5.31.   ■

Exercises

5.43 Construct a homomorphism from $\mathbb{Z}[i] \to \mathbb{Z}[i]$ that has i in its kernel. What is the entire kernel?

5.44 Find the kernel of the homomorphism $\mathbb{Q}[x] \to \mathbb{Q}[\sqrt{2}]$ defined by $f \mapsto f(\sqrt{2})$, where $\mathbb{Q}[\sqrt{2}] = \{a + b\sqrt{2} : a, b \in \mathbb{Q}\}$.

5.45 Show that the kernel of the evaluation homomorphism e_a in Corollary 5.21 is the set of polynomials in $R[x]$ that have a as a root.

5.46 Consider the set I of polynomials in $\mathbb{R}[x]$ that vanish on the set $\{3 \pm \sqrt{5}, 5 \pm \sqrt{7}\}$. Show that I is a principal ideal in $\mathbb{R}[x]$.

The notation $\subsetneq$ means "is a proper subset of" (in contrast to $\subseteq$ which indicates a subset which may or may not be proper).

5.47 * Find three ideals (a) in $\mathbb{Z}$ with the property that

$$(24) \subsetneq (a).$$

5.48 * Suppose a and b are integers. Show that $a \mid b$ if and only if $(b) \subseteq (a)$.

5.49 * If $a, b \in \mathbb{Z}$, prove that $(a) \cap (b) = (m)$, where $m = \mathrm{lcm}(a, b)$.

5.50 * Define the *sum* of ideals I and J in a commutative ring R by

$$I + J = \{u + v : u \in I \text{ and } v \in J\}.$$

   (i) Prove that $I + J$ is an ideal.

   (ii) If $a, b \in \mathbb{Z}$, prove that $(a) + (b) = (a, b) = (d)$, where $d = \gcd(a, b)$.

5.51 * Define the *product* of ideals I and J in a commutative ring R by

$$IJ = \{a_1 b_1 + \cdots + a_n b_n : a_i \in I, b_i \in J, n \geq 1\}.$$

   (i) Prove that IJ is an ideal in R.

   (ii) Prove that if I and J are principal ideals, then IJ is principal. More precisely, if $I = (a)$ and $J = (b)$, then $IJ = (ab)$.

   (iii) If $I = (a_1, \ldots, a_s)$ and $J = (b_1, \ldots, b_t)$, prove that

$$IJ = (a_i b_j : 1 \leq i \leq s \quad \text{and} \quad 1 \leq j \leq t).$$

5.52 * Let I, J, and Q be ideals in a commutative ring R.

   (i) Prove that $IJ = JI$.

   (ii) Prove that $RI = I$.

   (iii) Prove that $I(JQ) = (IJ)Q$.

5.53 * If $(I_j)_{j \in J}$ is a family of ideals in a commutative ring R, prove that $\bigcap_{j \in J} I_j$ is an ideal in R.

5.54 *

   (i) If R and S are commutative rings, show that their **direct product** $R \times S$ is also a commutative ring, where addition and multiplication in $R \times S$ are defined coordinatewise:

$$(r, s) + (r', s') = (r + r', s + s')$$

and

$$(r, s)(r', s') = (rr', ss').$$

   This construction generalizes that of $(\mathbb{Z}_2)^n$ in Exercise 5.40 on page 212.

   (ii) Show that $R \times S$ is not a domain.

   (iii) Show that $R \times (0)$ is an ideal in $R \times S$.

   (iv) Show that $R \times (0)$ is a ring isomorphic to R, but it is not a subring of $R \times S$.

   (v) Prove that $\mathbb{Z}_6 \cong \mathbb{Z}_2 \times \mathbb{Z}_3$.

   (vi) Show that $\mathbb{Z}_4 \not\cong \mathbb{Z}_2 \times \mathbb{Z}_2$.

   (vii) Prove that $\mathbb{Z}_{mn} \cong \mathbb{Z}_m \times \mathbb{Z}_n$ if m and n are relatively prime.

      Hint. Use the Chinese Remainder Theorem.

5.55 If $R_1, \ldots, R_n$ are commutative rings, define their **direct product** $R_1 \times \cdots \times R_n$ by induction on $n \geq 2$ (it is the set of all n-tuples $(r_1, \ldots, r_n)$ with $r_i \in R_i$ for all i). Prove that the ring $(\mathbb{Z})^X$ in Example 5.16, where X is a set with $|X| = n$, is the direct product of n copies of $\mathbb{Z}_2$.

5.56 (i) Give an example of a commutative ring R with nonzero ideals I and J such that $I \cap J = (0)$.

   (ii) If I and J are nonzero ideals in a domain R, prove that $I \cap J \neq (0)$.

5.57 Let F be the set of all 2×2 real matrices of the form

$$A = \begin{bmatrix} a & b \\ -b & a \end{bmatrix}.$$

   (i) Prove that F is a field (with operations matrix addition and matrix multiplication).

   (ii) Prove that $\varphi \colon F \to \mathbb{C}$, defined by $\varphi(A) = a + ib$, is an isomorphism.

5.4 Connections: Boolean Things

In some high school programs, *Boolean Algebra* is called the "algebra of sets;" it usually focuses on establishing set-theoretic identities like

$$A \cap (B \cup C) = (A \cap B) \cup (A \cap C)$$

for subsets A, B, and C of a set X. Such formulas are proved by showing that an element lies in the left-hand side if and only it lies in the right-hand side.

Exercises 4.68 through 4.74 on page 169 gave you practice in doing this sort of thing, but they actually showed more. Recall Example 4.47: if 2^X is the family of all the subsets of a set X, then 2^X is a commutative ring with addition defined as *symmetric difference*,

$$A + B = (A - B) \cup (B - A) = A \cup B - (A \cap B),$$

Recall: if U, V are subsets of X, then $U - V = \{x \in X : x \in U$ and $x \notin V\}$.

and multiplication defined as intersection,

$$AB = A \cap B.$$

It follows, for all subsets A of X, that

$$A^2 = A, \quad A + \varnothing = A, \quad A + A = \varnothing, \quad \text{and} \quad AX = A;$$

the identity element 1 is the subset X itself. It follows from $A + A = \varnothing$ that every $A \in 2^X$ is its own additive inverse; that is, $A = -A$. Indeed, Exercise 5.58 on page 226 says that $1 = -1$ in 2^X. Since we often pass back and forth between the commutative ring 2^X and set theory, we say out loud that a minus sign will be used in set theory, as in the definition of symmetric difference, but it shall never be used when we are working in 2^X viewed as a ring.

We are going to show that calculations in the ring 2^X give more satifisfying proofs of set-theoretic identities; thus, regarding all subsets as forming a commutative ring is a definite advantage. Another goal is to use the calculations to establish the ***inclusion-exclusion*** principle, a very useful technique in counting problems.

Venn diagrams are visual representations in the plane of relationships among subsets in X. They convert words into pictures. For example, symmetric difference and intersection are illustrated by the Venn diagram in Figure 5.2.

Some standard words occurring in set theory, actually in logic, are NOT, AND, OR, and EXCLUSIVE OR. If we picture a statement a as the inside of a region A in the plane, then the Venn diagram of "NOT a" is the outside of A; it is the ***complement***

$$A^c = \{x \in X : x \notin A\}.$$

In Figure 5.2, $A + B$ is the shaded region, AB is the unshaded region.

Exercise 4.69 on page 168 says that $A^c = X + A$. If a and b are statements, then the Venn diagram of the statement "a AND b" is the intersection $A \cap B$, while the diagram of "a OR b" is the union $A \cup B$. EXCLUSIVE OR is the symmetric difference $A + B$; it pictures the statement "a OR b but not both" (as in the statement "Take it or leave it!").

The next result is Exercise 4.73 on page 169; you probably solved this exercise then using elements, as we now do.

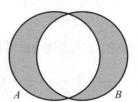

Figure 5.2. $A + B$ and AB.

Proposition 5.33 (De Morgan). *If A and B are subsets of a set X, then*

$$(A \cup B)^c = A^c \cap B^c.$$

Proof. We first show that $(A \cup B)^c \subseteq A^c \cap B^c$. If $x \in (A \cup B)^c$, then $x \notin A \cup B$. But $A \cup B$ consists of all elements in A or in B. So, $x \notin A \cup B$ implies $x \notin A$ and $x \notin B$; that is, $x \in A^c \cap B^c$.

For the reverse inclusion, take $x \in A^c \cap B^c$. Hence, $x \in A^c$ and $x \in B^c$; that is, $x \notin A$ and $x \notin B$. Thus, $x \notin A \cup B$, and $x \in (A \cup B)^c$.   ■

This proof is not very difficult, but it's also not very satisfying. The reasoning very much depends on the meanings of the connectives NOT, AND, and OR, as do the definitions of union and intersection. If feels as if we are just playing with words.

We are going to give a second proof, more in the spirit of commutative rings, that uses the binary operations in a special kind of commutative ring that distills the distinguishing feature of 2^X into one property.

Definition. A ***Boolean ring*** is a commutative ring R in which $a^2 = a$ for all $a \in R$.

Example 5.34.   (i) The ring 2^X of subsets of a set X, in Example 4.47, is a Boolean ring.

(ii) If X is a set, possibly infinite, then the family R of all finite subsets of X together with X itself is a Boolean ring with operations symmetric difference and intersection.   ▲

Let's extend familiar facts in 2^X to arbitrary Boolean rings R. Some of these calculations might look strange; just keep 2^X and Venn diagrams in the back of your mind as you work through them. For example, the following definitions, inspired by the particular Boolean ring 2^X, make sense in any Boolean ring.

Complement: $a' = 1 + a$ (see Exercise 4.69 on page 168)

Union: $a \vee b = a + b + ab$ (see Exercise 4.74(ii) on page 169)

Disjoint: $ab = 0$

Lemma 5.35. *Suppose that R is a Boolean ring and $a \in R$. Then:*

(i) $a + a = 0$

(ii) $aa' = 0$

(iii) $a' + a = 1$.

Proof.   (i)

$$a + a = (a + a)(a + a)$$
$$= a^2 + a^2 + a^2 + a^2$$
$$= a + a + a + a.$$

Now subtract $a + a$ from both sides to obtain $a + a = 0$.

(ii) $aa' = a(1 + a) = a + a^2 = a + a = 0$.

(iii) $a' + a = (1 + a) + a = 1 + (a + a) = 1 + 0 = 1$.   ■

Proposition 5.36. *Let R be a Boolean ring and $a, b \in R$.*

(i) *$a + b = ab' + a'b$, and the summands ab' and $a'b$ are disjoint.*

(ii) *$a \vee b = ab' + a'b + ab$, and the summands ab', $a'b$, and ab are pairwise disjoint.*

Proof.   (i) For all $x \in R$, $x + x = 0$, $xx' = 0$, and $x + x' = 1$. Hence,

$$a + b = a(b + b') + b(a + a') = ab + ab' + ab + a'b = ab' + a'b.$$

The summands are disjoint, because $(ab')(a'b) = 0$.

(ii)

$$a \vee b = a + b + ab = ab' + a'b + ab.$$

The summands are disjoint, for part (i) shows that ab' and $a'b$ are disjoint, while $(ab')ab = 0 = (a'b)ab$.   ∎

Work though these proofs for yourself, justifying each step. Notice how the particulars of 2^X are fading into the background.

Let's now see how working in an arbitrary Boolean ring reduces the proofs about facts in specific such rings like 2^X to algebraic calculations. Compare the set-theoretic proof of Proposition 5.33 with the following proof.

Proposition 5.37 (De Morgan = Proposition 5.33). *If A and B are subsets of a set X, then*

$$(A \cup B)^c = A^c \cap B^c.$$

Proof. We first work in a Boolean ring R and then pass to 2^X.
If $a, b \in R$, we want to show that $(a \vee b)' = a'b'$. But

$$(a \vee b)' = 1 + a \vee b = 1 + (a + b + ab),$$

The proof that $1 + (a + b + ab) = (1 + a)(1 + b)$ could be an exercise in any first-year high school algebra text.

which is equal to $a'b' = (1 + a)(1 + b)$.
Now interpret this general result in R in the particular Boolean ring 2^X, using the translations $A \vee B = A \cup B$, $AB = A \cap B$, and $1 + A = A^c$.   ∎

There's another De Morgan law in Exercise 4.73 on page 169. Algebra to the rescue.

Proposition 5.38 (De Morgan). *If A and B are subsets of a set X, then*

$$(A \cap B)^c = A^c \cup B^c.$$

Proof. Let R be a Boolean ring and $a, b \in R$. We want to show that

$$(ab)' = a' \vee b'.$$

The idea again is to first use "pure algebra," reducing everything to statements about addition and multiplication in R, and then translate the result into the language of 2^X. Now $(ab)' = 1 + ab$, and

$$a' \vee b' = a' + b' + a'b' = (1 + a) + (1 + b) + (1 + a)(1 + b).$$

Calculate:

$$(1 + a) + (1 + b) + (1 + a)(1 + b) = 1 + a + 1 + b + 1 + a + b + ab$$
$$= (1 + 1) + (a + a)$$
$$+ (b + b) + (1 + ab)$$
$$= 1 + ab$$
$$= (ab)'. \quad \blacksquare$$

We now solve an earlier exercise using this point of view.

Proposition 5.39 (= Exercise 4.70). *Let A, S be subsets of a set X. Then $S = A^c$ if and only if $A \cap S = \emptyset$ and $A \cup S = X$.*

Proof. It suffices to work in a Boolean ring and then to see what it says in the particular Boolean ring 2^X.

Assume that $s = a' = 1 + a$. Then

$$as = a(1 + a) = a + a^2 = a + a = 0,$$

and

$$a \vee s = a \vee (1 + a) = a + (1 + a) + a(1 + a) = a + 1 + a + a + a^2 = 1.$$

Conversely, if $as = 0$ and $a + s + as = 1$, then $a + s = 1$. But $-1 = 1$ in every Boolean ring, by Exercise 5.58 on page 226, and so $s = 1 + a = a'$. $\blacksquare$

The usual distributive law in a commutative ring is $a(b+c) = ab+ac$. The proof that the equation holds in 2^X essentially follows from the set-theoretic identity

$$A \cap (B \cup C) = (A \cap B) \cup (A \cap C).$$

We are now going to show that interchanging $\cap$ and $\cup$ gives another valid identity.

Proposition 5.40. *If A, B and C are subsets of a set X, then*

$$A \cup (B \cap C) = (A \cup B) \cap (A \cup C).$$

Proof. We must show that $a \vee bc = (a \vee b)(a \vee c)$; that is,

$$a + bc + abc = (a + b + ab)(a + c + ac).$$

Expand the right-hand side, remembering that $x^2 = x$ and $x + x = 0$ for all x:

$$(a + b + ab)(a + c + ac) = a^2 + ac + a^2c + ab + bc + abc$$
$$+ a^2b + abc + a^2bc$$
$$= a + bc + abc. \quad \blacksquare$$

Exercises

5.58 * Prove that $-1 = 1$ in every Boolean ring.

5.59 * Proposition 5.40 proves that if A, B, and C are subsets of a set X, then

$$A \cup (B \cap C) = (A \cup B) \cap (A \cup C).$$

Give another proof using set theory.

5.60 * If A, B_1, B_2, and B_3 are subsets of a set X, show that

$$A \cap (B_1 \cap B_2 \cap B_3) = (A \cap B_1) \cup (A \cap B_2) \cup (A \cap B_3).$$

Generalize to $A \cap \left(\bigcup_{i=1}^n B_i \right)$.

5.61 Let R be a ring in which multiplication is not assumed to be commutative (see the callout on page 156). If $a^2 = a$ for every $a \in R$, prove that R must be a commutative ring.

5.62 (i) If A, B are subsets of a set X, prove that $B - A = B \cap A^c$.

(ii) In any Boolean ring R, prove that $b + a = b(1 + a) + a(1 + b)$.

In Exercises 5.66 and 5.67, we use minus signs in a Boolean algebra. Since $-1 = +1$, all these signs are really +, but this notation invites you to compare these formulas with the statement of Inclusion-Exclusion.

5.63 Suppose that R is a Boolean ring and $a, b \in R$. Show that

$$a'b' = 1 - a - b + ab,$$

where $a' = 1 + a$.

5.64 Suppose that R is a Boolean ring and $a, b \in R$. Show that

$$1 - a'b' = a \vee b.$$

5.65 Suppose that R is a Boolean ring and $a, b \in R$. Show that

$$a \vee (b \vee c) = (a \vee b) \vee c.$$

5.66 Suppose that R is a Boolean ring and $(a_i)_{i=1}^n$ is a collection of n elements in R. Show that

$$1 - \prod_{i=1}^n a_i' = \bigvee_{i=1}^n a_i.$$

5.67 Suppose that R is a Boolean ring and $(a_i)_{i=1}^n$ is a collection of n elements in R. Show that

$$\prod_{i=1}^n a_i' = 1 - \sum_{1 \leq i \leq n} a_i + \sum_{1 \leq i < j \leq n} a_i a_j - \cdots + (-1)^n a_1 a_2 \dots a_n.$$

Hint. $a' = 1 - a$.

5.68 Suppose that R is a Boolean ring and $(a_i)_{i=1}^n$ is a collection of n elements in R. Show that

$$\bigvee_{i=1}^n a_i = \sum_{1 \leq i \leq n} a_i - \sum_{1 \leq i < j \leq n} a_i a_j + \cdots + (-1)^{n-1} a_1 a_2 \dots a_n.$$

5.69 In a Boolean ring, define $a \leq b$ to mean $a = ab$. Viewing 2^X as a Boolean ring, prove that $A \leq B$ in 2^X if and only if $A \subseteq B$.

5.70 An *atom* in a Boolean ring R is a nonzero element $a \in R$ with $x \leq a$ if and only if $x = 0$ or $x = a$. If R is a finite Boolean ring, prove that every $x \in R$ is a sum of atoms.

5.71 (i) If R is a finite Boolean ring, prove that $R \cong 2^X$, where X is the set of all atoms in R.

   (ii) **Take It Further.** Let R be the Boolean ring of all finite subsets of an infinite set X (see Example 5.34(ii)). Prove that $R \ncong 2^Y$ for any set Y.

   Hint. The simplest solution involves some set theory we have not discussed. If X is countable, then R is countable; however, if Y is any infinite set, then 2^Y is uncountable. Hence, there is no bijection $R \to 2^Y$.

Inclusion-Exclusion

Suppose you have a class of students, all of whom take either French or Spanish, but none of whom take both. If 15 students are studying French and 12 students are studying Spanish, you have $15 + 12 = 27$ students in your class. Denote the number of elements in a finite set A by

$$|A|.$$

Then one way to state the above fact is that if F is the set of students studying French and S is the set of students studying Spanish, then

$$|F \cup S| = |F| + |S|.$$

We make the above counting principle explicit.

Addition Principle. If A and B are *disjoint* finite subsets of a set X, then

$$|A \cup B| = |A| + |B|.$$

The Addition Principle extends, by induction, to any number of finite sets.

Lemma 5.41. *If $(A_i)_{i=1}^{n}$ is a family of pairwise disjoint finite sets, then*

$$\left| \bigcup_{i=1}^{n} A_i \right| = \sum_{i=1}^{n} |A_i|.$$

Proof. The proof is by induction on $n \geq 2$. The base step is the Addition Principle. For the inductive step,

$$\bigcup_{i=1}^{n} A_i = \left(\bigcup_{i=1}^{n-1} A_i \right) \cup A_n.$$

Now $\left(\bigcup_{i=1}^{n-1} A_i \right) \cap A_n = \varnothing$: Exercise 5.60 on page 226 gives

$$\left(\bigcup_{i=1}^{n-1} A_i \right) \cap A_n = (A_1 \cap A_n) \cup \cdots \cup (A_{n-1} \cap A_n),$$

and each $A_i \cap A_n = \varnothing$ because the subsets are pairwise disjoint. ∎

Let's return to your class of students, 15 of whom are studying French and 12 of whom are studying Spanish. What if 4 of them are studying both French and Spanish? You'd then have fewer than 27 students in the class because of double counting. A Venn diagram can help you figure out how to calculate the actual number. The goal of this subsection is to develop a general method of calculating the number of elements in the union of a finite collection of possibly overlapping finite sets.

As a Venn diagram illustrates, the Addition Principle no longer holds if A and B overlap, for elements in $A \cap B$ are counted twice in $|A| + |B|$. What is the formula giving a precise count of $|A \cup B|$? The number of things that get counted twice must be subtracted once.

Lemma 5.42. *If A and B are finite subsets of a set X, then*

$$|A \cup B| = |A| + |B| - |A \cap B|.$$

Proof. First note that $A \cup B$ is the disjoint union

$$A \cup B = (A - B) \cup (B - A) \cup (A \cap B),$$

so that Lemma 5.41 gives

$$|A \cup B| = |A \cap B^c| + |A^c \cap B| + |A \cap B|. \tag{5.1}$$

As usual, we first compute in a Boolean ring R, after which we specialize to 2^X. Recall Proposition 5.36(ii): if $a, b \in R$, then

$$a \vee b = ab' + a'b + ab,$$

where the summands on the right-hand side are pairwise disjoint. Hence, there are two more equations: factor out a to get

$$a \vee b = a(b' + b) + a'b = a + a'b,$$

or factor out b to get

$$a \vee b = ab' + b(a' + a) = ab' + b.$$

Since the summands on the right-hand side of each of the equations are pairwise disjoint, we can pass back to 2^X to obtain

$$|A \cup B| = |A| + |A^c \cap B|$$

and

$$|A \cup B| = |B| + |A \cap B^c|.$$

Add the equations:

$$2|A \cup B| = |A| + |B| + |A^c \cap B| + |A \cap B^c|. \tag{5.2}$$

Now Eq. (5.1) says that the last two terms on the right-hand side of Eq. (5.2) can be replaced by $|A \cup B| - |A \cap B|$, giving

$$2|A \cup B| = |A| + |B| + |A \cup B| - |A \cap B|.$$

Subtracting $|A \cup B|$ from both sides gives the desired result. ∎

Example 5.43. How many positive integers < 1000 are there that are not divisible by 5 or by 7? If the number of positive integers that are divisible by 5 or 7 is D, then the answer is $999 - D$. We compute D using Lemma 5.42.

Let

$$A = \{n \in \mathbb{Z} : 5 \mid n \text{ and } 0 < n < 1000\}$$

and

$$B = \{n \in \mathbb{Z} : 7 \mid n \text{ and } 0 < n < 1000\}.$$

The Division Algorithm gives $|A| = 199$, because $999 = 199 \cdot 5 + 4$; similarly, $|B| = 142$ and $|A \cap B| = 28$, where $A \cap B = \{n \in \mathbb{Z} : 35 \mid n \text{ and } 0 < n < 1000\}$. Hence,

$$|A \cup B| = |A| + |B| - |A \cap B|$$
$$= 199 + 142 - 28 = 313.$$

Therefore, there are exactly $999 - 313 = 686$ positive numbers < 1000 that are not divisible by 5 or by 7.   ▲

How to Think About It. You could probably convince yourself of the result in Lemma 5.42 with a Venn diagram accompanied by a few examples. While diagrams and examples can motivate insight, they are not substitutes for rigorous proof. The reason is that a picture can be misleading. For example, if you aren't careful about drawing a Venn diagram for the union of four or more regions, then some possible intersections might be overlooked.

Example 5.44. Let's look at the case of three finite subsets A, B, and C of a set X. Before reading on, what do you think the formula should be? The basic idea is to apply Lemma 5.42 twice.

$$
\begin{aligned}
|A \cup B \cup C| &= |(A \cup B) \cup C| \\
&= |A \cup B| + |C| - |(A \cup B) \cap C| \\
&= |A| + |B| - |A \cap B| + |C| - |(A \cap C) \cup (B \cap C)| \\
&= |A| + |B| + |C| - |A \cap B| \\
&\quad - \big(|A \cap C| + |B \cap C| - |A \cap B \cap C|\big) \\
&= |A| + |B| + |C| - \big(|A \cap B| + |A \cap C| + |B \cap C|\big) \\
&\quad + |A \cap B \cap C|.
\end{aligned}
$$

So, the number of elements in the union of three sets is the sum of the number of elements in each, minus the sum of the number of elements in the pairwise intersections, plus the number of elements that are common to all three.   ▲

   We want to generalize the formula in Example 5.44 to count the number of elements in a union of finitely many subsets. The difficulty in deriving such a formula by a brutal assault is that we must be careful that an element in the union is not counted several times, for it may occur in the intersection of several of the A_i. To illustrate, consider Figures 5.3 and 5.4, Venn diagrams depicting the various intersections obtained from three subsets and from four subsets.

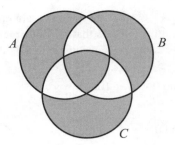

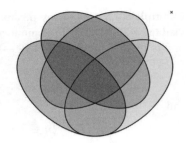

Figure 5.3. Subsets of three regions.   **Figure 5.4.** Subsets of four regions.

To count the number of elements in a union

$$A_1 \cup \cdots \cup A_r,$$

we proceed by induction, using the idea of Example 5.44 by shearing off A_r and treating the union of the rest as one set. The details are technical, so sharpen a pencil and follow along.

Given finite subsets $A_1, \ldots, A_r$ of a set X, let us write

$$A_{ij} = A_i \cap A_j, \text{ where } 1 \leq i < j \leq r,$$
$$A_{ijk} = A_i \cap A_j \cap A_k, \text{ where } 1 \leq i < j < k \leq r,$$
$$\vdots \qquad \vdots$$
$$A_{i_1 i_2 \cdots i_q} = A_{i_1} \cap A_{i_2} \cap \cdots \cap A_{1_q}, \text{ where } 1 \leq i_1 < i_2 < \cdots i_q \leq r,$$
$$\vdots \qquad \vdots$$
$$A_{12 \cdots r} = A_1 \cap \cdots \cap A_r.$$

Theorem 5.45 (Inclusion–Exclusion). *Given finite subsets $A_1, \ldots, A_r$ of a set X, we have*

$$|A_1 \cup \cdots \cup A_r| = \sum_{i \leq r} |A_i| - \sum_{i < j \leq r} |A_{ij}| + \sum_{i < j < k \leq r} |A_{ijk}| - \cdots + (-1)^{r-1}|A_{12 \cdots r}|.$$

Proof. The proof is by induction on $r \geq 2$. The base step is Lemma 5.42.

For the inductive step, the same Lemma gives

$$|A_1 \cup \cdots \cup A_{r-1}) \cup A_r| = |A_1 \cup \cdots \cup A_{r-1}| + |A_r| - |(A_1 \cup \cdots \cup A_{r-1}) \cap A_r|.$$

Now

$$|(A_1 \cup \cdots \cup A_{r-1}) \cap A_r| = |(A_1 \cap A_r) \cup \cdots \cup (A_{r-1} \cap A_r)|$$
$$= |A_{1r} \cup \cdots \cup A_{r-1\,r}|,$$

and the inductive hypothesis gives

$$|A_1 \cup \cdots \cup A_{r-1}| = \sum_{i \leq r-1} |A_i| - \sum_{i < j \leq r-1} |A_{ij}| + \cdots + (-1)^{r-2}|A_{12 \cdots (r-1)}|$$

as well as

$$|A_{1r} \cup \cdots \cup A_{r-1\,r}| = \sum_{i < r} |A_{ir}| - \sum_{i < j < r} |A_{ijr}| + \cdots + (-1)^{r-2}|A_{12 \cdots r}|.$$

Finally, collect terms, realizing that $-(-1)^{r-2} = (-1)^{r-1}$.   ∎

In Exercise 5.76 below, you will use Inclusion-Exclusion to give a formula computing the Euler ϕ-function $\phi(n)$.

Here is an interesting special case of Theorem 5.45 that applies when all intersections $A_{i_1} \cap \cdots \cap A_{i_q}$, for each $q \leq r$, have the same number of elements.

Corollary 5.46 (Uniform Inclusion-Exclusion). *If $A_1, \ldots, A_r$ are finite subsets of a set X such that, for each $q \leq r$, there is an integer s_q with*

$$|A_{i_1} \cap \cdots \cap A_{i_q}| = s_q,$$

then

$$|A_1 \cup \cdots \cup A_r| = rs_1 - \binom{r}{2}s_2 + \binom{r}{3}s_3 - \cdots + (-1)^{r-1}s_r.$$

Proof. By hypothesis, $|A_i| = s_1$ for all i, and so $\sum_i |A_i| = rs_1$. How many terms are there in the sum $\sum_{1 \leq i_1 < \ldots < i_q \leq r} |A_{i_1 \cdots i_q}|$? If $q = 2$, there is one term $|A_{ij}| = |A_i \cap A_j|$ for each pair of distinct A_i, A_j in $\{A_1, \ldots, A_r\}$; that is, there's one term for each choice of 2 of the r subsets. If $q = 3$, there is one term $|A_{ijk}| = |A_i \cap A_j \cap A_k|$ for each triple of distinct A_i, A_j, A_k in $\{A_1, \ldots, A_r\}$; that is, there's one term for each choice of 3 of the r subsets. In general, there are r choose q terms in the sum $\sum_{1 \leq i_1 < i_2 < \cdots < i_q \leq r} |A_{i_1 i_2 \cdots i_q}|$; thus, there are $\binom{r}{q}$ terms of the form $|A_{i_1 i_2 \cdots i_q}|$. Therefore, the sum $\sum_{1 \leq i_1 < \ldots < i_q \leq r} |A_{i_1 \cdots i_q}|$ in the Inclusion-Exclusion formula is here equal to $\binom{r}{q}s_q$. ∎

Example 5.47. Social Security numbers are 9-digit numbers of the form xxx-xx-xxxx (there are some constraints on the digits, but let's not worry about them here). How many Social Security numbers are there that contain all the odd digits?

As usual, it is easier to compute the size of the complement of a union. Let X be the set of all 9-digit numbers and, for $i = 1, 3, 5, 7, 9$, let

$$R_i = \{n \in X : i \text{ is not a digit in } n\}.$$

Thus, $R_1 \cup R_3 \cup R_5 \cup R_7 \cup R_9$ consists of all 9-digit numbers missing at least one odd digit. There are 10^9 Social Security numbers. For each i, we have $|R_i| = 9^9$ (for i does not occur). If $i < j$, then $|R_i \cap R_j| = 8^9$ (for i and j do not occur); if $i < j < k$, then $|R_i \cap R_j \cap R_k| = 7^9$, and so forth. By Corollary 5.46,

$$|R_1 \cup R_3 \cup R_5 \cup R_7 \cup R_9| = 5 \cdot 9^9 - \binom{5}{2}8^9 + \binom{5}{3}7^9 - \binom{5}{4}6^9 + 5^9.$$

Therefore, the answer is $10^9 - |R_1 \cup R_3 \cup R_5 \cup R_7 \cup R_9|$. ▲

You can compute this number explicitly if you really care to know it.

Exercises

5.72 There is a class of students, all of whom are taking French or Spanish. If 15 students are studying French, 12 are studying Spanish, and 4 are studying both, how many students are in the class? Notice that "or" is not "exclusive or."

Answer. 23.

5.73 There is a class of students, all of whom are taking either French, German, or Spanish. Suppose that 15 students are studying French, 12 students are studying German, and 10 students are studying Spanish; moreover, 4 students are studying French and German, 5 are studying German and Spanish, and 3 are studying French and Spanish. One brave soul is studying all three at once. How many students are in the class?

Answer.   26.

5.74 Is "Inclusion-Exclusion" an appropriate name for Theorem 5.45? Why?

5.75 Elvis is playing a game in which he tosses a fair coin and rolls a fair die. He wins if either the coin comes up heads or the die rolls a multiple of 3. What is the probability that Elvis wins the game?

5.76 * Recall that if p is a prime and ϕ is the Euler-ϕ function, then $\phi(p) = p - 1$ (see page 111).

(i) Suppose $n = p_1^{e_1} p_2^{e_2} p_3^{e_3}$ is a product of three prime powers. Show that

$$\phi(n) = n - \frac{n}{p_1} - \frac{n}{p_2} - \frac{n}{p_3} + \frac{n}{p_1 p_2} + \frac{n}{p_1 p_3} + \frac{n}{p_2 p_3} - \frac{n}{p_1 p_2 p_3}.$$

(ii) Generalize to show that if $n = p_1^{e_1} p_2^{e_2} \cdots p_n^{e_n}$, where $p_1, \ldots, p_n$ are distinct primes, then

$$\phi(n) = n \left(1 - \sum_i \frac{1}{p_i} + \sum_{i,j} \frac{1}{p_i p_j} - \sum_{i,j,l} \frac{1}{p_i p_j p_l} + \cdots + (-1)^k \frac{1}{p_1 \cdots p_k} \right).$$

(iii) Using the notation of part (ii), show that

$$\phi(n) = n \prod_{k=1}^{n} \left(1 - \frac{1}{p_k} \right).$$

6 Arithmetic of Polynomials

The two most important rings appearing in precollege mathematics are $\mathbb{Z}$ and $k[x]$ (where k is usually $\mathbb{Q}$, $\mathbb{R}$, or $\mathbb{C}$). The goal of this chapter is to show that these rings share some basic structural properties: both are domains, each has a division algorithm, and non-units in each are products, in essentially only one way, of *irreducibles* (primes in $\mathbb{Z}$, polynomials in $k[x]$ having no nontrivial factorizations); there are numerous other parallels as well. Our program is to take familiar results about $\mathbb{Z}$ and investigate their analogs in $k[x]$. Sometimes a translation from $\mathbb{Z}$ to $k[x]$ is quite simple—not only is the analog of a theorem in Chapter 1 true, but so is its proof, mutatis mutandis; in other cases, however, some modifications in proofs are necessary.

6.1 Parallels to $\mathbb{Z}$

Divisibility

Let's begin with a discussion of divisibility.

Definition. If R is a commutative ring and $a, b \in R$, then a is a ***divisor*** of b, denoted by

$$a \mid b,$$

if there is $r \in R$ with $b = ar$. We continue using the usual synonyms: a divides b or b is a *multiple* of a.

The next result, analogous to Lemma 1.13, will be very useful in what follows. It allows us to use *degree* in $k[x]$ as a proxy for absolute value in $\mathbb{Z}$.

Lemma 6.1. *Let k be a field and let $f(x), g(x) \in k[x]$. If $f \neq 0$ and $f \mid g$, then*

$$\deg(f) \leq \deg(g).$$

Proof. If $g = fq$, where $q(x) \in k[x]$, then Lemma 5.8(ii) gives $\deg(g) = \deg(fq) = \deg(f) + \deg(q)$. Since $\deg(q) \geq 0$, we have $\deg(f) \leq \deg(g)$. ∎

Recall that a *unit* in a commutative ring R is an element that has a multiplicitive inverse in R. The only units in $\mathbb{Z}$ are ± 1, but the polynomial ring $k[x]$, where k is a field, has many units, as the next proposition shows.

Sometimes, we'll denote a polynomial by $f(x)$; other times, we'll simply write f. Both conventions are commonly used in algebra.

Proposition 6.2. *If k is a field, then $u(x) \in k[x]$ is a unit if and only if u is a nonzero polynomial of degree 0; that is, u is a nonzero constant.*

Proof. If u is a unit, then there is a polynomial $v(x) \in k[x]$ with $uv = 1$. Thus, $u \mid 1$ and, by Lemma 6.1, we have $\deg(u) \leq \deg(1) = 0$. Hence, $\deg(u) = 0$.

Conversely, if $\deg(u) = 0$, then $u \in k$. Since k is a field and $u \neq 0$, there is an inverse u^{-1} in k; that is, u is a unit in k. A fortiori, u is a unit in $k[x]$. ∎

Describing the units in $k[x]$ when k is not a field is much more complicated. For example, a nonzero constant need not be a unit: 5 is not a unit in $\mathbb{Z}[x]$. And a unit need not be a constant: $(2x + 1)^2 = 4x^2 + 4x + 1 = 1$ in $\mathbb{Z}_4[x]$, so that $2x + 1$ is a unit in $\mathbb{Z}_4[x]$ (it is its own inverse).

Multiplying an element of a commutative ring by a unit doesn't change any of its essential algebraic properties. It's convenient to give a name to elements that are so related.

Definition. An *associate* of an element a in a commutative ring R is an element of the form ua for some unit $u \in R$.

Example 6.3. (i) Since the only units in $\mathbb{Z}$ are ± 1, the associates of an integer m are $\pm m$.

(ii) There are only four units in the Gaussian integers $\mathbb{Z}[i]$, by Proposition 4.42: namely ± 1 and $\pm i$. Hence, every nonzero Gaussian integer z has four associates: $z, -z, iz, -iz$.

(iii) There are exactly six units in the Eisenstein integers $\mathbb{Z}[\omega]$, where $\omega = \frac{1}{2}\left(-1 + i\sqrt{3}\right)$, by Exercise 4.45 on page 165. Hence, every Eisenstein integer z has exactly six associates: $\pm z, \pm \omega z, \pm \omega^2 z$.

(iv) If k is a field, Proposition 6.2 says that the associates of $f(x) \in k[x]$ are nonzero multiples uf for $u \in k$. ▲

Proposition 6.4. *If k is a field, every nonzero polynomial in $k[x]$ has a monic associate.*

Proof. If the leading coefficient of f is c, then c, being a nonzero element of k, is a unit, and so f is associate to $c^{-1}f$. ∎

In a commutative ring R, every element $a \in R$ is divisible by units u (for $a = u(u^{-1}a)$) and associates ua [for $a = u^{-1}(ua)$]. An element having only these obvious divisors is called *irreducible*.

Definition. An element a in a commutative ring R is *irreducible* in R if it is neither zero nor a unit and its only divisors are units and associates.

The definition of prime on page 22 says that primes are positive.

An integer n is irreducible in $\mathbb{Z}$ if and only if $n = \pm p$ for some prime p; that is, n is an associate of a prime. When k is a field, Proposition 6.2 implies that every associate uf of a polynomial $f(x) \in k[x]$ has the same degree as f, and it is easy to see that if f is irreducible, then uf is also irreducible.

How to Think About It. We have defined *irreducible in R*, not *irreducible*, for irreducibility depends on the ambient ring. In particular, irreducibility of a polynomial in $k[x]$ depends on the coefficient ring k, hence on $R = k[x]$. For example, the polynomial $x^2 + 1$, when viewed as lying in $\mathbb{R}[x]$, is irreducible. On the other hand, when $x^2 + 1$ is viewed as lying in the larger ring $\mathbb{C}[x]$, it is not irreducible, for $x^2 + 1 = (x + i)(x - i)$ and neither factor is a unit. Similarly, a prime p may factor in some larger commutative ring containing $\mathbb{Z}$. For example, in the Gaussian integers $\mathbb{Z}[i]$, the prime 5 in $\mathbb{Z}$ factors: $5 = (2 + i)(2 - i)$. Since the only units in $\mathbb{Z}[i]$ are ± 1 and $\pm i$, by Example 6.3(iii), the factors are neither units nor associates of 5 in $\mathbb{Z}[i]$.

In general, testing a polynomial for irreducibility is hard. Here is a criterion for irreducibility of polynomials over fields that uses degree to narrow the kinds of polynomials that need to be tested as factors.

Proposition 6.5. *Let k be a field and let $f(x) \in k[x]$ be a nonconstant polynomial. Then f is irreducible in $k[x]$ if and only if it has no factorization $f = gh$ in $k[x]$ with both factors having degree $< \deg(f)$.*

Proof. If f is irreducible in $k[x]$ and $f = gh$, then one factor, say g, is a unit (why?). By Proposition 6.2, we have $\deg(g) = 0 < \deg(f)$, for f is nonconstant.

Conversely, if $f = gh$ and f is not a product of polynomials of smaller degree, then one factor, say g, must have degree 0, hence it is a unit. Therefore f is irreducible in $k[x]$. ∎

And the other factor is $h = g^{-1}f$, an associate of f.

Every linear polynomial $a(x) = rx + s \in k[x]$, where k is a field, is irreducible in $k[x]$: if $a = fg$, then $1 = \deg(a) = \deg(f) + \deg(g)$. Hence, $\deg(f), \deg(g) \in \{0, 1\}$. It follows that one degree is 0 and the other is 1, and so a is irreducible, by Proposition 6.5. There are fields k whose only irreducible polynomials are linear; for example, the Fundamental Theorem of Algebra says that $\mathbb{C}$ is such a field.

See Corollary 6.15.

Proposition 6.5 need not be true if the ring of coefficients is not a field. Indeed, linear polynomials need not be irreducible. For example, $5x + 5 = 5(x + 1)$ is not irreducible in $\mathbb{Z}[x]$, even though one factor has degree 0 and the other degree 1, for 5 is not a unit in $\mathbb{Z}[x]$.

Proposition 6.6. *Let R be a domain and let $a, b \in R$.*

(i) $a \mid b$ and $b \mid a$ if and only if a and b are associates.

(ii) *Let k be a field and $a, b \in R = k[x]$ be monic polynomials. If $a \mid b$ and $b \mid a$, then $a = b$.*

Proof. (i) If $a \mid b$ and $b \mid a$, there are $r, s \in R$ with $b = ra$ and $a = sb$, and so $b = ra = rsb$. If $b = 0$, then $a = 0$ (because $b \mid a$); if $b \neq 0$, then we may cancel it (R is a domain) to obtain $1 = rs$. Hence, r and s are units, and a and b are associates. The converse is obvious (and it does not need the hypothesis that R be a domain).

(ii) Corollary 5.9 tells us that R is a domain, so, by part (i), there is a unit $u \in k[x]$ with $a = ub$. Now u is a nonzero constant, by Proposition 6.2. Because $a \mid b$ and $b \mid a$, a and b have the same degree (by Lemma 6.1), say m. Since they are monic, the leading coefficient of ub is u and the leading coefficient of a is 1. Hence $u = 1$ and $a = b$. ∎

The next example shows that we need the hypothesis in Proposition 6.6 that R be a domain.

Example 6.7 (Kaplansky). Let X be the interval $[0, 3]$. We claim that there are elements $a, b \in C(X)$ (see Example 4.31(ii)) each of which divides the other yet which are not associates. Define

$$a(t) = 1 - t = b(t) \qquad \text{for all } t \in [0, 1]$$
$$a(t) = 0 = b(t) \qquad \text{for all } t \in [1, 2]$$
$$a(t) = t - 2 \qquad \text{for all } t \in [2, 3]$$
$$b(t) = -t + 2 \qquad \text{for all } t \in [2, 3].$$

If $v \in C(X)$ satisfies $v(t) = 1$ for all $t \in [0, 1]$ and $v(t) = -1$ for all $t \in [2, 3]$, then it is easy to see that $b = av$ and $a = bv$ (same v); hence, a and b divide each other.

Suppose a and b are associates: there is a unit $u \in C(X)$ with $b = au$. As for v above, $u(t) = 1$ for all $t \in [0, 1]$ and $u(t) = -1$ for all $t \in [2, 3]$; in particular, $u(1) = 1$ and $u(2) = -1$. Since u is continuous, the Intermediate Value Theorem of calculus says that $u(t) = 0$ for some $t \in [1, 2]$. But this contradicts Exercise 4.41(ii) on page 164, which says that units in $C(X)$ are never 0.   ▲

The next result shows that irreducible polynomials over a field behave like primes in $\mathbb{Z}$; they are "building blocks" in the sense that every nonconstant polynomial can be expressed in terms of them.

Proposition 6.8. *If k is a field, then every nonconstant polynomial in $k[x]$ is a product of irreducibles.*

We continue to use the term *product* as we have in earlier chapters: a product can have only one factor. Thus, it's okay to say that a single irreducible is a product of irreducibles.

Proof. If the proposition is false, then the set

$$C = \{a(x) \in k[x] : a \text{ is neither a constant nor a product of irreducibles}\}$$

is nonempty. Let $h(x) \in C$ have least degree (the Least Integer Axiom guarantees h exists). Since $h \in C$, it is not a unit, and so $0 < \deg(h)$; since h is not irreducible, $h = fg$, where neither f nor g is a unit, and so, by Proposition 6.2, neither f nor g is constant. Hence, Lemma 6.1 gives $0 < \deg(f) < \deg(h)$ and $0 < \deg(g) < \deg(h)$. It follows that $f \notin C$ and $g \notin C$, for their degrees are too small (h has the smallest degree of polynomials in C). Thus, both f and g are products of irreducibles and, hence, $h = fg$ is a product of irreducibles, contradicting $h \in C$. Therefore, C is empty, and the proposition is true.   ∎

Corollary 6.9. *If k is a field, then every nonconstant $f(x) \in k[x]$ has a factorization*

$$f(x) = ap_1(x) \cdots p_n(x),$$

where a is a nonzero constant and the p_i are monic irreducibles.

Proof. Apply the result of Exercise 6.8 on page 243 to a factorization of f as in the proposition.   ∎

We continue showing that polynomials over fields behave very much like integers. Let's first do some long division.

$$
\begin{array}{r}
4x^3 - 14x^2 \hspace{3.5cm} \\
x^2 + 3x - 2 \overline{\smash{\big)}\, 4x^5 - 2x^4 + x^3 \cdots} \\
\underline{4x^5 + 12x^4 - 8x^3} \hspace{0.9cm} \\
-14x^4 + 9x^3 \hspace{0.9cm}
\end{array}
$$

$$\vdots$$

This process can be completed until we get 0 or a remainder of degree < 2 (which is it?). Generalizing, there is a Division Algorithm for $R[x]$, where R is any commutative ring: if $a(x), b(x) \in R[x]$ and a is monic, then there are $q(x), r(x) \in R[x]$ with $b = qa + r$, where $r = 0$ or $\deg(r) < \deg(a)$. The basic idea is to mimic what we've just done.

Proposition 6.10. *Let R be a commutative ring and $f(x), g(x) \in R[x]$. If f is monic, then there exist $q(x), r(x) \in R[x]$ with*

$$g = qf + r,$$

where $r = 0$ or $\deg(r) < \deg(f)$.

Proof. Let

$$f = x^n + a_{n-1}x^{n-1} + \cdots + a_0 \quad \text{and} \quad g = b_m x^m + b_{m-1}x^{m-1} + \cdots + b_0.$$

If $m = \deg(g) < \deg(f) = n$, then take $q = 0$ and $r = g$.

If $m \geq n$, the quotient begins with $b_m x^{m-n}$ multiplied by f; now subtract, getting a polynomial of degree less than m. The rest of the proof is by induction on $m = \deg(g) \geq n$. If

$$G(x) = g - b_m x^{m-n} f,$$

then either $G = 0$ or $\deg(G) < m = \deg(g)$. If $G = 0$, we are done: set $q = b_m x^{m-n}$ and $r = 0$. If $G \neq 0$, the inductive hypothesis gives polynomials q' and r with $G = q'f + r$, where either $r = 0$ or $\deg(r) < \deg(f)$. Therefore, $g - b_m x^{m-n} f = q'f + r$, and so

$$g = \left(b_m x^{m-n} + q' \right) f + r. \quad \blacksquare$$

In $\mathbb{Z}$, if $b < a$, then $b = 0a + b$; for example, $27 = 0 \cdot 35 + 27$. Similarly for polynomials: $x^2 + 1 = 0(x^3 + x^2 - 1) + (x^2 + 1)$.

When R is a field, we can divide by every nonzero polynomial, not merely by monic ones; moreover, the quotient and remainder are unique.

Theorem 6.11 (Division Algorithm). *Let k be a field and $f(x), g(x) \in k[x]$. If $f \neq 0$, then there exist unique $q(x), r(x) \in k[x]$ with*

$$g = qf + r,$$

where $r = 0$ or $\deg(r) < \deg(f)$.

Proof. We first prove the existence of q and r. Now $f = a_n x^n + \cdots + a_0$, where $a_n \neq 0$. Since k is a field, it contains the inverse a_n^{-1}. Hence, $a_n^{-1} f$ is monic, and Proposition 6.10 gives $q'(x), r(x) \in k[x]$ with

$$g = q'(a_n^{-1} f) + r,$$

where either $r = 0$ or $\deg(r) < \deg(a_n^{-1} f) = \deg(f)$. Therefore,

$$g = qf + r,$$

where $q = q'a_n^{-1}$.

To prove uniqueness of q and r, assume that $g = Qf + R$, where $R = 0$ or $\deg(R) < \deg(f)$. Then $qf + r = g = Qf + R$, and

$$(q - Q)f = R - r.$$

If $R \neq r$, then each side, being nonzero, has a degree. Since k is a field, $k[x]$ is a domain (Lemma 5.8), and so

$$\deg\big((q - Q)f\big) = \deg(q - Q) + \deg(f)$$
$$\geq \deg(f),$$

while $\deg(R - r) \leq \max\{\deg(R), \deg(r)\} < \deg(f)$, a contradiction. Hence, $R = r$ and $(q - Q)f = 0$. As $f \neq 0$, it can be canceled: thus, $q - Q = 0$ and $q = Q$. ∎

By Exercise 6.5 on page 243, Theorem 6.11 remains true if we weaken the hypothesis so that k is only a domain.

There is a two-step strategy to determine whether one integer divides another: first, use the Division Algorithm; then show that the remainder is zero. This same strategy can now be used for polynomials.

Example 6.12. This example shows that quotients and remainders may not be unique when the coefficients do not lie in a domain. In $\mathbb{Z}_4[x]$, let $b(x) = 2x^3 + 3$ and $a(x) = 2x^2 + 2x + 1$. Then

$$2x^3 + 3 = (x + 1)(2x^2 + 2x + 1) + (x + 2)$$
$$= (x + 3)(2x^2 + 2x + 1) + x.$$

The quotient and remainder in the first equation are $x + 1$ and $x + 2$, while the quotient and remainder in the second equation are $x + 3$ and x. Note that both $x + 2$ and x are linear, and hence

$$\deg(x + 2) = \deg(x)$$
$$= 1$$
$$< \deg(a)$$
$$= 2. \blacktriangle$$

In forthcoming investigations into roots of unity, we'll need to know whether $x^m - 1$ divides $x^n - 1$. Certainly this is true when $m \mid n$ because, if $n = mq$,

$$x^n - 1 = x^{mq} - 1$$
$$= (x^m)^q - 1$$
$$= (x^m - 1)\Big((x^m)^{q-1} + (x^m)^{q-2} + \cdots + \cdots (x^m)^2 + x^m + 1\Big).$$

The converse is also true, and the proof uses the Division Algorithms in $\mathbb{Z}$ and in $k[x]$.

Proposition 6.13. *If k is a field, then $x^m - 1$ divides $x^n - 1$ in $k[x]$ if and only if $m \mid n$.*

Proof. We've seen above that $x^m - 1$ divides $x^n - 1$ if $m \mid n$.

Conversely, suppose that $x^m - 1$ divides $x^n - 1$. If $n = mq + r$ where $0 \le r < m$, then

$$
\begin{aligned}
x^n - 1 &= x^{mq+r} - 1 \\
&= x^{mq+r} - x^r + x^r - 1 \\
&= x^r (x^{mq} - 1) + (x^r - 1).
\end{aligned}
$$

We're assuming that $x^m - 1$ divides $x^n - 1$ and, as in the discussion just preceding this proposition, $x^m - 1$ divides $x^{mq} - 1$. Hence, by the 2 out of 3 property for polynomials (Exercise 6.7 on page 243), $x^m - 1$ divides $x^r - 1$. Since $r < m$, we must have $r = 0$ (why?). ■

Roots

We are going to apply the preceding results to roots of polynomials. We've been using the word "root" all along; let's begin with a formal definition.

Definition. If $f(x) \in k[x]$, where k is a field, then a **root** of f **in k** is an element $a \in k$ with $f(a) = 0$.

How to Think About It. We have just defined "root in k," not "root." Often, a root of a polynomial $f(x) \in k[x]$ may live in a larger field K containing k, but we still call it a root of f. For example, $f(x) = x^2 - 2$ has its coefficients in $\mathbb{Q}$, but we usually say that $\sqrt{2}$ is a root of f even though $\sqrt{2}$ is irrational; that is, $\sqrt{2} \notin \mathbb{Q}$.

Etymology. Why is a root so called? Just as the Greeks called the bottom side of a triangle its base (as in the area formula $\frac{1}{2}$ altitude × base), they also called the bottom side of a square its base. A natural question for the Greeks was: given a square of area A, what is the length of its side? Of course, the answer is $\sqrt{A}$. Were we inventing a word for $\sqrt{A}$, we might have called it the *base* of A or the *side* of A. Similarly, consider the analogous three-dimensional question: given a cube of volume V, what is the length of its edge? The answer $\sqrt[3]{V}$ might be called the *cube base* of V, and $\sqrt{A}$ might then be called the *square base* of A. Why, then, do we call these numbers cube *root* and square *root*? What has any of this to do with plants?

Since tracing the etymology of words is not a simple matter, we only suggest the following explanation. Through 400 CE, most mathematics was written in Greek, but, by the fifth century, India had become a center of mathematics, and important mathematical texts were also written in Sanskrit. The Sanskrit term for square root is *pada*. Both Sanskrit and Greek are Indo-European languages, and the Sanskrit word *pada* is a cognate of the Greek word *podos*; both mean *base* in the sense of the foot of a pillar or, as above, the bottom of a square. In both languages, however, there is a secondary meaning "the root of a plant." In translating from Sanskrit, Arab mathematicians chose the secondary

This title can be translated from Arabic, but the words already had a technical meaning: both *jabr* and *muqabala* refer to operations akin to subtracting the same number from both sides of an equation.

meaning, perhaps in error (Arabic is not an Indo-European language), perhaps for some unknown reason. For example, the influential book by al-Khwarizmi, *Al-jabr w'al muqabala*, which appeared in the year 830 CE, used the Arabic word *jidhr*, meaning root of a plant. (The word "algebra" is a European version of the first word in the title of this book; the author's name has also come into the English language as the word "algorithm.") This mistranslation has since been handed down through the centuries; the term *jidhr* became standard in Arabic mathematical writings, and European translations from Arabic into Latin used the word *radix* (meaning root, as in *radish* or *radical*). The notation $r2$ for $\sqrt{2}$ occurs in European writings from about the twelfth century (but the square root symbol did not arise from the letter r; it evolved from an old dot notation). However, there was a competing notation in use at the same time, for some scholars who translated directly from the Greek denoted $\sqrt{2}$ by $l2$, where l abbreviates the Latin word *latus*, meaning "side." Finally, with the invention of logarithms in the 1500s, r won out over l, for the notation $l2$ was then commonly used to denote $\log 2$. The passage from square root to cube root to the root of a polynomial equation other than $x^2 - a$ and $x^3 - a$ is a natural enough generalization. Thus, as pleasant as it would be, there seems to be no botanical connection with roots of equations.

Proposition 6.14 (Remainder Theorem). *Let* $f(x) \in k[x]$, *where* k *is a field. If* $u \in k$, *then there is* $q(x) \in k[x]$ *with*

$$f(x) = q(x)(x - u) + f(u).$$

Proof. The Division Algorithm gives

$$f(x) = q(x)(x - u) + r,$$

where either $r = 0$ or $\deg(r) < \deg(x - u) = 1$; hence, the remainder r is a constant. By Corollary 5.21, evaluation at u is a homomorphism; hence, $f(u) = q(u)(u - u) + r$, and so $f(u) = r$. ∎

Proposition 6.14 is often paraphrased to say that $f(u)$ is the remainder after dividing $f(x)$ by $x - u$.

Here is a connection between roots and factoring.

Corollary 6.15 (Factor Theorem). *Let* $f(x) \in k[x]$, *where* k *is a field, and let* $a \in k$. *Then* a *is a root of* f *in* k *if and only if* $x - a$ *divides* f.

Proof. If a is a root of f in k, then $f(a) = 0$, and Proposition 6.14 gives $f(x) = q(x)(x - a)$. Conversely, if $f(x) = g(x)(x - a)$, then evaluating at a gives $f(a) = g(a)(a - a) = 0$; that is, a is a root of f in k. ∎

The next result turns out to be very important.

Theorem 6.16. *Let* k *be a field. If* $f(x) \in k[x]$ *has degree* n, *then* f *has at most* n *roots in* k.

Proof. We prove the statement by induction on $n \geq 0$. If $n = 0$, then f is a nonzero constant, and the number of its roots in k is zero. Now let $n > 0$. If f

has no roots in k, we are done, for $0 \leq n$. Otherwise, we may assume that f has a root a in k. By Corollary 6.15,

$$f(x) = q(x)(x - a);$$

moreover, $q(x) \in k[x]$ has degree $n - 1$. If there is another root of f in k, say b, where $b \neq a$, then evaluating at b gives

$$0 = f(b) = q(b)(b - a).$$

Since $b - a \neq 0$, we have $q(b) = 0$ (for k is a field, hence a domain); that is, b is a root of q. But $\deg(q) = n - 1$, so that the inductive hypothesis says that q has at most $n - 1$ roots in k. Therefore, f has at most n roots in k, namely a and the roots of q.   ∎

Example 6.17.   Theorem 6.16 is not true for polynomials with coefficients in an arbitrary commutative ring. For example, the quadratic polynomial $x^2 - 1$ in $\mathbb{Z}_8[x]$ has four roots in $\mathbb{Z}_8$, namely 1, 3, 5, and 7. On the other hand, Exercise 6.14 on page 247 says that Theorem 6.16 remains true if we assume that the coefficient ring is only a domain.   ▲

Recall that every polynomial $f(x) \in k[x]$ determines the polynomial function $f^{\#} \in \mathrm{Poly}(k)$, where $f^{\#} \colon k \to k$ is defined by $a \mapsto f(a)$ for all $a \in k$. On page 204, however, we saw that the nonzero polynomial $f(x) = x^p - x \in \mathbb{F}_p[x]$ determines the constant function zero; different polynomials can determine the same polynomial function. This pathology vanishes when the field k is infinite.

Proposition 6.18.   *Let k be an infinite field and $f(x), g(x) \in k[x]$. If f and g determine the same polynomial function (that is, $f^{\#} = g^{\#}$, so that $f(a) = g(a)$ for all $a \in k$), then $f = g$.*

Proof. If $f \neq g$, then the polynomial $h = f - g$, being nonzero, has a degree, say n. But every element of k is a root of h; since k is infinite, h has more than n roots, and this contradicts Theorem 6.16.   ∎

This proof yields a more general result.

Corollary 6.19.   *Let k be a (possibly finite) field, and let $f(x), g(x) \in k[x]$, where $\deg(f) \leq \deg(g) = n$. If $f(a) = g(a)$ for $n + 1$ elements $a \in k$, then $f = g$.*

Proof. If $f \neq g$, then $\deg(f - g)$ is defined; but $\deg(f - g) \leq n$, and so $f - g$ has too many roots.   ∎

We can now show that $k[x]$ and $\mathrm{Poly}(k)$ are structurally the same for the most familiar fields k.

Theorem 6.20.   *If k is an infinite field, then*

$$k[x] \cong \mathrm{Poly}(k).$$

Proof. In Example 5.14(ii), we saw that $\varphi\colon k[x] \to \mathrm{Poly}(k)$, sending $f \mapsto f^{\#}$, is a surjective homomorphism. Since k is infinite, Proposition 6.18 applies to show that φ is injective. Therefore, φ is an isomorphism.   ∎

We now generalize Proposition 6.18 to polynomials in several variables. Denote the n-tuple $(x_1, \ldots, x_n)$ by X.

Proposition 6.21. *Let k be an infinite field.*

(i) *If $f(X) \in k[X] = k[x_1, \ldots, x_n]$ is nonzero, then there are $a_1, \ldots, a_n \in k$ with $f(a_1, \ldots, a_n) \neq 0$.*

(ii) *If $f(X), g(X) \in k[X]$ and*

$$f(a_1, \ldots, a_n) = g(a_1, \ldots, a_n) \ \text{for all} \ (a_1, \ldots, a_n) \in k^n,$$

then $f = g$.

Proof. (i) The proof is by induction on $n \geq 1$. If $n = 1$, then the result is Proposition 6.18, because $f(a) = 0$ for all $a \in k$ implies $f = 0$. For the inductive step, assume that

$$f(x_1, \ldots, x_n) = B_0 + B_1 x_n + B_2 x_n^2 + \cdots + B_r x_n^r,$$

where all $B_i \in k[x_1, \ldots, x_{n-1}]$ and $B_r \neq 0$. By induction, there is $\alpha = (a_1, \ldots, a_{n-1}) \in k^{n-1}$ with $B_r(\alpha) \neq 0$. Hence, $f(\alpha, x_n) \in k[x_n]$, and

$$f(\alpha, x_n) = B_0(\alpha) + B_1(\alpha) x_n + \cdots + B_r(\alpha) x_n^r \neq 0.$$

Since $f(\alpha, x_n)$ and $g(\alpha, x_n)$ lie in $k[x_n]$, we have $f(\alpha, b), g(\alpha, b) \in k$.

By the base step, there is $a_n \in k$ with $f(\alpha, a_n) \neq 0$.

(ii) The proof is by induction on $n \geq 1$; the base step is Proposition 6.18. For the inductive step, write

$$f(X, y) = \sum_i p_i(X) y^i \quad \text{and} \quad g(X, y) = \sum_i q_i(X) y^i,$$

where X denotes $(x_1, \ldots, x_{n-1})$ (by allowing some p's and q's to be zero, we may assume that both sums involve the same indices i). Suppose that $f(\alpha, b) = g(\alpha, b)$ for every $\alpha \in k^{n-1}$ and every $b \in k$. For fixed $\alpha \in k^{n-1}$, define $F_\alpha(y) = \sum_i p_i(\alpha) y^i$ and $G_\alpha(y) = \sum_i q_i(\alpha) y^i$. Since both $F_\alpha(y)$ and $G_\alpha(y)$ are in $k[y]$, the base step gives $p_i(\alpha) = q_i(\alpha)$ for all $\alpha \in k^n$. By the inductive hypothesis, $p_i(X) = q_i(X)$ for all i, and hence

$$f(X, y) = \sum_i p_i(X) y^i = \sum_i q_i(X) y^i = g(X, y).   ∎$$

Exercises

6.1 Prove that the only units in $\mathbb{Z}[x]$ are ± 1, and that the only associates of a polynomial $f(x) \in \mathbb{Z}[x]$ are $\pm f$.

6.2 * Let R be a domain, and let $p(x), q(x) \in R[x]$.

(i) If p and q are irreducible, prove that $p \mid q$ if and only if there is a unit u with $q = up$.

(ii) If, in addition, both p and q are monic, prove that $p \mid q$ implies $u = 1$ and $p = q$.

6.3 (i) If R is a domain, prove that the only units in $R[x]$ are units in R.

   (ii) The domain $\mathbb{Z}_2$ has only one unit. Give an example of an infinite domain having only one unit.

6.4 Let R be a commutative ring and let $a(x), b(x) \in R[x]$, where $a \neq 0$. Prove that Proposition 6.10 generalizes: if the leading coefficient of a is a unit, then there exist $q(x), r(x) \in R[x]$ with $b = qa + r$, where either $r = 0$ or $\deg(r) < \deg(a)$.

6.5 * Let k be a domain and let $a(x), b(x) \in k[x]$, where $a \neq 0$. Prove that the uniqueness statement in the Division Algorithm generalizes: if there are q, r, Q, R in $k[x]$ with $qa + r = b = Qa + R$, where $r = 0$ or $\deg(r) < \deg(a)$, and where $R = 0$ or $\deg(R) < \deg(a)$, then $R = r$ and $Q = q$.

6.6 Let k be a domain and let $f(x) \in k[x]$. If $a(x)$ is an associate of f, prove that $\deg(f) = \deg(a)$. Give an example to show that the statement may be false if k is not a domain.

6.7 * Show that there is a "2 out of 3" result for polynomials, analogous to the one for integers: if k is a field and $f, g, h \in k[x]$ are polynomials such that $f = g + h$, then a polynomial that divides two of the three will divide the third.

6.8 * Let R be a domain and $f(x) \in R[x]$ be nonzero. If $f = g_1 \cdots g_n$, where $g_i(x) \in R[x]$ for all i, show that there exist a nonzero $a \in R$ and monic $g_i'(x) \in R[x]$ with $f = ag_1' \cdots g_n'$.

6.9 (i) Let $f(x), g(x) \in \mathbb{Q}[x]$ with f monic. Write a pseudocode (or a program in a CAS) implementing the Division Algorithm with input f, g and output $q(x), r(x)$, the quotient and remainder.

   (ii) Find the quotient and remainder by dividing $x^3 + 2x^2 - 8x + 6$ by $x - 1$ as you would in high school.

6.10 * If R is a commutative ring, define a relation $\equiv$ on R by $a \equiv b$ if they are associates. Prove that $\equiv$ is an equivalence relation on R.

6.11 A student claims that $x - 1$ is not irreducible in $\mathbb{Q}[x]$ because there is a factorization $x - 1 = (\sqrt{x} + 1)(\sqrt{x} - 1)$. Explain the error of his ways.

6.12 * Prove that the ideal (x, y) in $k[x, y]$, where k is a field, is not a principal ideal.

Greatest Common Divisors

We now introduce gcd's of polynomials $f(x), g(x) \in R[x]$. It doesn't make sense to say that $f \leq g$, even when $R = \mathbb{R}$, but it does make sense to say $\deg(f) \leq \deg(g)$. Although some of the coming definitions make sense for polynomial rings $R[x]$ over a commutative ring R, we will focus our attention on the rings $k[x]$ for fields k.

Definition. Let k be a field. A ***common divisor*** of polynomials $a(x), b(x) \in k[x]$ is a polynomial $c(x) \in k[x]$ with $c \mid a$ and $c \mid b$. If a and b are not both 0, define their ***greatest common divisor***, denoted by

$$\gcd(a, b),$$

to be a monic common divisor d of a and b of largest degree. If $a = 0 = b$, define $\gcd(0, 0) = 0$.

Note the convention that greatest common divisors are monic. We'll say more about this in a moment.

The next proposition shows that gcd's exist; it is true, but not obvious, that every pair $a, b \in k[x]$ has a unique gcd (Corollary 6.29).

Proposition 6.22. *If k is a field and $a(x), b(x) \in k[x]$, then a gcd of a, b exists.*

Proof. We saw, in Lemma 6.1, that if c and a are polynomials with $c \mid a$, then $\deg(c) \leq \deg(a)$. It follows that gcd's exist, for common divisors do exist (1 is always a common divisor), and there is an upper bound on the degrees of common divisors, namely, $\max\{\deg(a), \deg(b)\}$. Finally, a common divisor d of largest degree can be replaced by a monic associate.   ■

Defining gcd's of polynomials to be monic is just a normalization; after all, when we defined gcd's of integers, we insisted they should be positive. This will be needed to prove uniqueness of gcd's.

Example 6.23. Here is an easy computation of a gcd, generalizing Lemma 1.17. Let k be a field and $p(x) \in k[x]$ be a monic irreducible polynomial. If $b(x) \in k[x]$, then

$$\gcd(p, b) = \begin{cases} p & \text{if } p \mid b \\ 1 & \text{otherwise.} \end{cases}$$

A common divisor c of p and b is, in particular, a divisor of p. But the only monic divisors of p are p and 1, and so $\gcd(p, b) = p$ or 1; it is p if $p \mid b$, and it is 1 otherwise.   ▲

We are going to see that gcd's of polynomials are linear combinations. The proof of this fact for gcd's of integers essentially involved ideals in $\mathbb{Z}$, and so we now examine ideals in $k[x]$.

In any commutative ring R, associates a and b generate the same principal ideal (the converse may be false if R is not a domain).

Proposition 6.24. *Let R be a domain and $a, b \in R$. The principal ideals (a) and (b) are equal if and only if a and b are associates.*

Proof. If $(a) = (b)$, then $a \in (b)$; hence, $a = rb$ for some $r \in R$, and so $b \mid a$. Similarly, $b \in (a)$ implies $a \mid b$, and so Proposition 6.6 shows that a and b are associates.

Conversely, if $a = ub$, where u is a unit, then $a \in (b)$ and $(a) \subseteq (b)$. Similarly, $b = u^{-1}a$ implies $(b) \subseteq (a)$, and so $(a) = (b)$.   ■

Ideals in general commutative rings can be quite complicated, but we have seen, in Theorem 5.29, that every ideal in $\mathbb{Z}$ is principal. When k is a field, all the ideals in $k[x]$ are also principal.

Theorem 6.25. *If k is a field, then every ideal in $k[x]$ is a principal ideal. In fact, either $I = (0)$ or there is a unique monic $d(x)$ with $I = (d) = \{rd : r \in k\}$.*

Proof. If $I = (0)$, then I is a principal ideal with generator 0; that is, $I = (0)$. Otherwise, let $a(x)$ be a polynomial in I of least degree. Since $a \in k[x]$ is nonzero, its leading coefficient $c \neq 0$; since k is a field, c^{-1} exists, and $d = c^{-1}a$ is monic. By Proposition 6.24, $(a) = (d)$.

Clearly, $(d) \subseteq I$. For the reverse inclusion, let $f(x) \in I$. By the Division Algorithm, there are $q(x), r(x) \in k[x]$ with $f = qd + r$, where either $r = 0$ or $\deg(r) < \deg(d)$. But $r = f - qd \in I$, so that if $r \neq 0$, its existence contradicts d being a polynomial in I of least degree. Hence, $r = 0$, $d \mid f$, and $f \in (d)$. Therefore, $I \subseteq (d)$, and $I = (d)$.

To prove uniqueness, suppose that $d'(x) \in k[x]$ is a monic polynomial with $(d') = (d)$. By Proposition 6.24, d' and d are associates; there is a unit $u \in k[x]$ with $d' = ud$. Now $u \in k$, by Proposition 6.2. Since both d' and d are monic, we have $u = 1$ and $d' = d$. ∎

Recall Example 5.27(iv): the set I consisting of all polynomials $f(x) \in \mathbb{R}[x]$ having i as a root is an ideal in $\mathbb{R}[x]$ containing $(x^2 + 1)$. We can now say more.

Corollary 6.26. *The ideal $I \subseteq \mathbb{R}[x]$ consisting of all polynomials $f(x) \in \mathbb{R}[x]$ having i as a root is equal to $(x^2 + 1)$.*

Proof. Now $(x^2 + 1) \subseteq I$. For the reverse inclusion, we know that $I = (d)$, where d is the unique monic polynomial of least degree in I. But $x^2 + 1$ is a monic polynomial in I, and there can be no such polynomial of smaller degree lest i be a root of a linear polynomial in $\mathbb{R}[x]$. ∎

It is not true that ideals in arbitrary commutative rings are necessarily principal, as the next example shows.

Example 6.27. Let $R = \mathbb{Z}[x]$, the commutative ring of all polynomials over $\mathbb{Z}$. It is easy to see that the set I of all polynomials with even constant term is an ideal in $\mathbb{Z}[x]$. We show that I is *not* a principal ideal.

Suppose there is $d(x) \in \mathbb{Z}[x]$ with $I = (d)$. The constant $2 \in I$, so that there is $f(x) \in \mathbb{Z}[x]$ with $2 = df$. Since the degree of a product is the sum of the degrees of the factors, $0 = \deg(2) = \deg(d) + \deg(f)$. Since degrees are nonnegative, it follows that $\deg(d) = 0$; i.e., d is a nonzero constant. As constants here are integers, the candidates for d are ± 1 and ± 2. Suppose $d = \pm 2$; since $x \in I$, there is $g(x) \in \mathbb{Z}[x]$ with $x = dg = \pm 2g$. But every coefficient on the right side is even, while the coefficient of x on the left side is 1. This contradiction gives $d = \pm 1$. Thus, d is a unit and, by Example 5.30, $I = (d) = \mathbb{Z}[x]$, another contradiction. Therefore, no such d exists; that is, I is not a principal ideal. ▲

But see Exercise 6.22 on page 248. There is $h(x) \in \mathbb{Z}[x]$ with $I = (2, h)$

Recall that if R is any commutative ring and $a, b \in R$, then a *linear combination* of a, b is an element of R of the form $sa + tb$, where $s, t \in R$. Given a, b, the set I of all linear combinations of a, b is an ideal in R. The next theorem parallels Theorem 1.19.

Theorem 6.28. *If k is a field and $f(x)$, $g(x) \in k[x]$, then any gcd of f, g is a linear combination of f and g; that is, if $d(x)$ is a gcd, then there are $s(x), t(x) \in k[x]$ with*

$$d = sf + tg.$$

Proof. The set I of all linear combinations of f and g is an ideal in $k[x]$; by Theorem 6.25, there is $d(x) \in k[x]$ with $I = (d)$. If both f and g are 0, then

$d = 0$, and we are done; otherwise, we may assume that d is monic. We know that $d = sf + tg$ for some s and t, because d lies in I. We claim that d is a gcd. Now d is a common divisor, for $f, g \in I = (d)$. If h is a common divisor of f and g, then $f = f_1 h$ and $g = g_1 h$. Hence, $d = sf + tg = (sf_1 + tg_1)h$ and $h \mid d$. Therefore, $\deg(h) \leq \deg(d)$, and so d is a monic common divisor of largest degree. ∎

We can characterize gcd's in $k[x]$.

Corollary 6.29. *Let k be a field and let $f(x)$, $g(x) \in k[x]$. A monic common divisor $d(x)$ is a* gcd *of f, g if and only if d is divisible by every common divisor; that is, if h is any common divisor of f, g, then $h \mid d$.*

Proof. The end of the proof of Theorem 6.28 shows that if h is a common divisor, then $h \mid d$. Conversely, if $h \mid d$, then $\deg(h) \leq \deg(d)$, and so d is a monic common divisor of largest degree. ∎

Theorem 6.30. *Let $f(x)$, $g(x) \in k[x]$, where k is a field, and let $I = (f, g)$ be the ideal of all linear combinations of f and g.*

(i) *If $d(x) \in k[x]$ is monic, then $d = \gcd(f, g)$ if and only if $I = (d)$.*

(ii) *f and g have a unique* gcd.

> Recall that gcd's are required to be monic. That's essential to uniqueness.

Proof. (i) Suppose that $d = \gcd(f, g)$. We show that $(d) \subseteq I$ and $I \subseteq (d)$. Theorem 6.28 shows that $d \in I$; therefore, $(d) \subseteq I$ (for every multiple of d is also a linear combination). For the reverse inclusion, let $h = uf + vg \in I$. Now $d \mid f$ and $d \mid g$, because d is a common divisor, and so $d \mid h$. Hence, $h = rd \in (d)$; that is, $I \subseteq (d)$, and so $I = (d)$.

Conversely, suppose that $I = (d)$. Then $d = sf + tg$, and so every common divisor h of f, g is a divisor of d. Hence, Corollary 6.29 gives $d = \gcd(f, g)$.

(ii) If d and d' are gcd's of f and g, then $(d) = (d')$, by part (i). Since both d and d' are monic, we must have $d = d'$, by Theorem 6.25. ∎

How to Think About It. It's a good idea to stop and take stock of where we are in our program of displaying parallels between integers and polynomials. For polynomials over a field, we have, so far

- extended the notion of divisibility
- generalized "prime" to "irreducible"
- shown that factorizations into irreducibles exist
- established a division algorithm
- shown that the gcd of two polynomials exists and is unique
- shown that the gcd of two polynomials is a linear combination of them.

Thinking back to Chapter 1, what's next? There were two main paths we took then: one led to unique factorization—the Fundamental Theorem of Arithmetic; one led to Euclidean Algorithms. We'll follow both these paths for polynomials.

Exercises

6.13 Find the gcd of each pair (f, g) in $\mathbb{Q}[x]$ and write it as a linear combination of f and g.

    (i) $(x^3 - x^2 - x - 2, x^3 - 3x^2 + 3x - 2)$

    (ii) $(x^6 - 1, x^5 - 1)$

    (iii) $(x^3 - x^2 - x - 2, 2x^3 - 4x^2 + 2x - 4)$

    (iv) $(x^6 - 1, x^6 + x^5 - 2)$

    (v) $\left((2x + 1)(x^6 - 1), (2x + 1)(x^5 - 1)\right)$

    (vi) $(3x^6 - 3, 2x^5 - 2)$

6.14 * Let R be a domain. If $f(x) \in R[x]$ has degree n, prove that f has at most n roots in R.

Hint. Use Frac(R).

6.15 If k is a field in which $1 + 1 \neq 0$, prove that $\sqrt{1 - x^2}$ is not a rational function over k.

Hint. Mimic the classical proof that $\sqrt{2}$ is irrational.

6.16 In Exercise 6.10 on page 243, we saw that the relation $\equiv$ on a commutative ring R, defined by $a \equiv b$ if they are associates, is an equivalence relation. Prove that if R is a domain, then there is a bijection from the family of all equivalence classes to the family of all principal ideals in R.

6.17 *

    (i) If $f(x)$ and $g(x)$ are relatively prime in $k[x]$ (k a field) and each divides a polynomial h, prove that their product fg also divides h.

    (ii) If $p_1, p_2, \ldots, p_n$ are polynomials so that $\gcd(p_i, p_j) = 1$, and each p_i divides a polynomial h, prove that $p_1 p_2 \cdots p_n$ also divides h.

6.18 *

    (i) Find $\gcd(3x^3 - 2x^2 + 3x - 2, 3x^2 + x - 2)$ in $\mathbb{C}[x]$.

    (ii) Write a pseudocode (or a program in a CAS) implementing Euclidean Algorithm I.

    (iii) Write a pseudocode (or a program in a CAS) implementing Euclidean Algorithm II.

    Hint. Model your routine after the functions in Exercise 1.67 on page 36.

6.19 * Prove the converse of Euclid's Lemma. Let k be a field and let $f(x) \in k[x]$ be a nonconstant polynomial; if, whenever f divides a product of two polynomials, it necessarily divides one of the factors, then f is irreducible. (See Theorem 1.21.)

6.20 (i) Find two polynomials in $\mathbb{Q}[x]$ whose associated polynomial functions agree with this input-output table:

Input	Output
1	3
4	17
5	26

    (ii) Classify the set of all polynomials that agree on the table.

6.21 (i) Show that the set of polynomials in $\mathbb{Q}[x]$ that vanish on $\{1, 2, 3\}$ is an ideal in $\mathbb{Q}[x]$.

    (ii) What is a generator of this ideal?

6.22 * In Example 6.27, we saw that the ideal I in $\mathbb{Z}[x]$ consisting of all polynomials with even constant term is not a principal ideal. Find a polynomial $h(x) \in \mathbb{Z}[x]$ so that $I = (2, h)$; that is, I consists of all the linear combinations of 2 and h.

6.23 Let k be a field and $f(x), g(x) \in k[x]$. Generalize Exercises 5.49 and 5.50 on page 220: if $d(x) = \gcd(f, g)$ and $m(x) = \text{lcm}(f, g)$, prove that

$$(f) + (g) = (d) \quad \text{and} \quad (f) \cap (g) = (m).$$

6.24 Show, in $\mathbb{Z}_8[x]$, that $x^2 - 1$ has two distinct factorizations into irreducibles.

Hint. See Example 6.17.

Unique Factorization

The main result in this subsection is a generalization of the Fundamental Theorem of Arithmetic to polynomials: the factorization of every nonconstant polynomial over a field as a product of irreducibles is essentially unique.

We begin by proving Euclid's Lemma for polynomials. As for integers, it shows that irreducibility is a strong assumption when dealing with divisibility.

Theorem 6.31 (Euclid's Lemma). *Let k be a field and let $f(x), g(x) \in k[x]$. If $p(x)$ is an irreducible polynomial in $k[x]$ and $p \mid fg$, then*

$$p \mid f \quad or \quad p \mid g.$$

More generally, if $p \mid f_1 \cdots f_n$, then $p \mid f_i$ for some i.

Proof. Assume that $p \mid fg$ but that $p \nmid f$. Since p is irreducible, $\gcd(p, f) = 1$, and so $1 = sp + tf$ for some polynomials $s(x)$ and $t(x)$. Therefore,

$$g = spg + tfg.$$

But $p \mid fg$, by hypothesis, and so Exercise 6.7 on page 243 gives $p \mid g$. The last statement follows by induction on $n \geq 2$. ∎

The converse of Euclid's Lemma is true; see Exercise 6.19 on page 247.

Polynomial versions of arithmetic theorems in Chapter 1 now follow.

Definition. Two polynomials $f(x), g(x) \in k[x]$, where k is a field, are called *relatively prime* if their gcd is 1.

Corollary 6.32. *Let $f(x), g(x), h(x) \in k[x]$, where k is a field, and let h and f be relatively prime. If $h \mid fg$, then $h \mid g$.*

Proof. The proof of Theorem 6.31 works here. Since $\gcd(h, f) = 1$, we have $1 = sh + tf$, and so $g = shg + tfg$. But $fg = hh_1$ for some $h_1(x) \in k[x]$, and so $g = h(sg + th_1)$. ∎

Definition. If k is a field, then a rational function $f(x)/g(x) \in k(x)$ is in *lowest terms* if f and g are relatively prime.

Proposition 6.33. *If k is a field, every nonzero $f(x)/g(x) \in k(x)$ can be put in lowest terms.*

Proof. If $f = df'$ and $g = dg'$, where $d = \gcd(f, g)$, then f' and g' are relatively prime, and so f'/g' is in lowest terms. ∎

There is an analog of the Euclidean Algorithm in $\mathbb{Z}$ that can be applied to compute gcd's of polynomials.

Theorem 6.34 (Euclidean Algorithm I). *If k is a field and $f(x), g(x) \in k[x]$, then there is an algorithm computing $\gcd(f, g)$.*

Proof. The proof is essentially a repetition of the proof of the Euclidean Algorithm in $\mathbb{Z}$; just iterate the Divison Algorithm. Each line comes from the line above it by moving some terms "southwest."

$$g = q_1 f + r_1$$
$$f = q_2 r_1 + r_2$$
$$r_1 = q_3 r_2 + r_3$$
$$\vdots$$
$$r_{n-3} = q_{n-1} r_{n-2} + r_{n-1}$$
$$r_{n-2} = q_n r_{n-1} + r_n$$
$$r_{n-1} = q_{n+1} r_n.$$

Since the degrees of the remainders are strictly decreasing, the procedure must stop after at most $\deg(f)$ steps. The claim is that $d = r_n$ is the gcd, once it is made monic. We see that d is a common divisor of f and g by back substitution: repeated applications of "2 out of 3," working from the bottom up. To see that d is the gcd, work from the top down to show that if c is any common divisor of f and g, then $c \mid r_i$ for every i. ∎

The Euclidean Algorithm may not produce a monic last remainder. The gcd is the monic associate of the last nonzero remainder.

Example 6.35 (Good Example). Let

$$f(x) = 3x^3 - 2x^2 + 3x - 2 \quad \text{and} \quad g(x) = 3x^2 + x - 2;$$

we compute $\gcd(f, g)$.

$$3x^3 - 2x^2 + 3x - 2 = (x - 1)(3x^2 + x - 2) + (6x - 4)$$
$$3x^2 + x - 2 = \left(\tfrac{1}{2}x + \tfrac{1}{2}\right)(6x - 4) + 0.$$

Rewriting in simpler notation:

$$f = (x - 1)g + r$$
$$g = \left(\tfrac{1}{2}x + \tfrac{1}{2}\right)r.$$

The last remainder is $6x - 4$. As we warned, it's not monic, and we must make it so. Thus, we need to take its monic associate (multiplying by $\tfrac{1}{6}$):

$$\gcd(f, g) = x - \tfrac{2}{3}. \quad \blacktriangle$$

It's the same for ordinary long division—hand calculations for small integers are quite simple, but they get very messy when dividing two large integers.

Example 6.36 (Bad Example). The Euclidean Algorithm applied to integers is quite efficient, in terms of the number of steps it takes to get to the answer. It's the same for polynomials, but the steps get quite cumbersome when carried out by hand—the complexity comes from the computational overhead in the hand calculations, not in the efficiency of the algorithm itself. A CAS removes this obstacle.

The following steps calculate $\gcd(x^4 - x^3 - 5x^2 + 8x - 4, 3x^3 - 6x^2 + x - 2)$ via the Euclidean Algorithm; all the quotients and remainders were calculated with a CAS:

$$x^4 - x^3 - 5x^2 + 8x - 4 = \left(\tfrac{1}{3}x + \tfrac{1}{3}\right)(3x^3 - 6x^2 + x - 2)$$
$$+ \left(-\tfrac{10}{3}x^2 + \tfrac{25}{3}x - \tfrac{10}{3}\right)$$
$$3x^3 - 6x^2 + x - 2 = \left(-\tfrac{9}{10}x - \tfrac{9}{20}\right)\left(-\tfrac{10}{3}x^2 + \tfrac{25}{3}x - \tfrac{10}{3}\right) + \left(\tfrac{7}{4}x - \tfrac{7}{2}\right)$$
$$-\tfrac{10}{3}x^2 + \tfrac{25}{3}x - \tfrac{10}{3} = \left(-\tfrac{40}{21}x + \tfrac{20}{21}\right)\left(\tfrac{7}{4}x - \tfrac{7}{2}\right)$$

Multiplying by $\tfrac{4}{7}$ produces the gcd of $x - 2$. ▲

Here is an unexpected bonus from the Euclidean Algorithm.

Corollary 6.37. *Let k be a subfield of a field K, so that $k[x]$ is a subring of $K[x]$. If $f(x), g(x) \in k[x]$, then their gcd in $k[x]$ is equal to their gcd in $K[x]$.*

Proof. We may assume that $f \neq 0$, for $\gcd(0, g) = g$ (actually, g's monic associate). The Division Algorithm in $K[x]$ gives

$$g = Qf + R,$$

where $Q, R \in K[x]$ and either $R = 0$ or $\deg(R) < \deg(f)$; since $f, g \in k[x]$, the Division Algorithm in $k[x]$ gives

$$g = qf + r,$$

where $q, r \in k[x]$ and either $r = 0$ or $\deg(r) < \deg(f)$. But the equation $g = qf + r$ also holds in $K[x]$ because $k[x] \subseteq K[x]$, so that the uniqueness of quotient and remainder in the Division Algorithm in $K[x]$ gives $Q = q \in k[x]$ and $R = r \in k[x]$. Therefore, the list of equations occurring in the Euclidean Algorithm in $K[x]$ is exactly the same as the list occurring in the Euclidean Algorithm in the smaller ring $k[x]$. In particular, the gcd, being the last remainder (made monic), is the same in both polynomial rings. ∎

See Exercise 6.18(i) on page 247.

To illustrate, even though there are more divisors with complex coefficients, the gcd of $3x^3 - 2x^2 + 3x - 2$ and $3x^2 + x - 2$, computed in $\mathbb{R}[x]$, is equal to their gcd computed in $\mathbb{C}[x]$.

As in $\mathbb{Z}$, the Division Algorithm in $k[x]$ can also be used to compute coefficients occurring in an expression of the gcd as a linear combination.

Theorem 6.38 (Euclidean Algorithm II). *If k is a field and $f(x), g(x) \in k[x]$, then there is an algorithm finding a pair of polynomials $s(x)$ and $t(x)$ with $\gcd(f, g) = sf + tg$.*

Proof. Let $d = \gcd(f, g)$. To find s and t with $d = sf + tg$, again work from the last remainder back to f and g:

$$r_n = r_{n-2} - q_n r_{n-1}$$
$$= r_{n-2} - q_n(r_{n-3} - q_{n-1}r_{n-2})$$
$$= (1 + q_n q_{n-1})r_{n-2} - q_n r_{n-3}$$
$$\vdots$$
$$= sf + tg \quad \blacksquare$$

Example 6.39. Let's compute $\gcd(f, g)$, where $f(x) = x^3 - 2x^2 + x - 2$ and $g(x) = x^4 - 1$.

$$x^4 - 1 = (x + 2)(x^3 - 2x^2 + x - 2) + (3x^2 + 3)$$
$$x^3 - 2x^2 + x - 2 = (x + 2)(3x^2 + 3) + 15.$$

Rewriting in simpler notation:

$$g = (x + 2)f + r$$
$$f = (x + 2)r + 15.$$

We see that the last remainder, 15, is a constant. As we warned, it need not be monic, and we must make it so. Thus, $\gcd(f, g) = 1$; that is, f and g are relatively prime.

We now use Euclidean Algorithm II to find $s(x), t(x)$ with $d = sf + tg$. Using letters,

$$d = f - q'r$$
$$g = f - q'(g - qf)$$
$$= (1 + q'q)f - q'g.$$

Now set $r = 3x^2 + 3$, $q = x + 2$, and $q' = 3x + 6$. We have

$$15 = \big((1 + (3x + 6)(x + 2)\big)f - (x + 2)g$$
$$= (3x^2 + 12x + 13)f - (x + 2)g.$$

Since gcd's are monic,

$$1 = \tfrac{1}{15}(3x^2 + 12x + 13)f - \tfrac{1}{15}(x + 2)g.$$

A computer can be programmed to carry out Euclidean Algorithm II (see Exercise 6.18 on page 247). Once programmed, messy calculations are not a problem. Indeed, using the polynomials from Example 6.36, we have

$$\gcd(x^4 - x^3 - 5x^2 + 8x - 4, 3x^3 - 6x^2 + x - 2) = x - 2$$

Working as above (with the help of a CAS), we get

$$\left(\tfrac{9}{10}x + \tfrac{9}{20}\right)(x^4 - x^3 - 5x^2 + 8x - 4)$$
$$+ \left(-\tfrac{3}{10}x^2 - \tfrac{9}{20}x + \tfrac{17}{20}\right)(3x^3 - 6x^2 + x - 2)$$
$$= \tfrac{7}{4}x - \tfrac{7}{2}.$$

Multiplying both sides of this equation by $\tfrac{4}{7}$ gives the linear combination.   ▲

The next result, an analog for polynomials of the Fundamental Theorem of Arithmetic, shows that the factorization of a polynomial as a product of irreducible polynomials is essentially unique.

Theorem 6.40 (Unique Factorization). *If k is a field, then every nonconstant polynomial $f(x) \in k[x]$ is a product of a nonzero constant and monic irreducibles.*

Moreover, if f has two such factorizations,

$$f = ap_1 \cdots p_m \quad and \quad f = bq_1 \cdots q_n,$$

(that is, a and b are nonzero constants and the p's and q's are monic irreducibles), then $a = b$, $m = n$, and the q's may be re-indexed so that $q_i = p_i$ for all i.

Proof. We proved the existence of a factorization in Corollary 6.9.

To prove uniqueness, suppose that there is an equation

$$ap_1 \cdots p_m = bq_1 \cdots q_n$$

in which a and b are nonzero constants and the p's and q's are monic irreducibles. We prove, by induction on $M = \max\{m, n\} \geq 1$, that $a = b$, $m = n$, and the q's may be re-indexed so that $q_i = p_i$ for all i. For the base step $M = 1$, we have $ap_1 = bq_1$. Now a is the leading coefficient, because p_1 is monic, while b is the leading coefficient, because q_1 is monic. Therefore, $a = b$, and canceling gives $p_1 = q_1$. For the inductive step, the given equation shows that $p_m \mid q_1 \cdots q_n$. By Euclid's Lemma for polynomials, there is some i with $p_m \mid q_i$. But q_i, being monic irreducible, has no monic divisors other than 1 and itself, so that $q_i = p_m$. Re-indexing, we may assume that $q_n = p_m$. Canceling this factor, we have $ap_1 \cdots p_{m-1} = bq_1 \cdots q_{n-1}$. By the inductive hypothesis, $a = b$, $m - 1 = n - 1$ (hence $m = n$) and, after re-indexing, $q_i = p_i$ for all i. ∎

Here is another way to state uniqueness, using Proposition 6.24: after re-indexing, the ideals $(p_1), \ldots, (p_m)$ and $(q_1), \ldots, (q_m)$ are the same.

Collect like factors.

Definition. Let $f(x) \in k[x]$, where k is a field. A *prime factorization* of f is

$$f(x) = ap_1(x)^{e_1} \cdots p_m(x)^{e_m},$$

where a is a nonzero constant, the p_i's are *distinct* monic irreducible polynomials, and $e_i \geq 0$ for all i.

Theorem 6.40 shows that every nonconstant polynomial f has prime factorizations; moreover, if all the exponents $e_i > 0$, then the factors in it are unique. Let $f(x), g(x) \in k[x]$, where k is a field. As with integers, using zero exponents allows us to assume that the same irreducible factors occur in both prime factorizations:

$$f = p_1^{a_1} \cdots p_m^{a_m} \quad and \quad g = p_1^{b_1} \cdots p_m^{b_m}.$$

Definition. If f and g are elements in a commutative ring R, then a **common multiple** is an element $h \in R$ with $f \mid h$ and $g \mid h$. If f and g in R are not both 0, define their **least common multiple**, denoted by

$$\operatorname{lcm}(f, g),$$

to be a monic common multiple c of them with $c \mid h$ for every common multiple h. If $f = 0 = g$, define their lcm to be 0.

We now use prime factorizations having zero exponents.

Proposition 6.41. *Let $f(x), g(x) \in k[x]$, where k is a field, have prime factorizations $f = p_1^{a_1} \cdots p_n^{a_n}$ and $g = p_1^{b_1} \cdots p_n^{b_n}$ in $k[x]$, where $a_i, b_i \geq 0$ for all i.*

(i) *$f \mid g$ if and only if $a_i \leq b_i$ for all i.*

(ii) *If $m_i = \min\{a_i, b_i\}$ and $M_i = \max\{a_i, b_i\}$, then*

$$\gcd(f, g) = p_1^{m_1} \cdots p_n^{m_n} \quad \text{and} \quad \operatorname{lcm}(f, g) = p_1^{M_1} \cdots p_n^{M_n}.$$

Proof. (i) If $f \mid g$, then $g = fh$, where $h = p_1^{c_1} \cdots p_n^{c_n}$ and $c_i \geq 0$ for all i. Hence,

$$g = p_1^{b_1} \cdots p_n^{b_n} = \left(p_1^{a_1} \cdots p_m^{a_m}\right)\left(p_1^{c_1} \cdots p_n^{c_n}\right) = p_1^{a_1 + c_1} \cdots p_n^{a_n + c_n}.$$

By uniqueness, $a_i + c_i = b_i$; hence, $a_i \leq a_i + c_i = b_i$. Conversely, if $a_i \leq b_i$, then there is $c_i \geq 0$ with $b_i = a_i + c_i$. It follows that $h = p_1^{c_1} \cdots p_n^{c_n} \in k[x]$ and $g = fh$.

(ii) Let $d = p_1^{m_1} \cdots p_n^{m_n}$. Now d is a common divisor, for $m_i \leq a_i, b_i$. If $D = p_1^{e_1} \cdots p_n^{e_n}$ is any other common divisor, then $0 \leq e_i \leq \min\{a_i, b_i\} = m_i$, and so $D \mid d$. Therefore, $\deg(D) \leq \deg(d)$, and d is the gcd (for it is monic). The argument for lcm is similar. ∎

Corollary 6.42. *If k is a field and $f(x), g(x) \in k[x]$ are monic polynomials, then*

$$\operatorname{lcm}(f, g) \gcd(f, g) = fg.$$

Proof. The result follows from Proposition 6.41, for $m_i + M_i = a_i + b_i$. ∎

Since the Euclidean Algorithm computes the gcd in $k[x]$ when k is a field, Corollary 6.42 computes the lcm.

$$\operatorname{lcm}(f, g) = \frac{fg}{\gcd(f, g)}.$$

We can use roots to detect whether two polynomials are relatively prime.

Corollary 6.43. *If $f(x), g(x) \in \mathbb{R}[x]$ have no common root in $\mathbb{C}$, then f, g are relatively prime in $\mathbb{R}[x]$.*

Proof. Assume that $d = \gcd(f, g) \neq 1$, where $d \in \mathbb{R}[x]$. By the Fundamental Theorem of Algebra, d has a complex root α. By Corollary 6.37, $d = \gcd(f, g)$ in $\mathbb{C}[x]$. Since $(x - \alpha) \mid d$ in $\mathbb{C}[x]$, we have $(x - \alpha) \mid f$ and $(x - \alpha) \mid g$. By Corollary 6.15, α is a common root of f and g. ∎

How to Think About It. There is nothing magic about $\mathbb{R}$ and $\mathbb{C}$. It can be proved that every field k has an ***algebraic closure*** $\overline{k}$; that is, there is a field $\overline{k}$ containing k as a subfield, and every $f(x) \in \overline{k}[x]$ is a product of linear factors. In particular, since $k[x] \subseteq \overline{k}[x]$, every $f(x) \in k[x]$ is a product of linear factors in $\overline{k}[x]$; that is, $\overline{k}$ contains all the roots of f. Thus, Corollary 6.43 can be generalized by replacing $\mathbb{R}$ and $\mathbb{C}$ by k and $\overline{k}$.

We know that $\mathbb{C}$ can be viewed as a vector space over its subfield $\mathbb{R}$, and $\dim_{\mathbb{R}}(\mathbb{C}) = 2$. But things are not so simple for algebraic closures $\overline{k}$ of other fields k. It is always true that $\overline{k}$ is a vector space over k, but its dimension need not be 2. In fact, $\dim_k(\overline{k})$ need not even be finite: for example, $\dim_{\mathbb{Q}}(\overline{\mathbb{Q}}) = \infty$ and, if k is finite, then $\dim_k(\overline{k}) = \infty$.

Let k be a field, and assume that all the roots of a polynomial $f(x) \in k[x]$ lie in k: there are $a, r_1, \ldots, r_n \in k$ with

$$f(x) = a \prod_{i=1}^{n} (x - r_i).$$

If $r_1, \ldots, r_s$, where $s \le n$, are the distinct roots of f, then a prime factorization of f is

$$f(x) = a(x - r_1)^{e_1} (x - r_2)^{e_2} \cdots (x - r_s)^{e_s}.$$

We call e_j the ***multiplicity*** of the root r_j. As linear polynomials are always irreducible, unique factorization shows that multiplicities of roots are well-defined.

Exercises

6.25 Let $f(x), g(x) \in k[x]$, where k is a field. If fg is a square, must f or g be a square? What if $\gcd(f, g) = 1$?

6.26 Let $f(x), g(x) \in k[x]$, where k is a field, be relatively prime. If $h(x) \in k[x]$ and $h^2 \mid fg$, prove that $h^2 \mid f$ or $h^2 \mid g$.

6.27 Let $k = \mathbb{F}_2(x)$. Prove that $f(t) = t^2 - x \in k[t]$ is an irreducible polynomial. (We shall see later that there is a field K containing k and an element u with $u^2 = x$, so that $f(t) = (t - u)^2$ in $K[t]$.)

6.28 In $\mathbb{Z}_p[x]$, show that if f is an irreducible factor of $x^{p^n} - x$, then f^2 does not divide $x^{p^n} - x$.

6.29 Determine, for each of the following polynomials in $\mathbb{Q}[x]$ whether or not it is irreducible in $\mathbb{Q}[x]$, in $\mathbb{R}[x]$, or in $\mathbb{C}[x]$.
  (i) $x^2 - 7x + 6$.
  (ii) $x^2 + 2x - 1$.
  (iii) $x^2 + x + 1$.

6.30 * Show that $f(x) = x^3 + 5x^2 - 10x + 15$ is irreducible in $\mathbb{Q}[x]$.

In Section 6.2, we will give different criteria for determining whether polynomials are irreducible (in particular, we will discuss f on page 267). However, we ask you to solve this problem now so you will appreciate the theorems to be proved.

Principal Ideal Domains

There are other classes of domains that enjoy an analog of the Fundamental Theorem of Arithmetic; one such is the following.

Definition. A domain R is a ***principal ideal domain***, usually abbreviated by the acronym **PID**, if every ideal in R is a principal ideal.

We have already seen examples of PIDs.

Example 6.44. (i) Theorem 5.29 shows that $\mathbb{Z}$ is a PID.

(ii) Theorem 6.25 shows that $k[x]$ is a PID when k is a field.

(iii) Every field k is a PID, for its only ideals are $k = (1)$ and (0).

(iv) Not every domain is a PID. In Example 6.27, we saw that there are ideals in $\mathbb{Z}[x]$ that are not principal ideals.

(v) Exercise 6.12 on page 243 shows that $k[x, y]$, polynomials in two variables over a field k, is not a PID. ▲

(vi) In Chapter 8, we shall see that the rings of Gaussian integers and of Eisenstein integers are PIDs.

PIDs enjoy many of the properties shared by $\mathbb{Z}$ and $k[x]$ (k a field). In particular, they have a fundamental theorem of arithmetic, and the proof of this fact parallels the program we developed for $\mathbb{Z}$ and $k[x]$.

We begin by defining gcd's in a general commutative ring R. We can't use $\leq$ as we did in $\mathbb{Z}$, nor can we use degrees as we did in $k[x]$, but we can use the idea in Corollaries 1.20 and 6.29.

Definition. Let R be a commutative ring. If $a, b \in R$, then a gcd of a, b is a common divisor $d \in R$ that is divisible by every common divisor: if $c \mid a$ and $c \mid b$, then $c \mid d$.

Just defining a term doesn't guarantee it always exists—we could define *unicorn* if we were asked to do so, and there are rings, even domains, containing elements having no gcd (see Exercise 6.33 on page 259). Even if a gcd does exist, there is the question of uniqueness. In $\mathbb{Z}$, uniqueness of a gcd follows from our assuming, as part of the definition, that gcd's are positive; in $k[x]$, uniqueness follows from our assuming, as part of the definition, that gcd's are monic. Neither assumption makes sense in a general commutative ring; however, we do have a measure of uniqueness in domains.

Proposition 6.45. *Let R be a domain. If d and d' are gcd's of a, b in R, then d and d' are associates and $(d) = (d')$.*

Proof. By definition, both d and d' are common divisors of a, b; moreover, $d \mid d'$ and $d' \mid d$. Since R is a domain, Proposition 6.6 applies, and d and d' are associates. The second statement follows from Proposition 6.24. ■

Although there are domains with elements not having a gcd, we now show gcd's always exist in PIDs.

Theorem 6.46. *Let R be a PID. If $a, b \in R$, then a gcd of a, b exists and it is a linear combination of a, b.*

Proof. As every ideal in R, the ideal $(a, b) = \{ua + vb : u, v \in R\}$ is a principal ideal: there is $d \in R$ with $(a, b) = (d)$. Of course, d is a linear combination of a, b, say, $d = sa + tb$ for some $s, t \in R$, and it suffices to prove d is a gcd. Now d is a common divisor: $a \in (a, b) = (d)$, so that $a = rd$ for some $r \in R$; hence, $d \mid a$; similarly, $d \mid b$. Finally, if $c \mid a$ and $c \mid b$, then $c \mid d = sa + tb$. . ∎

We can now show that Euclid's Lemma holds in PIDs.

Theorem 6.47 (Euclid's Lemma). *Let R be a PID and $p \in R$ be irreducible. If $p \mid ab$, where $a, b \in R$, then $p \mid a$ or $p \mid b$.*

This proof should look quite familiar to you.

Proof. If $p \nmid a$, then 1 is a gcd of p, a, for the only divisors of p are units and associates. Thus, Theorem 6.46 says that there exist $s, t \in R$ with $1 = sp + ta$. Hence, $b = spb + tab$. But $ab = pr$, for some $r \in R$, and so $p \mid b$, as desired. ∎

How to Think About It.

To prove the unique factorization theorem in $\mathbb{Z}$ and in $k[x]$, we first proved that every element can be factored into irreducibles. After that, we showed that such factorizations are essentially unique. Let's carry on with that development for arbitrary PIDs.

To prove factorization into irreducibles in a PID, we need an abstract property of principal ideal domains, one that was previewed in Exercises 5.47 and 5.48 on page 220.

Suppose that R is a PID and that $r \in R$ is neither zero nor a unit. Must r be a product of irreducibles? If not, then r is not irreducible (recall that we allow products to have only one factor); thus, r factors: say, $r = ab$, where neither a nor b is a unit. If both a and b are products of irreducibles, then so is r, and we're done. So, suppose one of them, say a, is not a product of irreducibles. Thus, a is not irreducible, and $a = cd$. where neither c nor d is a unit. If both c and d are products of irreducibles, then so is a, and this is a contradiction. We've got a tiger by the tail! We can keep repeating this argument ad infinitum.

Let's rephrase these factorizations of r in terms of ideals; after all, $a \mid r$ says that $r = r'a$ for some $r' \in R$; that is, $(r) \subseteq (a)$. But this inclusion must be strict: $(r) \subsetneq (a)$, lest r and a be associates (they're not, because b is not a unit). The tiger tells us that there is an infinite strictly increasing sequence of ideals.

Lemma 6.48. *If R is a PID, then every ascending chain of ideals*

$$I_1 \subseteq I_2 \subseteq \cdots \subseteq I_n \subseteq I_{n+1} \subseteq \cdots$$

stops; that is, there is N with $I_n = I_N$ for all $n \geq N$.

Proof. Suppose there is an ascending chain of ideals that does not stop. Throwing away any repetitions $I_n = I_{n+1}$ if necessary, we may assume that there is a strictly ascending chain of ideals

$$I_1 \subsetneq I_2 \subsetneq \cdots \subsetneq I_n \subsetneq I_{n+1} \subsetneq \cdots .$$

By Exercise 6.31 on page 258, $J = \bigcup_{n \geq 1} I_n$ is an ideal in R. And since R is a PID, J is principal; there is $d \in J$ with $J = (d)$. Now d got into J by being in I_m for some m; that is, $(d) \subseteq I_m$. Hence,

$$J = (d) \subseteq I_m \subsetneq I_{m+1} \subseteq J.$$

This is a contradiction. $\blacksquare$

Lemma 6.48 gives us factorization into irreducibles.

Proposition 6.49. *If R is a PID, then every nonzero, non-unit $r \in R$ is a product of irreducibles.*

Proof. If a is a divisor of r, then $(r) \subseteq (a)$, as we saw above. Call a divisor a of r a ***proper divisor*** if a is neither a unit nor an associate of r. If a is a proper divisor of r, then $(r) \subsetneq (a)$: if the inclusion is not strict, then $(a) = (r)$, and this forces a and r to be associates, by Proposition 6.6.

Call a nonzero non-unit $r \in R$ ***sweet*** if it is a product of irreducibles; call it ***sour*** otherwise. We must show that there are no sour elements. So, suppose r is a sour element. Now r is not irreducible, so $r = ab$, where both a and b are proper divisors. But the product of sweet elements is sweet, so that at least one of the factors, say, a, is sour. As we observed in the first paragraph, we have $(r) \subsetneq (a)$. Repeat this for a instead of r. It follows by induction that there exists a sequence $a_1 = r, a_2 = a, a_3, \dots, a_n, \dots$ of sour elements with each a_{n+1} a proper divisor of a_n. But this sequence yields a strictly ascending chain of ideals

$$(a_1) \subsetneq (a_2) \subsetneq (a_3) \subsetneq \cdots ,$$

contradicting Lemma 6.48. $\blacksquare$

Proposition 6.49 gives existence. The next theorem gives a fundamental theorem of arithmetic for PIDs: every nonzero non-unit has a unique factorization as a product of irreducibles.

Theorem 6.50. *Let R be a PID. Every $r \in R$, neither 0 nor a unit, has a factorization as a product of irreducibles which is unique in the following sense*: *if*

$$p_1 \cdots p_n = r = q_1 \cdots q_m,$$

where the p's and q's are irreducible, then $m = n$ and the q's can be reindexed so that q_i and p_i are associates for all i.

Proof. Proposition 6.48(iii) shows that every $r \in R$, neither 0 nor a unit, is a product of irreducibles.

To prove uniqueness, suppose that r is a nonzero non-unit and

$$p_1 \cdots p_n = r = q_1 \cdots q_m,$$

where the p's and q's are irreducible. By Euclid's Lemma, p_n irreducible implies $p_n \mid q_i$ for some i. Since q_i is irreducible, we have p_n and q_i are associates: there is a unit $u \in R$ with $q_i = up_n$. Re-index the q's so that q_i is now $q_m = up_n$, cancel p_n from both sides, and replace q_1 by uq_1. Thus,

$$p_1 \cdots p_{n-1} = r = (uq_1) \cdots q_{m-1}.$$

Note that uq_1 is irreducible (for it is an associate of an irreducible). The proof is completed, as in Theorem 6.40, by induction on $\max\{n, m\}$. ∎

So, every PID has a fundamental theorem of arithmetic. It turns out that there are other domains occurring in nature, not PIDs, which also enjoy such a theorem.

Definition. A domain R is a ***unique factorization domain***, usually abbreviated **UFD**, if

(i) every $a \in R$ that is neither 0 nor a unit is a product of irreducibles;

(ii) this factorization is unique in the following sense: if

$$p_1 \cdots p_n = a = q_1 \cdots q_m,$$

where the p's and q's are irreducible, then $n = m$ and, after re-indexing, p_i and q_i are associates for all i.

Further Results. We've just seen that every PID is a UFD. The converse is false: there are UFDs that are not principal ideal domains. A theorem of Gauss states that if a domain A is a UFD, then $A[x]$ is also a UFD. For example, $\mathbb{Z}[x]$ is a UFD (this is not a PID). If k is a field, then it follows by induction on $n \geq 1$ that $R = k[x_1, \ldots, x_n]$, polynomials in several variables, is a UFD (R is not a PID if $n \geq 2$).

As we've mentioned earlier, the erroneous assumption that every domain is a UFD was behind many incorrect "proofs" of Fermat's Last Theorem. The ring $\mathbb{Z}[\sqrt{-5}]$ is not a UFD: we'll see in Chapter 8 that

$$3 \cdot 2 = 6 = (1 + \sqrt{-5})(1 - \sqrt{-5})$$

are two different factorizations of 6 into irreducibles in $\mathbb{Z}[\sqrt{-5}]$ (and $1 + \sqrt{-5}$ is not an associate of 2 or of 3). Another example: $\mathbb{Z}[\zeta_{23}]$ is not a UFD, and 23 is the smallest prime p for which $\mathbb{Z}[\zeta_p]$ is not a UFD (see [23] Chapter 1, p. 7).

Exercises

6.31 *

(i) Let I and J be ideals in a commutative ring R. Prove that their union $I \cup J$ is an ideal if and only if $I \subseteq J$ or $J \subseteq I$.

(ii) Let $I_1 \subseteq I_2 \subseteq \cdots \subseteq I_n \subseteq \cdots$ be an ascending chain of ideals in a commutative ring R. Prove that

$$\bigcup_{n \geq 1}^{\infty} I_n$$

is an ideal in R.

6.32 Consider ascending chains of ideals in $\mathbb{Z}$:

(i) Find two different ascending chains of ideals in which $I_1 = (24)$.

(ii) Show that every ascending chain of ideals has only finitely many distinct terms.

(iii) Find the longest strictly ascending chain of ideals that starts with (72) (an ascending chain of ideals is **strictly ascending** if all inclusions $I_j \subseteq I_{j+1}$ are strict inclusions $I_j \subsetneqq I_{j+1}$).

(iv) Find the longest strictly ascending chain of ideals that starts with (101).

6.33 * Let R be the subset of $k[x]$ (where k is a field) consisting of all polynomials $f(x)$ having no linear term; that is,

$$f(x) = a_0 + a_2 x^2 + a_3 x^3 + \cdots.$$

(i) Prove that R is a subring of $k[x]$.

(ii) Prove that x^5 and x^6 do not have a gcd in R.

6.34 Recall that $\mathbb{R}^{\mathbb{R}}$, the set of all real valued functions of a real variable, is a commutative ring under pointwise addition and multiplication. Let $n \geq 0$ be an integer, and let I_n be the set of all functions in $\mathbb{R}^{\mathbb{R}}$ vanishing on integer multiples of n.

(i) Show that I_n is an ideal in $\mathbb{R}^{\mathbb{R}}$.

(ii) Find a function that is in I_8 but not in I_4.

(iii) Show that

$$I_2 \subsetneqq I_4 \subsetneqq I_8 \subsetneqq \cdots \subsetneqq I_{2^j} \subsetneqq \cdots$$

(iv) Show that this ascending chain of ideals does not stop.

(v) Conclude that there are ideals in $\mathbb{R}^{\mathbb{R}}$ that are not principal.

6.2   Irreducibility

Although there are some techniques to help decide whether an integer is prime, the general problem is open and is very difficult (indeed, this is precisely why RSA public key codes are secure). Similarly, it is very difficult to determine whether a polynomial is irreducible, but there are some useful techniques that frequently work. Most of our attention will be on $\mathbb{Q}[x]$ and $\mathbb{Z}[x]$, but some of the results do generalize to other rings of coefficients.

For polynomials of low degree, we have a simple and useful irreducibility criterion.

Proposition 6.51. *Let k be a field and let $f(x) \in k[x]$ be a quadratic or cubic polynomial. Then f is irreducible in $k[x]$ if and only if f has no root in k.*

Proof. An irreducible polynomial f of degree > 1 has no roots in k, by Corollary 6.15, for if $r \in k$ is a root, then $f(x) = (x - r)g(x)$ in $k[x]$. Conversely, if f is not irreducible, then $f = gh$, where neither g nor h is constant; thus, neither g nor h has degree 0. Since $\deg(f) = 2$ or 3 and $\deg(f) = \deg(g) + \deg(h)$, at least one of the factors has degree 1 and, hence, f has a root in k.   ∎

Proposition 6.51 is no longer true for polynomials of degree ≥ 4; for example, $f(x) = x^4 + 2x^2 + 1 = (x^2 + 1)(x^2 + 1)$ obviously factors in $\mathbb{R}[x]$, so it's not irreducible, yet f has no real roots.

A polynomial $f(x)$ is reducible if has a linear factor $x - a$, and there's a simple test for that; see whether a is a root of f. But to check whether f has a root a, we need a candidate for a.

Theorem 6.52 (Rational Root Theorem). *If $f(x) = a_0 + a_1 x + \cdots + a_n x^n \in \mathbb{Z}[x] \subseteq \mathbb{Q}[x]$, then every rational root of f has the form b/c, where $b \mid a_0$ and $c \mid a_n$. In particular, if f is monic, then every rational root of f is an integer.*

Proof. We may assume that a root b/c is in lowest terms; that is, $\gcd(b, c) = 1$. Evaluating gives $0 = f(b/c) = a_0 + a_1 b/c + \cdots + a_n b^n / c^n$, and multiplying through by c^n gives

$$0 = a_0 c^n + a_1 bc^{n-1} + \cdots + a_n b^n.$$

Reducing this mod b shows that $b \mid a_0 c^n$; since $\gcd(b, c) = 1$, Corollary 1.22 gives $b \mid a_0$. Similarly, reducing mod c gives $c \mid a_n b^n$. Since $\gcd(b, c) = 1$, we have $c \mid a_n$. ∎

It follows from the second statement that if an integer a is not the nth power of an integer, then $x^n - a$ has no rational roots; that is, $\sqrt[n]{a}$ is irrational. In particular, $\sqrt{2}$ is irrational. Thus, Theorem 6.52 is a vast generalization of Proposition 1.26.

Had we known Theorem 6.52 earlier, we could have easily dealt with the "bad cubic" $f(x) = x^3 - 7x + 6$ in Example 3.5. Since the candidates for its rational roots are $\pm 1, \pm 2, \pm 3, \pm 6$, we would have quickly found the factorization $f(x) = (x - 1)(x - 2)(x + 3)$.

If $f(x) \in \mathbb{Q}[x]$ happens to be in $\mathbb{Z}[x]$, there is a useful theorem of Gauss comparing the factorizations of f in $\mathbb{Z}[x]$ and in $\mathbb{Q}[x]$ that concludes that f is irreducible over $\mathbb{Q}$. Our proof involves Example 5.23: the homomorphism $r_p : \mathbb{Z} \to \mathbb{Z}_p$, sending $j \mapsto [j]$, gives a homomorphism $r_p^* : \mathbb{Z}[x] \to \mathbb{Z}_p[x]$, called ***reduction mod p***. If $f = a_0 + a_1 x + \cdots + a_n x^n \in \mathbb{Z}[x]$, then

$$r_p^* : f \mapsto \overline{f}, \text{ where } \overline{f}(x) = [a_0] + [a_1]x + \cdots + [a_n]x^n \in \mathbb{Z}_p[x].$$

Thus, r_p^* merely reduces all coefficients mod p.

Theorem 6.53 (Gauss's Lemma). *Let $f(x) \in \mathbb{Z}[x]$. If there are $G(x), H(x) \in \mathbb{Q}[x]$ with $f = GH$, then there are $g(x), h(x) \in \mathbb{Z}[x]$ with $\deg(g) = \deg(G)$, $\deg(h) = \deg(H)$, and $f = gh$.*

Proof. Clearing denominators in the equation $f = GH$, there are positive integers n', n'' so that $g = n'G$ and $h = n''H$, where both g, h lie in $\mathbb{Z}[x]$. Setting $n = n'n''$, we have

$$nf = (n'G)(n''H) = gh \text{ in } \mathbb{Z}[x]. \tag{6.1}$$

Let p be a prime divisor of n, and reduce the coefficients mod p. Eq. (6.1) becomes

$$0 = \overline{g}(x)\overline{h}(x).$$

Recall that $\mathbb{F}_p$ is another notation for $\mathbb{Z}_p$; we use it when we want to regard $\mathbb{Z}_p$ as a field.

But $\mathbb{F}_p[x]$ is a domain, because $\mathbb{F}_p$ is a field, and so at least one of the factors, say $\overline{g}$, is 0; that is, all the coefficients of g are multiples of p. Therefore, we may write $g = pg'$, where all the coefficients of g' lie in $\mathbb{Z}$. If $n = pm$, then

$$(pm)f = nf = gh = (pg')h \text{ in } \mathbb{Z}[x].$$

Cancel p, and continue canceling primes until we reach a factorization $f = g^*h^*$ in $\mathbb{Z}[x]$. Note that $\deg(g^*) = \deg(g)$ and $\deg(h^*) = \deg(h)$. ∎

The contrapositive of Gauss's Lemma is more convenient to use.

Corollary 6.54. *If $f(x) \in \mathbb{Z}[x]$ is irreducible in $\mathbb{Z}[x]$, then f is irreducible in $\mathbb{Q}[x]$.*

How to Think About It. We agree that Gauss's Lemma, though very useful, is rather technical. Gauss saw that the ideas in the proof could be generalized to apply to polynomials in several variables over a field.

The basic use of reduction mod p was previewed on page 216 when we proved that -1 is not a square in $\mathbb{Z}$ by showing that it's not a square in $\mathbb{Z}_3$. Reduction mod p gives a criterion for irreducibility of f in $\mathbb{Z}[x]$ by testing the irreducibility of $\overline{f}$ in $\mathbb{F}_p[x]$. The precise statement is:

Proposition 6.55. *Let $f(x) = a_0 + a_1x + \cdots + x^n \in \mathbb{Z}[x]$ be monic. If p is prime and $\overline{f} \in \mathbb{F}_p[x]$ is irreducible in $\mathbb{F}_p[x]$, then f is irreducible in $\mathbb{Z}[x]$ and, hence, in $\mathbb{Q}[x]$.*

> The hypothesis that f is monic can be relaxed; we may assume instead that p does not divide its leading coefficient.

Proof. Suppose f factors in $\mathbb{Z}[x]$; say $f = gh$, where $0 < \deg(g) < \deg(f)$ and $0 < \deg(h) < \deg(f)$. By Exercise 6.8, we may assume that both g and h are monic. Now $\overline{f} = \overline{g}\,\overline{h}$ (for r_p^* is a homomorphism), so that $\deg(\overline{f}) = \deg(\overline{g}) + \deg(\overline{h})$. And $\overline{f}$, $\overline{g}$, and $\overline{h}$ are monic, because f, g, and h are, so $\deg(\overline{f}) = \deg(f)$, $\deg(\overline{g}) = \deg(g)$, and $\deg(\overline{h}) = \deg(h)$; this contradicts the irreducibility of $\overline{f}$ in $\mathbb{F}_p[x]$. Therefore, f is irreducible in $\mathbb{Z}[x]$. Finally, f is irreducible in $\mathbb{Q}[x]$, by Gauss's Lemma. ∎

For example, $x^2 + 1$ is irreducible in $\mathbb{Q}[x]$ because $\overline{x^2 + 1}$ is irreducible in $\mathbb{Z}_3[x]$.

Theorem 6.55 says that if one can find a prime p with $\overline{f}$ irreducible in $\mathbb{F}_p[x]$, then f is irreducible in $\mathbb{Q}[x]$. The finiteness of $\mathbb{F}_p$ is a genuine advantage, for there are only a finite number of polynomials in $\mathbb{F}_p[x]$ of any given degree. In principle, then, we can test whether a polynomial of degree n in $\mathbb{F}_p[x]$ is irreducible by looking at *all* possible factorizations of it.

The converse of Theorem 6.55 is false: $x^2 - 2$ is irreducible in $\mathbb{Q}[x]$ (it has no rational root), but it factors mod 2 (as x^2); you can check, however, that $x^2 - [2]$ is irreducible in $\mathbb{F}_3[x]$. But Theorem 6.55 may not apply at all: we'll see in Example 6.67 that $x^4 + 1$ is irreducible in $\mathbb{Q}[x]$, but it factors in $\mathbb{F}_p[x]$ for every prime p (see [26], p. 304).

In order to use Theorem 6.55, we will need an arsenal of irreducible polynomials over finite fields.

Example 6.56. We determine the irreducible polynomials in $\mathbb{F}_2[x]$ of small degree.

As always, the linear polynomials x and $x + 1$ are irreducible.

There are four quadratics: $x^2, x^2 + x, x^2 + 1, x^2 + x + 1$ (more generally, there are p^n monic polynomials of degree n in $\mathbb{F}_p[x]$, for there are p choices for each of the n coefficients $a_0, \ldots, a_{n-1}$). Since each of the first three has a root in $\mathbb{F}_2$, there is only one irreducible quadratic, namely, $x^2 + x + 1$.

There are eight cubics, of which four are reducible because their constant term is 0 (so that x is a factor). The remaining polynomials are

$$x^3 + 1, \qquad x^3 + x + 1, \qquad x^3 + x^2 + 1, \qquad x^3 + x^2 + x + 1.$$

Since 1 is a root of the first and fourth, the middle two are the only irreducible cubics. Proposition 6.51 now applies.

There are sixteen quartics, of which eight are reducible because their constant term is 0. Of the eight with nonzero constant term, those having an even number of nonzero coefficients have 1 as a root. There are now only four surviving polynomials f, and each has no roots in $\mathbb{F}_2$; that is, they have no linear factors. The only possible factorization for any of them is $f = gh$, where both g and h are irreducible quadratics. But there is only one irreducible quadratic, namely, $x^2 + x + 1$. Therefore, $x^4 + x^2 + 1 = (x^2 + x + 1)^2$ factors, and the other three quartics are irreducible.

Irreducible Polynomials of Low Degree over $\mathbb{F}_2$

Degree 2: $x^2 + x + 1$.

Degree 3: $x^3 + x + 1, \quad x^3 + x^2 + 1$.

Degree 4: $x^4 + x^3 + 1, \quad x^4 + x + 1, \quad x^4 + x^3 + x^2 + x + 1$. ▲

Example 6.57. Here is a list of the monic irreducible quadratics and cubics in $\mathbb{F}_3[x]$. You can verify that the list is correct by first enumerating all such polynomials; there are six monic quadratics having nonzero constant term, and there are eighteen monic cubics having nonzero constant term. It must then be checked which of these have 1 or -1 as a root, for Proposition 6.51 applies.

Note that $-1 = 2$ in $\mathbb{Z}_3$.

Monic Irreducible Quadratics and Cubics over $\mathbb{F}_3$

Degree 2: $x^2 + 1, \qquad\qquad x^2 + x - 1, \qquad\qquad x^2 - x - 1$.

Degree 3: $x^3 - x + 1, \qquad\quad x^3 + x^2 - x + 1, \qquad x^3 - x^2 + 1,$

$x^3 - x^2 + x + 1, \quad x^3 - x - 1, \qquad\qquad x^3 + x^2 - 1,$

$x^3 + x^2 + x - 1, \quad x^3 - x^2 - x - 1$. ▲

Example 6.58. Here are some applications of Theorem 6.55.

(i) The polynomial $f(x) = 3x^3 - 3x + 1$ is irreducible in $\mathbb{Q}[x]$, for $\overline{f} = x^3 + x + 1$ is irreducible in $\mathbb{F}_2[x]$.

(ii) We show that $f(x) = x^4 - 5x^3 + 2x + 3$ is irreducible in $\mathbb{Q}[x]$. By Theorem 6.52, the only candidates for rational roots of f are ± 1 and ± 3, and you can check that none is a root. Since f is a quartic, we cannot yet conclude that it is irreducible, for it might be a product of (irreducible) quadratics.

The criterion of Theorem 6.55 works like a charm. Since $\overline{f} = x^4 + x^3 + 1$ in $\mathbb{F}_2[x]$ is irreducible, by Example 6.56, it follows that f is irreducible in $\mathbb{Q}[x]$. (It wasn't necessary to check that f has no rational roots; irreducibility of $\overline{f}$ is enough to conclude irreducibility of f. In spite of this, it is a good habit to first check for rational roots.)

(iii) Let $\Phi_5(x) = x^4 + x^3 + x^2 + x + 1 \in \mathbb{Q}[x]$. In Example 6.56, we saw that $\overline{\Phi_5}(x) = x^4 + x^3 + x^2 + x + 1$ is irreducible in $\mathbb{F}_2[x]$, and so Φ_5 is irreducible in $\mathbb{Q}[x]$. ▲

Further Results.

We can count the number N_n of irreducible polynomials of degree n in $\mathbb{F}_p[x]$. In [17], pp. 83–84, it is shown that

$$p^n = \sum_{d|n} d\, N_d, \tag{6.2}$$

where the sum is over the positive divisors d of n.

This equation can be solved for N_n. If $m = p_1^{e_1} \cdots p_n^{e_n}$, define the **_Möbius function_** by

$$\mu(m) = \begin{cases} 1 & \text{if } m = 1; \\ 0 & \text{if any } e_i > 1; \\ (-1)^n & \text{if } 1 = e_1 = e_2 = \cdots = e_n. \end{cases}$$

It turns out that Eq. (6.2) is equivalent to

$$N_n = \frac{1}{n} \sum_{d|n} \mu(d)\, p^{n/d}.$$

One application of this formula is that, for every $n \geq 1$, there exists an irreducible polynomial in $\mathbb{F}_p[x]$ of degree n.

The definition of μ seems to come out of nowhere, but it occurs in many problems at the intersection of combinatorics and number theory. See A.Cuoco, "Searching for Möbius," *College Mathematics Journal*, 37:2, (148–153), 2009.

Exercises

6.35 Let $f(x) = x^2 + x + 1 \in \mathbb{F}_2[x]$. Prove that f is irreducible in $\mathbb{F}_2[x]$, but that f has a root $\alpha \in \mathbb{F}_4$. Use the construction of $\mathbb{F}_4$ in Exercise 4.55 on page 165 to display α explicitly.

6.36 Show that $x^4 + x + 1$ is not irreducible in $\mathbb{R}[x]$ even though it has no roots in $\mathbb{R}$.

6.37 (i) If k is a field and each of $f(x), g(x) \in k[x]$ has a root α in k, show that α is a root of $\gcd(f, g)$.

(ii) How does this apply to the polynomials in Examples 6.35 and 6.36?

6.38 If p is a prime, show that, in $\mathbb{Z}_p[x]$,

$$x^p - x = \prod_{i=0}^{p-1} (x - i) \quad \text{and} \quad x^{p-1} - 1 = \prod_{i=1}^{p-1} (x - i).$$

6.39 (**Wilson's Theorem**). Suppose that p is a prime in $\mathbb{Z}$. Show that

$$(p - 1)! \equiv -1 \bmod p.$$

6.40 *

(i) Let $f(x) = (x - a_1) \cdots (x - a_n) \in k[x]$, where k is a field. Show that f has **_no repeated roots_** (i.e., all the a_i are distinct) if and only if $\gcd(f, f') = 1$, where f' is the derivative of f.

Hint. Use Exercise 5.17 on page 203.

(ii) Prove that if $p(x) \in \mathbb{Q}[x]$ is an irreducible polynomial, then p has no repeated roots in $\mathbb{C}$.

Hint. Use Corollary 6.37.

6.41 If p is prime, prove that there are exactly $\frac{1}{3}(p^3 - p)$ monic irreducible cubic polynomials in $\mathbb{F}_p[x]$.

6.42 Determine whether the following polynomials are irreducible in $\mathbb{Q}[x]$.

(i) $f(x) = x^5 - 4x + 2$.

(ii) $f(x) = x^4 + x^2 + x + 1$.

Hint. Show that f has no roots in $\mathbb{F}_3$ and that a factorization of f as a product of quadratics would force impossible restrictions on the coefficients.

(iii) $f(x) = x^4 - 10x^2 + 1$.

Hint. Show that f has no rational roots and that a factorization of f as a product of quadratics would force impossible restrictions on the coefficients.

6.43 Is $x^5 + x + 1$ irreducible in $\mathbb{F}_2[x]$?

Hint. Use Example 6.56.

6.44 Let $f(x) = (x^p - 1)/(x - 1)$, where p is prime. Using the identity

$$f(x + 1) = x^{p-1} + pq(x),$$

where $q(x) \in \mathbb{Z}[x]$ has constant term 1, prove that

$$\Phi_p(x^{p^n}) = x^{p^n(p-1)} + \cdots + x^{p^n} + 1$$

is irreducible in $\mathbb{Q}[x]$ for all $n \geq 0$.

6.45 Let k be a field, and let $f(x) = a_0 + a_1 x + \cdots + a_n x^n \in k[x]$ have degree n and nonzero constant term a_0. If f is irreducible, prove that $a_n + a_{n-1}x + \cdots + a_0 x^n$ is irreducible.

Roots of Unity

In Chapter 3, we defined an nth root of unity ζ to be **primitive** if every nth root of unity is a power of ζ. For example, i is a primitive 4th root of unity. Note that i is also an 8th root of unity, for $i^8 = 1$, but it's not a primitive 8th root of unity; $\frac{\sqrt{2}}{2}(1 + i)$ is a primitive 8th root of unity.

Lemma 6.59. *Every nth root of unity $\zeta \in \mathbb{C}$ is a primitive dth root of unity for a unique divisor d of n.*

Proof. We know that $\zeta^n = 1$; let d be the smallest positive integer for which $\zeta^d = 1$. By the Division Algorithm, there are integers q and r with $n = qd + r$, where $0 \leq r < d$. Now

$$1 = \zeta^n = \zeta^{qd+r} = \zeta^{dq}\zeta^r = \zeta^r,$$

because $\zeta^d = 1$. But $r < d$ and $\zeta^r = 1$; if $r > 0$, then we contradict d being the smallest positive such exponent. Therefore, $r = 0$ and $d \mid n$. This shows that ζ is a dth root of unity. Its first d powers,

$$1, \ \zeta, \ \zeta^2, \ \ldots, \ \zeta^{d-1},$$

are all distinct (Exercise 6.51 on page 269). Since there are exactly d dth roots of unity, they are all powers of ζ, and so ζ is primitive. $\blacksquare$

Definition. If d is a positive integer, then the dth *cyclotomic polynomial* is defined by

$$\Phi_d(x) = \prod(x - \zeta),$$

where ζ ranges over all the *primitive dth roots of unity*.

Proposition 6.60. *Let n be a positive integer and regard $x^n - 1 \in \mathbb{Z}[x]$. Then*

(i)

$$x^n - 1 = \prod_{d \mid n} \Phi_d(x),$$

This proposition sheds light on your discovery in Exercise 3.59 on page 116.

where d ranges over all the positive divisors d of n (in particular, both $\Phi_1(x)$ and $\Phi_n(x)$ are factors).

(ii) $\Phi_n(x)$ *is a monic polynomial in $\mathbb{Z}[x]$.*

Proof. (i) For each divisor d of n, collect all terms in the equation $x^n - 1 = \prod(x - \zeta)$ with ζ a primitive dth root of unity. Thus,

$$x^n - 1 = \prod_{d \mid n} h_d(x),$$

where $h_d(x) = \prod(x - \zeta)$ with ζ an nth root of unity that is also a primitive dth root of unity. But every such ζ must be an nth root of unity: by Lemma 6.59, $n = dq$ for some integer d, and $1 = \zeta^n = \zeta^{dq}$. Therefore, $h_d(x) = \Phi_d(x)$.

(ii) The proof is by strong induction on $n \geq 1$. The base step is true, for $\Phi_1(x) = x - 1$. For the inductive step $n > 1$, write

$$x^n - 1 = \Phi_n(x)F(x),$$

where $F(x) = \prod \Phi_d(x)$ with $d \mid n$ and $d < n$. The inductive hypothesis says that all the factors Φ_d of F are monic polynomials in $\mathbb{Z}[x]$; hence, F is a monic polynomial in $\mathbb{Z}[x]$. By Proposition 6.10, $\Phi_n(x) = (x^n - 1)/F(x)$ is a monic polynomial in $\mathbb{Z}[x]$, as desired. ∎

Example 6.61. The formula in Proposition 6.60(i) can be used to calculate $\Phi_n(x)$ for any n. Indeed, solving for $\Phi_n(x)$ in

$$x^n - 1 = \prod_{d \mid n} \Phi_d(x)$$

we have

$$\Phi_n(x) = \frac{x^n - 1}{\prod_{d \mid n, \, d < n} \Phi_d(x)}$$

Using the fact that $\Phi_1(x) = x - 1$, we have a recursively defined function:

$$\Phi_n(x) = \begin{cases} x - 1 & \text{if } n = 1 \\ (x^n - 1)/\prod_{d \mid n, \, d < n} \Phi_d(x) & \text{if } n > 1. \end{cases}$$

n	$\Phi_n(x)$
1	$x - 1$
2	$x + 1$
3	$x^2 + x + 1$
4	$x^2 + 1$
5	$x^4 + x^3 + x^2 + 1$
6	$x^2 - x + 1$
7	$x^6 + x^5 + x^4 + x^3 + x^2 + 1$
8	$x^4 + 1$
9	$x^6 + x^3 + 1$
10	$x^4 - x^3 + x^2 - x + 1$
11	$x^{10} + x^9 + x^8 + x^7 + x^6 + x^5 + x^4 + x^3 + x^2 + 1$
12	$x^4 - x^2 + 1$

Figure 6.1. Cyclotomic polynomials.

You should verify that $x^{12} - 1 = \prod_{d \in \{1,2,3,4,6,12\}} \Phi_d(x)$. The recursive definition can be programmed into a CAS (see Exercise 6.55 on page 270); Figure 6.1 displays the first dozen cyclotomic polynomials. There's no simple pattern to these polynomials, but calculating a good number of them gives you food for thought and leads to interesting conjectures. For example, can you conjecture anything about $\deg(\Phi_n)$? All the coefficients of the cyclotomic polynomials displayed in Figure 6.1 are 0 and ± 1, but your guess that this is always true is wrong [see Exercise 6.55(iii) on page 270]. Do any of the $\Phi_n(x)$ factor in $\mathbb{Z}[x]$?   ▲

When $p \leq 11$ is prime, $\Phi_p(x)$ is $x^{p-1} + x^{p-2} + \cdots + x^2 + x + 1$. We now prove this is true for every prime p.

Proposition 6.62. *If p is prime,*

$$\Phi_p(x) = x^{p-1} + x^{p-2} + \cdots + x^2 + x + 1.$$

Proof. By Proposition 6.60,

$$x^p - 1 = \Phi_1(x)\Phi_p(x) = (x - 1)\Phi_p(x),$$

and the Division Algorithm gives

$$\Phi_p(x) = \frac{x^p - 1}{x - 1} = x^{p-1} + x^{p-2} + \cdots + x^2 + x + 1. \quad \blacksquare$$

Recall that the **Euler ϕ-function** $\phi(n)$ is defined by

$$\phi(n) = \text{number of } k \text{ with } 1 \leq k \leq n \text{ and } \gcd(k, n) = 1.$$

The next proposition shows that $\phi(n)$ is intimately related to $\Phi_n(x)$, and this leads to a simple proof of a fact from number theory.

Proposition 6.63. (i) $\phi(n) = \deg(\Phi_n)$.

(ii) *For every integer $n \geq 1$, we have $n = \sum_{d|n} \phi(d)$.*

Proof. (i) This follows at once from Corollary 3.30, which says that there are $\phi(n)$ primitive nth roots of unity.

(ii) Immediate from Proposition 6.60(i) and part (i), for

$$n = \sum_{d|n} \deg(\Phi_d) = \sum_{d|n} \phi(d). \quad \blacksquare$$

Proposition 6.63(ii) is often proved in number theory courses without mentioning cyclotomic polynomials; the resulting proof is much more difficult.

We've shown that $\Phi_n(x) \in \mathbb{Z}[x]$, and we'll finish this section by showing that Φ_p is irreducible in $\mathbb{Q}[x]$ when p is prime. It turns out that Φ_n is actually irreducible in $\mathbb{Q}[x]$ for every n, but the proof is more difficult (see [17] p.195).

As any linear polynomial over a field, the cyclotomic polynomial $\Phi_2 = x + 1$ is irreducible in $\mathbb{Q}[x]$; $\Phi_3 = x^2 + x + 1$ is irreducible in $\mathbb{Q}[x]$ because it has no rational roots; we saw, in Example 6.58, that Φ_5 is irreducible in $\mathbb{Q}[x]$. We'll next introduce another irreducibility criterion, useful in its own right, that will allow us to prove that Φ_p is irreducible in $\mathbb{Q}[x]$ for all primes p. An example will motivate the criterion.

Example 6.64. Exercise 6.30 on page 254 asked you to show that $f(x) = x^3 + 5x^2 - 10x + 15$ is irreducible in $\mathbb{Z}[x]$. You now have machinery that makes this easy. For example, you could invoke Theorem 6.52 (the Rational Root Theorem) to show that f has no root in $\mathbb{Q}$ (or $\mathbb{Z}$) and then use Proposition 6.51.

But let's use another technique that shows the power of reducing coefficients. Suppose that $f(x), g(x), h(x) \in \mathbb{Z}[x]$ and $f = gh$, where neither g nor h is constant; reduce the coefficients mod 5. Because reduction mod 5 is a homomorphism, we have $\overline{f} = \overline{g}\,\overline{h}$. But all the coefficients of f (except the leading one) are divisible by 5, so we have

$$x^3 = \overline{g}\overline{h} \quad \text{in } \mathbb{Z}_5[x].$$

Since x is irreducible (it's a linear polynomial), we can apply unique factorization in $\mathbb{Z}_5[x]$ to conclude that both $\overline{g}$ and $\overline{h}$ are of the form ux^m where u is a unit in $\mathbb{Z}_5$. Pulling this back to $\mathbb{Z}$, we see that all the coefficients of g and h, except their leading coefficients, are divisible by 5. Hence the constant term of gh (which is the product of the constant terms of g and h) is divisible by 25. But $gh = f$ and the constant term of f is 15, which is not divisible by 25. Hence no non-trivial factorization of f exists. ▲

Theorem 6.65 (Eisenstein Criterion). *Let $f(x) = a_0 + a_1 x + \cdots + a_n x^n \in \mathbb{Z}[x]$. If there is a prime p dividing a_i for all $i < n$ but with $p \nmid a_n$ and $p^2 \nmid a_0$, then f is irreducible in $\mathbb{Q}[x]$.*

Usually, Kadiddlehopper was the first to discover Kadiddlehopper's Theorem, but not always. For example, the Eisenstein Criterion is in a paper of Eisenstein of 1850, but it appeared in a paper of Schönemann in 1845.

Proof. (**R. Singer**). Let $r_p^* : \mathbb{Z}[x] \to \mathbb{F}_p[x]$ be reduction mod p, and let $\overline{f}$ denote $r_p^*(f)$. If f is not irreducible in $\mathbb{Q}[x]$, then Gauss's Lemma gives polynomials $g(x), h(x) \in \mathbb{Z}[x]$ with $f = gh$, where $g(x) = b_0 + b_1 x + \cdots + b_m x^m$, $h(x) = c_0 + c_1 x + \cdots + c_k x^k$, and $m, k > 0$. There is thus an equation $\overline{f} = \overline{g}\overline{h}$ in $\mathbb{F}_p[x]$.

Since $p \nmid a_n$, we have $\overline{f} \neq 0$; in fact, $\overline{f} = ux^n$ for some unit $u \in \mathbb{F}_p$, because all its coefficients, aside from its leading coefficient, are 0. By

Theorem 6.40, unique factorization in $k[x]$ where k is a field, we must have $\overline{g} = vx^m$ and $\overline{h} = wx^k$ (for units v, w in $\mathbb{F}_p$), so that each of $\overline{g}$ and $\overline{h}$ has constant term 0. Thus, $[b_0] = 0 = [c_0]$ in $\mathbb{F}_p$; equivalently, $p \mid b_0$ and $p \mid c_0$. But $a_0 = b_0 c_0$, and so $p^2 \mid a_0$, a contradiction. Therefore, f is irreducible in $\mathbb{Q}[x]$.   ∎

Let's see that $\Phi_p(x) = x^{p-1} + x^{p-2} + \cdots + x + 1$ is irreducible in $\mathbb{Q}[x]$ when p is prime. Gauss showed how to transform $\Phi_p(x)$ so that the Eisenstein Criterion applies.

Lemma 6.66. *Let $g(x) \in \mathbb{Z}[x]$. If there is $c \in \mathbb{Z}$ with $g(x + c)$ irreducible in $\mathbb{Z}[x]$, then g is irreducible in $\mathbb{Q}[x]$.*

Proof. By Exercise 5.42 on page 216, the function $\varphi \colon \mathbb{Z}[x] \to \mathbb{Z}[x]$, given by

$$f(x) \mapsto f(x + c),$$

is an isomorphism (its inverse is $f(x) \mapsto f(x - c)$). If $g(x) = s(x)t(x)$, then

$$\varphi(g) = \varphi(st) = \varphi(s)\varphi(t). \tag{6.3}$$

But $\varphi(g) = g(x + c)$, so that Eq. (6.3) is a forbidden factorization of $g(x + c)$. Hence, Corollary 6.54 says that g is irreducible in $\mathbb{Q}[x]$.   ∎

Example 6.67. Consider $f(x) = x^4 + 1 \in \mathbb{Q}[x]$. Now

$$f(x + 1) = (x + 1)^4 + 1 = x^4 + 4x^3 + 6x^3 + 4x + 2.$$

The Eisenstein Criterion, using the prime $p = 2$, shows that $f(x + 1)$ is irreducible in $\mathbb{Q}[x]$, and Lemma 6.66 shows that $x^4 + 1$ is irreducible in $\mathbb{Q}[x]$.   ▲

Theorem 6.68 (Gauss). *For every prime p, the pth cyclotomic polynomial $\Phi_p(x)$ is irreducible in $\mathbb{Q}[x]$.*

Proof. Since $\Phi_p(x) = (x^p - 1)/(x - 1)$, we have

$$\Phi_p(x + 1) = [(x + 1)^p - 1]/x = x^{p-1} + \binom{p}{1}x^{p-2} + \binom{p}{2}x^{p-3} + \cdots + p.$$

Since p is prime, we have $p \mid \binom{p}{i}$ for all i with $0 < i < p$ (Proposition 2.26); hence, the Eisenstein Criterion applies, and $\Phi_p(x + 1)$ is irreducible in $\mathbb{Q}[x]$. By Lemma 6.66, Φ_p is irreducible in $\mathbb{Q}[x]$.   ∎

It's not true that $x^{n-1} + x^{n-2} + \cdots + x + 1$ is irreducible when n is not prime. For example, when $n = 4$, $x^3 + x^2 + x + 1 = (x + 1)(x^2 + 1)$.

Further results. Gauss used Theorem 6.68 to prove that a regular 17-gon can be constructed with ruler and compass (ancient Greek mathematicians did not know this). He also constructed regular 257-gons and 65537-gons. We will look at ruler–compass constructions in Chapter 7.

Exercises

6.46 * Let $\zeta = e^{2\pi i/n}$ be a primitive nth root of unity.

    (i) Prove, for all $n \geq 1$, that
$$x^n - 1 = (x-1)(x-\zeta)(x-\zeta^2)\cdots(x-\zeta^{n-1}),$$
    and, if n is odd, that
$$x^n + 1 = (x+1)(x+\zeta)(x+\zeta^2)\cdots(x+\zeta^{n-1}).$$

    (ii) For numbers a and b, prove that
$$a^n - b^n = (a-b)(a-\zeta b)(a-\zeta^2 b)\cdots(a-\zeta^{n-1}b),$$
    and, if n is odd, that
$$a^n + b^n = (a+b)(a+\zeta b)(a+\zeta^2 b)\cdots(a+\zeta^{n-1}b).$$

    Hint. Set $x = a/b$ if $b \neq 0$.

6.47 * Let k be a field and $a \in k$. Show that, in $k[x]$,
$$x^n - a^n = (x-a)\big(x^{n-1} + x^{n-1}a + x^{n-2}a^2 + \ldots a^{n-1}x + a^n\big).$$

6.48 If k is a field, $a \in k$, and $f(x) = c_n x^n + c_{n-1}x^{n-1} + \cdots + c_0 \in k[x]$, then rewrite
$$f(x) - f(a) = \big(c_n x^n + c_{n-1}x^{n-1} + \cdots + c_0\big) - \big(c_n a^n + c_{n-1}a^{n-1} + \cdots + c_0\big)$$
and use Exercise 6.47 to give another proof of Corollary 6.15.

6.49 Determine whether the following polynomials are irreducible in $\mathbb{Q}[x]$.

    (i) $f(x) = 3x^2 - 7x - 5$.   (ii) $f(x) = 2x^3 - x - 6$.

    (iii) $f(x) = 8x^3 - 6x - 1$.   (iv) $f(x) = x^3 + 6x^2 + 5x + 25$.

    (v) $f(x) = x^4 + 8x + 12$.

    Hint. In $\mathbb{F}_5[x]$, $f(x) = (x+1)g(x)$, where g is irreducible.

6.50 Use the Eisenstein Criterion to prove that if a is a squarefree integer, then $x^n - a$ is irreducible in $\mathbb{Q}[x]$ for every $n \geq 1$. Conclude that there are irreducible polynomials in $\mathbb{Q}[x]$ of every degree $n \geq 1$.

6.51 * In the proof of Lemma 6.59, we claimed that the first d powers of ζ are distinct, where ζ is an nth root of unity and d is the smallest positive integer with $\zeta^d = 1$. Prove this claim.

6.52 * Let ζ be an nth root of unity. Lemma 6.59 shows that ζ is a primitive dth root of unity for some divisor d of n. Show that the divisor is unique.

6.53 Consider a finite table of data:

Input	Output
a_1	b_1
a_2	b_2
a_3	b_3
⋮	⋮
a_n	b_n

Show that two polynomial functions (defined over $\mathbb{Q}$) agree on the table if and only if their difference is divisible by
$$\prod_{i=1}^{n}(x - a_i).$$

6.54 (i) Show that the set of polynomials in $\mathbb{Q}[x]$ that vanish on $\{\alpha_1, \ldots, \alpha_n\}$ is an ideal in $\mathbb{Q}[x]$.

(ii) What is a generator of this ideal?

6.55 *

(i) Implement the recursively defined function for $\Phi_n(x)$ given in Example 6.61 in a CAS.

(ii) Use it to generate $\Phi_n(x)$ for, say, $1 \leq n \leq 50$.

(iii) Use the CAS to find the smallest value of n for which a coefficient of $\Phi_n(x)$ is not 0, 1, or -1.

6.3   Connections: Lagrange Interpolation

A popular activity in high school mathematics is finding a polynomial that agrees with a table of data. For example, students are often asked to find a polynomial agreeing with a table like this:

There are several methods for fitting polynomial functions to data; see Chapter 1 of [7].

Input	Output
-3	12
2	22
3	72
-4	-26

On the surface, the problem of fitting data seems to have little to do with the ideas in this chapter. However, by placing it in a more abstract setting, we'll see that it yields Lagrange Interpolation, a result useful in its own right. Thus, this problem fits right into the theory of commutative rings; in fact, it's really the Chinese Remainder Theorem! But isn't the Chinese Remainder Theorem about solving some congruences? Well, yes, and we'll see that we can make the notion of congruence apply here. But first, let's find a polynomial $f(x)$ by hand that fits the table.

$f(-3) = 12$ $\Leftrightarrow$ the remainder when $f(x)$ is divided by $(x + 3)$ is 12

$f(2) = 22$ $\Leftrightarrow$ the remainder when $f(x)$ is divided by $(x - 2)$ is 22

$f(3) = 72$ $\Leftrightarrow$ the remainder when $f(x)$ is divided by $(x - 3)$ is 72

$f(-4) = -26$ $\Leftrightarrow$ the remainder when $f(x)$ is divided by $(x + 4)$ is -26.

The statement about remainders looks structurally similar to the solution we constructed to the problem from Qin Jiushao on page 146. We now make this similarity precise.

Recall that $a \equiv b \bmod m$ in $\mathbb{Z}$ means that $m \mid (a - b)$. We can define congruence in $k[x]$, where k is a field: given $m(x) \in k[x]$, then $f(x), g(x)$ are ***congruent*** mod m, denoted by

$$f \equiv g \bmod m,$$

if $m \mid (f - g)$.

Take It Further. Throwing caution to the winds, here's a fantastic generalization. Rephrase congruence mod m in $\mathbb{Z}$ in terms of ideals: since $m \mid (a-b)$ if and only if $a - b \in (m)$, we have $a \equiv b$ mod m if and only if $a - b \in (m)$. Let R be a commutative ring and I be any, not necessarily principal, ideal in R. If $a, b \in R$, define

$$a \equiv b \bmod I$$

to mean $a - b \in I$ (we'll actually use this generalization when we discuss quotient rings in Chapter 7).

We can now rephrase the Division Algorithm in $k[x]$ using congruence. The statement: given $m(x), f(x) \in k[x]$ with $m(x) \neq 0$, there exist $q(x), r(x) \in k[x]$ with $f = qm + r$, where $r = 0$ or $\deg(r) < \deg(m)$, can be rewritten to say

$$f \equiv r \bmod m.$$

And Proposition 6.14, the Remainder Theorem, says that if $m(x) = x - a$, then $f \equiv f(a) \bmod (x - a)$. Thus, the constraints on f can be rewritten

$$f \equiv 12 \bmod (x + 3)$$
$$f \equiv 22 \bmod (x - 2)$$
$$f \equiv 72 \bmod (x - 3)$$
$$f \equiv -26 \bmod (x + 4).$$

Notice that the four linear polynomials are pairwise relatively prime.

Let's push the similarity a little further, using the localization idea on page 146. Suppose we can find polynomials g, h, k, and ℓ satisfying

$g(-3) = 1$	$h(-3) = 0$	$k(-3) = 0$	$\ell(-3) = 0$
$g(2) = 0$	$h(2) = 1$	$k(2) = 0$	$\ell(2) = 0$
$g(3) = 0$	$h(3) = 0$	$k(3) = 1$	$\ell(3) = 0$
$g(-4) = 0$	$h(-4) = 0$	$k(-4) = 0$	$\ell(-4) = 1.$

Setting $f = 12g + 22h + 72k - 26\ell$, we have a polynomial that fits the original table (why?). Now Proposition 6.15, the Factor Theorem, shows that g is divisible by the linear polynomials $x - 2$, $x - 3$, and $x + 4$. Since they are irreducible in $\mathbb{Q}[x]$, they are pairwise relatively prime; hence, Exercise 6.17(ii) on page 247 says that g is divisible by their product: there is $A(x)$ such that

$$g(x) = A(x - 2)(x - 3)(x + 4).$$

In fact, we can choose A to be a constant and have $g(-3) = 1$: set

$$1 = g(-3) = A(-3 + 2)(-3 - 3)(-3 + 4);$$

that is, $A = 1/30$ and $g(x) = \frac{1}{30}(x - 2)(x - 3)(x + 4)$. Similarly,

- $h(x) = B(x + 3)(x - 3)(x + 4)$ and $h(2) = 1$ implies $B = -1/30$, so

$$h(x) = -\frac{1}{30}(x + 3)(x - 3)(x + 4)$$

- $k(x) = C(x + 3)(x - 2)(x + 4)$ and $k(3) = 1$ implies $C = 1/42$, so

$$k(x) = \tfrac{1}{42}(x + 3)(x - 2)(x + 4)$$

- $\ell(x) = D(x + 3)(x - 2)(x - 3)$ and $\ell(-4) = 1$ implies $D = -1/42$, so

$$\ell(x) = -\tfrac{1}{42}(x + 3)(x - 2)(x - 3).$$

Now putting $f = 12g + 22h + 72k - 26\ell$, we have, after simplification (carried out by a CAS):

$$f(x) = 2x^3 + 4x^2 - 8x + 6.$$

You can check that f matches the table.

This method is called **_Lagrange Interpolation_**, and it applies to any finite set of input-output pairs. As we just saw, it's the same method used in the proof of the Chinese Remainder Theorem, but applied to polynomials rather than integers. Since our goal is merely to display connections, we leave the proof to the reader.

Theorem 6.69 (Chinese Remainder Theorem for Polynomials). *Let k be a field. If $m_1, \ldots, m_r \in k[x]$ are pairwise relatively prime and $b_1, \ldots, b_r \in k[x]$, then the simultaneous congruences*

$$f \equiv b_1 \bmod m_1$$
$$f \equiv b_2 \bmod m_2$$
$$\vdots \quad \vdots$$
$$f \equiv b_r \bmod m_r$$

have an explicit solution, namely,

$$f = b_1\,(s_1 M_1) + b_2\,(s_2 M_2) + \cdots + b_r\,(s_r M_r),$$

where

$$M_i = m_1 m_2 \cdots \widehat{m}_i \cdots m_r \quad and \quad s_i M_i \equiv 1 \bmod m_i \; for \; 1 \leq i \leq r.$$

Furthermore, any solution to this system is congruent to f mod $m_1 m_2 \cdots m_r$.

Proof. The proof of Theorem 4.27 can be easily adapted to prove this.   ∎

Example 6.70. The calculations we have just made can be used to illustrate the theorem. The table gives

$$b_1 = 12, \qquad b_2 = 22, \qquad b_3 = 72, \qquad b_4 = -26;$$

applying the Remainder Theorem to the table entries,

$$m_1 = x + 3, \qquad m_2 = x - 2, \qquad m_3 = x - 3, \qquad m_4 = x + 4.$$

Then

$$M_1 = (x - 2)(x - 3)(x + 4),$$
$$M_2 = (x + 3)(x - 3)(x + 4),$$
$$M_3 = (x + 3)(x - 2)(x + 4),$$
$$M_4 = (x + 3)(x - 2)(x - 3),$$

and

$$s_1 = 1/30, \qquad s_2 = -1/30, \qquad s_3 = 1/42, \qquad s_4 = -1/42.$$

Note, for example, that

$$
\begin{aligned}
s_1 M_1 - 1 &= \tfrac{1}{30}(x-2)(x-3)(x+4) - 1 \\
&= \frac{x^3 - x^2 - 14x + 24}{30} - 1 \\
&= \frac{x^3 - x^2 - 14x - 6}{30} \\
&= (x+3)\frac{x^2 - 4x - 2}{30},
\end{aligned}
$$

so that $s_1 M_1 \equiv 1 \bmod m_1$. This is not magic; if you look carefully at how s_1 is calculated (we called it A on the previous page), you'll see that it is none other than

$$1/M_1(-3).$$

And the Remainder Theorem (again) says that the remainder when $g(x) = M_1(x)/M_1(-3)$ is divided by $x + 3$ is

$$g(-3) = M_1(-3)/M_1(-3) = 1.$$

Similarly, we have $s_i M_i \equiv 1 \bmod m_i$ for the other values of i.

Finally, the statement about any other solution to the system follows from Exercise 6.53 on page 269. ▲

Compare the statement of Theorem 6.69 to the more typical statement of Lagrange Interpolation.

Theorem 6.71 (Lagrange Interpolation). *Let k be a field. An explicit way of writing the polynomial $f(x) \in k[x]$ of minimal degree that takes the values b_i at distinct points a_i, for $1 \le i \le r$, is*

$$f(x) = b_1 \frac{M_1(x)}{M_1(a_1)} + b_2 \frac{M_2(x)}{M_2(a_2)} + b_3 \frac{M_3(x)}{M_3(a_3)} + \cdots + b_r \frac{M_r(x)}{M_r(a_r)},$$

where $M_i(x)$ is the polynomial defined by

$$M_i(x) = (x - a_1)(x - a_2) \cdots (x - a_{i-1})\widehat{(x - a_i)}(x - a_{i+1}) \cdots (x - a_r).$$

Some things to note:

(i) Theorem 6.69 is more general than Theorem 6.71, for it allows the moduli m_i to be any finite set of relatively prime polynomials (Lagrange Interpolation only considers moduli of the form $x - a$).

(ii) On the other hand, Theorem 6.71 is more explicit: it implies that s_i (in the statement of the Chinese Remainder Theorem) is $1/M_i(a_i)$ (in the statement of Lagrange Interpolation).

(iii) The statement of Lagrange Interpolation goes on to say that the polynomial obtained by this method is the one of lowest degree that fits the conditions.

You'll verify these last two items in Exercises 6.60 and 6.63 below.

Exercises

6.56 Find a polynomial that agrees with the table

Input	Output
0	3
1	4
2	7
3	48
4	211

6.57 A radio show offered a prize to the first caller who could predict the next term in the sequence

$$1, 2, 4, 8, 16.$$

(i) What would you get if you used "common sense?"

(ii) What would you get if you used Lagrange Interpolation?

6.58 Another radio show offered a prize to the first caller who could predict the next term in the sequence

$$14, 3, 26, 8, 30.$$

After no one got it for a few days, the host announced that these are the first five numbers that were retired from the Mudville Sluggers baseball team. Use Lagrange Interpolation to predict the next number that was retired in Mudville.

6.59 The following table fits the quadratic $f(x) = x^2 - 3x + 5$; that is, $f(0) = 5$, $f(1) = 3$, etc. Now forget about f and use Lagrange Interpolation to find a polynomial that fits the table.

The result of Exercise 6.53 allows you to fool many standardized tests.

Input	Output
0	5
1	3
2	3
3	5
4	9

It seems that this table should fool Lagrange Interpolation, which produces a degree 4 polynomial. Does it?

6.60 * Show that Lagrange Interpolation produces a polynomial of smallest degree that agrees with a given input-output table.

6.61 (i) Find a polynomial $g(x)$ that agrees with the table

Input	Output
4	24
5	60
6	120
7	210
8	336

(ii) Factor g into irreducibles.

6.62 It is known that there's a cubic polynomial function $f(x) \in \mathbb{Q}[x]$ such that, for positive integers n,

$$f(n) = \sum_{k=0}^{n-1} k^2.$$

   (i) Find f.

   Hint. A cubic is determined by four inputs.

   (ii) Prove that $f(n) = \sum_{k=0}^{n-1} k^2$ for all positive integers n.

6.63 * Using the notation of Theorem 6.71, show that

$$M_i(x)/M_i(a_i) \equiv 1 \mod (x - a_i).$$

7 Quotients, Fields, and Classical Problems

In Chapter 4, we introduced the idea of congruence modulo an integer m as a way to "ignore" multiples of m in calculations by concentrating on remainders. This led to an arithmetic of congruence classes and the construction of the commutative ring $\mathbb{Z}_m$ in which multiples of m are set equal to 0. You now know that the multiples of m form an ideal (m) in $\mathbb{Z}$, and so $\mathbb{Z}_m$ can be thought of as a commutative ring obtained from $\mathbb{Z}$ in which all the elements of (m) are set equal to 0.

In this chapter, we introduce *quotient rings*, a generalization of this construction. Given a commutative ring R and an ideal $I \subseteq R$, we will produce a new commutative ring R/I that forces all the elements of I to be 0.

In particular, beginning with the commutative ring $k[x]$ and the ideal (f), where k is a field and $f(x) \in k[x]$, we shall see that identifying (f) with 0 produces an element α in the quotient ring $k[x]/(f)$ that is a root of f: if $f(x) = c_0 + c_1 x + \cdots + c_n x^n$, then $f(\alpha) = c_0 + c_1\alpha + \cdots + c_n\alpha^n = 0$. Moreover, the complex number field is a special case: $\mathbb{C}$ is the quotient ring arising from $\mathbb{R}[x]$ and the ideal $(x^2 + 1)$. Another byproduct of the quotient ring construction is the existence and classification of all finite fields (there are others beside $\mathbb{F}_p$ and $\mathbb{F}_4$).

In the last section, we will apply fields to settle classical geometric problems that arose over two millenia ago: using only ruler and compass, can we duplicate the cube, trisect an angle, square the circle, or construct regular n-gons?

7.1 Quotient Rings

In Chapter 3, we said that the approach of many Renaissance mathematicians to the newly invented complex numbers was to consider them as polynomials or rational functions in i, where calculations are carried out as usual with the extra simplification rule $i^2 = -1$. In constructing $\mathbb{Z}_m$ from $\mathbb{Z}$, we ignored multiples of m (that is, we set them all equal to 0), and we saw that this idea is compatible with addition and multiplication. Let's see if we can mimic this idea, starting with $\mathbb{R}[x]$, and apply it to $\mathbb{C}$, as our Renaissance ancestors wished. Can we replace the symbol x in a polynomial $f(x)$ by a new symbol i that satisfies $i^2 = -1$? Well, if we make $x^2 = -1$, then we are setting $x^2 + 1 = 0$. So, as with constructing $\mathbb{Z}_m$, let's set all the multiples of $x^2 + 1$ in $\mathbb{R}[x]$ equal to 0.

This analogy with $\mathbb{Z}_m$ looks promising: let the commutative ring $\mathbb{R}[x]$ correspond to $\mathbb{Z}$, and let the (principal) ideal $(x^2 + 1)$ in $\mathbb{R}[x]$ correspond to the ideal (m) in $\mathbb{Z}$. Push this analogy further. Elements in $\mathbb{Z}_m$ are congruence classes $[a]$, where $a \in \mathbb{Z}$. Let's invent new elements $[f]$ corresponding to polynomials $f(x)$. In more detail, $[a]$ denotes $\{a + qm : q \in \mathbb{Z}\}$, and we forced $qm = 0$ (actually, using the Division Algorithm in $\mathbb{Z}$, this allowed us to focus on remainders after dividing by m). Defining $[f] = \{f(x) + q(x)(x^2 + 1) : q(x) \in \mathbb{R}[x]\}$ would allow us to focus on the remainder after dividing $f(x)$ by $x^2 + 1$. Indeed, the Division Algorithm in $\mathbb{R}[x]$ writes

$$f(x) = q(x)(x^2 + 1) + r(x),$$

where $r(x) = 0$ or $\deg(r) < 2$. In other words, we could write $[f] = [r]$, where $r(x) = a + bx$ for $a, b \in \mathbb{R}$. Hold it! If the bracket notation makes $x^2 + 1 = 0$, then $x^2 = -1$, and we may as well write i instead of $[x]$. Looks a lot like $\mathbb{C}$ to us! Now it turns out that this idea is also compatible with addition and multiplication, as we shall see when we introduce *quotient rings* precisely. The construction makes sense for any commutative ring R and any ideal I in R; moreover, it constructs not only $\mathbb{C}$ but many other important systems as well.

Definition. Let I be an ideal in a commutative ring R. We say that $a, b \in R$ are **congruent** **mod** I, written

$$a \equiv b \bmod I,$$

if $a - b \in I$.

This does generalize our earlier definition of congruence when $R = \mathbb{Z}$, $m \geq 0$, and $I = (m)$. If $a, b \in \mathbb{Z}$ and $a \equiv b \bmod (m)$, then $a - b \in (m)$. But $a - b \in (m)$ if and only if $m \mid a - b$; that is, $a \equiv b \bmod m$ in the old sense.

We now note that congruence mod I is an equivalence relation on R.

Proposition 7.1. *Let a, b, c be elements in a commutative ring R. If I is an ideal in R, then*

(i) $a \equiv a \bmod I$

(ii) *if $a \equiv b \bmod I$, then $b \equiv a \bmod I$*

(iii) *if $a \equiv b \bmod I$ and $b \equiv c \bmod I$, then $a \equiv c \bmod I$.*

Proof. Just modify the proof of Proposition 4.3. ∎

The next result shows that the new notion of congruence is compatible with addition and multiplication of elements in R.

Proposition 7.2. *Let I be an ideal in a commutative ring R.*

(i) *If $a \equiv a' \bmod I$ and $b \equiv b' \bmod I$, then*

$$a + b \equiv a' + b' \bmod I.$$

More generally, if $a_i \equiv a_i' \bmod I$ for $i = 1, \ldots, k$, then

$$a_1 + \cdots + a_k \equiv a_1' + \cdots + a_k' \bmod I.$$

(ii) *If $a \equiv a' \bmod I$ and $b \equiv b' \bmod I$, then*

$$ab \equiv a'b' \bmod I.$$

More generally, if $a_i \equiv a_i' \bmod I$ for $i = 1, \ldots, k$, then

$$a_1 \cdots a_k \equiv a_1' \cdots a_k' \bmod I.$$

(iii) *If $a \equiv b$ mod I, then*

$$a^k \equiv b^k \text{ mod } I \text{ for all } k \geq 1.$$

Proof. This is a straightforward modification of the proof of Proposition 4.5. For example, here is the proof of Proposition 4.5(i). If $m \mid (a - a')$ and $m \mid (b-b')$, then $m \mid (a+b)-(a'+b')$, because $(a+b)-(a'+b') = (a-a')+(b - b')$. Rewrite this here by changing "$m \mid (a - a')$" to "$a - a' \in I$." ∎

We now mimic the construction of the commutative rings $\mathbb{Z}_m$ by first generalizing the idea of a congruence class.

Definition. Let I be an ideal in a commutative ring R. If $a \in R$, then the ***coset*** $a + I$ is the subset

$$a + I = \{a + z : z \in I\} \subseteq R.$$

Thus, the coset $a + I$ is the set of all those elements in R that are congruent to a mod I. Cosets generalize the notion of congruence class and so, by analogy, the coset $a + I$ is often called a mod I.

Proposition 7.3. *If $R = \mathbb{Z}$, $I = (m)$, and $a \in \mathbb{Z}$, then the coset*

$$a + I = a + (m) = \{a + km : k \in \mathbb{Z}\}$$

is equal to the congruence class $[a] = \{n \in \mathbb{Z} : n \equiv a \text{ mod } m\}$.

Proof. If $u \in a + (m)$, then $u = a + km$ for some $k \in \mathbb{Z}$. Hence, $u - a = km$, $m \mid (u - a)$, $u \equiv a$ mod m, and $u \in [a]$; that is, $a + (m) \subseteq [a]$.

For the reverse inclusion, if $v \in [a]$, then $v \equiv a$ mod m, $m \mid (v - a)$, $v - a = \ell m$ for some $\ell \in \mathbb{Z}$, and $v = a + \ell m \in a + (m)$. Therefore, $[a] \subseteq a + (m)$, and so $a + (m) = [a]$. ∎

In Proposition 7.1, we saw that congruence mod I is an equivalence relation on R; in Exercise 7.6 on page 285, you will prove that if $a \in R$, then its equivalence class is the coset $a + I$. It follows that the family of all cosets is a partition of R (see Proposition A.17); that is, cosets are nonempty, R is the union of the cosets, and distinct cosets are disjoint: if $a + I \neq b + I$, then $(a + I) \cap (b + I) = \varnothing$.

When are two cosets mod I the same? In Proposition 4.2, we answered this question by proving that $a \equiv b$ mod m if and only if each of a and b has the same remainder after dividing by m.

Proposition 7.4. *Let I be an ideal in a commutative ring R. If $a, b \in R$, then $a + I = b + I$ if and only if $a \equiv b$ mod I. In particular, $a + I = I$ if and only if $a \in I$.*

Proof. Note first that $a \in a + I$, for $0 \in I$ and $a = a + 0$. If $a + I = b + I$, then $a \in b + I$; hence, $a = b + i$ for some $i \in I$, and so $a - b \in I$ and $a \equiv b$ mod I.

Conversely, assume that $a - b \in I$; say $a - b = i$. To see whether $a + I \subseteq b + I$, we must show that if $a + i' \in a + I$, where $i' \in I$, then $a + i' \in b + I$. But $a + i' = (b + i) + i' = b + (i + i') \in b + I$ (for ideals are closed under addition). The reverse inclusion, $b + I \subseteq a + I$, is proved similarly. Therefore, $a + I = b + I$. ∎

We have now generalized congruence mod m to congruence mod an ideal and congruence classes to cosets. The next step is to assemble the cosets and make a commutative ring with them.

Definition. If I is an ideal in a commutative ring R, we denote the set of all its cosets by R/I:

$$R/I = \{a + I : a \in R\}.$$

Once the set $\mathbb{Z}_m$ was built, we equipped it with the structure of a commutative ring by defining addition and multiplication of congruence classes. We carry out that program now for R/I.

Definition. Let I be an ideal in a commutative ring R.
   Define addition $\alpha \colon R/I \times R/I \to R/I$ by

$$\alpha \colon (a + I, b + I) \mapsto a + b + I$$

and define multiplication $\mu \colon R/I \times R/I \to R/I$ by

$$\mu \colon (a + I, b + I) \mapsto ab + I.$$

Example 7.5. Suppose that $R = \mathbb{Z}[x]$ and I is the principal ideal $(x^2 + x + 1)$. If $a = 3 + 2x$ and $b = 4 + 3x$, then

$$(a + I)(b + I) = ab + I = (3 + 2x)(4 + 3x) + I = 12 + 17x + 6x^2 + I.$$

But, by Exercise 7.4(ii) on page 285, $12 + 17x + 6x^2 \equiv 6 + 11x \bmod I$ (in fact, $(12 + 17x + 6x^2) - (6 + 11x) = 6(x^2 + x + 1)$), so that

$$(3 + 2x + I)(4 + 3x + I) = 6 + 11x + I. \quad \blacktriangle$$

Lemma 7.6. *Addition and multiplication $R/I \times R/I \to R/I$ are well-defined functions.*

Proof. Let $a + I = a' + I$ and $b + I = b' + I$; that is, $a - a' \in I$ and $b - b' \in I$.
   To see that addition is well-defined, we must show that $a' + b' + I = a + b + I$. This is true:

$$(a + b) - (a' + b') = (a - a') + (b - b') \in I.$$

To see that multiplication $R/I \times R/I \to R/I$ is well-defined, we must show that $(a' + I)(b' + I) = a'b' + I = ab + I$; that is, $ab - a'b' \in I$. But this is true:

$$ab - a'b' = ab - a'b + a'b - a'b' = (a - a')b + a'(b - b') \in I. \quad \blacksquare$$

The proof of Theorem 4.32, which shows that $\mathbb{Z}_m$ is a commutative ring, generalizes to show that R/I is a commutative ring. Here are the details.

Theorem 7.7. *If I is an ideal in a commutative ring R, then R/I is a commutative ring.*

Proof. Each of the eight axioms in the definition of commutative ring must be verified; all the proofs are routine, for they are inherited from the corresponding property in R. If $a, b, c \in R$, then we have

(i) Commutativity of addition:

$$(a + I) + (b + I) = a + b + I = b + a + I = (b + I) + (a + I).$$

(ii) The zero element is $I = 0 + I$, for $I + (a + I) = 0 + a + I = a + I$.

(iii) The negative of $a + I$ is $-a + I$, for $(a + I) + (-a + I) = 0 + I = I$.

(iv) Associativity of addition:

$$[(a + I) + (b + I)] + (c + I) = (a + b + I) + (c + I)$$
$$= [(a + b) + c] + I = [a + (b + c)] + I$$
$$= (a + I) + (b + c + I) = (a + I) + [(b + I) + (c + I)].$$

(v) Commutativity of multiplication:

$$(a + I)(b + I) = ab + I = ba + I = (b + I)(a + I).$$

(vi) The multiplicative identity is $1 + I$, for $(1 + I)(a + I) = 1a + I = a + I$.

(vii) Associativity of multiplication:

$$[(a + I)(b + I)](c + I) = (ab + I)(c + I)$$
$$= [(ab)c] + I = [a(bc)] + I$$
$$= (a + I)(bc + I) = (a + I)[(b + I)(c + I)].$$

(viii) Distributivity:

$$(a + I)\big[(b + I) + (c + I)\big] = (a + I)(b + c + I)$$
$$= [a(b + c)] + I = (ab + ac) + I$$
$$= (ab + I) + (ac + I)$$
$$= (a + I)(b + I) + (a + I)(c + I). \quad \blacksquare$$

Definition. The commutative ring R/I constructed in Theorem 7.7 is called the *quotient ring* of R modulo I (it is usually pronounced "R mod I").

We said that quotient rings generalize the construction of $\mathbb{Z}_m$. Let's show that the commutative rings $\mathbb{Z}/(m)$ and $\mathbb{Z}_m$ are not merely isomorphic, they are identical.

We have already seen, in Proposition 7.3, that they have the same elements: for every $a \in \mathbb{Z}$, the coset $a + (m)$ and the congruence class $[a]$ are subsets of $\mathbb{Z}$, and they are equal. But the operations coincide as well. They have the same addition:

$$\big(a + (m)\big) + \big(b + (m)\big) = a + b + (m) = [a + b] = [a] + [b]$$

and they have the same multiplication:

$$\big(a + (m)\big)\big(b + (m)\big) = ab + (m) = [ab] = [a][b].$$

Thus, quotient rings truly generalize the integers mod m.

If $I = R$, then R/I consists of only one coset, and so R/I is the zero ring (in Chapter 4, we said that the zero ring does arise occasionally). Since the zero ring is not very interesting, we usually assume, when forming quotient rings, that ideals are proper ideals. Recall, in constructing $\mathbb{Z}_m$, that we usually assumed that $m \geq 2$.

The definitions of addition and multiplication in R/I involve an interplay between reducing modulo the ideal I and the operations of addition and multiplication in R/I. In the special case of $\mathbb{Z}_m$, we called this interplay "reduce as you go." But that's just an informal way of describing a homomorphism. More precisely, if we define a function $\pi: R \to R/I$ by $\pi: a \mapsto a + I$, then we can rewrite $a + b + I = (a + I) + (b + I)$ as $\pi(a + b) = \pi(a) + \pi(b)$; similarly, $ab + I = (a + I)(b + I)$ can be rewritten as $\pi(ab) = \pi(a)\pi(b)$.

The word "map" is often used as a synonym for *function* or *homomorphism*.

Definition. If I is an ideal in a commutative ring R, then the ***natural map*** is the function $\pi: R \to R/I$ given by

$$a \mapsto a + I;$$

that is, $\pi(a) = a + I$.

Proposition 7.8. *If I is an ideal in a commutative ring R, then the natural map $\pi: R \to R/I$ is a surjective homomorphism, and* $\ker \pi = I$.

Proof. We have just seen that $\pi(a+b) = \pi(a)+\pi(b)$ and $\pi(ab) = \pi(a)\pi(b)$. Since $\pi(1) = 1 + I$, the multiplicative identity in R/I, we see that π is a homomorphism.

Now π is surjective: if $a + I \in R/I$, then $a + I = \pi(a)$. Finally, by definition, $\ker \pi = \{a \in R \mid \pi(a) = 0 + I\}$. But $\pi(a) = a + I$, and $a + I = 0 + I$ if and only if $a \in I$ (Proposition 7.4). The result follows. $\blacksquare$

Here is the converse of Proposition 5.25: Every ideal is the kernel of some homomorphism.

Corollary 7.9. *Given an ideal I in a commutative ring R, there exists a commutative ring A and a homomorphism $\varphi: R \to A$ with $I = \ker \varphi$.*

Proof. If we set $A = R/I$, then the natural map $\pi: R \to R/I$ is a homomorphism with $I = \ker \pi$. $\blacksquare$

We know that isomorphic commutative rings are essentially the same, being "translations" of one another; that is, if $\varphi: R \to S$ is an isomorphism, we may think of $r \in R$ as being in English while $\varphi(r) \in S$ is in French. The next theorem shows that quotient rings are essentially images of homomorphisms. It also shows how to modify a homomorphism to make it an isomorphism.

There are second and third isomorphism theorems, but they are less useful (see Exercise 7.15 on page 286).

Theorem 7.10 (First Isomorphism Theorem). *Let R and A be commutative rings. If $\varphi: R \to A$ is a homomorphism, then $\ker \varphi$ is an ideal in R, $\operatorname{im} \varphi$ is a subring of A, and*

$$R/\ker \varphi \cong \operatorname{im} \varphi.$$

Proof. Let $I = \ker \varphi$. We have already seen, in Proposition 5.25, that I is an ideal in R and $\operatorname{im} \varphi$ is a subring of A.

Define $\widetilde{\varphi}: R/I \to \operatorname{im} \varphi$ by

$$\widetilde{\varphi}(r + I) = \varphi(r).$$

We claim that $\widetilde{\varphi}$ is an isomorphism. First, $\widetilde{\varphi}$ is well-defined. If $r + I = s + I$, then $r - s \in I = \ker \varphi$, $\varphi(r - s) = 0$, and $\varphi(r) = \varphi(s)$. Hence

$$\widetilde{\varphi}(r + I) = \varphi(r) = \varphi(s) = \widetilde{\varphi}(s + I).$$

Next, $\widetilde{\varphi}$ is a homomorphism because φ is.

$$\begin{aligned}
\widetilde{\varphi}\big((r + I) + (s + I)\big) &= \widetilde{\varphi}(r + s + I) \\
&= \varphi(r + s) = \varphi(r) + \varphi(s) \\
&= \widetilde{\varphi}(r + I) + \widetilde{\varphi}(s + I).
\end{aligned}$$

Similarly, $\widetilde{\varphi}\big((r + I)(s + I)\big) = \widetilde{\varphi}(r + I)\widetilde{\varphi}(s + I)$ (Exercise 7.7 on page 285). As $\widetilde{\varphi}(1 + I) = \varphi(1) = 1$, we see that $\widetilde{\varphi}$ a homomorphism.

We show that $\widetilde{\varphi}$ is surjective. If $a \in \operatorname{im} \varphi$, then there is $r \in R$ with $a = \varphi(r)$; plainly, $a = \varphi(r) = \widetilde{\varphi}(r + I)$.

Finally, we show that $\widetilde{\varphi}$ is injective. If $\widetilde{\varphi}(r + I) = 0$, then $\varphi(r) = 0$, and $r \in \ker \varphi = I$. Hence, $r + I = I$; that is, $\ker \widetilde{\varphi} = \{I\}$ and $\widetilde{\varphi}$ is injective, by Proposition 5.31. Therefore, $\widetilde{\varphi}$ is an isomorphism. ∎

We can illustrate this last proof with a picture; such a picture is often called a **commutative diagram** if composites of maps having same domain and same target are equal. Here, $i: \operatorname{im} \varphi \to A$ is the inclusion, and $\varphi = i\widetilde{\varphi}\pi$.

$$
\begin{array}{ccc}
R & \xrightarrow{\ \varphi\ } & A \\
{\scriptstyle \pi}\big\downarrow & & \big\uparrow{\scriptstyle i} \\
R/I & \xrightarrow[\widetilde{\varphi}]{} & \operatorname{im} \varphi
\end{array}
$$

Here's a trivial example. If R is a commutative ring, then (0) is an ideal. The identity $1_R: R \to R$ is a surjective homomorphism with $\ker 1_R = (0)$, so that the First Isomorphism Theorem gives the isomorphism $\widetilde{1}_R: R/(0) \to R$; that is, $R/(0) \cong R$.

Theorem 7.10 has more interesting applications than showing that $R/(0) \cong R$. For example, it gives us the tools needed to tighten up the discussion of the alternate construction of $\mathbb{C}$ that began this section.

Theorem 7.11. *The quotient ring $\mathbb{R}[x]/(x^2 + 1)$ is a field isomorphic to the complex numbers $\mathbb{C}$.*

Proof. Consider the evaluation $\varphi: \mathbb{R}[x] \to \mathbb{C}$ (as in Corollary 5.21) with $\varphi(x) = i$ and $\varphi(a) = a$ for all $a \in \mathbb{R}$; that is,

$$\varphi: f(x) = a_0 + a_1 x + a_2 x^2 + \cdots \mapsto f(i) = a_0 + a_1 i + a_2 i^2 + \cdots.$$

Now φ is surjective, for $a + ib = \varphi(a + bx)$, and so the First Isomorphism Theorem gives an isomorphism $\widetilde{\varphi}: \mathbb{R}[x]/\ker \varphi \to \mathbb{C}$, namely $f(x) + \ker \varphi \mapsto f(i)$. But Corollary 6.26 gives $\ker \varphi = (x^2 + 1)$; therefore, $\mathbb{R}[x]/(x^2 + 1) \cong \mathbb{C}$ as commutative rings, by the First Isomorphism Theorem. We know that $\mathbb{C}$ is a field, and any commutative ring isomorphic to a field must, itself, be a field. Thus, the quotient ring $\mathbb{R}[x]/(x^2 + 1)$ is another construction of $\mathbb{C}$. ∎

Hence, the high school approach to complex numbers contains the germ of a correct idea.

How to Think About It. Because every element of $\mathbb{R}[x]$ is congruent to a linear polynomial $a + bx \bmod (x^2 + 1)$, every element of $\mathbb{R}[x]/(x^2 + 1)$ can be written as $a + bx + (x^2 + 1)$ for some real numbers a and b.

Example 7.12. (i) Since $\mathbb{R}[x]/(x^2 + 1)$ is a field, every nonzero element in it has a multiplicative inverse. Let's find the inverse of an element $a + bx + (x^2 + 1)$ by "pulling back" the formula in $\mathbb{C}$,

$$\frac{1}{a + bi} = \frac{a - bi}{a^2 + b^2},$$

to $\mathbb{R}[x]/(x^2 + 1)$, using the inverse of the isomorphism φ in the proof of Theorem 7.11. Now $\varphi^{-1}(a + bi) = a + bx + (x^2 + 1)$, so that

$$\frac{1}{a + bx + (x^2 + 1)} = \frac{a - bx}{a^2 + b^2} + (x^2 + 1).$$

(ii) Euclidean Algorithm II gives another way of finding the inverse, writing $\gcd(a + bx, x^2 + 1)$ as a linear combination of $a + bx$ and $x^2 + 1$. The algorithms in Exercise 6.18(iii) produce the linear combination in $\mathbb{R}[x]$:

$$\left(\frac{a - bx}{b^2}\right)(a + bx) + 1(x^2 + 1) = \frac{a^2 + b^2}{b^2}.$$

Dividing both sides by $(a^2 + b^2)/b^2$, we have

$$\frac{a - bx}{a^2 + b^2}(a + bx) + \frac{b^2}{a^2 + b^2}(x^2 + 1) = 1.$$

Moving to $\mathbb{R}[x]/(x^2 + 1)$, this implies again that

$$\left(a + bx + (x^2 + 1)\right)^{-1} = \frac{a - bx}{a^2 + b^2} + (x^2 + 1) \quad \blacktriangle$$

We end this section with a generalization of Theorem 7.11. If you chase back the arguments to their source, you'll see that all we needed is that $\mathbb{R}[x]$ is a PID and $x^2 + 1$ is an irreducible element in $\mathbb{R}[x]$.

Proposition 7.13. *If R is a PID and p is an irreducible element in R, then $R/(p)$ is a field.*

Proof. It suffices to show that every nonzero element $a + (p)$ in the commutative ring $R/(p)$ has a multiplicative inverse. Since $a + (p) \neq 0$, we have $a \notin (p)$; that is, $p \nmid a$. Since R is a PID, Theorem 6.46 says that gcd's exist and are linear combinations. In particular, $\gcd(a, p) = 1$, so there are $s, t \in R$ with $sa + tp = 1$. Thus,

$$1 + (p) = sa + (p) = \left(sa + (p)\right) = \left(s + (p)\right)\left(a + (p)\right)$$

in $R/(p)$, and $\left(a + (p)\right)^{-1} = s + (p)$. Therefore, $R/(p)$ is a field.   ∎

Exercises

7.1 Are any cosets of (5) in $\mathbb{Z}$ ideals?

7.2 Prove Proposition 7.1.

7.3 Prove Proposition 7.2.

7.4 * In $\mathbb{Q}[x]/(x^2 + x + 1)$, write each term in the form $a + bx$ with $a, b \in \mathbb{Q}$.

  (i) $(3 + 2x)(4 + 3x)$ (ii) $12 + 17x + 6x^2$

  (iii) x^2 (iv) x^3

  (v) $(1 - x)^2$ (vi) $(1 - x)(1 - x^2)$

  (vii) $(a + bx)(a + bx^2)$ (viii) $(a + bx)^2$.

7.5 In $\mathbb{Q}[x]/(x^4 + x^3 + x^2 + x + 1)$, write each term in the form $a + bx + cx^2 + dx^3$ with a, b, c, d rational numbers.

  (i) x^5

  (ii) $(1 - x)(1 - x^2)(1 - x^3)(1 - x^4)$

  (iii) $(1 + x)(1 + x^2)(1 + x^3)(1 + x^4)$.

7.6 * In Proposition 7.1, we saw that if I is an ideal in a commutative ring R, then congruence mod I is an equivalence relation on R. Prove that the equivalence classes are the cosets mod I.

7.7 In the notation of Theorem 7.10, show that

$$\widetilde{\varphi}\big((r + I)(s + I)\big) = \widetilde{\varphi}(r + I)\widetilde{\varphi}(s + I).$$

7.8 * Let $\varphi \colon R \to S$ be an isomorphism of commutative rings. Assume that $I \subseteq R$ and $J \subseteq S$ are ideals and that $\varphi(I) = J$, where $\varphi(I) = \{\varphi(a) : a \in I\}$. Prove that $\widetilde{\varphi} \colon R/I \to S/J$, given by $\widetilde{\varphi} \colon r + I \mapsto \varphi(r) + J$, is an isomorphism.

7.9 Let I be an ideal in a commutative ring R.

  (i) If S is a subring of R and $I \subseteq S$, prove that

  $$S/I = \{r + I : r \in S\}$$

  is a subring of R/I.

  (ii) If J is an ideal in R and $I \subseteq J$, prove that

  $$J/I = \{r + I : r \in J\}$$

  is an ideal in R/I.

7.10 Show that the subring $\mathbb{Z}[x]/(x^2 + 1)$ of $\mathbb{R}[x]/(x^2 + 1)$ is isomorphic to the Gaussian integers $\mathbb{Z}[i]$.

7.11 Show that there is an isomorphism of fields:

$$\mathbb{R}[x]/(x^2 + 1) \cong \mathbb{R}[x]/(x^2 + x + 1)$$

Hint. Both are isomorphic to $\mathbb{C}$.

7.12 Show that

$$\mathbb{Q}[x]/(x^2 + x + 1) \cong \mathbb{Q}[\omega] = \{u + v\omega : u, v \in \mathbb{Q}\},$$

where $\omega = \frac{1}{2}\left(-1 + i\sqrt{3}\right)$.

7.13 Show that the subring $\mathbb{Z}[x]/(x^2 + x + 1)$ of $\mathbb{R}[x]/(x^2 + x + 1)$ is isomorphic to the Eisenstein integers $\mathbb{Z}[\omega]$.

7.14 For each element of $\mathbb{Q}[x]/(x^2 + x + 1)$, find the multiplicative inverse.

   (i) $3 + 2x + (x^2 + x + 1)$

   (ii) $5 - x + (x^2 + x + 1)$

   (iii) $15 + 7x - 2x^2 + (x^2 + x + 1)$

   (iv) $a + bx + (x^2 + x + 1)$ (in terms of a and b).

7.15 * Prove the ***Third Isomorphism Theorem***: If R is a commutative ring having ideals $I \subseteq J$, then J/I is an ideal in R/I, and there is an isomorphism

$$(R/I)/(J/I) \cong R/J.$$

Hint. Show that the function $\varphi: R/I \to R/J$, given by $a + I \mapsto a + J$, is a homomorphism, and apply the First Isomorphism Theorem.

7.16 For every commutative ring R, prove that $R[x]/(x) \cong R$.

7.17 An ideal I in a commutative ring R is called a ***prime ideal*** if I is a proper ideal such that $ab \in I$ implies $a \in I$ or $b \in I$.

   (i) If p is a prime number, prove that (p) is a prime ideal in $\mathbb{Z}$.

   Hint. Euclid's Lemma.

   (ii) Prove that if an ideal (m) in $\mathbb{Z}$ is a prime ideal, then $m = 0$ or $|m|$ is a prime number.

7.18 Let I be a proper ideal in $k[x]$, where k is a field.

   (i) If p is an irreducible polynomial, prove that (p) is a prime ideal in $k[x]$.

   (ii) Prove that if an ideal (f) in $k[x]$ is a prime ideal, then $f = 0$ or f is an irreducible polynomial.

7.19 Let I be a proper ideal in a commutative ring R.

   (i) Prove that (0) is a prime ideal in R if and only if R is a domain.

   (ii) Prove that I is a prime ideal if and only if $a \notin I$ and $b \notin I$ imply $ab \notin I$.

   (iii) Prove that I is a prime ideal if and only if R/I is a domain.

7.20 Prove that (x) is a prime ideal in $\mathbb{Z}[x]$.

Hint. Is $\mathbb{Z}[x]/(x)$ a domain?

7.21 An ideal I in a commutative ring R is called a ***maximal ideal*** if I is a proper ideal for which there is no proper ideal J with $I \subsetneq J$.

   (i) If p is a prime number, prove that (p) is a maximal ideal in $\mathbb{Z}$.

   (ii) Prove that if an ideal (m) in $\mathbb{Z}$ is a maximal ideal, then $|m|$ is a prime number.

7.22 Let I be a proper ideal in $k[x]$, where k is a field.

   (i) If p is an irreducible polynomial, prove that (p) is a maximal ideal in $k[x]$.

   (ii) Prove that if an ideal (f) in $k[x]$ is a maximal ideal, then f is an irreducible polynomial.

7.23 * Let I be a proper ideal in a commutative ring R.

   (i) Prove that (0) is a maximal ideal in R if and only if R is a field.

   (ii) Prove that I is a maximal ideal if and only if R/I is a field. Conclude that if k is a field and $p(x) \in k[x]$ is irreducible, then $k[x]/(p)$ is a field.

   (iii) Prove that every maximal ideal is a prime ideal.

7.24 (i) Prove that J is a maximal ideal in $\mathbb{Z}[x]$, where J consists of all polynomials with even constant term.

   Hint. Prove that $\mathbb{Z}[x]/J \cong \mathbb{F}_2$.

   (ii) Prove that the prime ideal (x) in $\mathbb{Z}[x]$ is not a maximal ideal.

7.2   Field Theory

General results about quotient rings R/I have a special character when R enjoys extra hypotheses. In this section, we investigate properties of fields with an eye to using the ideas behind the isomorphism

$$\mathbb{R}[x]/(x^2 + 1) \cong \mathbb{C}.$$

We are going to apply quotient rings to prove some interesting results: for every polynomial $f(x) \in k[x]$, where k is a field, there exists a field extension E/k containing all the roots of f; we will also be able to prove the existence of finite fields other than $\mathbb{F}_p$.

Characteristics

Contemplating "any field" seems quite daunting, and so it makes sense for us to begin classifying fields. First of all, fields come in two types: those that contain a subfield isomorphic to $\mathbb{Q}$, and those that contain a subfield isomorphic to $\mathbb{F}_p$ for some prime p.

Recall the definition of na on page 160, where $n \in \mathbb{Z}$ and a is an element of a commutative ring R. For example, $3a$ means $a + a + a$ and $(-3)a$ means $-a - a - a$. More generally, if n is a nonnegative integer, then na means

$$\underbrace{a + a + \cdots + a}_{n \text{ times}},$$

$0a = 0$, and $-na$ is the sum of $|n|$ copies of $-a$. Note that na is a hybrid in the sense that it is the product of an integer and an element of R, not the product of two ring elements. However, na can be viewed as the product of two elements in R, for if e is the multiplicative identity in R, then $ne \in R$ and $na = (ne)a$. In particular, $3a = a + a + a = (e + e + e)a = (3e)a$. This "action" of $\mathbb{Z}$ on R is really a homomorphism.

Lemma 7.14. *If R is a commutative ring with multiplicative identity e, then the function $\chi: \mathbb{Z} \to R$, given by*

$$\chi(n) = ne,$$

is a homomorphism.

Proof. Exercise 7.25 on page 293.   ∎

Proposition 7.15. *If k is a field and $\chi: \mathbb{Z} \to k$ is the map $\chi: n \mapsto ne$, where e is the multiplicative identity in R, then either $\operatorname{im} \chi \cong \mathbb{Z}$ or $\operatorname{im} \chi \cong \mathbb{F}_p$ for some prime p.*

Proof. Since every ideal in $\mathbb{Z}$ is principal, $\ker \chi = (m)$ for some integer $m \geq 0$. If $m = 0$, then χ is an injection, and $\operatorname{im} \chi \cong \mathbb{Z}$. If $m \neq 0$, the First Isomorphism Theorem gives $\mathbb{Z}_m = \mathbb{Z}/(m) \cong \operatorname{im} \chi \subseteq k$. Since k is a field, $\operatorname{im} \chi$ is a domain, and so m is prime (Exercise 5.3 on page 195). Writing p instead of m, we have $\operatorname{im} \chi \cong \mathbb{Z}_p = \mathbb{F}_p$.   ∎

Corollary 7.16. *Every field k contains a subfield isomorphic to either $\mathbb{Q}$ or $\mathbb{F}_p$.*

Proof. Proposition 7.15 shows that k contains a subring isomorphic to $\mathbb{Z}$ or to $\mathbb{F}_p$ for some prime p. If the subring is $\mathbb{Z}$, then, because the field k contains multiplicative inverses for all of its non-zero elements, it contains an isomorphic copy of $\mathbb{Q} = \mathrm{Frac}(\mathbb{Z})$. More precisely, Exercise 5.38(ii) on page 212 says that a field containing an isomorphic copy of $\mathbb{Z}$ as a subring must contain an isomorphic copy of $\mathbb{Q}$.   ■

By Exercise 7.28(i) on page 293, k can't contain an isomorphic copy of *both* $\mathbb{Q}$ and $\mathbb{F}_p$; by Exercise 7.28(ii) on page 293, k can't contain copies of $\mathbb{F}_p$ and $\mathbb{F}_q$ for distinct primes p and q.

Definition. A field has *characteristic* 0 if $\ker \chi = (0)$; it has *characteristic* p if $\ker \chi = (p)$ for some prime p.

This distinction is the first step in classifying different types of fields.

The fields $\mathbb{Q}$, $\mathbb{R}$, $\mathbb{C}$, and $\mathbb{C}(x)$ have characteristic 0, as do any of their subfields. Every finite field has characteristic p for some prime p (after all, if $\ker \chi = (0)$, then $\mathrm{im}\, \chi \cong \mathbb{Z}$ is infinite); $\mathbb{F}_p(x)$, the field of all rational functions over $\mathbb{F}_p$, is an infinite field of characteristic p.

Proposition 7.17. *Let k be a field of characteristic $p > 0$.*

(i) $pa = 0$ *for all* $a \in k$.

Exercise 5.36 on page 212 proves a congruence version of (ii) for $a, b \in \mathbb{Z}$.

(ii) *If* $q = p^n$, *then* $(a + b)^q = a^q + b^q$ *for all* $a, b \in k$.

(iii) *If k is finite, then* $\varphi : k \to k$, *given by*

$$\varphi : a \mapsto a^p,$$

is an isomorphism.

Proof.   (i) Since k has characteristic p, we have $\ker(\chi) = (p)$; that is, $\chi(p) = p1 = 0$ (we have reverted to our usual notation, so that 1 denotes the multiplicative identity). But the hybrid product pa can be viewed as a product of two ring elements: $pa = (p1)a = 0a = 0$.

(ii) Expand $(a + b)^p$ by the Binomial Theorem, and note that $p \mid \binom{p}{j}$ for all $1 \le j \le p - 1$. By (i), all the inside terms vanish. The argument is completed by induction on $n \ge 1$.

(iii) It is obvious that $\varphi(1) = 1$ and

$$\varphi(ab) = (ab)^p = a^p b^p = \varphi(a)\varphi(b).$$

By (ii), $\varphi(a + b) = \varphi(a) + \varphi(b)$. Therefore, φ is a homomorphism. Since $\ker \varphi$ is a proper ideal in k (for $1 \notin \ker \varphi$), we have $\ker \varphi = (0)$, because k is a field, and so φ is an injection. Finally, since k is finite, the Pigeonhole Principle applies, and φ is an isomorphism.   ■

We have seen finite fields $\mathbb{F}_p$ with p elements, for every prime p, and in Exercise 4.55 on page 165, we saw a field $\mathbb{F}_4$ with exactly four elements. The next result shows that the number of elements in a finite field must be a prime power; there is no field having exactly 15 elements. Theorem 7.38 will show, for every prime p and every integer $n \ge 1$, that there exists a field having exactly p^n elements.

Example A.20(iv) in the Appendices shows that if a commutative ring R contains a subring k that is a field, then R is a vector space over k: vectors are elements $r \in R$, while scalar multiplication by $a \in k$ is the given multiplication ar of elements in R. The vector space axioms are just some of the axioms in the definition of commutative ring.

Proposition 7.18. *If K is a finite field, then $|K| = p^n$ for some prime p and some $n \geq 1$.*

Proof. The prime field of K is isomorphic to $\mathbb{F}_p$ for some prime p, by Proposition 7.15. As we remarked, K is a vector space over $\mathbb{F}_p$; as K is finite, it is obviously finite-dimensional. If $\dim_{\mathbb{F}_p}(K) = n$, then $|K| = p^n$, by Corollary A.34 in the Appendix. ■

> If K is a vector space over k, its dimension is denoted by $\dim_k(K)$ or, more briefly, by $\dim(K)$.

Extension Fields

The Fundamental Theorem of Algebra states that every nonconstant polynomial in $\mathbb{C}[x]$ is a product of linear polynomials in $\mathbb{C}[x]$; that is, $\mathbb{C}$ contains all the roots of every polynomial in $\mathbb{C}[x]$. Using ideas similar to those allowing us to view $\mathbb{C}$ as a quotient ring, we'll prove Kronecker's Theorem, a *local* analog of the Fundamental Theorem of Algebra for polynomials over an arbitrary field k: given $f(x) \in k[x]$, there is some field E containing k as a subfield that also contains all the roots of f. (We call this a *local* analog, for even though the larger field E contains all the roots of the polynomial f, it may not contain roots of some other polynomials in $k[x]$.) In fact, we'll see how to construct such an E making basic use of quotient rings of the form $k[x]/I$, where k is a field.

> Theorem 4.43 says that $\mathbb{Z}_m$ is a field if and only if m is a prime in $\mathbb{Z}$; Proposition 7.19 is the analog for $k[x]$.

Proposition 7.19. *If k is a field and $I = (f)$, where $f(x) \in k[x]$ is nonconstant, then the following are equivalent:*

(i) *f is irreducible*

(ii) *$k[x]/I$ is a field*

(iii) *$k[x]/I$ is a domain.*

Proof. (i) $\Rightarrow$ (ii) Since $k[x]$ is a PID, this follows at once from Proposition 7.13.

(ii) $\Rightarrow$ (iii) Every field is a domain.

(iii) $\Rightarrow$ (i) Assume that $k[x]/I$ is a domain. If f is not irreducible, then there are $g(x), h(x) \in k[x]$ with $f = gh$, where $\deg(g) < \deg(f)$ and $\deg(h) < \deg(f)$. Recall that the zero in $k[x]/I$ is $0 + I = I$. Thus, if $g + I = I$, then $g \in I = (f)$ and $f \mid g$, contradicting $\deg(g) < \deg(f)$. Similarly, $h + I \neq I$. However, the product $(g + I)(h + I) = f + I = I$ is zero in the quotient ring, which contradicts $k[x]/I$ being a domain. Therefore, f is irreducible. ■

The structure of general quotient rings R/I can be complicated, but for special choices of R and I, the commutative ring R/I can be easily described. For example, when k is a field and $p(x) \in k[x]$ is an irreducible polynomial, the following proposition gives a complete description of the field $R/I = k[x]/(p)$, and it shows how to build a field K in which $p(x)$ has a root.

> This section will be using various facts about dimension, and you may wish to look in Appendix A.3 to refresh your memory.

Proposition 7.20. *Let k be a field and $K = k[x]/(p)$, where $p(x) \in k[x]$ is a monic irreducible polynomial of degree d and $I = (p)$.*

(i) *K is a field, and*

$$k' = \{a + I : a \in k\}$$

is a subfield of K isomorphic to k. If k' is identified with k via $a \mapsto a + I$, then k is a subfield of K.

If we view k as a subfield of K, then it makes sense to speak of a root of p in K.

(ii) *$z = x + I$ is a root of p in K.*

(iii) *If $g(x) \in k[x]$ and z is a root of g in K, then $p \mid g$ in $k[x]$.*

(iv) *p is the unique monic irreducible polynomial in $k[x]$ having z as a root.*

(v) *K is a vector space over k, the list $1, z, z^2, \dots, z^{d-1}$ is a basis, and*

$$\dim_k(K) = d.$$

Proof.   (i) Since p is irreducible, Proposition 7.19 says that the quotient ring $K = k[x]/I$ is a field, while Corollary 5.32 on page 220 says that the natural map $a \mapsto a + I$ restricts to an isomorphism $k \to k'$.

(ii) Let $p(x) = a_0 + a_1 x + \cdots + a_{d-1}x^{d-1} + x^d$, where $a_i \in k$ for all i. In light of the identification of k and k' in (i), we may view $p(x)$ as $\sum_j (a_j + I)x^j$. Hence, since $z = x + I$,

$$\begin{aligned}
p(z) &= (a_0 + I) + (a_1 + I)z + \cdots + (1 + I)z^d \\
&= (a_0 + I) + (a_1 + I)(x + I) + \cdots + (1 + I)(x + I)^d \\
&= (a_0 + I) + (a_1 x + I) + \cdots + (x^d + I) \\
&= a_0 + a_1 x + \cdots + x^d + I \\
&= p + I = I,
\end{aligned}$$

because $I = (p)$. But $I = 0 + I$ is the zero element of $K = k[x]/I$; thus, $p(z) = 0$ and z is a root of p.

(iii) If $p \nmid g$ in $k[x]$, then $\gcd(g, p) = 1$ because p is irreducible. Therefore, there are $s, t \in k[x]$ with $1 = sp(x) + tg(x)$. Since $k[x] \subseteq K[x]$, we may regard this as an equation in $K[x]$. Setting $x = z$ gives the contradiction $1 = 0$.

(iv) Let $h(x) \in k[x]$ be a monic irreducible polynomial having z as a root. By part (iii), we have $p \mid h$. Since h is irreducible, we have $h = cp$ for some constant c; since h and p are monic, we have $c = 1$ and $h = p$.

(v) Example A.20 in the Appendices shows that K is a vector space over k. Every element of K has the form $f(x) + I$, where $f \in k[x]$. By the Division Algorithm, there are polynomials $q, r \in k[x]$ with $f = qp + r$ and either $r = 0$ or $\deg(r) < d = \deg(p)$. Since $f - r = qp \in I$, it follows that $f + I = r + I$. Let $r(x) = b_0 + b_1 x + \cdots + b_{d-1}x^{d-1}$, where $b_i \in k$ for all i. As in (ii), we see that $r + I = b_0 + b_1 z + \cdots + b_{d-1}z^{d-1}$. Therefore, $1, z, z^2, \dots, z^{d-1}$ spans K.

To see that the list is linearly independent, suppose that

$$\sum_{i=0}^{d-1} c_i z^i = 0 \quad \text{in } K = k[x]/(p);$$

lifting to $k[x]$, this says that

$$\sum_{i=0}^{d-1} c_i x^i \equiv 0 \bmod (p) \quad \text{in } k[x],$$

so that $p \mid \sum_{i=0}^{d-1} c_i x^i$ in $k[x]$. But $\deg(p) = d$, so that all $c_i = 0$. ∎

Definition. If K is a field containing k as a subfield, then K is called an ***extension field*** of k, and we write "K/k is an extension field." An extension field K/k is a ***finite extension*** if K is a finite-dimensional vector space over k. The dimension of K, denoted by

$$[K : k],$$

is called the ***degree*** of K/k.

This notation should not be confused with the notation for a quotient ring, for a field K has no interesting ideals; in particular, if $k \subsetneq K$, then k is not an ideal in K.

Corollary 7.21. *If $k(z)/k$ is a field extension, where z is a root of an irreducible polynomial $p(x) \in k[x]$, then*

$$[k(z) : k] = \deg(p).$$

Proof. This is just a restatement of Proposition 7.20(v). ∎

Corollary 7.21 shows why $[K : k]$ is called the degree of K/k.

How to Think About It. At first glance, many people see Proposition 7.20 as a cheat: we cook up a field that contains a root of $p(x)$ by reducing mod p. The root is thus a *coset*, not a "number." But, just as mathematicians gradually came to see the elements of $\mathbb{C}$ as numbers (through their constant use in calculations), one can develop a feel for arithmetic in $k[x]/(p)$ in which the cosets become concrete objects in their own right, as in the next example.

Example 7.22. Suppose that $k = \mathbb{Q}$ and

$$p(x) = x^3 + x^2 - 2x - 1.$$

You can check that p is irreducible (it's a cubic without a rational root (why?), so it can't factor). Now

$$K = \mathbb{Q}[x]/(x^3 + x^2 - 2x - 1)$$

is a field, $[K : \mathbb{Q}] = 3$, and a basis for K over $\mathbb{Q}$ is

$$1 + (p), \ x + (p), \ x^2 + (p).$$

Hence, every element of K can be represented by a quadratic expression

$$a + bx + cx^2 + (p),$$

where $a, b, c \in \mathbb{Q}$. The expression is unique, because p is a cubic: if two quadratics $f, g \in \mathbb{Q}[x]$ are congruent mod (p), then either $f - g = 0$ or $\deg(f - g) < 2$; the latter cannot occur, and so $f = g$. So, let's drop the "$+(p)$" decoration and just represent an element of K by the unique quadratic

Eventually, we dropped the bracket notation for congruence classes, abbreviating $[a]$ to a.

polynomial in its congruence class mod p. Using this convention, the elements of K are thus named by quadratic polynomials in $\mathbb{Q}[x]$.

What about the arithmetic? Just as in $\mathbb{C} = \mathbb{R}[x]/(x^2+1)$, calculations in K are carried out by calculating in $\mathbb{Q}[x]$, dividing by p, and taking the remainder. Indeed, because $x^3 + x^2 - 2x - 1 = 0$ in K, we have an equation in K,

$$x^3 = -x^2 + 2x + 1.$$

Hence, to calculate in K, we calculate in $\mathbb{Q}[x]$ with the additional simplification rule that x^3 is replaced by $-x^2 + 2x + 1$. Beginning to sound familiar?

So, for example, here are some calculations in K:

(i) Addition looks just the same as in $\mathbb{Q}[x]$ because addition doesn't increase degree:

$$\left(a + bx + cx^2\right) + \left(d + ex + fx^2\right)$$
$$= (a + d) + (b + e)x + (c + f)x^2.$$

(ii) Multiplication requires a simplification. For example, in K,

$$(3 + 2x + 4x^2)(-1 + 5x + 7x^2) = 3 + 53x + 77x^2,$$

a fact that you can verify (by hand or CAS) by expanding the left-hand side and reducing mod (p).

In general, expand

$$\left(a + bx + cx^2\right)\left(d + ex + fx^2\right)$$

as

$$cfx^4 + (bf + ce)x^3 + (af + be + cd)x^2 + (ae + bd)x + ad,$$

and then simplify, replacing occurrences of x^3 by $-x^2 + 2x + 1$:

$$= cfx\left(x^3\right) + (bf + ce)x^3 + (af + be + cd)x^2 + (ae + bd)x + ad$$
$$= cfx\left(-x^2 + 2x + 1\right) + (bf + ce)\left(-x^2 + 2x + 1\right)$$
$$\quad + (af + be + cd)x^2 + (ae + bd)x + ad$$

$$= \text{etc.}$$

> CAS environments will do all of this work for you—just ask for the remainder when a product is divided by p.

A little practice with such calculations gives you the feeling that you are indeed working with "numbers" in a system and, if K had any use, you'd soon become very much at home in it just as our Renaissance predecessors became at home in $\mathbb{C}$. ▲

> You could use Cardano's formula to find expressions for the roots of $p(x)$. Why not try it?

While we've constructed a field extension K/k in which $p(x) \in k[x]$ has a root α, we have little idea about what that root *is*, even when $k = \mathbb{R}$ and α is a complex number. For example, if $p(x) = x^2 + 1$, we can't tell whether $\alpha = i$ or $\alpha = -i$. Proposition 7.20 doesn't give you a way to find roots— it just gives you a way to construct an extension field containing k and in which the operations behave as if $p(x)$ is 0. Playing with these operations might actually give you some ideas about the three complex numbers that make $p(x) = x^3 + x^2 - 2x - 1$ equal to 0 in $\mathbb{C}$; see Example 7.22 above and Exercises 7.29 and 7.30 below.

Exercises

7.25 * Prove Lemma 7.14.

7.26 * If X is a subset of a field k, then $\langle X \rangle$, the **subfield generated by** X, is the intersection of all the subfields containing X (by Exercise 4.61(iii) on page 168, the intersection of any family of subfields of k is itself a subfield of k).

  (i) Prove that $\langle X \rangle$ is the smallest such subfield in the sense that any subfield F containing X must contain $\langle X \rangle$.

  (ii) Define the **prime field** of a field k to be the intersection of all the subfields of k. Prove that the prime field of k is the subfield generated by 1.

  (iii) Prove that the prime field of a field is isomorphic to either $\mathbb{Q}$ or $\mathbb{F}_p$.

7.27 * If k is a field of characteristic $p > 0$ and $a \in k$, prove that

$$(x + a)^p = x^p + a^p.$$

Hint. Use Proposition 7.17 and the Binomial Theorem.

7.28 Let R be a commutative ring, and let p, q be distinct primes.

  (i) Prove that R cannot have subrings A and B with $A \cong \mathbb{Q}$ and $B \cong \mathbb{F}_p$.

  (ii) Prove that R cannot have subrings A and B with $A \cong \mathbb{F}_p$ and $B \cong \mathbb{F}_q$.

  (iii) Why doesn't the existence of $R = \mathbb{F}_p \times \mathbb{F}_q$ contradict part (ii)? (Exercise 5.54 on page 221 defines the *direct product* of rings.)

7.29 * As in Example 7.22, let $p(x) = x^3 + x^2 - 2x - 1$ and let $K = \mathbb{Q}[x]/(p)$. Let

$$\alpha = x, \beta = x^2 - 2, \text{ and } \gamma = x^3 - 3x = -x^2 - x + 1.$$

Calculate in K, writing each result as $a + bx + cx^2$:

  (i) $\alpha + \beta + \gamma$

  (ii) $\alpha\beta + \alpha\gamma + \beta\gamma$

  (iii) $\alpha\beta\gamma$.

7.30 * As in Example 7.22, let $p(x) = x^3 + x^2 - 2x - 1$ and let $K = \mathbb{Q}[x]/(p)$. Show that, in K,

$$p(x^2 - 2) = p(x^3 - 3x) = 0.$$

Hence the three roots of p in K are $x + (p)$, $x^2 - 2 + (p)$, and $x^3 - 3x + (p)$.

Algebraic Extensions

The first step in classifying fields is by their characteristics. Here is the second step: we define *algebraic extensions*.

Definition. Let K/k be an extension field. An element $z \in K$ is **algebraic** over k if there is a nonzero polynomial $f(x) \in k[x]$ having z as a root; otherwise, z is **transcendental** over k. A field extension K/k is **algebraic** if every $z \in K$ is algebraic over k.

When a real number is called transcendental, it usually means that it is transcendental over $\mathbb{Q}$. For example, it was proved by Lindemann that π is a transcendental number (see [15], pp. 47–57 or [3], p. 5); there is no nonzero $f(x) \in \mathbb{Q}[x]$ with $f(\pi) = 0$.

Proposition 7.23. *If K/k is a finite extension field, then K/k is an algebraic extension.*

Proof. By definition, K/k finite means that K has finite dimension n as a vector space over k. Suppose that z is an element of K. By Corollary A.39 in Appendix A.3, the list of $n + 1$ vectors $1, z, z^2, \ldots, z^n$ is linearly dependent: there are $c_0, c_1, \ldots, c_n \in k$, not all 0, with $\sum c_i z^i = 0$. Thus, the polynomial $f(x) = \sum c_i x^i$ is not the zero polynomial, and z is a root of f. Therefore, z is algebraic over k.   ∎

The converse of the last proposition is not true; the field $\mathbb{A}$ of all complex numbers algebraic over $\mathbb{Q}$ is an algebraic extension of $\mathbb{Q}$ that is not a finite extension. (The fact that $\mathbb{A}$ is a field is not obvious, but it is true (see [17], Chapter 6).)

Definition. If K/k is an extension field and $z \in K$, then $k(z)$, the subfield of K obtained by *adjoining* z to k, is the intersection of all those subfields of K containing k and z.

More generally, if A is a subset of K, define $k(A)$ to be the intersection of all the subfields of K containing $k \cup A$; we call $k(A)$ the subfield of K obtained by *adjoining* A to k. In particular, if $A = \{z_1, \ldots, z_n\}$ is a finite subset, then we may denote $k(A)$ by $k(z_1, \ldots, z_n)$.

In Exercise 7.43 on page 308, you'll show that $k(A)$ is the smallest subfield of K containing k and A; that is, if E is any subfield of K containing k and A, then $k(A) \subseteq E$.

Proposition 7.20 starts with an irreducible polynomial $p(x) \in k[x]$ and constructs an extension K/k in which p has a root. Suppose we start with the root; that is, suppose that z is algebraic over k. Can we find a polynomial p so that $k(z)$ (the smallest extension of k that contains z) can be realized as $k[x]/(p)$? Let's look at an example.

Example 7.24. Suppose that $K = \mathbb{R}$, $k = \mathbb{Q}$, and $z = \sqrt{2} + \sqrt{3}$. First of all, z is algebraic over $\mathbb{Q}$. To see this, proceed as you would in high school algebra.

$$z^2 = 5 + 2\sqrt{6},$$
$$(z^2 - 5)^2 = 24,$$
$$z^4 - 10z^2 + 1 = 0.$$

Hence, z is a root of $h(x) = x^4 - 10x^2 + 1$, so it is algebraic over $\mathbb{Q}$.

Consider the evaluation homomorphism $\psi : \mathbb{Q}[x] \to \mathbb{R}$ (provided by Theorem 5.19) given by

$$\psi : f(x) \mapsto f(z).$$

The First Isomorphism Theorem suggests that we look at im ψ and ker ψ.

- im ψ contains $\mathbb{Q}$ (because $\psi(a) = a$ for all $a \in \mathbb{Q}$) and z (because $\psi(x) = z$). It follows that any subfield of $\mathbb{R}$ that contains $\mathbb{Q}$ and z contains im ψ. In other words,

$$\text{im } \psi = \mathbb{Q}(z).$$

- If $I = \ker \psi$, then the First Isomorphism Theorem gives

$$\mathbb{Q}[x]/I \cong \operatorname{im} \psi.$$

- $\operatorname{im} \psi$ is a subring of $\mathbb{R}$, so it is a domain. And I is an ideal in $\mathbb{Q}[x]$, so it is principal, say $I = (p)$, where $p(x) \in \mathbb{Q}[x]$.
- Furthermore, since $\mathbb{Q}[x]/I$ is a domain, p is an irreducible polynomial in $\mathbb{Q}[x]$, which we can take to be monic.

Thus, we have an isomorphism:

$$\Psi \colon \mathbb{Q}[x]/(p) \cong \operatorname{im} \psi,$$

namely $\Psi \colon f(x) + (p) \mapsto f(z)$. Since $\operatorname{im} \psi = \mathbb{Q}(z)$, we have

$$\mathbb{Q}[x]/(p) \cong \mathbb{Q}(z).$$

There it is: $\mathbb{Q}(z)$ is realized as a quotient of $\mathbb{Q}[x]$ by an irreducible polynomial. And, because we started with a specific z, we can do better: we can find p. We'll see later, in Example 7.32, that $p(x) = h(x)$. ▲

Example 7.24 contains most of the ideas of the general result.

Theorem 7.25. (i) *If K/k is an extension field and $z \in K$ is algebraic over k, then there is a unique monic irreducible polynomial $p(x) \in k[x]$ having z as a root. Moreover, if $I = (p)$, then $k[x]/I \cong k(z)$; indeed, there exists an isomorphism*

$$\Psi : k[x]/I \to k(z)$$

with $\Psi(x + I) = z$ and $\Psi(c + I) = c$ for all $c \in k$.
(ii) *If $z' \in K$ is another root of $p(x)$, then there is an isomorphism*

$$\theta : k(z) \to k(z')$$

with $\theta(z) = z'$ and $\theta(c) = c$ for all $c \in k$.

Proof. (i) As in Example 7.24, consider the evaluation homomorphism $\psi \colon k[x] \to K$, given by Theorem 5.19:

$$\psi \colon f \mapsto f(z).$$

Now $\operatorname{im} \psi$ is the subring of K consisting of all polynomials in z, that is, all elements of the form $f(z)$ with $f \in k[x]$, while $\ker \psi$ is the ideal in $k[x]$ consisting of all those $g(x) \in k[x]$ having z as a root. Since every ideal in $k[x]$ is a principal ideal, we have $\ker \psi = (p)$ for some monic polynomial $p(x) \in k[x]$. But the First Isomorphism Theorem says that $k[x]/(p) \cong \operatorname{im} \psi$, which is a domain, and so p is irreducible, by Proposition 7.19. The same proposition says that $k[x]/(p)$ is a field, and so there is an isomorphism $\Psi \colon k[x]/(p) \cong \operatorname{im} \psi$; namely $\Psi \colon f(x) + I \mapsto \psi(f) = f(z)$. Hence, $\operatorname{im} \Psi$ is a subfield of K containing k and z. But every such subfield of K must contain $\operatorname{im} \psi$, so that $\operatorname{im} \Psi = \operatorname{im} \psi = k(z)$. We have proved everything in the statement except the uniqueness of p; but this follows from Proposition 7.20(iv).

(ii) By (i), there are isomorphisms $\Psi: k[x]/I \rightarrow k(z)$ and $\Psi': k[x]/I \rightarrow k(z')$ with $\Psi(c+I) = c = \Psi'(c+I)$ for all $c \in k$; moreover, $\Psi: x+I \mapsto z$ and $\Psi': x + I \mapsto z'$. The composite $\theta = \Psi' \circ \Psi^{-1}$ is the desired isomorphism; it satisfies $\theta(c) = c$ for all $c \in k$, and

$$\theta: f(z) \mapsto f(z'),$$

for all $f(x) \in k[x]$. ∎

The proof of Theorem 7.25(ii) is described by the following diagram.

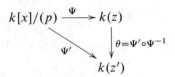

Definition. If K/k is an extension field and $z \in K$ is algebraic over k, then the unique monic irreducible polynomial $p(x) \in k[x]$ having z as a root is called the ***minimal polynomial*** of z over k, and it is denoted by

$$p(x) = \text{irr}(z, k).$$

The minimal polynomial $\text{irr}(z, k)$ depends on k. For example, $\text{irr}(i, \mathbb{R}) = x^2 + 1$, while $\text{irr}(i, \mathbb{C}) = x - i$.

Example 7.26. We know that $i \in \mathbb{C}$ is algebraic over $\mathbb{R}$, and $\text{irr}(i, \mathbb{R}) = x^2 + 1$. Now $-i$ is another root of $x^2 + 1$. The isomorphism $\theta: \mathbb{C} = \mathbb{R}(i) \rightarrow \mathbb{R}(-i) = \mathbb{C}$ with $\theta(i) = -i$ and $\theta(c) = c$ for all $c \in \mathbb{R}$ is, of course, complex conjugation.

The adjunction of one root of $\text{irr}(i, \mathbb{R})$ also adjoins the other root, for the minimal polynomial here is quadratic. But this doesn't always happen. For example, $z = \sqrt[3]{5}$ is algebraic over $\mathbb{Q}$ with minimal polynomial $x^3 - 5$. Theorem 7.25 tells us that

$$\Psi: \mathbb{Q}[x]/(x^3 - 5) \rightarrow \mathbb{Q}(z),$$

given by $f(x) + (x^3 - 5) \mapsto f(\sqrt[3]{5})$, is an isomorphism. But the roots of $x^3 - 5$ are not all contained in $\mathbb{Q}(z)$; indeed, the other two roots are not real, while $\mathbb{Q}[z] \subseteq \mathbb{R}$; in fact, in $\mathbb{C}[x]$,

$$x^3 - 5 = (x - z)(x - \omega z)(x - \omega^2 z),$$

where ω is our old friend $\frac{1}{2}\left(-1 + i\sqrt{3}\right)$. Theorem 7.25(i) tells us that the fields

$$\mathbb{Q}(z), \quad \mathbb{Q}(\omega z), \quad \mathbb{Q}(\omega^2 z)$$

are all isomorphic via isomorphisms that fix $\mathbb{Q}$ pointwise. One of these is

$$\theta : \mathbb{Q}(z) \rightarrow \mathbb{Q}(\omega z),$$

defined as follows: every element of $\mathbb{Q}(z)$ is of the form $f(z)$ where $f(x) + (x^3 - 5)$ is a coset in $\mathbb{Q}[x]/(x^3 - 5)$. Then

$$\theta\left(f(z)\right) = f(\omega z).$$

Again, a diagram illustrates the work just done.

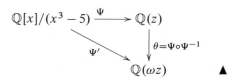

The following formula is quite useful, especially when proving a theorem by induction on degree. Before reading the proof, you may want to refresh your memory by looking at Appendix A.3 on linear algebra.

Theorem 7.27. *Let $k \subseteq E \subseteq K$ be fields, with E/k and K/E finite extension fields. Then K/k is a finite extension field, and*

$$[K : k] = [K : E][E : k].$$

Proof. If $A = a_1, \ldots, a_n$ is a basis of E over k and $B = b_1, \ldots, b_m$ is a basis of K over E, then it suffices to prove that the list X of all $a_i b_j$ is a basis of K over k.

To see that X spans K, take $u \in K$. Since B is a basis of K over E, there are scalars $\lambda_j \in E$ with $u = \sum_j \lambda_j b_j$. Since A is a basis of E over k, there are scalars $\mu_{ji} \in k$ with $\lambda_j = \sum_i \mu_{ji} a_i$. Therefore, $u = \sum_{ij} \mu_{ji} a_i b_j$, and X spans K over k. (Check that this makes sense in the special case $A = a_1, a_2, a_3$ and $B = b_1, b_2$.)

To prove that X is linearly independent over k, assume that there are scalars $\mu_{ji} \in k$ with $\sum_{ij} \mu_{ji} a_i b_j = 0$. If we define $\lambda_j = \sum_i \mu_{ji} a_i$, then $\lambda_j \in E$ and $\sum_j \lambda_j b_j = 0$. Since B is linearly independent over E, it follows that

$$0 = \lambda_j = \sum_i \mu_{ji} a_i$$

for all j. Since A is linearly independent over k, it follows that $\mu_{ji} = 0$ for all j and i, as desired. ∎

There are several classical geometric problems, such as trisecting an arbitrary angle with ruler and compass in which Theorem 7.27 plays a critical role (see Section 7.3).

Example 7.28. We now show how Theorem 7.27, the multiplicativity of degree in a tower of extension fields, can be used to calculate degrees; we also show, given an extension field E/k, that an explicit basis of E over k can sometimes be constructed. We urge you to work though this example carefully; it will help make the preceding development much more concrete, and you will see how all these ideas come together.

Let's return to Exercise 3.56 on page 116 (if you haven't attempted this exercise, you should try it now). It involves $\zeta = \cos(2\pi/7) + i \sin(2\pi/7)$, a primitive 7th root of unity; note that the powers of ζ are the vertices of a regular 7-gon in the complex plane. Using Proposition 6.62 and the language we have since introduced, we can now say that

$$\mathrm{irr}(\zeta, \mathbb{Q}) = \Phi_7(x) = x^6 + x^5 + x^4 + x^3 + x^2 + x + 1,$$

because $\Phi_7(x)$ is irreducible (Theorem 6.68), and so $[\mathbb{Q}(\zeta) : \mathbb{Q}] = 6$. For any nonnegative integer k, we have

$$\left(\zeta^k\right)^{-1} = \overline{\zeta^k} = \zeta^{7-k},$$

by Theorem 3.32(ii). We defined α, β, and γ in Exercise 3.56 on page 116 by

$$\alpha = \zeta + \zeta^6 = 2\cos(2\pi/7)$$
$$\beta = \zeta^2 + \zeta^5 = 2\cos(4\pi/7)$$
$$\gamma = \zeta^3 + \zeta^4 = 2\cos(6\pi/7),$$

and we saw that

$$\alpha + \beta + \gamma = -1$$
$$\alpha\beta + \alpha\gamma + \beta\gamma = -2$$
$$\alpha\beta\gamma = 1.$$

It follows that α, β, and γ are roots of

$$x^3 + x^2 - 2x - 1.$$

Ah, but this is precisely the irreducible $p(x)$ in Example 7.22. There, you constructed a field in which p has a root, but you didn't know what the roots are. Now you know: they are α, β, and γ, all real numbers, determined by expressions involving cosines. Furthermore, the construction in Theorem 7.25 gives a field isomorphic to $\mathbb{Q}(\alpha)$. But $\mathbb{Q}(\alpha)$ contains *all* the roots of p; for example,

$$\begin{aligned} \alpha^2 &= \left(\zeta + \zeta^6\right)^2 \\ &= \zeta^2 + 2\zeta^7 + \zeta^{12} \\ &= \zeta^2 + 2 + \zeta^5 \\ &= \beta + 2, \end{aligned}$$

and, hence,

$$\beta = \alpha^2 - 2 \in \mathbb{Q}(\alpha).$$

In the same way, you can expand α^3 to see that

$$\gamma = \alpha^3 + 3\alpha,$$

so that γ is also an element of $\mathbb{Q}(\alpha)$.

Since $\alpha = \zeta + \zeta^{-1}$, we see that $\mathbb{Q}(\alpha)$ is a subfield of $\mathbb{Q}(\zeta)$. And, since

$$[\mathbb{Q}(\alpha) : \mathbb{Q}] = \deg p = 3,$$

Theorem 7.27 gives

$$[\mathbb{Q}(\zeta) : \mathbb{Q}(\alpha)][\mathbb{Q}(\alpha) : \mathbb{Q}] = [\mathbb{Q}(\zeta) : \mathbb{Q}].$$

Hence, $[\mathbb{Q}(\zeta) : \mathbb{Q}(\alpha)] \cdot 3 = 6$, and $[\mathbb{Q}(\zeta) : \mathbb{Q}(\alpha)] = 2$. Therefore, the extension $\mathbb{Q}(\zeta)/\mathbb{Q}$ decomposes into a cubic extension of $\mathbb{Q}$ followed by a quadratic extension of $\mathbb{Q}(\alpha)$. This implies that

$$\deg\big(\mathrm{irr}\,(\zeta, \mathbb{Q}(\alpha))\big) = 2;$$

This "tower of fields" is sometimes illustrated with a diagram that displays the degrees:

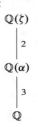

that is, ζ is a root of a quadratic polynomial with coefficients in $\mathbb{Q}(\alpha)$. Finding this quadratic is deceptively easy. We have

$$\zeta + \zeta^6 = \alpha \quad \text{and} \quad \zeta\zeta^6 = \zeta^7 = 1.$$

Thus, ζ is a root of

$$x^2 - \alpha x + 1 \in \mathbb{Q}(\alpha)[x].$$

There's another tower, $\mathbb{Q} \subseteq E \subseteq \mathbb{Q}(\zeta)$, with $E/\mathbb{Q}$ a quadratic extension and $\mathbb{Q}(\zeta)/E$ a cubic extension (i.e., writing $6 = 2 \cdot 3$ rather than $6 = 3 \cdot 2$). We constructed α, β, γ by breaking up the roots of Φ_7 into three sums, each a pair of complex conjugates. Instead, let's try to break the roots up into two sums, say δ and ϵ, each having three terms, and each sum containing just one member of every conjugate pair $\{\zeta^j, \overline{\zeta^j} = \zeta^{7-j}\}$; further, we'd like both $\delta + \epsilon$ and $\delta\epsilon$ rational. A little experimenting (and a CAS) leads to defining them like this:

$$\delta = \zeta + \zeta^2 + \zeta^4$$
$$\epsilon = \zeta^6 + \zeta^5 + \zeta^3.$$

Note that $\epsilon + \delta = -1$, so that $\epsilon \in \mathbb{Q}(\delta)$. We can now form the tower

$$\mathbb{Q} \subseteq \mathbb{Q}(\delta) \subseteq \mathbb{Q}(\zeta).$$

You can also check that $\epsilon\delta = 2$, so that ϵ and δ are the roots of the quadratic polynomial in $\mathbb{Q}[x]$:

$$x^2 + x + 2.$$

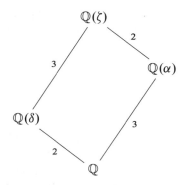

$(x - \epsilon)(x - \delta)$
$x^2 - \underbrace{(\epsilon + \delta)}_{-1} x + \dfrac{\epsilon\delta}{2}$

The roots of this polynomial are, by the quadratic formula, $\frac{1}{2} \pm \frac{\sqrt{7}}{2}$. See Exercise 7.35 on page 300.

Hence, $[\mathbb{Q}(\delta) : \mathbb{Q}] = 2$ and $[\mathbb{Q}(\zeta) : \mathbb{Q}(\delta)] = 3$. We now have two ways to decompose the extension $\mathbb{Q}(\zeta)/\mathbb{Q}$, which we draw in the following diagram:

$$
\begin{array}{ccc}
 & \mathbb{Q}(\zeta) & \\
{}^3 \diagup & & \diagdown {}^2 \\
\mathbb{Q}(\delta) & & \mathbb{Q}(\alpha) \\
{}_2 \diagdown & & \diagup {}_3 \\
 & \mathbb{Q} &
\end{array}
$$

Finally, ζ must be a root of a cubic polynomial with coefficients in $\mathbb{Q}(\delta)$. Again, the calculations are deceptively simple:

$$\zeta + \zeta^2 + \zeta^4 = \delta,$$
$$\zeta\zeta^2 + \zeta\zeta^4 + \zeta^2\zeta^4 = \zeta^3 + \zeta^5 + \zeta^6 = \epsilon,$$
$$\zeta\zeta^2\zeta^4 = \zeta^7 = 1,$$

so that ζ, ζ^2, and ζ^4 are roots of

$$x^3 - \delta x^2 + \epsilon x - 1 \in \mathbb{Q}(\delta)[x]. \quad \blacktriangle$$

Exercises

[handwritten margin notes: look @ ps. 266d what props it, Field exty diagrams]

7.31 As usual, let $\zeta_n = \cos(2\pi/n) + i \sin(2\pi/n)$.

(i) Find the minimal polynomial of ζ_n over $\mathbb{Q}$ for all n between 1 and 10.

(ii) What is minimal polynomial of ζ_p over $\mathbb{Q}$ if p is prime?

(iii) What is minimal polynomial of ζ_{p^2} over $\mathbb{Q}$ if p is prime?

7.32 If p is a prime, and $\zeta_p = \cos(2\pi/p) + i \sin(2\pi/p)$, show that

$$[\mathbb{Q}(\zeta_p) : \mathbb{Q}] = p - 1.$$

7.33 Show that $x^2 - 3$ is irreducible in $\mathbb{Q}(\sqrt{2})[x]$.

7.34 Let $k \subseteq K \subseteq E$ be fields. Prove that if E is a finite extension of k, then E is a finite extension of K, and K is a finite extension of k.

7.35 Show that

(i) $\cos(2\pi/7) + \cos(4\pi/7) + \cos(8\pi/7) = -\frac{1}{2}$.

(ii) $\left(\sin(2\pi/7) + \sin(4\pi/7) + \sin(8\pi/7) \right)^2 = \frac{7}{2}$.

7.36 Let $k \subseteq F \subseteq K$ be a tower of fields, and let $z \in K$. Prove that if $k(z)/k$ is finite, then $[F(z) : F] \leq [k(z) : k]$. Conclude that $[F(z) : F]$ is finite.

Hint. Use Proposition 7.20 to obtain an irreducible polynomial $p(x) \in k[x]$; the polynomial p may factor in $K[x]$.

7.37 Let K/k be an extension field. If $A \subseteq K$ and $u \in k(A)$, prove that there are only finitely many $a_1, \ldots, a_n \in A$ with $u \in k(a_1, \ldots, a_n)$.

7.38 Let E/k be a field extension. If $v \in E$ is algebraic over k, prove that v^{-1} is algebraic over k.

Splitting Fields

We now prove a result of Kronecker that says that if $f(x) \in k[x]$ is not constant, where k is a field, then there is some extension field K/k containing all the roots of f.

Theorem 7.29 (Kronecker). *If k is a field and $f(x) \in k[x]$ is nonconstant, there exists an extension field K/k with f a product of linear polynomials in $K[x]$.*

Proof. The proof is by induction on $\deg(f) \geq 1$. If $\deg(f) = 1$, then f is linear and we can take $K = k$. If $\deg(f) > 1$, write $f = pg$ in $k[x]$, where $p(x)$ is irreducible. Now Proposition 7.20 provides an extension field F/k containing a root z of p. Hence, $p = (x - z)h$, and so $f = pg = (x - z)hg$ in $F[x]$. By induction (since $\deg(hg) < \deg(f)$), there is an extension field K/F (so that K/k is also an extension field) with hg, and hence f, a product of linear factors in $K[x]$. ∎

How to Think About It. For the familiar fields $\mathbb{Q}$, $\mathbb{R}$, and $\mathbb{C}$, Kronecker's Theorem offers nothing new. The Fundamental Theorem of Algebra says that every nonconstant $f(x) \in \mathbb{C}[x]$ has a root in $\mathbb{C}$; it follows, by induction on the degree of f, that all the roots of f lie in $\mathbb{C}$; that is, $f(x) = a(x - z_1) \cdots (x - z_n)$, where $a \in \mathbb{C}$ and $z_j \in \mathbb{C}$ for all j. On the other hand, if $k = \mathbb{F}_p$ or

$k = \mathbb{C}(x) = \text{Frac}(\mathbb{C}[x])$, the Fundamental Theorem does not apply. However, Kronecker's Theorem does apply to tell us, for any $f(x) \in k[x]$, that there is always some larger field K containing all the roots of f; for example, there is an extension field $K/\mathbb{C}(x)$ containing $\sqrt{x}$, and there is an extension field $E/\mathbb{F}_3$ containing the roots of $x^2 - x - 1 \in \mathbb{F}_3[x]$.

A field F is called *algebraically closed* if *every* nonconstant polynomial $f(x) \in F[x]$ has a root in F (for example, $\mathbb{C}$ is algebraically closed). In contrast, extension fields K/k constructed in Kronecker's Theorem (that give roots of only one polynomial at a time) are usually not algebraically closed. Every field k does have an *algebraic closure*: there is an algebraic extension F/k that is algebraically closed (Kronecker's Theorem is one ingredient of the proof; see [25], p. 328).

The extension field K/k in Kronecker's Theorem need not be unique. Indeed, if f is a product of linear factors in K, then it is so in any extension of K. Therefore, let's consider the "smallest" field in which f is a product of linear factors. But the lack of uniqueness is not necessarily a consequence of K being too large, as we shall see in Example 7.33.

Definition. If K/k is an extension field and $f(x) \in k[x]$ is nonconstant, then f **splits over** K if $f(x) = a(x - z_1) \cdots (x - z_n)$, where $z_1, \ldots, z_n$ are in K and $a \in k$.

An extension field E/k is called a *splitting field* of f *over* k if f splits over E, but f does not split over any extension field F/k such that $k \subseteq F \subsetneq E$.

Consider $f(x) = x^2 + 1 \in \mathbb{Q}[x]$. The roots of f are $\pm i$, and so f splits over $\mathbb{C}$; that is, $f(x) = (x - i)(x + i)$ is a product of linear polynomials in $\mathbb{C}[x]$. However, $\mathbb{C}$ is not a splitting field of f over $\mathbb{Q}$: there are proper subfields of $\mathbb{C}$ containing $\mathbb{Q}$ and all the roots of f. For example, $\mathbb{Q}(i)$ is such a subfield; in fact, it is the splitting field of f over $\mathbb{Q}$.

A splitting field of a polynomial $g(x) \in k[x]$ depends on k as well as on g. A splitting field of $x^2 + 1$ over $\mathbb{Q}$ is $\mathbb{Q}(i)$, while a splitting field of $x^2 + 1$ over $\mathbb{R}$ is $\mathbb{R}(i) = \mathbb{C}$.

Corollary 7.30. *If k is a field and $f(x) \in k[x]$, then a splitting field of f over k exists.*

Proof. By Kronecker's Theorem, there is an extension field K/k such that f splits in $K[x]$; say $f(x) = a(x - z_1) \cdots (x - z_n)$. The subfield $E = k(z_1, \ldots, z_n)$ of K is a splitting field of f over k, because a proper subfield of E must omit some z_i. ∎

Example 7.31. (i) Let $f(x) = x^n - 1 \in \mathbb{Q}[x]$, and let $E/\mathbb{Q}$ be a splitting field. If $\zeta = e^{2\pi i/n}$ is a primitive nth root of unity, then $\mathbb{Q}(\zeta) = E$ is a splitting field of f, for every nth root of unity is a power of ζ, and $\zeta^j \in \mathbb{Q}(\zeta)$ for all j.

(ii) There are n distinct nth roots of unity in $\mathbb{C}$, but there may be fewer roots of unity over fields of characteristic p. For example, let $f(x) = x^3 - 1 \in \mathbb{Z}_3[x]$. Since $x^3 - 1 = (x - 1)^3$, by Exercise 7.27 on page 293, we see that there is only one cube root of unity here. ▲

How to Think About It. When we defined the field $k(A)$ obtained from a field k by adjoining a set A, we assumed that $A \subseteq K$ for some extension field K/k. But suppose no larger field K is given at the outset. For example, can the roots of $f(x) = x^2 - x - 1 \in \mathbb{F}_3[x]$ be adjoined to $\mathbb{F}_3$? Yes. In light of Kronecker's Theorem, there is some field extension $K/\mathbb{F}_3$ containing the roots of f, say α, β; now we do have the larger field, and so $\mathbb{F}_3(\alpha, \beta)$ makes sense; we can adjoin the roots of f to $\mathbb{F}_3$. Such an extension field K may not be unique, but we shall see that any two of them are isomorphic.

Example 7.32. Let's return to Example 7.24, where we saw that

$$z = \sqrt{2} + \sqrt{3}$$

is a root of $h(x) = x^4 - 10x^2 + 1$, and that $\mathbb{Q}(\sqrt{2} + \sqrt{3})$ can be realized as a quotient $\mathbb{Q}[x]/(p)$ for some irreducible monic polynomial $p(x) \in \mathbb{Q}[x]$ having z as a root.

As promised, we can now do better: we can show that $p = h$ and that $E = \mathbb{Q}(\sqrt{2} + \sqrt{3})$ is a splitting field of $h(x) = x^4 - 10x^2 + 1$, as well as a splitting field of $g(x) = (x^2 - 2)(x^2 - 3)$.

Because $x^4 - 10x^2 + 1$ is a quadratic in x^2, we can apply the quadratic formula to see that if w is any root of h, then $w^2 = 5 \pm 2\sqrt{6}$. But the identity $\left(\sqrt{a} + \sqrt{b}\right)^2 = a + 2\sqrt{ab} + b$ gives $w = \pm(\sqrt{2} + \sqrt{3})$. Similarly, $5 - 2\sqrt{6} = \left(\sqrt{2} - \sqrt{3}\right)^2$, so that h has distinct roots, namely

$$z = \sqrt{2} + \sqrt{3}, \quad -\sqrt{2} - \sqrt{3}, \quad \sqrt{2} - \sqrt{3}, \quad -\sqrt{2} + \sqrt{3}.$$

Note that $x^2 = w$, where w is a root of $w^2 - 10w + 1$; that is, $w = \frac{1}{2}\left(10 \pm \sqrt{96}\right) = 5 \pm 2\sqrt{6}$.

By Theorem 6.52, the only possible rational roots of h are ± 1, and so we have just proved that all these roots are irrational.

We claim that h is irreducible in $\mathbb{Q}[x]$ (so, $p = h$ after all). It suffices to show that h has no quadratic factor $q(x) \in \mathbb{Q}[x]$ (why?). If, on the contrary, $h = qq'$ for two monic quadratic polynomials in $\mathbb{Q}[x]$, then the roots of h are paired up, two for q and two for q'. Suppose $q(z) = 0$. Then the other root of q, call it z', is one of

$$\sqrt{2} - \sqrt{3}, \quad -\sqrt{2} - \sqrt{3}, \quad -\sqrt{2} + \sqrt{3}.$$

Now, if $q(x) = x^2 + bx + c$, then $-b = z + z'$ and $c = zz'$. But you can check, for each choice of z', that either $z + z'$ or zz' is irrational. Since $q \in \mathbb{Q}[x]$, this is a contradiction, and so h is irreducible.

We now know that $[E : \mathbb{Q}] = 4$. Let $F = \mathbb{Q}(\sqrt{2}, \sqrt{3})$, so that we have a tower of fields $\mathbb{Q} \subseteq E \subseteq F$. Theorem 7.27 tells us that

$$[F : \mathbb{Q}] = [F : E][E : \mathbb{Q}].$$

On the other hand,

$$[F : \mathbb{Q}] = [F : \mathbb{Q}(\sqrt{2})][\mathbb{Q}(\sqrt{2}) : \mathbb{Q}].$$

Now $[\mathbb{Q}(\sqrt{2}) : \mathbb{Q}] = 2$, because $\sqrt{2}$ is a root of the irreducible quadratic $x^2 - 2$ in $\mathbb{Q}[x]$. We claim that $[F : \mathbb{Q}(\sqrt{2})] \leq 2$. The field F arises by adjoining $\sqrt{3}$ to $\mathbb{Q}(\sqrt{2})$; either $\sqrt{3} \in \mathbb{Q}(\sqrt{2})$, in which case the degree is 1,

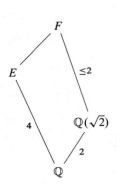

or $x^2 - 3$ is irreducible in $\mathbb{Q}(\sqrt{2})[x]$, in which case the degree is 2 (by Exercise 7.33 on page 300, it is 2). It follows that $[F : \mathbb{Q}] \leq 4$, and so the equation $[F : \mathbb{Q}] = [F : E][E : \mathbb{Q}]$ gives $[F : E] = 1$; that is, $F = E$, so that F not only arises from $\mathbb{Q}$ by adjoining all the roots of h, but it also arises from $\mathbb{Q}$ by adjoining all the roots of $g(x) = (x^2 - 2)(x^2 - 3)$. ▲

A *splitting field* of $f(x) \in k[x]$ is a *smallest* field extension E/k containing all the roots of f. We say "a" splitting field instead of "the" splitting field because splitting fields of f over k are not unique. Corollary 7.30 constructed a splitting field of $f(x) \in k[x]$ as a subfield of a field extension K/k, where f splits in K. But there may be distinct such field extensions K/k.

Example 7.33. Consider $f(x) = x^2 + x + 1 \in \mathbb{F}_2[x]$. Now f is irreducible (for it is a quadratic with no roots in $\mathbb{F}_2$), and f is a product of linear polynomials in $K = \mathbb{F}_2[x]/(f)$, by Proposition 7.20: if $z = x + (f) \in K$, then $f(x) = (x - z)^2$ in $K[x]$ (remember that $-1 = 1$ here). On the other hand, in Exercise 4.55 on page 165, we constructed a field K' with elements the four matrices $\begin{bmatrix} a & b \\ b & a+b \end{bmatrix}$ (where $a, b \in \mathbb{F}_2$) and operations matrix addition and matrix multiplication. You can check that if $u = \begin{bmatrix} 1 & 1 \\ 1 & 0 \end{bmatrix}$, then $u \in K'$ and $f(x) = (x - u)^2$ in $K'[x]$. ▲

Our next goal is to show that splitting fields are unique up to isomorphism. We paraphrase Theorem 7.25(ii).

Let K/k be an extension field, and let $z, z' \in K$ be roots of some irreducible $p(x) \in k[x]$. Then there is an isomorphism $\theta \colon k(z) \to k(z')$ with $\theta(z) = z'$ and $\theta(c) = c$ for all $c \in k$.

We need a generalization. Suppose that $f(x) \in k[x]$ is a polynomial, not necessarily irreducible, and let $E = k(z_1, \ldots, z_t)$ and $E' = k(z_1', \ldots, z_t')$ be splitting fields of f. Is there an isomorphism $\theta \colon E \to E'$ that carries the roots $z_1, \ldots, z_t$ to the roots $z_1', \ldots, z_t'$ and that fixes all the elements $c \in k$? The obvious way to proceed is by induction on $\deg(f)$ (making use of the fact that f has an irreducible factor, which will let us use Theorem 7.25). Think about proving the inductive step. We'll have an isomorphism $\varphi \colon k(z_1) \to k(z_1')$ that we'll want to extend to an isomorphism $\widetilde{\varphi} \colon k(z_1, z_2) \to k(z_1', z_j')$ for some j; that is, $\widetilde{\varphi} \colon k(z_1)(z_2) \to k(z_1')(z_j')$. The base fields $k(z_1)$ and $k(z_1')$ are no longer equal; they are only isomorphic. The upshot is that we have to complicate the statement of what we are going to prove in order to take account of this.

First, recall Corollary 5.22:

If R and S are commutative rings and $\varphi \colon R \to S$ is a homomorphism, then there is a unique homomorphism $\varphi^* \colon R[x] \to S[x]$ given by

$$\varphi^* \colon r_0 + r_1 x + r_2 x^2 + \cdots \mapsto \varphi(r_0) + \varphi(r_1)x + \varphi(r_2)x^2 + \cdots.$$

Moreover, φ^* is an isomorphism if φ is.

As we said, we are forced to complicate our earlier result.

Lemma 7.34. *Let $\varphi: k \to k'$ be an isomorphism of fields and $\varphi^*: k[x] \to k'[x]$ the isomorphism of Corollary 5.22; let $p(x) \in k[x]$ be irreducible, and let $p' = \varphi^*(p)$.*

(i) *p' is irreducible in $k'[x]$, and the map $\Phi: k[x]/(p) \to k'[x]/(p')$, defined by $\Phi: f + (p) \mapsto \varphi^*(f) + (p')$, is an isomorphism of fields.*

(ii) *Let K/k be a field extension, let $z \in K$ be algebraic over k, and let $p(x) = \mathrm{irr}(z, k)$. If $p' = \varphi^*(p) \in k'[x]$ and z' is a root of p' in some extension of k', then φ can be extended to an isomorphism $k(z) \to k'(z')$ that maps z to z'.*

Proof. (i) This is straightforward, for φ carries the ideal (p) in $k[x]$ onto the ideal (p') in $k'[x]$, and Exercise 7.8 on page 285 applies. Exercise 7.44 on page 308 asks you to give the details.

(ii) By (i), there are isomorphisms

$$\psi : k[x]/(p) \to k(z) \quad \text{and} \quad \psi' : k'[x]/(p') \to k'(z').$$

By Lemma 7.34, there is an isomorphism

$$\Phi : k[x]/(p) \to k'(x)/(p'),$$

and the composite $\psi' \circ \Phi \circ \psi^{-1}$ is the desired isomorphism. ∎

Here is a picture of the Lemma's proof.

$$
\begin{array}{ccc}
k[x]/(p) & \xrightarrow{\ \psi\ } & k(z) \\
\Big\downarrow{\scriptstyle\Phi} & & \Big\downarrow{\scriptstyle\psi'\circ\Phi\circ\psi^{-1}} \\
k'[x]/(p') & \xrightarrow{\ \psi'\ } & k'(z')
\end{array}
$$

We now give the version we need.

Theorem 7.35. *Let $\varphi: k \to k'$ be an isomorphism of fields and $\varphi^*: k[x] \to k'[x]$ the isomorphism $k[x]/(p) \to k'[x]/(p')$ in Lemma 7.34. Let $f(x) \in k[x]$ and $f^*(x) = \varphi^*(f) \in k'[x]$. If E is a splitting field of f over k and E' is a splitting field of f^* over k', then there is an isomorphism $\Phi: E \to E'$ extending φ.*

Proof. The proof is by induction on $d = [E : k]$. If $d = 1$, then f is a product of linear polynomials in $k[x]$, and it follows easily that f^* is also a product of linear polynomials in $k'[x]$. Therefore, $E' = k'$, and we may set $\Phi = \varphi$.

For the inductive step, choose a root z of f in E that is not in k, and let $p(x) = \mathrm{irr}(z, k)$ be the minimal polynomial of z over k. Now $\deg(p) > 1$, because $z \notin k$; moreover, $[k(z) : k] = \deg(p)$, by Proposition 7.20. Let z' be a root of p^* in E', so that $p^* = \mathrm{irr}(z', k')$.

By Lemma 7.34(ii), there is an isomorphism $\widetilde{\varphi}: k(z) \to k'(z')$ extending φ with $\widetilde{\varphi}(z) = z'$. We may regard f as a polynomial with coefficients in $k(z)$, for $k \subseteq k(z)$ implies $k[x] \subseteq k(z)[x]$. We claim that E is a splitting field of f over $k(z)$; that is,

$$E = k(z)(z_1, \ldots, z_n),$$

where $z_1, \ldots, z_n$ are the roots of $f(x)/(x - z)$. After all,

$$E = k(z, z_1, \ldots, z_n) = k(z)(z_1, \ldots, z_n).$$

Similarly, E' is a splitting field of f^* over $k'(z')$. But $[E : k(z)] < [E : k]$, by Theorem 7.27, so the inductive hypothesis gives an isomorphism $\Phi: E \to E'$ that extends $\widetilde{\varphi}$ and, hence, φ.   ■

Corollary 7.36. *If k is a field and $f(x) \in k[x]$, then any two splitting fields of f over k are isomorphic via an isomorphism that fixes k pointwise.*

Proof. Let E and E' be splitting fields of $f(x)$ over k. If φ is the identity, then Theorem 7.35 applies at once.   ■

Classification of Finite Fields

We know, thanks to Proposition 7.18, that every finite field has p^n elements for some prime p and integer $n \geq 1$. We also know that a field k with p^n elements must have characteristic p, so that $pa = 0$ for all $a \in k$, by Proposition 7.17. In this section, we show that fields with exactly p^n elements exist, and that any two having the same number of elements are isomorphic.

First, we show that every nonzero element a in a finite field with q elements is a $(q - 1)$st root of unity (of course, a is not a complex root of unity). We have seen the idea of the next proof in the proof of Theorem 4.63.

Lemma 7.37. *Let k be a finite field having q elements. If $a \in k$ is nonzero, then $a^{q-1} = 1$.*

Proof. Let $k^{\#} = \{a_1, a_2, \ldots, a_{q-1}\}$ be the nonzero elements of k. We claim, for any $a \in k^{\#}$, that the function $\mu_a: a_i \mapsto aa_i$ takes values in $k^{\#}$: since k is a field, it is a domain, and so $aa_i \neq 0$. We now claim $\mu_a: k^{\#} \to k^{\#}$ is injective: if $aa_i = aa_j$, then the cancellation law gives $a_i = a_j$. Finally, since $k^{\#}$ is finite, the Pigeonhole Principle shows that μ_a is a bijection. It follows that $aa_1, aa_2, \ldots, aa_{q-1}$ is just a rearrangement of $a_1, a_2, \ldots, a_{q-1}$. Hence,

$$a_1 a_2 \cdots a_{q-1} = (aa_1)(aa_2) \cdots (aa_{q-1}) = a^{q-1} a_1 a_2 \cdots a_{q-1}.$$

Now cancel $a_1 a_2 \cdots a_{q-1}$ to obtain $1 = a^{q-1}$.   ■

We now show, given a prime power $q = p^n$, that there exists a field with p^n elements. Our guess is that Galois realized that $\mathbb{C}$ can be constructed by adjoining a root of $x^2 + 1$ to $\mathbb{R}$, so that it was natural for him (but not for any of his contemporaries!) to adjoin a root of a polynomial to $\mathbb{F}_p$. However, Kronecker's Theorem was not proved until a half century after Galois's death.

Theorem 7.38 (Galois). *If p is prime and n is a positive integer, then there exists a field having exactly p^n elements.*

Proof. Write $q = p^n$. In light of Lemma 7.37, it is natural to consider roots of the polynomial

$$g(x) = x^q - x \in \mathbb{F}_p[x].$$

By Kronecker's Theorem, there is a field extension $K/\mathbb{F}_p$ with g a product of linear factors in $K[x]$. Define

$$E = \{z \in K : g(z) = 0\};$$

that is, E is the *set* of all the roots of g. We claim that all the roots of g are distinct. Since the derivative $g'(x) = qx^{q-1} - 1 = p^n x^{q-1} - 1 = -1$ (by Proposition 7.17), we have $\gcd(g, g') = 1$. By Exercise 6.40 on page 263, all the roots of g are, indeed, distinct; that is, E has exactly $q = p^n$ elements.

The theorem will follow if E is a subfield of K. Of course, $1 \in E$. If a, $b \in E$, then $a^q = a$ and $b^q = b$. Hence, $(ab)^q = a^q b^q = ab$, and $ab \in E$. By Proposition 7.17, $(a + b)^q = a^q + b^q = a + b$, so that $a + b \in E$. Therefore, E is a subring of K. Finally, if $a \neq 0$, then Lemma 7.37 says that $a^{q-1} = 1$, and so the inverse of a is a^{q-2} (which lies in E because E is closed under multiplication). ∎

Proposition 7.39. *If k is a finite field having $q = p^n$ elements, then every $a \in k$ is a root of $x^q - x$.*

Proof. This follows directly from Lemma 7.37. ∎

It is remarkable that the next theorem was not proved until the 1890s, 60 years after Galois discovered finite fields.

E. H. Moore was an algebraist who later did research in geometry and foundations of analysis.

Corollary 7.40 (Moore). *Any two finite fields having exactly p^n elements are isomorphic.*

Proof. By Proposition 7.39, every element of E is a root of $g(x) = x^q - x \in \mathbb{F}_p[x]$, and so E is a splitting field of g over $\mathbb{F}_p$. ∎

Finite fields are often called **Galois fields** in honor of their discoverer. In light of Corollary 7.40, we may speak of *the* field with q elements, where $q = p^n$ is a power of a prime p, and we denote it by

$$\mathbb{F}_q.$$

The next example displays different finite fields with the same number of elements; by Moore's Theorem, they are isomorphic.

Example 7.41. (i) In Exercise 4.55 on page 165, we constructed the field $\mathbb{F}_4$ with four elements:

$$\mathbb{F}_4 = \left\{ \left[\begin{smallmatrix} a & b \\ b & a+b \end{smallmatrix} \right] : a, b \in \mathbb{F}_2 \right\}.$$

On the other hand, since $f(x) = x^2 + x + 1 \in \mathbb{F}_2[x]$ is irreducible, the quotient $K = \mathbb{F}_2[x]/(f)$ is a field. By Proposition 7.20, F consists of all $a + bz$, where $z = x + (f)$ is a root of f and $a, b \in \mathbb{F}_2$. Hence K also is a field with four elements.

(ii) According to the table in Example 6.57, there are three monic irreducible quadratics in $\mathbb{F}_3[x]$, namely

$$p(x) = x^2 + 1, \quad q(x) = x^2 + x - 1, \quad \text{and} \quad r(x) = x^2 - x - 1;$$

each gives rise to a field with $9 = 3^2$ elements, namely quotient rings of $\mathbb{F}_3[x]$. Let us look at the first two in more detail. Proposition 7.20 says that $E = \mathbb{F}_3[x]/(p)$ is given by

$$E = \{a + bz : \text{ where } z^2 + 1 = 0\}.$$

Similarly, if $F = \mathbb{F}_3[x]/(q)$, then

$$F = \{a + bu : \text{ where } u^2 + u - 1 = 0\}.$$

Without Moore's Theorem, it is not instantly obvious that the two fields are isomorphic. You can check that the map $\varphi \colon E \to F$ (found by trial and error), defined by $\varphi(a + bz) = a + b(1 - u)$, is an isomorphism.

Now $\mathbb{F}_3[x]/(x^2 - x - 1)$ is another field with nine elements; Exercise 7.46 asks for an explicit isomorphism with E.

(iii) In Example 6.57, we exhibited eight monic irreducible cubics $p(x) \in \mathbb{F}_3[x]$; each gives rise to a field $\mathbb{F}_3[x]/(p)$ having $27 = 3^3$ elements, and Moore's Theorem says that they are all isomorphic to one another.   ▲

The following result is known.

Theorem 7.42 (Primitive Element). *Let K/k be a finite field extension; that is, $[K : k] < \infty$. If either k has characteristic 0 or K is a finite field, then there exists $\alpha \in K$ such that $K = k(\alpha)$.*

Proof. [26], p. 301. ■

Actually, more is known when K is finite: it can be shown that every nonzero element of K is a power of α (not merely a linear combination of powers of α).

Corollary 7.43. *For every integer $n \geq 1$, there exists an irreducible polynomial in $\mathbb{F}_p[x]$ of degree n.*

Proof. Let $h(x) = \mathrm{irr}(\alpha, \mathbb{F}_p)$ be the minimal polynomial of α. Since h is irreducible, Corollary 7.21 gives $\dim_k(K) = \deg(h)$. But if $|K| = p^n$, then $\dim_k(K) = n$. Therefore, since there exists a finite field with exactly p^n elements, there exists an irreducible polynomial of degree n. ■

Exercises

7.39 Let $f(x), g(x) \in k[x]$ be monic polynomials, where k is a field. If g is irreducible and every root of f (in an appropriate splitting field) is also a root of g, prove that $f = g^m$ for some integer $m \geq 1$.

Hint. Use induction on $\deg(h)$.

7.40 Determine whether any of the following pairs of fields are isomorphic.

(i) $\mathbb{Q}(i)$ and $\mathbb{Q}(\frac{1}{2}(1 + i))$

(ii) $\mathbb{Q}(i)$ and $\mathbb{Q}(\sqrt{3})$

(iii) $\mathbb{Q}(\sqrt{2})$ and $\mathbb{Q}(\sqrt{3})$

(iv) $\mathbb{Q}(\sqrt{2})$ and $\mathbb{Q}(\sqrt{6})$

7.41 Let $f(x) = s_0 + s_1 x + \cdots + s_{n-1} x^{n-1} + x^n \in k[x]$, where k is a field, and suppose that $f(x) = (x-z_1)(x-z_2) \cdots (x-z_n)$, where the z_i lie in some splitting field. Prove that $s_{n-1} = -(z_1 + z_2 + \cdots + z_n)$ and $s_0 = (-1)^n z_1 z_2 \cdots z_n$. Conclude that the sum and product of all the roots of f lie in k.

7.42 (i) Show that $[\mathbb{Q}(\cos(2\pi/7)) : \mathbb{Q}] = 3$.

(ii) Find the minimal polynomial for $\cos \frac{2\pi}{7}$ over $\mathbb{Q}$.

(iii) Find all the roots of this polynomial.

Hint. See Exercise 3.56 on page 116.

7.43 Suppose that K/k is an extension field and if $A \subseteq K$. Show that $k(A)$ is the smallest subfield of K containing k and A; that is, if E is any subfield of K containing k and A, then $k(A) \subseteq E$.

7.44 Prove Lemma 7.34.

7.45 Using the setup from Example 7.41(i), show that the map $\varphi : \mathbb{F}_4 \to K$, defined by $\varphi \left(\begin{bmatrix} a & b \\ b & a+b \end{bmatrix} \right) = a + bz$, is an isomorphism.

7.46 Using the setup from Example 7.41(ii), show that

$$\mathbb{F}_3/(x^2 + 1) \cong \mathbb{F}_3[x]/(x^2 - x - 1)$$

without using Corollary 7.40.

7.47 Prove that $\mathbb{F}_3[x]/(x^3 - x^2 + 1) \cong \mathbb{F}_3[x]/(x^3 - x^2 + x + 1)$ without using Corollary 7.40.

7.48 Write addition and multiplication tables for the field $\mathbb{F}_8$ with eight elements using an irreducible cubic over $\mathbb{F}_2$.

7.49 (i) Is $\mathbb{F}_4$ isomorphic to a subfield of $\mathbb{F}_8$?

(ii) For a prime p, prove that if $\mathbb{F}_{p^n}$ is isomorphic to a subfield of $\mathbb{F}_{p^m}$, then $n \mid m$ (the converse is also true).

Hint. View $\mathbb{F}_{p^m}$ as a vector space over $\mathbb{F}_{p^n}$.

7.3 Connections: Ruler–Compass Constructions

There are myths in several ancient civilizations in which the gods demand precise solutions of mathematical problems in return for granting relief from catastrophes. We quote from van der Waerden [35].

> In the dialogue *Platonikos* of Eratosthenes, a story was told about the problem of doubling the cube. According to this story, as Theon of Smyrna recounts it in his book *Exposition of mathematical things useful for the reading of Plato*, the Delians asked for an oracle in order to be liberated from a plague. The god (Apollo) answered through the oracle that they had to construct an altar twice as large as the existing one without changing its shape. The Delians sent a delegation to Plato, who referred them to the mathematicians Eudoxus and Helikon of Kyzikos.

The altar was cubical in shape, and so the problem involves constructing $\sqrt[3]{2}$ (the volume of a cube with edges of length ℓ is ℓ^3). The gods were cruel, for although there is a geometric construction of $\sqrt{2}$ (it is the length of the diagonal of a square with sides of length 1), we are going to prove that it is impossible to construct $\sqrt[3]{2}$ by the methods of Euclidean geometry — that is, by using only ruler and compass. (Actually, the gods were not so cruel, for the Greeks did use other methods. Thus, Menaechmus constructed $\sqrt[3]{2}$ as the intersection of the parabolas $y^2 = 2x$ and $x^2 = y$; this is elementary for us, but it was an ingenious feat when there was no analytic geometry and no algebra. There was also a solution found by Nicomedes.)

There are several other geometric problems handed down from the Greeks. Can one trisect every angle? Can one construct a regular n-gon? More precisely, can one inscribe a regular n-gon in the unit circle? Can one "square the circle;" that is, can one construct a square whose area is equal to the area of a given circle? Since the disk with radius 1 has area π, can one construct a square with sides of length $\sqrt{\pi}$?

If we are not careful, some of these problems appear ridiculously easy. For example, a 60° angle can be trisected using a protractor: just find 20° and draw the angle. Thus, it is essential to state the problems carefully and to agree on certain ground rules. The Greek problems specify that only two tools, ruler and compass, are allowed, and each must be used in only one way. The goal of this section is to determine exactly what can be constructed using the two "Euclidean tools." The answer will involve some surprising applications of ideas from this chapter.

How to Think About It. In many geometry classes, constructions are now taught using *dynamic geometry* software. These environments can be used in the same way that one uses physical rulers and compasses; the principles are the same, and what's possible in them is what's possible with pencil and paper. This brings up an important point. Constructions made in dynamic geometry environments are likely to be more accurate than those carried out with pencil and paper, but the goal here is not approximation—we are not content with constructing $\sqrt[3]{2}$ correct to 100 decimal places; the goal is to find $\sqrt[3]{2}$ exactly, just as we can find $\sqrt{2}$ exactly as the length of the diagonal of the unit square. We now seek to determine just what constructions are possible, and so we must use precise definitions.

Notation. Let P and Q be points in the plane; we denote the line segment with endpoints P and Q by $\overline{PQ}$, and we denote its length by PQ. If P and Q are points, we'll let $L(P, Q)$ denote the line through P and Q, and $C(P, Q)$ the circle with center P and radius PQ. We'll also denote the circle with center P and radius r (for a positive number r) by $C(P, r)$.

The formal discussion begins with defining the tools by saying exactly what each is allowed to do.

Definition. A *ruler* is a tool that can be used to draw the line $L(P, Q)$ determined by points P and Q.

A *compass* is a tool that can be used to draw circles; given two points P and Q, it can draw $C(P, Q)$ and $C(Q, P)$.

What we are calling a *ruler*, others call a *straightedge*. For them, a *ruler* can be used not only to draw lines but to measure distances as well.

In many high school texts, $L(P, Q)$ is written as $\overleftrightarrow{PQ}$. Of course, we can't physically draw the infinite line $L(P, Q)$, but $\overline{PQ}$ has endpoints and $L(P, Q)$ does not.

How to Think About It. Just to show you how fussy we are, let us point out a subtlety about what a compass cannot do. Suppose we are given three points: P, Q, and R. We are allowed to draw the circle $C(P, Q)$ with center P and radius $r = PQ$. But we are *not* allowed to draw the circle $C(R, r)$ with center R and radius r. Reason: a compass is allowed to draw a circle only if two points are given at the outset; but the circle $C(R, r)$ cannot be drawn (using the compass as in the definition) because only one point, namely R, is given at the outset. Our compass is called a **collapsible compass** as compared to the more versatile compass that's allowed to draw $C(R, r)$. We mention this now only because the proof of Theorem 7.48(ii) may appear more complicated than necessary (we'll say something more there).

Constructions with ruler and compass are carried out in the plane. Since every construction has only a finite number of steps, we shall be able to define *constructible points* inductively. Once this precise definition is given, we will be able to show that it is impossible to double the cube or to trisect arbitrary angles using only a ruler and compass. Angles such as $90°$ and $45°$ can be trisected using a ruler and compass (for we can construct a $30°$ angle, which can then be bisected), but we shall see that a $60°$ angle is impossible to trisect. When we say *impossible*, we mean what we say; we do not mean that it is merely very difficult. You should ponder how anything can be proved to be impossible. This is an important idea, and we recommend letting students spend an evening trying to trisect a $60°$ angle by themselves as one step in teaching them the difference between hard and impossible.

About 425 BCE, Hippias of Elis was able to square the circle by drawing a certain curve as well as lines and circles. We shall see that this construction is impossible using only ruler and compass.

Given the plane, we establish a coordinate system by first choosing two distinct points, A and A'; call the line they determine the *x-axis*. Use a compass to draw the two circles $C(A, A')$ and $C(A', A)$ of radius AA' with centers A and A', respectively (see Figure 7.1). These two circles intersect in two points P_1 and P_2; the line $L(P_1, P_2)$ they determine is called the *y-axis*; it is the perpendicular-bisector of $\overline{AA'}$, and it intersects the x-axis in a point O, called the *origin*. We define the distance OA to be 1. We have introduced coordinates into the plane; of course, $O = (0, 0)$, $A = (1, 0)$, and $A' = (-1, 0)$. Consider the point P_1 in Figure 7.1. Now $\triangle OAP_1$ is a right triangle with legs $\overline{OA}$ and $\overline{OP_1}$. The hypotenuse $\overline{AP_1}$ has length $2 = AA'$ (for this is the radius of $C(A, A')$). Since $OA = 1$, the Pythagorean Theorem gives $P_1 = (0, \sqrt{3})$. Similarly, $P_2 = (0, -\sqrt{3})$.

Informally, we construct a new point Q from old points E, F, G, and H by using the first pair $E \neq F$ to draw a line or circle, the second pair $G \neq H$ to draw a line or circle, and then obtaining Q as one of the points of intersection

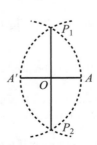

Figure 7.1. The first constructible points.

of the two lines, of the line and the circle, or of the two circles. More generally, a point is called *constructible* if it is obtained from A and A' by a finite number of such steps.

Given a pair of constructible points, we do *not* assert that every point on the line or the circle they determine is constructible. For example, we can draw the x-axis $L(A', A)$, but, as we'll see, not every point on it is constructible.

We now begin the formal discussion. Our goal is Theorem 7.52 which gives an algebraic characterization of constructibility. Recall, given distinct points P and Q in the plane, that $L(P, Q)$ is the line they determine and $C(P, Q)$ is the circle with center P and radius $\overline{PQ}$.

Definition. Let $E \neq F$ and $G \neq H$ be points in the plane. A point Q is *constructible from* E, F, G, and H if either

(i) $Q \in L(E, F) \cap L(G, H)$, where $L(E, F) \neq L(G, H)$;

(ii) $Q \in L(E, F) \cap C(G, H)$;

(iii) $Q \in C(E, F) \cap C(G, H)$, where $C(E, F) \neq C(G, H)$.

A point Q is *constructible* if $Q = A$, $Q = A'$, or there are points $P_1, \dots, P_n$ with $Q = P_n$ such that every point P_{j+1} (for $1 \leq j$) is constructible from points E, F, G, H in $\{A, A', P_1, \dots, P_j\}$.

If $L(E, F) \neq L(G, H)$ and $L(E, F)$ is not parallel to $L(G, H)$, then $L(E, F) \cap L(G, H)$ is a single point (there are at most two points comprising any of these intersections).

We illustrate the formal definition of constructibility by showing that every angle can be bisected with ruler and compass.

Lemma 7.44. (i) *The perpendicular-bisector of a given line segment $\overline{AB}$ can be drawn.*

(ii) *If A and B are constructible points, then the midpoint of $\overline{AB}$ is constructible.*

(iii) *If a point $P = (\cos\theta, \sin\theta)$ is constructible, then $Q = (\cos(\theta/2), \sin(\theta/2))$ is constructible.*

Proof. (i) The construction is the same as in Figure 7.1. Here, there are two points of intersection of the circles $C(A, B)$ and $C(B, A)$, say P_1 and P_2, and $L(P_1, P_2)$ is the perpendicular-bisector of $\overline{AB}$.

(ii) The midpoint is the intersection of $\overline{AB}$ and its perpendicular-bisector.

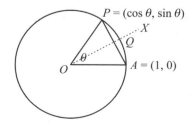

Figure 7.2. Bisecting an angle.

(iii) The point $A = (1, 0)$ is constructible. By (i), the perpendicular-bisector $L(O, X)$ of the chord $\overline{PA}$ can be drawn. The point Q lies in the intersection of $L(O, X)$ and the unit circle, and so Q is constructible. (See Figure 7.2.) ■

We know that the Parallel Postulate is not true in non-Euclidean geometry. What hidden hypotheses of Euclidean geometry are we using to make this construction?

Here is the (tricky) constructible version of the Parallel Postulate.

Lemma 7.45. *If U, V, P are distinct constructible points with $P \notin L(U, V)$, then there is a constructible point Q with $L(P, Q)$ parallel to $L(U, V)$.*

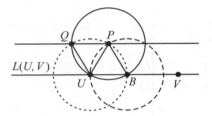

Figure 7.3. Parallel postulate.

Proof. The proof refers to Figure 7.3. Choose U so that $L(P, U)$ is not perpendicular to $L = L(U, V)$. Thus, L is not tangent to the circle $C(P, U)$, and so $C(P, U)$ meets L in another point, say B (of course, B is constructible). Let $Q \in C(P, U) \cap C(U, P)$. Clearly, Q is constructible, and we claim that $L(P, Q)$ is parallel to L. Indeed, we claim that the quadrilateral $PBUQ$ is a rhombus and hence it is a parallelogram. Now PQ is a radius of $C(P, U)$, PU is a radius of both $C(P, U)$ and $C(B, U)$, and BU is a radius of both $C(B, U)$ and $C(U, P)$. Hence, $PQ = PU = PB$, as we want. ■

In high school geometry, the goal is to construct certain figures with ruler and compass. We are about to shift the focus, considering instead the notion of *constructible numbers*. "Numbers?" Well, analytic geometry equips points with coordinates, and we have seen how to regard points as complex numbers.

Definition. A *complex number $z = x + iy$ is **constructible** if the point (x, y) is a constructible point.

Exercise 7.51 on page 326 shows that every element of $\mathbb{Z}[i]$ is a constructible number. So is our old friend $\omega = -\frac{1}{2} + i\frac{\sqrt{3}}{2}$.

How to Think About It. We asked you earlier to contemplate how we could prove that something is impossible. The basic strategy is an elaborate indirect proof: assuming a certain point Y is constructible, we will reach a contradiction. The first step is essentially analytic geometry: replace points by their coordinates, as we have just done by defining constructible complex numbers. The next step involves modern algebra; don't just consider one constructible number; consider the set K of *all* constructible numbers, for the totality of them may have extra structure that we can exploit. In fact, we will see that K is a subfield of $\mathbb{C}$. Not only can we translate points to numbers, we can also translate the definition of constructibility into algebra. If a point P_{n+1} is constructible from (constructible) points $P_0, P_1, \ldots, P_n$, we shall see that its complex brother P_{n+1} is algebraic over the subfield $F = \mathbb{Q}(P_0, P_1, \ldots, P_n)$. Given that lines have linear equations and circles have quadratic equations, it is not surprising that $[F(P_{n+1}) : F] \leq 2$. The ultimate criterion that a point Y be constructible is essentially that $[\mathbb{Q}(Y) : \mathbb{Q}]$ is a power of 2.

It follows, for example, that the classical geometric problem of duplicating the cube corresponds to the algebraic problem of checking whether $[\mathbb{Q}(\sqrt[3]{2}) : \mathbb{Q}]$ is a power of 2. Since this degree is 3, the assumption that we can duplicate the cube leads to the contradiction in arithmetic that $3 = 2^k$ for some integer k.

We continue our discussion of constructibility.

Lemma 7.46. *A complex number $z = a + ib$ is constructible if and only if its real and imaginary parts are constructible.*

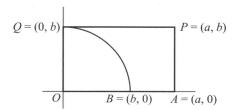

Figure 7.4. Real and imaginary parts.

Proof. If $z = a + ib$ is constructible, then construct lines through $P = (a, b)$ parallel to each axis (Lemma 7.45). The intersection of the vertical line and the x-axis is $A = (a, 0)$, so that A is constructible, and hence $a = a + 0i$ is a constructible real number. Similarly, the point $Q = (0, b)$, the intersection of the horizontal line and the y-axis, is constructible. It follows that $B = (b, 0)$ is constructible, for it is an intersection point of the x-axis and $C(O, Q)$. Hence, $b = b + 0i$ is a constructible real number.

Conversely, assume that a and b are constructible real numbers; that is, $(a, 0)$ and $B = (b, 0)$ are constructible points. The point $Q = (0, b)$ is constructible, being the intersection of the y-axis and $C(O, B)$. By Lemma 7.45, the vertical line through $(a, 0)$ and the horizontal line through $(0, b)$ can be drawn, and (a, b) is their intersection. Therefore, (a, b) is a constructible point, and so $z = a + ib$ is a constructible number. ■

Definition. Denote the subset of $\mathbb{C}$ of all *constructible numbers* by K.

The next lemma allows us to focus on *real* constructible numbers.

Lemma 7.47. (i) *If $K \cap \mathbb{R}$ is a subfield of $\mathbb{R}$, then K is a subfield of $\mathbb{C}$.*

(ii) *If $K \cap \mathbb{R}$ is a subfield of $\mathbb{R}$ and if $\sqrt{a} \in K$ whenever $a \in K \cap \mathbb{R}$ is positive, then K is closed under square roots.*

Proof. (i) If $z = a + ib$ and $w = c + id$ are constructible, then $a, b, c, d \in K \cap \mathbb{R}$, by Lemma 7.46. Hence, $a + c, b + d \in K \cap \mathbb{R}$, because $K \cap \mathbb{R}$ is a subfield, and so $(a+c)+i(b+d) \in K$, by Lemma 7.46. Similarly, $zw = (ac - bd) + i(ad + bc) \in K$. If $z \neq 0$, then $z^{-1} = (a/z\overline{z}) - i(b/z\overline{z})$. Now $a, b \in K \cap \mathbb{R}$, by Lemma 7.46, so that $z\overline{z} = a^2 + b^2 \in K \cap \mathbb{R}$, because $K \cap \mathbb{R}$ is a subfield of $\mathbb{C}$. Therefore, $z^{-1} \in K$.

(ii) If $z = a + ib \in K$, then $a, b \in K \cap \mathbb{R}$, by Lemma 7.46, and so $r^2 = a^2 + b^2 \in K \cap \mathbb{R}$, as in part (i). Since r^2 is nonnegative, the hypothesis gives $r \in K \cap \mathbb{R}$ and $\sqrt{r} \in K \cap \mathbb{R}$.

Now $z = r(\cos\theta + i\sin\theta)$, so that $\cos\theta + i\sin\theta = r^{-1}z \in K$, because K is a subfield of $\mathbb{C}$ by part (i). By Lemma 7.44, $\cos\frac{\theta}{2} + i\sin\frac{\theta}{2}$ can be constructed, and hence is in K. But $\sqrt{z} = \sqrt{r}\left(\cos\frac{\theta}{2} + i\sin\frac{\theta}{2}\right) \in K$, as desired.   ∎

Theorem 7.48. *The set of all constructible numbers K is a subfield of $\mathbb{C}$ that is closed under square roots and complex conjugation.*

Proof. It suffices to prove that the properties of $K \cap \mathbb{R}$ in Lemma 7.47 hold. Let a and b be constructible real numbers.

(i) $-a$ *is constructible.*

If $P = (a, 0)$ is a constructible point, then $(-a, 0)$ is the other intersection of the x-axis and $C(O, P)$.

(ii) $a + b$ *is constructible.*

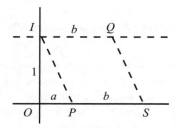

Figure 7.5. $a + b$.

> You are tempted to use a compass with center I and radius b to draw $C(P, b)$. But this is illegal. Remember: we're using a collapsible compass that requires two points given at the outset; here, only one is available, namely P.

Assume that a and b are positive. Let $I = (0, 1)$, $P = (a, 0)$, and $Q = (b, 1)$. Now Q is constructible: it is the intersection of the horizontal line through I and the vertical line through $(b, 0)$, both of which can be drawn by Lemma 7.45 (the latter point is constructible, by hypothesis). The line through Q parallel to IP intersects the x-axis in $S = (a+b, 0)$, as desired.

To construct $b - a$, let $P = (-a, 0)$ in Figure 7.5. Thus, $a + b$ and $-a + b$ are constructible; by part (i), $-a - b$ and $a - b$ are also constructible. Thus, $a + b$ is constructible, no matter whether a and b are both positive, both negative, or have opposite sign.

(iii) ab *is constructible.*

By part (i), we may assume that both a and b are positive. In Figure 7.6, $A = (1, 0)$, $B = (1 + a, 0)$, and $C = (0, b)$. Define D to be the intersection of the y-axis and the line through B parallel to AC. Since the triangles $\triangle OAC$ and $\triangle OBD$ are similar,

$$OB/OA = OD/OC;$$

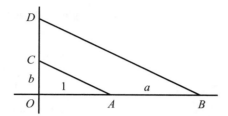

Figure 7.6. *ab*.

hence $(a + 1)/1 = (b + CD)/b$, and $CD = ab$. Therefore, $b + ab$ is constructible. Since $-b$ is constructible, by part (i), we have $ab = (b + ab) - b$ constructible, by part (ii).

(iv) *If $a \neq 0$, then a^{-1} is constructible.*

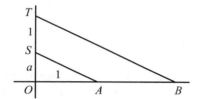

Figure 7.7. a^{-1}.

Let $A = (1, 0)$, $S = (0, a)$, and $T = (0, 1 + a)$. Define B as the intersection of the x-axis and the line through T parallel to AS; thus, $B = (1 + u, 0)$ for some u. Similarity of the triangles $\triangle OSA$ and $\triangle OTB$ gives

$$OT/OS = OB/OA.$$

Hence, $(1 + a)/a = (1 + u)/1$, and so $u = a^{-1}$. Therefore, $1 + a^{-1}$ is constructible, and so $(1 + a^{-1}) - 1 = a^{-1}$ is constructible.

(v) *If $a \geq 0$, then $\sqrt{a}$ is constructible.*

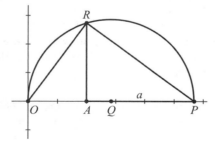

Figure 7.8. $\sqrt{a}$.

Let $A = (1, 0)$ and $P = (1 + a, 0)$; construct Q, the midpoint of $\overline{OP}$ (if U, V are constructible points, then the midpoint of the segment $\overline{UV}$ is its intersection with the perpendicular-bisector, constructed as in Figure 7.1). Define R as the intersection of the circle $C(Q, O)$ with the vertical line through A. The (right) triangles $\triangle AOR$ and $\triangle ARP$ are similar, so that

$$OA/AR = AR/AP,$$

and hence $AR = \sqrt{a}$.

(vi) *If $z = a + ib \in K$, then $\bar{z} = a - ib$ is constructible.*

By Lemma 7.47, K is a subfield of $\mathbb{C}$. Now $a, b \in K$, by Lemma 7.46, and $i \in K$, as we saw on page 310. It follows that $-bi \in K$, and so $a - ib \in K$. ∎

Corollary 7.49. *If a, b, c are constructible, then the roots of the quadratic $ax^2 + bx + c$ are constructible.*

Proof. This follows from Theorem 7.48 and the quadratic formula. ∎

We now consider subfields of $\mathbb{C}$ to enable us to prove an inductive step in the upcoming theorem.

Lemma 7.50. *Let F be a subfield of $\mathbb{C}$ containing i that is closed under complex conjugation. Let $z = a + ib, w = c + id \in F$, and let $P = (a, b)$ and $Q = (c, d)$.*

(i) *If $a + ib \in F$, then $a \in F$ and $b \in F$.*

(ii) *If the equation of $L(P, Q)$ is $y = mx + q$, where $m, q \in \mathbb{R}$, then $m, q \in F$.*

(iii) *If the equation of $C(P, Q)$ is $(x-a)^2 + (y-b)^2 = r^2$, where $a, b, r \in \mathbb{R}$, then $r^2 \in F$.*

Proof. (i) If $z = a + ib \in F$, then $a = \frac{1}{2}(z + \bar{z}) \in F$ and $ib = \frac{1}{2}(z - \bar{z}) \in F$; since we are assuming that $i \in F$, we have $b = -i(ib) \in F$.

(ii) By (i), the numbers a, b, c, d lie in F. Hence, $m = (d - b)/(c - a) \in F$ and $q = b - ma \in F$.

(iii) The circle $C(P, Q)$ has equation $(x - a)^2 + (y - b)^2 = r^2$, and $r^2 \in F$ because $r^2 = (c - a)^2 + (d - b)^2$. ∎

As we said earlier, the next result is intuitively obvious, for the equation of a line is linear and the equation of a circle is quadratic. However, the coming proof involves some calculations.

Lemma 7.51. *Let F be a subfield of $\mathbb{C}$ containing i and which is closed under complex conjugation. Let P, Q, R, S be points whose coordinates lie in F, and let $\alpha = u + iv \in \mathbb{C}$. If either of the following is true,*

$$\alpha \in L(P, Q) \cap L(R, S), \quad where \; L(P, Q) \neq L(R, S),$$
$$\alpha \in L(P, Q) \cap C(R, S),$$
$$\alpha \in C(P, Q) \cap C(R, S), \quad where \; C(P, Q) \neq C(R, S),$$

then $[F(\alpha) : F] \leq 2$.

Proof. If $L(P, Q)$ is not vertical, then Lemma 7.50(ii) says that $L(P, Q)$ has equation $y = mx + b$, where $m, b \in F$. If $L(P, Q)$ is vertical, then its equation is $x = b$ because $P = (a, b) \in L(P, Q)$, and so $b \in F$, by Lemma 7.50(i). Similarly, $L(R, S)$ has equation $y = nx + c$ or $x = c$, where $m, b, n, c \in F$. Since these lines are not parallel, one can solve the pair of linear equations for (u, v), the coordinates of $\alpha \in L(P, Q) \cap L(R, S)$, and they also lie in F. In this case, therefore, $[F(\alpha) : F] = 1$.

See Exercises 7.52 and 7.53 on page 326.

Let $L(P, Q)$ have equation $y = mx + b$ or $x = b$, and let $C(R, S)$ have equation $(x - c)^2 + (y - d)^2 = r^2$; by Lemma 7.50, we have $m, q, r^2 \in F$. Since $\alpha = u + iv \in L(P, Q) \cap C(R, S)$,

$$
\begin{aligned}
r^2 &= (u - c)^2 + (v - d)^2 \\
&= (u - c)^2 + (mu + q - d)^2,
\end{aligned}
$$

so that u is a root of a quadratic polynomial with coefficients in $F \cap \mathbb{R}$. Hence, $[F(u) : F] \leq 2$. Since $v = mu + q$, we have $v \in F(u)$, and, since $i \in F$, we have $\alpha \in F(u)$. Therefore, $\alpha = u + iv \in F(u)$, and so $[F(\alpha) : F] \leq 2$.

Let $C(P, Q)$ have equation $(x - a)^2 + (y - b)^2 = r^2$ and let $C(R, S)$ have equation $(x - c)^2 + (y - d)^2 = s^2$. By Lemma 7.50, we have $r^2, s^2 \in F \cap \mathbb{R}$. Since $\alpha \in C(P, Q) \cap C(R, S)$, there are equations

$$
(u - a)^2 + (v - b)^2 = r^2 \text{ and } (u - c)^2 + (v - d)^2 = s^2.
$$

After expanding, both equations have the form $u^2 + v^2 + something = 0$. Setting the *something*'s equal gives an equation of the form $tu + t'v + t'' = 0$, where $t, t', t'' \in F$. Coupling this with the equation of one of the circles returns us to the situation of the second paragraph. ■

Here is the criterion we have been seeking: an algebraic characterization of a geometric idea; it is an exact translation from geometry into algebra.

Theorem 7.52. *A complex number z is constructible if and only if there is a tower of fields*

$$
\mathbb{Q} = K_0 \subseteq K_1 \subseteq \cdots \subseteq K_n \subseteq \mathbb{C},
$$

where $z \in K_n$ and $[K_{j+1} : K_j] \leq 2$ for all j.

Proof. Let $z = a + ib$, and let $P = (a, b)$ be the corresponding point in the plane. If z is a constructible number, then P is a constructible point, and so there is a sequence of points $A, A', P_1, \ldots, P_n = P$ with each P_{j+1} obtainable from $\{A, A', P_1, \ldots, P_j\}$; since i is constructible, we may assume that $P_1 = (0, 1)$. Define

$$
K_1 = \mathbb{Q}(z_1) \text{ and } K_{j+1} = K_j(z_{j+1}),
$$

where z_j corresponds to the point P_j and there are points E, F, G, H lying in $\{A, A', P_1, \ldots, P_j\}$ with one of the following:

$$
\begin{aligned}
P_{j+1} &\in L(E, F) \cap L(G, H), \\
P_{j+1} &\in L(E, F) \cap C(G, H), \\
P_{j+1} &\in C(E, F) \cap C(G, H).
\end{aligned}
$$

We may assume, by induction on $j \geq 1$, that K_j is closed under complex conjugation, so that Lemma 7.51 applies to show that $[K_{j+1} : K_j] \leq 2$. Finally, K_{j+1} is also closed under complex conjugation, for if z_{j+1} is a root of a quadratic $f(x) \in K_j[x]$, then $\overline{z}_{j+1}$ is the other root of f.

Conversely, given a tower of fields as in the statement, then Theorem 7.48 and Lemma 7.50 show that z is constructible. ∎

Corollary 7.53. *If a complex number z is constructible, then $[\mathbb{Q}(z) : \mathbb{Q}]$ is a power of 2.*

Proof. This follows from Theorems 7.52 and 7.27: If $k \subseteq E \subseteq K$ are fields with E/k and K/E finite extension fields, then $[K : k] = [K : E][E : k]$. ∎

The converse of Corollary 7.53 is false; it can be shown that there are non-constructible numbers z with $[\mathbb{Q}(z) : \mathbb{Q}] = 4$ (see [27], p. 136).

Corollary 7.54. (i) *The real number $\cos(2\pi/7)$ is not constructible.*

(ii) *The complex 7th root of unity ζ_7 is not constructible.*

Proof. (i) We saw in Example 7.28 that $[\mathbb{Q}(\cos(2\pi/7)) : \mathbb{Q}] = 3$.

(ii) $\zeta_7 = \cos(2\pi/7) + i\sin(2\pi/7)$. ∎

We'll soon have more to say about constructibility of roots of unity.

We can now deal with the Greek problems, two of which were solved by Wantzel in 1837. The notion of dimension of a vector space was not known at that time; in place of Theorem 7.52, Wantzel proved that if a number is constructible, then it is a root of an irreducible polynomial in $\mathbb{Q}[x]$ of degree 2^n for some n.

Nicomedes solved the Delian problem of doubling the cube using a marked ruler and compass.

Theorem 7.55 (Wantzel). *It is impossible to duplicate the cube using only ruler and compass.*

Proof. The question is whether $z = \sqrt[3]{2}$ is constructible. Since $x^3 - 2$ is irreducible, $[\mathbb{Q}(z) : \mathbb{Q}] = 3$, by Theorem 7.20; but 3 is not a power of 2. ∎

Consider how ingenious this proof is. At the beginning of this section, you were asked to ponder how we can prove impossibility. As we said when we outlined this argument, the constructibility of a point was translated into algebra, and the existence of a geometric construction produces an arithmetic contradiction. This is a spectacular use of the idea of modeling!

A student in one of our classes, imbued with the idea of continual progress through technology, asked, "Will it ever be possible to duplicate the cube with ruler and compass?" *Impossible* here is used in its literal sense.

Theorem 7.56 (Wantzel). *It is impossible to trisect a 60° angle using only ruler and compass.*

Proof. We may assume that one side of the angle is on the x-axis, and so the question is whether $z = \cos(20°) + i\sin(20°)$ is constructible. If z were constructible, then Lemma 7.46 would show that $\cos(20°)$ is constructible.

The triple angle formula on page 110 gives

$$\cos(3\alpha) = 4\cos^3\alpha - 3\cos\alpha.$$

Setting $\alpha = 20°$, we have $\cos 3\alpha = \frac{1}{2}$, so that $z = \cos(20°)$ is a root of $4x^3 - 3x - \frac{1}{2}$; equivalently, $\cos(20°)$ is a root of $f(x) = 8x^3 - 6x - 1 \in \mathbb{Z}[x]$. A cubic is irreducible in $\mathbb{Q}[x]$ if and only if it has no rational roots. By Theorem 6.52, the only candidates for rational roots are ± 1, $\pm\frac{1}{2}$, $\pm\frac{1}{4}$, and $\pm\frac{1}{8}$; since none of these is a root, as one easily checks, it follows that f is irreducible. (Alternatively, we can prove irreducibility using Theorem 6.55, for $\overline{f}(x) = x^3 + x - 1$ is irreducible in $\mathbb{Z}_7[x]$.) Therefore, $3 = [\mathbb{Q}(z) : \mathbb{Q}]$, by Theorem 7.20(ii), and so $z = \cos(20°)$ is not constructible because 3 is not a power of 2. ∎

Theorem 7.57 (Lindemann). *It is impossible to square the circle with ruler and compass.*

Proof. The problem is whether we can construct a square whose area is π, the area of the unit circle; If the side of the square has length z, we are asking whether $z = \sqrt{\pi}$ is constructible. Now $\mathbb{Q}(\pi)$ is a subfield of $\mathbb{Q}(\sqrt{\pi})$. We have already mentioned that Lindemann proved that π is transcendental (over $\mathbb{Q}$), so that $[\mathbb{Q}(\pi) : \mathbb{Q}]$ is infinite. It follows from Corollary A.41 in Appendix A.3 that $[\mathbb{Q}(\sqrt{\pi}) : \mathbb{Q}]$ is also infinite. Thus, $[\mathbb{Q}(\sqrt{\pi}) : \mathbb{Q}]$ is surely not a power of 2, and so $\sqrt{\pi}$ is not constructible. ∎

Other construction tools

If a ruler is allowed not only to draw a line but to measure distance using marks on it (as most of our rulers are used nowadays), then the added function makes it a more powerful instrument. Both Nicomedes and Archimedes were able to trisect arbitrary angles using a marked ruler and a compass; we present Archimedes' proof here.

Theorem 7.58 (Archimedes). *Every angle can be trisected using a marked ruler and compass.*

Proof. It is easy to construct $\gamma = 30°$, $\gamma = 60°$, and $\gamma = 90°$. The trigonometric Addition Formula shows that if $z = \cos\beta + i\sin\beta$ and $z' = \cos\gamma + i\sin\gamma$ can be found, so can $zz' = \cos(\beta + \gamma) + i\sin(\beta + \gamma)$. Now if $3\beta = \alpha$, then $3(\beta + 30°) = \alpha + 90°$, $3(\beta + 60°) = \alpha + 180°$, and $3(\beta + 90°) = \alpha + 270°$.

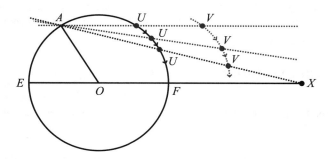

Figure 7.9. Sliding ruler.

Thus, it suffices to trisect an acute angle α, for if $\alpha = 3\beta$ and β can be found, then so can $\beta + 30°$, $\beta + 60°$, and $\beta + 90°$ be found.

Draw the given angle $\alpha = \angle AOE$, where the origin O is the center of the unit circle. Take a ruler on which the distance 1 has been marked; that is, there are points U and V on the ruler with $UV = 1$. There is a chord through A parallel to $L(E, F)$; place the ruler so that the chord is $\overline{AU}$. Since α is acute, U lies in the first quadrant. Keeping A on the sliding ruler, move the point U down the circle; the ruler intersects the extended diameter $L(E, F)$ in some point X with $UX > 1$. Continue moving U down the circle, keeping A on the sliding ruler, until the ruler intersects $L(E, F)$ in the point V.

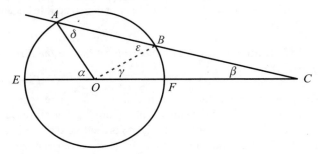

Figure 7.10. Trisecting α.

We claim that $\beta = \angle UVO = \frac{1}{3}\alpha$. Now

$$\alpha = \delta + \beta,$$

because α is an exterior angle of $\triangle AOV$, and hence it is the sum of the two opposite internal angles. Since $\triangle OAU$ is isosceles ($\overline{OA}$ and $\overline{OU}$ are radii), $\delta = \varepsilon$, and so

$$\alpha = \varepsilon + \beta.$$

But $\varepsilon = \gamma + \beta = 2\beta$, for it is an exterior angle of the isosceles triangle $\triangle UVO$; therefore,

$$\alpha = 2\beta + \beta = 3\beta. \quad \blacksquare$$

In addition to investigating more powerful tools, one can look at what can be accomplished with fewer tools. It was proved by Mohr in 1672 and, independently, by Mascheroni in 1797, that every geometric construction carried out by ruler and compass can be done without the ruler. There is a short proof of the theorem given by Hungerbühler in *American Mathematical Monthly*, 101 (1994), pp. 784–787.

Constructing Regular n-gons

High school geometry students are often asked to construct various regular polygons. In light of our present discussion, we can phrase such problems more carefully.

Which regular polygons can be inscribed in the unit circle using only ruler and compass?

Because they can construct 90° and 60° angles, high school students can construct squares and hexagons (just make right and 60° central angles), and they can connect every other vertex of their hexagon to inscribe an equilateral triangle. Also, by using the perpendicular-bisector construction, they can inscribe a regular polygon with twice as many sides as an already constructed one, so they can inscribe regular polygons with $3 \cdot 2^n$ and $4 \cdot 2^n$ sides for any positive integer n. Archimedes knew that π is the area of the unit circle, and he approximated it by inscribing and circumscribing a regular 96-gon (he began with a regular hexagon and then doubled the number of sides four times).

This is about as far as most high school programs get, although some treat polygons with $5 \cdot 2^n$ sides (using Exercise 3.48 to construct the decagon and then connecting every other vertex). This is also about as far as Greek geometers got, although they also were able to show (see Exercise 7.67) that if a regular m-gon and and a regular n-gon are inscribable in a circle (again, with only ruler and compass), then so is a regular nm-gon; for example, a regular 15-gon can be inscribed. However, it was unknown whether all regular polygons could be so inscribed.

About 2000 years later, around 1796, Gauss—still in his teens—essentially invented the main results in this section, and he applied them to the problem of determining whether a regular polygon could be inscribed in a circle with ruler and compass (he wrote that his main result on this problem led to his decision to become a mathematician). We'll develop his methods here.

Theorem 3.28 tells us that the vertices of a regular n-gon inscribed in the unit circle can be realized in the complex plane as the set of roots to $x^n - 1$, and that these roots are all powers of

> We'll revisit the construction of the pentagon in just a minute, putting it a more general setting.

$$\zeta_n = \cos(2\pi/n) + i \sin(2\pi/n).$$

So, we can recast our question about inscribability and ask:

For which values of n is ζ_n a constructible number?

Well, we can hit this question with Theorem 7.52:

> Given the development so far, you may already see that the problem can be translated to the algebra of constructible complex numbers, but this was a huge leap for mathematicians of Gauss's time and certainly out of reach for Greek geometers.

Corollary 7.59. *A regular n-gon can be inscribed in the unit circle with ruler and compass if and only if there is a tower of fields*

$$\mathbb{Q} = K_0 \subseteq K_1 \subseteq \cdots \subseteq K_n \subseteq \mathbb{C},$$

where $\zeta_n = e^{2\pi i/n}$ lies in K_n and $[K_{j+1} : K_j] \leq 2$ for all j.

Proof. Indeed, a regular n-gon can be so inscribed if and only if ζ_n and, hence, all its powers, are constructible numbers. ∎

Gauss showed how to construct such a tower when $n = 17$, and his method was general in principle, leading to a complete classification of inscribable regular polygons. Before we state the main result, let's work through two examples as Gauss did (all laid out in detail by him in [14], Section VII).

Example 7.60. In Example 3.34 on page 113, we showed how to find explicit formulas for the vertices of a regular pentagon inscribed in the unit circle.

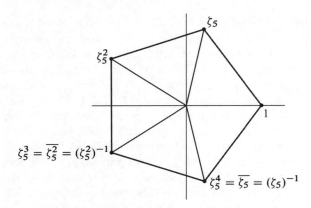

Figure 7.11. Unit 5-gon.

Let's look at this from the perspective of this chapter. Write ζ instead of ζ_5. The nonreal roots of $x^5 - 1$, namely ζ, ζ^2, ζ^3, ζ^4, are the roots of the irreducible polynomial

$$\Phi_5(x) = x^4 + x^3 + x^2 + x + 1.$$

It follows that

$$[\mathbb{Q}(\zeta) : \mathbb{Q}] = 4,$$

so Corollary 7.53 tells us that there's a chance that ζ is constructible. In Example 3.34, without using this language, we actually constructed the tower of quadratic extensions necessary to guarantee that ζ is, in fact, constructible. We showed that if $g = \zeta + \zeta^4 = 2\cos(2\pi/5)$ and $h = \zeta^2 + \zeta^3 = \cos(4\pi/5)$, then g and h are roots of the quadratic equation $x^2 + x - 1$, so that

$$[\mathbb{Q}(\zeta + \zeta^4) : \mathbb{Q}] = 2.$$

By Theorem 7.27,

$$[\mathbb{Q}(\zeta) : \mathbb{Q}(\zeta + \zeta^4)][\mathbb{Q}(\zeta + \zeta^4) : \mathbb{Q}] = [\mathbb{Q}(\zeta) : \mathbb{Q}] = 4,$$

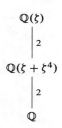

so that

$$[\mathbb{Q}(\zeta) : \mathbb{Q}(\zeta + \zeta^4)] = 2,$$

and we have our tower of quadratic extensions:

$$\mathbb{Q} \subseteq \mathbb{Q}(\zeta + \zeta^4) \subseteq \mathbb{Q}(\zeta) \quad \blacktriangle$$

Gauss's construction of the 17-gon

Stepping back a bit, we can describe what we did with the pentagon: the non-real roots are

$$\{\zeta, \zeta^2, \zeta^3, \zeta^4\}$$

(we are still writing ζ instead of ζ_5). There are four roots. The first story of our tower is $\mathbb{Q}(\zeta + \zeta^4)$, generated by the sums of pairs of the roots: $\zeta + \zeta^4$ and $\zeta^2 + \zeta^3$. The top story, $\mathbb{Q}(\zeta)$, is generated by the individual roots themselves.

This is the basic idea behind Gauss's insight into the 17-gon, but the situation here is more complicated. Change notation again; now let

$$\zeta = \zeta_{17} = \cos(2\pi/17) + i \sin(2\pi/17).$$

Because the minimal polynomial of ζ over $\mathbb{Q}$ is

$$\Phi(x) = x^{16} + x^{15} + \cdots + x^2 + x + 1,$$

we have $[\mathbb{Q}(\zeta) : \mathbb{Q}] = 16$. There are sixteen roots of Φ_{17}: $\{\zeta^k : 0 \le k \le 15\}$. Together with 1, these points on the unit circle are the vertices of our regular 17-gon.

For each factorization $16 = ef$, Gauss divided the roots into e sums of $f = 16/e$ roots each: $\eta_{e,0}, \eta_{e,1}, \ldots, \eta_{e,e-1}$, where each $\eta_{e,k}$ is a sum of f roots. For example, he divided the sixteen roots into two sums of eight each, which we can call $\eta_{2,0}$ and $\eta_{2,1}$, as follows:

> We'll see what method Gauss used to partition the roots into these sums in just a minute.

$$\eta_{2,0} = \zeta + \zeta^9 + \zeta^{13} + \zeta^{15} + \zeta^{16} + \zeta^8 + \zeta^4 + \zeta^2$$
$$\eta_{2,1} = \zeta^3 + \zeta^{10} + \zeta^5 + \zeta^{11} + \zeta^{14} + \zeta^7 + \zeta^{12} + \zeta^6.$$

There are also four sums of four each:

$$\eta_{4,0} = \zeta + \zeta^{13} + \zeta^{16} + \zeta^4$$
$$\eta_{4,1} = \zeta^3 + \zeta^5 + \zeta^{14} + \zeta^{12}$$
$$\eta_{4,2} = \zeta^9 + \zeta^{15} + \zeta^8 + \zeta^2$$
$$\eta_{4,3} = \zeta^{10} + \zeta^{11} + \zeta^7 + \zeta^6.$$

And there are eight sums of two each:

> Each period of length > 1 is a real number; you can check that if ζ^k occurs in $\eta_{e,k}$, so does
>
> $$\zeta^{17-k} = \zeta^{-k} = \overline{\zeta^k},$$
>
> so that each period is a sum of terms of form $z + \overline{z}$, and hence a real number.

$$\eta_{8,0} = \zeta + \zeta^{16}$$
$$\eta_{8,1} = \zeta^3 + \zeta^{14}$$
$$\eta_{8,2} = \zeta^9 + \zeta^8$$
$$\eta_{8,3} = \zeta^{10} + \zeta^7$$
$$\eta_{8,4} = \zeta^{13} + \zeta^4$$
$$\eta_{8,5} = \zeta^5 + \zeta^{12}$$
$$\eta_{8,6} = \zeta^{15} + \zeta^2$$
$$\eta_{8,7} = \zeta^{11} + \zeta^6.$$

Finally, there are sixteen "sums" of one each, namely

$$\left\{ \eta_{16,k} = \zeta^k \mid 0 \le k \le 15 \right\}.$$

Gauss called each of the $\eta_{e,k}$ a **period of length** $f = 16/e$. The plan is to show that the periods of length eight lie in a quadratic extension K_1 of $\mathbb{Q}$, the periods of length four lie in a quadratic extension K_2 of K_1, and so on, building a tower of quadratic extensions ending with $\mathbb{Q}(\zeta)$.

The calculations will sometimes be involved so, once again, pull out your pencil or computer.

> If you use a CAS, you can perform all of these calculations in $\mathbb{Q}[x]/(\Phi(x))$.

- Because $\Phi(\zeta) = 0$, we see that $\eta_{2,0} + \eta_{2,1} = -1$.

- With a little patience and care (or a CAS), you can check that

$$\eta_{2,0}\eta_{2,1} = 4(\eta_{2,0} + \eta_{2,1}) = -4.$$

Hence $\eta_{2,0}$ and $\eta_{2,1}$ are roots of $x^2 + x - 4$, and so

$$[\mathbb{Q}(\eta_{2,1}) : \mathbb{Q}] \leq 2.$$

You can also check that
$\eta_{2,0}^2 = \eta_{n,1}$ (Exercise 7.60
on page 327).

The first step in our tower is $\mathbb{Q} \subseteq \mathbb{Q}(\eta_{2,1})$. Note that $\eta_{2,0} = -4/\eta_{2,1}$, so that $\eta_{2,1} \in \mathbb{Q}(\eta_{2,0})$. Next, we move up to the periods of length 4. You can check (Exercise 7.61 on page 327) that

$$\eta_{4,1} + \eta_{4,3} = \eta_{2,1}$$
$$\eta_{4,1}\eta_{4,3} = -1.$$

Hence, $\eta_{4,1}$ and $\eta_{4,3}$ are roots of $x^2 - \eta_{2,1}x - 1$, and

$$[\mathbb{Q}(\eta_{4,1}) : \mathbb{Q}(\eta_{2,1})] \leq 2.$$

$\mathbb{Q}(\eta_{4,1})/\mathbb{Q}(\eta_{2,1})$ is the second story in our tower.

$$\mathbb{Q} \subseteq \mathbb{Q}(\eta_{2,1}) \subseteq \mathbb{Q}(\eta_{4,1}).$$

Note that $\eta_{4,3} = -1/\eta_{4,1}$, so that $\eta_{4,3} \in \mathbb{Q}(\eta_{4,1})$.

Up one more story—the periods of length eight: you can check (Exercise 7.62 on page 327) that

$$\eta_{8,1} + \eta_{8,5} = \eta_{4,1}$$
$$\eta_{8,1}\eta_{8,5} = \eta_{4,2}.$$

So, $\eta_{8,1}$ and $\eta_{8,5}$ are roots of $x^2 - \eta_{4,1}x + \eta_{4,2}$. This says that

$$[\mathbb{Q}(\eta_{8,1}) : \mathbb{Q}(\eta_{4,1}, \eta_{4,2})] \leq 2.$$

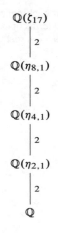

But, by Exercise 7.63 on page 327,

$$\mathbb{Q}(\eta_{4,1}, \eta_{4,2}) = \mathbb{Q}(\eta_{4,1}),$$

so that

$$[\mathbb{Q}(\eta_{8,1}) : \mathbb{Q}(\eta_{4,1})] \leq 2.$$

Assemble what we have built:

$$\mathbb{Q} \subseteq \mathbb{Q}(\eta_{2,1}) \subseteq \mathbb{Q}(\eta_{4,1}) \subseteq \mathbb{Q}(\eta_{8,1}) \subseteq \mathbb{Q}(\zeta).$$

The degree of each extension is at most 2; since

$$\mathbb{Q}(\zeta) = 16,$$

all the degrees are equal to 2 (Theorem 7.27). Hence, we have constructed a tower of fields, each quadratic over the one below, starting with $\mathbb{Q}$ and ending with $\mathbb{Q}(\zeta)$. We have proved that ζ is constructible.

Theorem 7.61. *A regular 17-gon can be inscribed in the unit circle with ruler and compass.*

One detail that remains is to see what method Gauss used to assign different powers of ζ to each period; how did Gauss decide which powers of ζ should occur in each $\eta_{e,k}$? The answer comes from Galois theory (a subject we only briefly touch on in Chapter 9). He employed an ingenious method using the fact that 3 is a *primitive element* in $\mathbb{F}_{17}$; that is, every nonzero element in $\mathbb{F}_{17}$ is a power of 3.

k	0	1	2	3	4	5	6	7	8	9	10	11	12	13	14	15
3^k	1	3	9	10	13	5	15	11	16	14	8	7	4	12	2	6

Gauss used this special property of 3 to define the periods: if $ef = 16$, there are k periods of length f defined by

$$\eta_{e,k} = \sum_{j=0}^{f-1} \zeta^{3^{k+je}}.$$

So, for example,

$$\eta_{8,4} = \zeta^{3^{4+0\cdot8}} + \zeta^{3^{4+1\cdot8}}$$
$$= \zeta^{81} + \zeta^{531441}$$
$$= \zeta^4 + \zeta^{13}.$$

Sufficiency of the following theorem, a feat the Greeks would have envied, was discovered by Gauss around 1796. He claimed necessity as well, but none of his published papers contains a complete proof of it. The first published proof of necessity is due to Wantzel, in 1837.

Theorem 7.62 (Gauss-Wantzel). *If p is an odd prime, then a regular p-gon is constructible if and only if p is a **Fermat prime**; that is, a prime of the form $p = 2^{2^t} + 1$ for some $t \geq 0$.*

Proof. We only prove necessity. The problem is whether $z = e^{2\pi i/p}$ is constructible. Now z is a root of the cyclotomic polynomial $\Phi_p(x)$, which is an irreducible polynomial in $\mathbb{Q}[x]$ of degree $p - 1$, by Corollary 6.68.

Since z is constructible, $p - 1 = 2^s$ for some s, so that

$$p = 2^s + 1.$$

We claim that s itself is a power of 2. Otherwise, there is an odd number $k > 1$ with $s = km$. Now k odd implies that -1 is a root of $x^k + 1$; in fact, there is a factorization in $\mathbb{Z}[x]$:

$$x^k + 1 = (x + 1)(x^{k-1} - x^{k-2} + x^{k-3} - \cdots + 1).$$

Thus, setting $x = 2^m$ gives a forbidden factorization of p in $\mathbb{Z}$:

$$p = 2^s + 1 = (2^m)^k + 1$$
$$= [2^m + 1][(2^m)^{k-1} - (2^m)^{k-2} + (2^m)^{k-3} - \cdots + 1]. \quad \blacksquare$$

The only known Fermat primes are 3, 5, 17, 257, and 65537. It follows from Theorem 7.62, for example, that it is impossible to construct regular 7-gons, 11-gons, or 13-gons.

One of the reasons has already been mentioned: each period should contain a sum of terms, each of form $\zeta^k + \zeta^{-k}$. This ensures that every story in our tower except the last is contained in $\mathbb{R}$, so we "save the complex step" for the end (why is this a good thing?).

Gauss established sufficiency by generalizing the construction of the 17-gon, giving explicit formulas of the $\eta_{e,f}$ for all pairs (e, f) with $ef = p - 1$ (see [33], pp. 200–206).

Further results. Fermat conjectured that all numbers of the form $F_m = 2^{2^m} + 1$ are prime, but Euler factored

$$F_5 = 2^{32} + 1 = 4,294,967,297 = 641 \times 6700417.$$

We now know that F_m is composite for $5 \le m \le 32$, but it is unknown whether F_{33} is prime. The largest F_m that has been shown to be composite (as of 2011) is $F_{2543548}$, and the latest conjecture is that there are only finitely many Fermat primes.

The strongest known result is:

Theorem 7.63. *A regular n-gon is constructible with ruler and compass if and only if $n = 2^k p_1 \cdots p_t$, where $k \ge 0$ and the p_i's are distinct Fermat primes.*

Proof. See [15], page 97. ∎

Exercises

7.50 Explain how to carry out each of the following constructions with ruler and compass. Prove that your method works.

(i) Copy a segment.

(ii) Copy an angle.

(iii) Construct a line parallel to a given one through a given point not on the line.

(iv) Construct a line perpendicular to a given one through a given point either on or off the line.

7.51 *

(i) Prove that every lattice point (m, n) in the plane is constructible. Conclude that every Gaussian integer is constructible.

(ii) Prove that every Eisenstein integer is constructible.

7.52 Suppose that ℓ and ℓ' are lines with equations $ax + by = c$ and $dx + ey = f$, and suppose that the coefficients of the equations are all in a field k.

(i) What condition on the coefficients guarantees that ℓ and ℓ' intersect in a unique point?

(ii) If ℓ and ℓ' intersect in a unique point P, show that P is point whose coordinates are in k.

7.53 If the quadratic polynomial $ax^2 + bx + c$ has coefficients in a field k, show that its roots are a quadratic extension of k.

7.54 Given a segment of length a, show how to construct a segment of length $a/5$.

7.55 Given a segment of length a, show how to construct a segment of length a/n, where n is any positive integer.

7.56 Show how to construct segments of length

(i) $\sqrt{5}$

(ii) $3 + \sqrt{5}$

(iii) $\sqrt{3 + \sqrt{5}}$

(iv) $\frac{\sqrt{3 + \sqrt{5}}}{3}$

(v) $\frac{\sqrt{3 + \sqrt{5}}}{3 - \sqrt{5}}$.

7.57 Show how to construct the complex numbers

(i) $1 + i$

(ii) $\frac{\sqrt{2}}{2}(1 + i)$

(iii) $\frac{\sqrt{2}}{2}(1 + i)$

(iv) $\cos 15° + i \sin 15°$

(v) $\cos 22.5° + i \sin 22.5°$

(vi) $\cos 36° + i \sin 36°$.

7.58 Show that the side of a regular decagon inscribed in the unit circle is constructible.

7.59 Show that if the side-length of a regular n-gon inscribed in the unit circle is constructible, so is the side-length of a regular $2n$-gon inscribed in the unit circle.

7.60 * Using the notation of this section, show that

$$\eta_{2,0}\eta_{2,1} = -4 \quad \text{and} \quad \eta_{2,0}^2 = \eta_{n,1}.$$

7.61 * Using the notation of this section, show that

$$\eta_{4,1} + \eta_{4,3} = \eta_{2,1} \quad \text{and} \quad \eta_{4,1}\eta_{4,3} = -1.$$

7.62 * Using the notation of this section, show that

$$\eta_{8,1} + \eta_{8,5} = \eta_{4,1} \quad \text{and} \quad \eta_{8,1}\eta_{8,5} = \eta_{4,2}.$$

7.63 * Using the notation of this section, show that

$$\eta_{4,1}^3 - 6\eta_{4,1} + 3 = -2\eta_{4,2},$$

so that $\eta_{4,2} \in \mathbb{Q}(\eta_{4,1})$

7.64 Let $\zeta = \zeta_{17}$ and let k be a nonnegative integer. Show that

$$\zeta^k + \zeta^{-k} = 2\cos(2k\pi/17).$$

7.65 Show that 3 is a primitive element for $\mathbb{Z}_5$, and apply this to Example 7.60.

7.66 Find the minimal polynomial over $\mathbb{Q}$ of $\eta_{8,k}$ for all $0 \le k \le 7$.

7.67 Show that if $(m, n) = 1$, and if ζ_n and ζ_m are constructible, so is ζ_{mn}.
Hint. Use Theorem 7.63.

8

Cyclotomic Integers

After proving Corollary 1.8, the special case of Fermat's Last Theorem for exponent $n = 4$, we observed that the full theorem would follow if we could prove, for every *odd* prime p, that there are no positive integers a, b, c with $a^p + b^p = c^p$. It is natural to factor this expression as in Exercise 3.50(ii) on page 115: for odd p, we have

$$(a + b)(a + \zeta b) \cdots (a + \zeta^{p-1} b) = c^p, \tag{8.1}$$

where $\zeta = e^{2\pi i/p}$ is a pth root of unity. We didn't have the language of rings at the time but, later you showed, in Exercise 4.65 on page 168, that the cyclotomic integers $\mathbb{Z}[\zeta]$ is a domain. How could we begin to use this observation? Recall Exercise 2.14 on page 59: if $ab = c^n$ in $\mathbb{Z}$, where n is a positive integer and a, b are relatively prime, then both a and b are nth powers. If $\mathbb{Z}[\zeta]$ behaved like $\mathbb{Z}$ and the factors on the left-hand side of Eq. (8.1) are pairwise relatively prime, then all the factors $a + \zeta^j b$ would be pth powers; that is,

$$a + \zeta^j b = d_j^p$$

for some $d_j \in \mathbb{Z}[\zeta]$. For example, consider the case $p = 3$, so that $\mathbb{Z}[\zeta] = \mathbb{Z}[\omega]$ is the ring of Eisenstein integers. The factorization is

$$(a + b)(a + \omega b)(a + \omega^2 b) = c^3.$$

If the factors on the left-hand side are pairwise relatively prime, then each of them is a cube of an Eisenstein integer. We can even say something if the factors are not relatively prime: assuming that $\mathbb{Z}[\omega]$ has unique factorization, any prime divisor of c must occur three times on the left-hand side (*prime* now means a prime in $\mathbb{Z}[\omega]$; that is, an element whose only divisors are units and associates—we will use the term *prime* here instead of *irreducible*).

But does the ring of cyclotomic integers behave as $\mathbb{Z}$? To solve Exercise 2.14, we need to use the Fundamental Theorem of Arithmetic: factorization into primes exists and is essentially unique. It turns out that some rings of cyclotomic integers do enjoy unique factorization into primes, but some do not. Indeed, it is known (see [36], p. 7) that $\mathbb{Z}[\zeta_{23}]$ does not have unique factorization.

It's clear that we need a more thorough investigation of the arithmetic in rings of cyclotomic integers. In particular, we already know the units in $\mathbb{Z}[i]$ and $\mathbb{Z}[\omega]$ (Example 6.3); what are the primes?

In Section 8.1, we retrace the by-now-familiar developments in Chapters 1 and 6 to establish division algorithms in $\mathbb{Z}[i]$ and $\mathbb{Z}[\omega]$ (using norm to measure size). Even though these are the easiest rings of cyclotomic integers, this will give us a clue how to proceed with $\mathbb{Z}[\zeta]$ for other roots of unity ζ. There will be a bonus: we'll be able to prove Fermat's Two-Square Theorem that characterizes all primes in $\mathbb{Z}$ which are sums of two squares.

As is our custom (because it is so useful), we'll generalize from these and the earlier examples of $\mathbb{Z}$ and $k[x]$, where k is a field, to define a *Euclidean domain*—a domain having a generalized division algorithm. We'll show that every Euclidean domain is a PID, so that, by Theorem 6.50, Euclidean domains are UFDs and thus have unique factorization.

In Section 8.2, we'll see that there are primes in $\mathbb{Z}[i]$ and $\mathbb{Z}[\omega]$ that are not ordinary integers. On the other hand, some primes in $\mathbb{Z}$ remain prime in the larger rings, while some split into non-unit factors. We'll then make a complete analysis of this phenomenon for $\mathbb{Z}[i]$ and sketch the analogous theory for $\mathbb{Z}[\omega]$.

In Section 8.3, we'll prove Fermat's theorem for exponent 3. The fact that there are no non-trivial integer solutions to $x^3 + y^3 = z^3$ is attributed to Euler; we'll prove the result making basic use of the arithmetic of $\mathbb{Z}[\omega]$.

In Section 8.4, we'll briefly sketch how the proof for exponent 3 generalizes to prime exponent p when the ring of cyclotomic integers $\mathbb{Z}[\zeta_p]$ is a UFD, where $\zeta = e^{2\pi i/p}$. But there are primes p for which $\mathbb{Z}[\zeta_p]$ does not have unique factorization. What then? We'll finish this section with a brief discussion about how Kummer's construct of *ideal numbers* (which Dedekind recognized as equivalent to what we now call *ideals*, which is why they are so-called) could be used to restore a kind of unique factorization to $\mathbb{Z}[\zeta_p]$.

Finally, in Section 8.5 we develop the machinery to prove a lovely theorem of Fermat that determines the number of ways a positive integer can be written as the sum of two perfect squares.

8.1   Arithmetic in Gaussian and Eisenstein Integers

We begin by showing that $\mathbb{Z}[i]$ and $\mathbb{Z}[\omega]$ have generalized division algorithms. Actually, we'll show that long division is possible in these rings; that is, there are quotients and remainders. However, quotients and remainders are not necessarily unique; stay tuned.

Given two Gaussian integers z and w, can we find Gaussian integers q and r so that $w = qz + r$, where r is "smaller than" z? The obvious way to compare size in $\mathbb{C}$ is with absolute values, but it's easier to calculate norms (squares of absolute value); by Exercise 3.62 on page 127, $|r| < |z|$ if and only if $N(r) < N(z)$. Let's start with an example.

Example 8.1. Take $z = -19 + 48i$ and $w = -211 + 102i$. Can we find q and r so that $w = qz + r$, where $N(r) < N(z)$? We can certainly divide z into w in $\mathbb{Q}[i]$; it's just that w/z may not be a Gaussian integer. In fact,

$$\frac{-211 + 102i}{-19 + 48i} = \frac{(-211 + 102i)(-19 - 48i)}{(-19 + 48i)(-19 - 48i)}$$

$$= \frac{8905 + 8190i}{2665}$$

$$= \frac{137}{41} + \frac{126}{41}i.$$

The idea is to take q to be the Gaussian integer closest to w/z in the complex plane, and then to find an r that makes up the difference. Since

$$\frac{w}{z} = \frac{137}{41} + \frac{126}{41}i \approx 3.34 + 3.07i,$$

we'll take $q = 3 + 3i$. What about r? There's no choice; since we want $w = qz + r$, set $r = w - qz$:

$$r = w - qz = (-211 + 102i) - (3 + 3i)(-19 + 48i) = -10 + 15i.$$

By construction, $w = qz + r$. What's more, that q is the closest Gaussian integer to w/z implies, as we'll see in the proof of the next proposition, that $N(r) < N(z)$. Indeed, $N(z)$ is much bigger that $N(r)$ in this example, because w/z is so close to q.

$$N(r) = (-10)^2 + (15)^2 = 325 \quad \text{and} \quad N(z) = (-19)^2 + (48)^2 = 2665. \quad \blacktriangle$$

This method for choosing q and r works in general.

Proposition 8.2 (Generalized Division Algorithm). *If z and w are Gaussian integers with $z \neq 0$, then there exist Gaussian integers q and r such that*

$$w = qz + r \quad \text{and} \quad N(r) < N(z).$$

Proof. Suppose that $w/z = a + bi$, where a and b are rational numbers (but not necessarily integers). As in Example 8.1, take q to be a Gaussian integer closest to w/z in the complex plane; more precisely, choose integers m and n so that

$$|a - m| \leq \tfrac{1}{2} \quad \text{and} \quad |b - n| \leq \tfrac{1}{2},$$

and let $q = m + ni$. Now define r to be the difference:

$$r = w - qz.$$

Clearly, $w = qz + r$, so the only thing to check is whether $N(r) < N(z)$. To this end, we have

$$N(r) = N(w - qz) = N\left(z\left(\frac{w}{z} - q\right)\right) = N(z)N\left(\frac{w}{z} - q\right).$$

But $w/z - q = (a - m) + (b - n)i$, so that

$$N\left(\frac{w}{z} - q\right) = (a - m)^2 + (b - n)^2 \leq \tfrac{1}{4} + \tfrac{1}{4} < 1.$$

It follows that $N(r) < N(z)$. $\blacksquare$

How to Think About It.   The earlier statements of the division algorithms for $\mathbb{Z}$ (Theorem 1.15) and for $k[x]$, where k is a field (Theorem 6.11), differ from that in Proposition 8.2; the latter statement does not assert uniqueness of quotient and remainder.

In fact, the way q and r are constructed shows that there may be several choices for q and r—locate w/z inside a unit square in the complex plane whose vertices are Gaussian integers, and then pick a closest vertex. There may be several of these, as the next example shows. Luckily, we won't need uniqueness of quotients and remainders to get unique factorization into primes.

Example 8.3. If $z = 2 + 4i$ and $w = -9 + 17i$, then

$$\frac{w}{z} = \frac{5}{2} + \frac{7}{2}i = 2.5 + 3.5i.$$

In contrast to Example 8.1, w/z sits smack in the middle of a unit square whose vertices are Gaussian integers and, hence, there are four choices for q, namely

$$2 + 3i, \quad 3 + 3i, \quad 3 + 4i, \quad 2 + 4i;$$

and there are four corresponding divisions, namely

$$-9 + 17i = (2 + 3i)(2 + 4i) + (-1 + 3i);$$
$$-9 + 17i = (3 + 3i)(2 + 4i) + (-3 - i);$$
$$-9 + 17i = (3 + 4i)(2 + 4i) + (1 - 3i);$$
$$-9 + 17i = (2 + 4i)(2 + 4i) + (3 + i).$$

All of these work. In fact, all the remainders have (the same) norm $10 < 20 = N(z)$. Even more: the remainders are all associates. Is this an accident? See Exercises 8.1 through 8.4 on page 336.   ▲

Alas, there are other rings $\mathbb{Z}[\zeta]$ of cyclotomic integers which do not have a generalized division algorithm.

There is an analogous result for the Eisenstein integers $\mathbb{Z}[\omega]$, and its proof is almost identical to that for the Gaussian integers. Recall that $\overline{c + d\omega} = c + d\omega^2 = c + d(-1 - \omega)$, and that

$$N(c + d\omega) = c^2 - cd + d^2.$$

Proposition 8.4 (Generalized Division Algorithm). *If z and w are Eisenstein integers with $z \neq 0$, then there exist Eisenstein integers q and r such that*

$$w = qz + r \quad and \quad N(r) < N(z).$$

Proof. Suppose that $w/z = a + b\omega$, where a and b are rational numbers (but not necessarily integers). Take q to be a Eisenstein integer closest to w/z in the complex plane (with respect to the norm); more precisely, choose integers m and n so that

$$|a - m| \leq \tfrac{1}{2} \quad and \quad |b - n| \leq \tfrac{1}{2},$$

and let $q = m + n\omega$. Now define r to be the difference

$$r = w - qz.$$

Clearly, $w = qz + r$, so the only thing to check is whether $N(r) < N(z)$. To this end, we have

$$N(r) = N(w - qz) = N\left(z\left(\frac{w}{z} - q\right)\right) = N(z)N\left(\frac{w}{z} - q\right).$$

But $w/z - q = (a - m) + (b - n)\omega$, so that

$$N\left(w/z - q\right) = (a - m)^2 - (a - m)(b - n) + (b - n)^2 \leq \tfrac{1}{4} + \tfrac{1}{4} + \tfrac{1}{4} < 1.$$

It follows that $N(r) < N(z)$.   ■

Why can't we modify the proof of Proposition 8.4 to prove the result for every ring of cyclotomic integers? The short answer is that there are counter-examples. But the reason the proof fails to generalize is that we can't verify $N(r) < N(z)$ in every $\mathbb{Z}[\zeta]$.

Example 8.5. Let's divide $w = 91 + 84\omega$ by $z = 34 + 53\omega$. First calculate w/z in $\mathbb{Q}[\omega]$:

$$\frac{91 + 84\omega}{34 + 53\omega} = \frac{(91 + 84\omega)(34 + 53\omega^2)}{(34 + 53\omega)(34 + 53\omega^2)} = \frac{2723 - 1967\omega}{2163}$$

$$= \frac{389}{309} - \frac{281}{309}\omega \approx 1.26 - .91\omega.$$

Now set $q = 1 - \omega$:

$$r = w - zq = 4 + 12\omega.$$

You can check that

$$N(z) = 34^2 - 34 \cdot 53 + 53^2 = 2163 \quad \text{and} \quad N(r) = 4^2 - 4 \cdot 12 + 12^2 = 112. \quad \blacktriangle$$

As with $\mathbb{Z}[i]$, there may be several choices for q when dividing Eisenstein integers (see Exercise 8.5 on page 337).

How to Think About It. If z, w are either in $\mathbb{Z}[i]$ or in $\mathbb{Z}[\omega]$, we could iterate the respective generalized division algorithms, as we did in $\mathbb{Z}$ or $k[x]$, to obtain Euclidean algorithms giving a greatest common divisor d of z and w; moreover, d can be expressed as a linear combination of z and w ensuring, as in earlier instances of this argument, that any common divisor of z and w is a factor of d. We'll give an example of such a calculation shortly.

Along the way, there may be choices to be made for quotients and remainders, possibly resulting in different "greatest" common divisors d. We ran into this situation before: the Euclidean Algorithm in $k[x]$ produces a gcd up to a unit factor. The same is true in $\mathbb{Z}[i]$ and $\mathbb{Z}[\omega]$, although it may not be obvious at this point because of the twists and turns that the Euclidean Algorithm might take. This is another example in which a more abstract setting can make things clearer (one reason for the added clarity is that abstraction casts away noise, allowing you to focus on the heart of a problem).

Euclidean Domains

Looking at our main examples—$\mathbb{Z}$, $k[x]$, $\mathbb{Z}[i]$, $\mathbb{Z}[\omega]$—we see that one key to a division algorithm is a measure of size: absolute value for $\mathbb{Z}$, degree for $k[x]$, norm for $\mathbb{Z}[i]$ and $\mathbb{Z}[\omega]$. Now we generalize.

Definition. A *Euclidean domain* is a domain R equipped with a *size function*

$$\partial : R - \{0\} \to \mathbb{N}$$

such that, for all $a, b \in R$ with $a \neq 0$, there exist q and r in R with

$$b = qa + r,$$

where either $r = 0$ or $\partial(r) < \partial(a)$.

∂ is defined on the nonzero elements of R and takes nonnegative integer values.

Some size functions have extra properties. For example, when R is a domain, then $R[x]$ is a domain, and degree (which is a size function on $R[x]$) satisfies $\deg(fg) = \deg(f) + \deg(g)$, while norm N (which is a size function on $\mathbb{Z}[i]$ and $\mathbb{Z}[\omega]$) satisfies $N(\alpha\beta) = N(\alpha)N(\beta)$. On the other hand, if ∂ is a size function of a Euclidean domain R, then so is ∂', where $\partial'(a) = \partial(a) + 1$ for all $a \in R - \{0\}$. It follows that a size function may have no algebraic properties; moreover, there may be no elements in R having size 0.

Euclidean domains have nice properties. The proof of the next proposition is essentially the same as that of Theorem 1.19.

Proposition 8.6. *Every Euclidean domain is a principal ideal domain.*

Proof. Suppose that R is a Euclidean domain with size function ∂. We want to show that every ideal I in R is principal. If $I = \{0\}$, then I is principal, and so we can assume that I contains nonzero elements. The set

$$S = \{\partial(z) : z \in I\}$$

is a set of nonnegative integers and, hence, it has a least element; call it m. Choose d to be any element of I of size m. We claim that $I = (d)$.

Clearly, $(d) \subseteq I$. To get the reverse inclusion, suppose that $z \in I$ is not 0; we must show that $z \in (d)$. Now there are q and r such that

$$z = qd + r,$$

where either $r = 0$ or $\partial(r) < \partial(d)$. But $r \in I$, because $r = z - dq$ and both z and d are in I. But $\partial(d)$ is the smallest size among elements of I; hence, $r = 0$, and $z = qd \in (d)$.   ■

Corollary 8.7 (Euclid's Lemma). *Let R be a Euclidean domain with $a, b \in R$. If $p \in R$ is irreducible and $p \mid ab$, then $p \mid a$ or $p \mid b$.*

Proof. This is a direct consequence of Theorem 6.47.   ■

Corollary 8.8. *Every Euclidean domain is a unique factorization domain; that is, every nonzero non-unit has a factorization into irreducibles that is essentially unique.*

Proof. This is a direct consequence of Theorem 6.50: every PID is a unique factorization domain.   ■

Corollary 8.8 probably piques your curiosity about what primes look like in Euclidean domains. We'll consider this question for $\mathbb{Z}[i]$ and $\mathbb{Z}[\omega]$ in the next section.

How to Think About It. Points about the development so far.

- Euclidean domains have generalized division *algorithms*, but they are not necessarily algorithms in the technical sense. They are procedures for computing quotients and remainders, but the division procedures, even for $\mathbb{Z}[i]$

and $\mathbb{Z}[\omega]$, are not deterministic: there is a choice about how to calculate quotients and remainders.

- In Section 1.3, we studied a direct path in Euclid's *Elements* from the Division Algorithm in $\mathbb{Z}$ to the Fundamental Theorem of Arithmetic. This path can be followed in a much more general setting. We just saw that every Euclidean domain enjoys Euclid's Lemma and a fundamental theorem.

- One way to show that a domain is a PID is to show that it is Euclidean—indeed, this is one of the most important uses of this notion. On the other hand, it's hard from first principles to show that a domain is *not* Euclidean (you have to show that no size function exists). Often, the easiest way to show that a domain is not Euclidean is to show that it's not a PID. So, for example, $\mathbb{Z}[x]$ is not a Euclidean domain.

- There are PIDs that are not Euclidean, so that the converse of Proposition 8.6 is false. Motzkin found a property of Euclidean domains that can be defined without mentioning its size function. He called an element d in an arbitrary domain R a **universal side divisor** if d is not a unit and, for every $r \in R$, either $d \mid r$ or there is some unit $u \in R$ with $d \mid (r + u)$. He then proved that every Euclidean domain contains a universal side divisor, namely any non-unit of smallest size. Now it was known that if $\alpha = \frac{1}{2}(1 + \sqrt{-19})$, then the ring $\mathbb{Z}[\alpha]$ is a PID. Motzkin then showed that $\mathbb{Z}[\alpha]$ has no universal side divisors, and he concluded that $\mathbb{Z}[\alpha]$ is a PID that is not a Euclidean domain (see Wilson, A principal ideal ring that is not a Euclidean ring. *Math. Magazine* 46 (1973), 34–38 and Williams, Note on non-Euclidean principal ideal domains, *Math. Magazine* 48 (1975), 176–177).

The fact that a Euclidean domain is a PID allows us a to talk about gcd's, thanks to Theorem 6.46. Using exactly the same logic as in Chapter 1, we can iterate division, creating a Euclidean algorithm that finds a gcd for us: just move factors on one line southwest on the next line (as in the next example).

Example 8.9. Building on the calculation in Example 8.5, let's find a gcd for $91 + 84\omega$ and $34 + 53\omega$ in $\mathbb{Z}[\omega]$. We'll use the algorithm outlined in Proposition 8.4 to carry out the divisions (a CAS is very useful here). There are four equations, which we present in "southwestern style:" if a row has the form $f = qh + r$, then the next row moves h and r southwest and looks like $h = q'r + r'$.

$$91 + 84\omega = (1 - \omega)(34 + 53\omega) + (4 + 12\omega)$$

$$34 + 53\omega = (3 - \omega)(4 + 12\omega) + (10 + 9\omega)$$

$$4 + 12\omega = (1 + \omega)(10 + 9\omega) + (3 + 2\omega)$$

$$10 + 9\omega = (4 + \omega)(3 + 2\omega).$$

Here is a second format, arranging the calculations as we did in $\mathbb{Z}$ in Chapter 1, that shows more detail.

$$
\begin{array}{r}
1-\ \omega \\
34+53\omega \overline{\smash{\big)}\ 91+84\omega} \\
\underline{87+72\omega}
\end{array}
$$

$$
\begin{array}{r}
3-\ \omega \\
4+12\omega \overline{\smash{\big)}\ 34+53\omega} \\
\underline{24+44\omega}
\end{array}
$$

$$
\begin{array}{r}
1+\ \omega \\
10+\ 9\omega \overline{\smash{\big)}\ 4+12\omega} \\
\underline{1+10\omega}
\end{array}
$$

$$
\begin{array}{r}
4+\omega \\
3+\ 2\omega \overline{\smash{\big)}\ 10+9\omega} \\
\underline{10+9\omega} \\
0
\end{array}
$$

Recall that in a general PID, the gcd of two elements a and b is a generator d of the principal ideal consisting of all linear combinations of a and b; in symbols, $(a,b) = (d)$.

So, we end with $3 + 2\omega$. Repeated application of Exercise 8.6 on page 337 shows that there is a chain of equalities of ideals:

$$(91 + 84\omega,\ 34 + 53\omega) = (34 + 53\omega,\ 4 + 12\omega) = (4 + 12\omega,\ 10 + 9\omega)$$
$$= (10 + 9\omega,\ 3 + 2\omega) = (3 + 2\omega);$$

that is,

$$(91 + 84\omega,\ 34 + 53\omega) = (3 + 2\omega).$$

Thus, $3 + 2\omega$ is a gcd of $91 + 84\omega$ and $34 + 53\omega$.

While the calculations are a bit tedious, you can work the four equations above backwards, as we did in Chapters 1 and 6, to write $3 + 2\omega$ as a linear combination of $91 + 84\omega$ and $34 + 53\omega$. Using a CAS, we found that

Appendix A.6 outlines a package for a CAS that allows you to calculate in $\mathbb{Z}[\omega]$.

$$3 + 2\omega = (5 + 3\omega)(91 + 84\omega) - (9 + 2\omega)(34 + 53\omega) \quad \blacktriangle$$

How to Think About It. We've seen, in $\mathbb{Z}[i]$ or $\mathbb{Z}[\omega]$, that there are sometimes choices for quotients and corresponding remainders in the generalized division algorithms. Hence, there may be more than one way to implement the Euclidean Algorithm and, so, more than one end result. But, thanks to Proposition 6.45 and the fact that $\mathbb{Z}[i]$ and $\mathbb{Z}[\omega]$ are PIDs, any two gcd's are associates. See Exercise 8.7 below for an example.

Exercises

8.1 Prove or Disprove and Salvage if Possible. Two Gaussian integers are associates if and only if they have the same norm.

8.2 How many possible numbers of "closest Gaussian integers" to a complex number are there? For each number, give an example.

8.3 Let z and w be Gaussian integers, and suppose that q and q' are Gaussian integers equidistant from w/z in the complex plane. Show that

$$N(\frac{w}{z} - q) = N(\frac{w}{z} - q').$$

8.4 Let z and w be Gaussian integers, and suppose that q and q' are Gaussian integers equidistant from w/z in the complex plane. Are $w/z - q$ and $w/zz - q'$ associates? If so, prove it; if not, give a counterexample.

8.5 If $z \neq 0$ and w are Eisenstein integers, how many possible numbers of quotients w/z are there in $\mathbb{Z}[\omega]$ satisfying the conditions of the division algorithm? For each number, give an example.

8.6 * Let R be a commutative ring. If a, b, c, and d are elements of R such that $b = da + c$, show that there is equality of ideals $(b, a) = (a, c)$.

8.7 If $z = 6 + 12i$ and $w = -13 + 74i$, show that
$$\frac{w}{z} = \frac{9}{2} + \frac{10}{3}i.$$

(i) Show that there are two q's with $w = qz + r$ in the generalized division algorithm, namely $q = 4 + 3i$ and $q = 5 + 3i$.

(ii) Apply the Euclidean Algorithm to find a gcd of z and w starting with $q = 4 + 3i$.

(iii) Apply the Euclidean Algorithm to find a gcd of z and w starting with $q = 5 + 3i$.

(iv) Are the two gcd's associates in $\mathbb{Z}[i]$?

8.2   Primes Upstairs and Primes Downstairs

We saw in the last chapter that an irreducible polynomial in $k[x]$ (for some field k) may factor in $K[x]$ for some extension field K/k. For example, $x^2 + 1$ is irreducible in $\mathbb{R}[x]$ but it factors in $\mathbb{C}[x]$. A similar phenomenon occurs for primes in $\mathbb{Z}$. Every ring of cyclotomic integers R has $\mathbb{Z}$ as a subring, and a prime $p \in \mathbb{Z}$ may factor in R.

Our goal in this section is to investigate primes in $\mathbb{Z}[i]$ and in $\mathbb{Z}[\omega]$, and the obvious way to begin doing this is by studying primes **downstairs**, that is, in $\mathbb{Z}$, and look at their behavior **upstairs**, that is, in rings of cyclotomic integers.

Corollary 8.8, the Fundamental Theorem for Euclidean domains, tells us that every element in such a domain has an essentially unique factorization into primes.

Lemma 8.10. *Let $R = \mathbb{Z}[\zeta_p]$ for any prime p. If $u \in R$, then u is a unit in R if and only if $N(u) = 1$.*

Proof. If u is a unit, there is $v \in R$ with $uv = 1$. Hence, $1 = N(uv) = N(u)N(v)$. As $N(u)$ and $N(v)$ are positive integers, we must have $N(u) = 1 = N(v)$.

Conversely, suppose that $N(u) = 1$. Since $N(u) = u\overline{u}$, we have u a unit in R (with inverse $\overline{u}$). ∎

The actual factorization of Gaussian integers or of Eisenstein integers into primes can be a tricky task, but here is a useful tool.

Proposition 8.11. *Let $R = \mathbb{Z}[i]$ or $\mathbb{Z}[\omega]$. If $z \in R$ and $N(z)$ is prime in $\mathbb{Z}$, then z is prime in R.*

Proof. We prove the contrapositive of the statement of the Proposition. If $z \in R$ and $z = wv$ for non-units w and v, then Lemma 8.10 gives $N(w) > 1$ and $N(v) > 1$. Hence,
$$N(z) = N(wv) = N(w)\,N(v),$$
and $N(z)$ is not prime in $\mathbb{Z}$. ∎

The converse of Proposition 8.11 is false; we'll soon see that 7 is prime in $\mathbb{Z}[i]$, but that $N(7) = 49$.

Example 8.12. (i) In Example 8.1, we divided $w = -211 + 102i$ by $z = -19 + 48i$ and got quotient $3 + 3i$ and remainder $-10 + 15i$. Let's carry out the rest of the Euclidean Algorithm to get a gcd of z and w.

We use two formats for the Euclidean Algorithm finding the gcd. Here's the southwestern version.

$$w = (3 + 3i)z + (-10 + 15i)$$
$$3 + 3i = (3 - i)(-10 + 15i) + (-4 - 7i)$$
$$-10 + 15i = (-1 - 2i)(-4 - 7i).$$

Hence, $\gcd(z, w) = 4 + 7i$ (we take an associate of $-4 - 7i$).

Here is a more detailed version of this calculation.

$$
\begin{array}{r}
3 + 3i \\
-19{+}48i\overline{)-211{+}102i} \\
-201{+}87i \\
\hline
-10{+}15i
\end{array}
\quad
\begin{array}{r}
3 - i \\
\overline{)-19{+}48i} \\
-15{+}55i \\
\hline
-4 - 7i
\end{array}
\quad
\begin{array}{r}
-1 - 2i \\
\overline{)-10{+}15i} \\
-10{+}15i \\
\hline
0
\end{array}
$$

Again we see that $\gcd(z, w) = 4 + 7i$. So $4 + 7i$ is a factor of both z and w.

(ii) Is the gcd $4 + 7i$ prime? If not, can we factor it explicitly? Since

$$N(4 + 7i) = 65 = 13 \cdot 5,$$

we claim that the norm of any prime factor π of $4 + 7i$ is either 13 or 5. If $4 + 7i = \pi\alpha$, where $\alpha \in \mathbb{Z}[i]$ is not a unit, then $N(\alpha) > 1$, by Lemma 8.10. Hence, as $N(4 + 7i) = N(\pi)N(\alpha)$, we must have $N(\pi) = 13$ or $N(\pi) = 5$. Well, 5 is small enough to do a direct check, and the only Gaussian integers with norm 5 are $2 \pm i$ and their associates. And we're in luck; $2 + i$ is a factor:

$$\frac{4 + 7i}{2 + i} = 3 + 2i.$$

Note that $N(3 + 2i) = 13$, so that $3 + 2i$ is prime, by Proposition 8.11.

(iii) To factor $w = -211 + 102i$, divide by the gcd:

$$\frac{-211 + 102i}{4 + 7i} = -2 + 29i.$$

Since $N(-2 + 29i) = 845 = 5 \cdot 13^2$, the same process as in part (ii) shows that

$$-2 + 29i = (2 + i)(5 + 12i) = (2 + i)(3 + 2i)^2.$$

Putting it all together, we have the prime factorization of w:

$$
\begin{aligned}
w &= -211 + 102i \\
&= (4 + 7i)(-2 + 29i) \\
&= (2 + i)(3 + 2i)(2 + i)(3 + 2i)^2 \\
&= (2 + i)^2(3 + 2i)^3.
\end{aligned}
$$

In Section 3.4, we used the fact that $5 + 12i = (3 + 2i)^2$ when we generated Pythagorean triples with Gaussian integers.

We leave it to you to find the prime factorization of z and to show that $\gcd(z, w) \operatorname{lcm}(z, w) = zw$. ▲

How to Think About It. How do we factor a positive rational integer m into primes? First of all, there is an algorithm determining whether m is prime. Use the Division Algorithm to see whether $2 \mid m$. If $2 \nmid m$, use the Division Algorithm to see whether $3 \mid m$. And so forth. Now if d is a divisor of m, then $d \leq m$, and so there are only finitely many candidates for divisors; hence, this process must stop. Of course, if we have any extra information about m, we may use it to cut down on the number of candidates. We must say that this algorithm is useful only for small numbers m; after all, the difficulty in factoring large numbers is the real reason that public key codes are secure.

A variation of this algorithm can be used to factor nonzero Gaussian integers. If $d, w \in \mathbb{Z}[i]$ and $d \mid w$, then $N(d) \leq N(w)$; hence, there are only finitely many Gaussian integers z which are candidates for being divisors of w. If $N(w)$ is prime, then Proposition 8.11 says that w is prime; if $N(z)$ is composite, we can proceed as in the last part of Example 8.12.

Laws of Decomposition

We now describe the primes in $\mathbb{Z}[i]$ (there will be a similar story for $\mathbb{Z}[\omega]$). The next lemma lets us concentrate on how primes downstairs in $\mathbb{Z}$ behave when they are viewed as elements upstairs in $\mathbb{Z}[i]$.

Notation. It gets tedious to keep saying "let p be a prime in $\mathbb{Z}$." From now on, let's call primes in $\mathbb{Z}$ *rational primes* to distinguish them from primes in other rings. Remember that a *prime* (or irreducible) element in a commutative ring is one whose only divisors are units and associates. We may also say *rational integer* to distinguish an ordinary integer in $\mathbb{Z}$ from a Gaussian integer, an Eisenstein integer or, more generally, a cyclotomic integer.

Lemma 8.13. *Every prime π in $\mathbb{Z}[i]$ divides a rational prime.*

Proof. Every Gaussian integer z divides its norm in $\mathbb{Z}[i]$, for $N(z) = z\overline{z}$. In particular, π divides a rational integer, namely its norm. Now $N(\pi)$ factors into primes in $\mathbb{Z}$:

$$
\pi\overline{\pi} = N(\pi) = p_1 p_2 \ldots p_k,
$$

and so

$$
\pi \mid p_1 p_2 \ldots p_k.
$$

The primes on the right-hand side are elements of $\mathbb{Z}[i]$ as well as of $\mathbb{Z}$.

But π is a prime in $\mathbb{Z}[i]$; hence, by Euclid's Lemma in $\mathbb{Z}[i]$, it divides one of the (prime) factors p_j on the right. ■

Example 8.14. We have seen that $\pi = 3 + 2i$ is a prime in $\mathbb{Z}[i]$; note that π divides 13, for

$$N(\pi) = (3 + 2i)(3 - 2i) = 13.$$

We sometimes say that $3 + 2i$ **lies above** 13. ▲

As we said earlier, there are primes in $\mathbb{Z}$ that remain prime in $\mathbb{Z}[i]$ and others that factor into new primes; the same is true for $\mathbb{Z}[\omega]$. It turns out that there's a beautiful theory, going back to Gauss, for how primes decompose in these rings, a theory that brings together many of the ideas you've studied so far. For example, here are some factorizations of rational primes when viewed as elements in $\mathbb{Z}[i]$:

$$5 = (2 + i)(2 - i); \quad 13 = (3 + 2i)(3 - 2i); \quad 29 = (5 + 2i)(5 - 2i).$$

In each of these cases, the rational prime decomposes as a norm: the product of a Gaussian integer and its conjugate. This is always the case.

Lemma 8.15. *Let p be a rational prime. If p is not prime in $\mathbb{Z}[i]$, then there exists some prime z in $\mathbb{Z}[i]$ with*

$$p = z\bar{z} = N(z).$$

Proof. Suppose $p = zw$, where z and w are non-unit Gaussian integers. Then

$$p^2 = N(p) = N(zw) = N(z)\,N(w),$$

And, in fact, z and w must be associates.

where neither $N(z)$ nor $N(w)$ is 1. But this is an equation in $\mathbb{Z}$, and so unique factorization in $\mathbb{Z}$ gives $p = N(z) = z\bar{z}$. Finally, z is prime in $\mathbb{Z}[i]$, by Proposition 8.11, because $N(z)$ is a rational prime. ■

Lemma 8.15 narrows the situation quite a bit. It says that if a rational prime factors in $\mathbb{Z}[i]$, it factors into exactly two conjugate Gaussian integers, each prime in $\mathbb{Z}[i]$. We say that such a rational prime **splits** in $\mathbb{Z}[i]$. We can state the result of Lemma 8.15 using only the arithmetic of $\mathbb{Z}$. Since $N(a + bi) = a^2 + b^2$, the lemma says that a prime splits if it can be written as the sum of two perfect squares. And the converse is also true.

Proposition 8.16. *A rational prime p splits in $\mathbb{Z}[i]$ if and only if p is a sum of two squares in $\mathbb{Z}$.*

Proof. If p splits in $\mathbb{Z}[i]$, then Lemma 8.15 says that there is a Gaussian integer $z = a + bi$ such that $p = N(z) = a^2 + b^2$.

Conversely, if $p = a^2 + b^2$, then $p = (a + bi)(a - bi)$. But $a + bi$ is prime in $\mathbb{Z}[i]$, by Proposition 8.11, because its norm, $N(a + bi) = p$, is a rational prime. ■

The question of which rational primes split in $\mathbb{Z}[i]$ thus comes down to the question of which primes are sums of two squares. Not every rational prime is a sum of two squares; for example, it's easy to see that 11 is not. Here is a nice (and perhaps surprising) connection to modular arithmetic. A quick example gives the idea. The prime 29 is the sum of two squares:

$$29 = 2^2 + 5^2.$$

As an equation in $\mathbb{F}_{29}$, this says that $2^2 + 5^2 = 0$. Multiply both sides by 6^2, for $6 = 5^{-1}$ in $\mathbb{F}_{29}$:

$$0 = 2^2 6^2 + 5^2 6^2$$
$$= (2 \cdot 6)^2 + (5 \cdot 6)^2$$
$$= (2 \cdot 6)^2 + 1,$$

so that $2 \cdot 6 = 12$ is a root of $x^2 + 1$ in $\mathbb{F}_{29}$. More generally, suppose that p is a prime and $p = a^2 + b^2$. We can assume that $0 < a, b < p$, so that both a and b are units in $\mathbb{F}_p$. We can write this as an equation: $a^2 + b^2 = 0$ in $\mathbb{F}_p$. Multiplying both sides by $(b^{-1})^2$, we get:

$$(ab^{-1})^2 + 1 = 0;$$

that is, ab^{-1} is a root of $x^2 + 1$ in $\mathbb{F}_p$. And the converse is true as well:

> Recall that $\mathbb{F}_p$ is another notation for $\mathbb{Z}_p$, the field of integers modulo p.

Proposition 8.17. *A rational prime p is a sum of two squares if and only if $x^2 + 1$ has a root in $\mathbb{F}_p$.*

Proof. We've just seen that an expression of p as the sum of two squares leads to a root of $x^2 + 1$ in $\mathbb{F}_p$.

> Another way to say this: p is the sum of two squares if and only if -1 is a square in $\mathbb{F}_p$.

Going the other way, suppose there is an integer n whose congruence class satisfies

$$n^2 + 1 = 0 \quad \text{in } \mathbb{F}_p.$$

Then, moving back to $\mathbb{Z}$, we see that n satisfies

$$p \mid (n^2 + 1).$$

Now go upstairs to $\mathbb{Z}[i]$. We have

$$p \mid (n + i)(n - i).$$

But p divides neither $n + i$ nor $n - i$ in $\mathbb{Z}[i]$ (otherwise, $n/p \pm i/p$ would be Gaussian integers). Euclid's Lemma says that p is not prime in $\mathbb{Z}[i]$ and, hence, by Lemma 8.15, p is the norm of some Gaussian integer z—that is, p is a sum of two squares. ∎

Corollary 8.18. *A rational prime p factors in $\mathbb{Z}[i]$ if and only if $x^2 + 1$ has a root in $\mathbb{F}_p$.*

> So, p factors in $\mathbb{Z}[i]$ if and only if $x^2 + 1$ factors in $\mathbb{F}_p[x]$.

Proof. Apply Proposition 8.17 and the Factor Theorem (Corollary 6.15). ∎

Let's summarize these various equivalent statements about a rational prime p.

- p factors in $\mathbb{Z}[i]$.
- $p = N(z)$ for some z in $\mathbb{Z}[i]$.
- $p = a^2 + b^2$ in $\mathbb{Z}$.
- $x^2 + 1$ has a root in $\mathbb{F}_p$.

The last criterion may seem the most remote, but it is actually the easiest to use—you have to check at most $(p-1)/2$ possible solutions to $x^2 + 1 = 0$ (because if α is a solution, so is $-\alpha$). If you try a few numerical cases, a pattern begins to emerge—the primes satisfying the last criterion all seem to be congruent to 1 mod 4. That's quite a beautiful and elegant result, which adds one more equivalent statement to the summarizing list above.

Theorem 8.19. *If p is an odd prime, then $x^2 + 1$ has a root in $\mathbb{F}_p$ if and only if $p \equiv 1 \bmod 4$.*

Proof. Assume that $p = 4k + 1$. Since p is prime, Theorem 4.9 (Fermat's Little Theorem) gives $a^{p-1} = 1$ for all nonzero a in $\mathbb{F}_p$. Thus, we have the factorization in $\mathbb{F}_p[x]$:

See Exercise 6.38 on page 263.

$$x^{p-1} - 1 = (x - 1)(x - 2)(x - 3)\ldots(x - (p-1)). \tag{8.2}$$

Hence,

$$\begin{aligned}
x^{p-1} - 1 &= x^{4k} - 1 \\
&= \left(x^4\right)^k - 1 \\
&= (x^4 - 1)\left[\left(x^4\right)^{k-1} + \left(x^4\right)^{k-2} + \left(x^4\right)^{k-3} + \cdots + 1\right] \\
&\qquad \text{by Exercise 6.47 on page 269} \\
&= (x^2 + 1)\left[(x^2 - 1)\left(x^4\right)^{k-1} + \left(x^4\right)^{k-2} + \left(x^4\right)^{k-3} + \cdots + 1\right] \\
&= (x^2 + 1)h(x). \tag{8.3}
\end{aligned}$$

Comparing Eqs. (8.2) and (8.3), the two factorizations of $x^{p-1} - 1$, and using unique factorization in $\mathbb{F}_p[x]$, we see that $x^2 + 1 = (x - \alpha)(x - \beta)$ for some $\alpha, \beta \in \mathbb{F}_p$.

Conversely, if p is odd and $p \not\equiv 1 \bmod 4$, then $p \equiv 3 \bmod 4$ (it can't be congruent to 0 or 2). But, by Proposition 8.17, if $x^2 + 1$ has a root in $\mathbb{F}_p$, then p is the sum of two squares in $\mathbb{Z}$. However, the sum of two squares in $\mathbb{Z}$ is never congruent to 3 mod 4: If $a = 0, 1, 2, 3$, then $a^2 \equiv 0, 1, 0, 1 \bmod 4$, and so $a^2 + b^2 \equiv 0, 1, 2 \bmod 4$. ■

Proposition 8.16, when combined with Theorem 8.19, gives us a nice fact of arithmetic, first established by Gauss.

The name of a theorem may not coincide with the name of the first person who proved it.

Corollary 8.20 (Fermat's Two-Square Theorem). *An odd rational prime p is a sum of two squares if and only if $p \equiv 1 \bmod 4$.*

Theorem 8.19 tells the whole story for odd primes: primes that are congruent to 1 mod 4 split into two conjugate factors, and primes that are congruent to 3 mod 4 stay prime (we call primes downstairs that stay prime upstairs ***inert***). There is one prime we haven't yet considered: $p = 2$. Now 2 factors in $\mathbb{Z}[i]$, because $x^2 + 1$ factors in $\mathbb{Z}_2[x]$. In fact

$$2 = (1 + i)(1 - i).$$

Are there other primes in $\mathbb{Z}[i]$ that are associate to their conjugates? That question is Exercise 8.10 on page 343.

But note that these two factors are associates:

$$1 + i = i(1 - i),$$

and so

$$2 = i(1 - i)^2.$$

Thus, 2 splits in a special way: it is associate to the *square* of a prime. We say that 2 ***ramifies*** in $\mathbb{Z}[i]$. Hence, our discussion gives a complete classification of how rational primes decompose in the Gaussian integers.

Theorem 8.21 (Law of Decomposition in Gaussian Integers). *Every rational prime p decomposes in $\mathbb{Z}[i]$ in one of three ways.*

(1) *p splits into two conjugate prime factors if $p \equiv 1$ mod 4*

(2) *p is inert if $p \equiv 3$ mod 4*

(3) *$p = 2$ ramifies: $2 = i(1-i)^2$.*

Corollary 8.22 (Classification of Gaussian Primes). *The primes π in $\mathbb{Z}[i]$ are of three types:*

(1) *$\pi = a + bi$, which lies above a rational prime p with $p \equiv 1$ mod 4; in this case, $N(\pi) = a^2 + b^2$*

(2) *$\pi = p$, where p is a rational prime with $p \equiv 3$ mod 4; in this case, $N(p) = p^2$*

(3) *$\pi = 1 - i$ and its associates; in this case, $N(1-i) = 2$.*

Proof. Now π divides some rational prime p, by Lemma 8.13. If $p \equiv 1$ mod 4, then π is of the first type; if $p \equiv 3$ mod 4, then p is inert and $\pi = p$; if $p \equiv 2$ mod 4, then p ramifies and $\pi = 1 - i$. ∎

As we mentioned earlier, 7 is a prime in $\mathbb{Z}[i]$, for $7 \equiv 3$ mod 4, but its norm 49 is not a rational prime. Thus, the converse of Proposition 8.11 is false.

How to Think About It. The fact that $2 = i(1-i)^2$ can be stated in terms of ideals in $\mathbb{Z}[i]$: there is equality of ideals

$$(2) = \left((1-i)^2\right).$$

In fact, if we use the definition of the product of ideals from Exercise 5.51 on page 220, the above equation of ideals can be written as

$$(2) = (1-i)(1-i) = (1-i)^2.$$

Exercises

8.8 (i) In Example 8.12 we found a gcd of $z = -19 + 48i$ and $w = -211 + 102i$ to be $4 + 7i$. Write $4 + 7i$ as a linear combination of z and w.

  (ii) Use part (i) to find the prime factorization of z.

  (iii) Show that $\gcd(z, w) \operatorname{lcm}(z, w) = zw$.

8.9 Show that if two Gaussian integers z and w have relatively prime norms in $\mathbb{Z}$, then z and w are relatively prime in $\mathbb{Z}[i]$. Is the converse true?

8.10 * Which primes in $\mathbb{Z}[i]$ are associate to their conjugates?

8.11 How many non-associate primes in $\mathbb{Z}[i]$ lie above 5?

8.12 In $\mathbb{Z}[i]$, show that every associate of $a + bi$ is conjugate to an associate of $b + ai$.

8.13 Show that every Gaussian integer is associate to one in the first quadrant of the complex plane. (We define the first quadrant to include the nonnegative x-axis but not the positive y-axis.)

8.14 Show that if two integers a and b can each be written as the sum of two squares, so can ab.

8.15 Factor each of these into primes in $\mathbb{Z}[i]$.

(i) 101   (ii) 31   (iii) 37

(iv) $7 + 4i$ (v) $8 + i$ (vi) 65

(vii) $7 + 3i$ (viii) $40 + 42i$ (ix) $154 + 414i$

8.16 Find the number of elements in each of the quotient rings.

(i) $\mathbb{Z}[i]/(1 - i)$ (ii) $\mathbb{Z}[i]/(2 + i)$

(iii) $\mathbb{Z}[i]/(3 + 2i)$ (iv) $\mathbb{Z}[i]/(5 + 12i)$

8.17 Take It Further. If z is a Gaussian integer, show that

$$|\mathbb{Z}[i]/(z)| = N(z).$$

Eisenstein Primes

The whole theory just given for $\mathbb{Z}[i]$ carries over to $\mathbb{Z}[\omega]$. Of course, the statements have to be modified slightly, but the proofs are almost identical to the corresponding results in $\mathbb{Z}[i]$. If you think about it, this shouldn't be a surprise: a proof using only algebraic properties of norm (for example, it is multiplicative) and properties of PIDs (unique factorization and Euclid's Lemma) should carry over mutatis mutandis.

We summarize the results for Eisenstein integers, providing sketches of proofs where we think it's necessary, but we leave the details to you. And these are important exercises, because they will help you digest the ideas in both rings.

Lemma 8.23. *Every prime in $\mathbb{Z}[\omega]$ divides a rational prime.*

Proof. Mimic the proof of Lemma 8.13. ∎

How about a law of decomposition for Eisenstein integers? Some rational primes factor in $\mathbb{Z}[\omega]$; for example:

> Try some other primes in $\mathbb{Z}[\omega]$. Any conjectures about which ones split?

$$7 = (3 + \omega)(3 + \omega^2); \quad 31 = (5 - \omega)(5 - \omega^2); \quad 97 = (3 - 8\omega)(3 - 8\omega^2).$$

In each of these cases, the prime in $\mathbb{Z}$ decomposes in $\mathbb{Z}[\omega]$ into a norm: the product of an Eisenstein integer and its conjugate. This is always the case.

Lemma 8.24. *Let p be a rational prime. If p is not prime in $\mathbb{Z}[\omega]$, then $p = z\bar{z} = N(z)$ for some prime z in $\mathbb{Z}[\omega]$.*

Proof. Mimic the proof of Lemma 8.15. ∎

As happened in $\mathbb{Z}[i]$, we can restate the result of Lemma 8.24 completely in terms of the arithmetic of $\mathbb{Z}$. Since $N(a + b\omega) = a^2 - ab + b^2$, the lemma says that a prime splits if it can be written in this form. The converse is also true.

Proposition 8.25. *A rational prime p splits in $\mathbb{Z}[\omega]$ if and only if p can be expressed as $a^2 - ab + b^2$ for integers $a, b \in \mathbb{Z}$.*

Proof. Mimic the proof of Lemma 8.16. ∎

Proposition 8.17 says that a rational prime p is a norm of some prime in $\mathbb{Z}[i]$ if and only if $x^2 + 1$ has its roots in $\mathbb{F}_p$. What might be an analog of this result for $\mathbb{Z}[\omega]$? Well, $x^2 + 1$ is the minimal polynomial for i; the minimal polynomial for ω is $x^2 + x + 1$. A careful look at the proof of Proposition 8.17 shows that we can use $x^2 + x + 1$ and modify the proof slightly to obtain another lovely result.

Proposition 8.26. *A rational prime p can be expressed as $a^2 - ab + b^2$ for integers $a, b \in \mathbb{Z}$ if and only if $x^2 + x + 1$ has a root in $\mathbb{F}_p$.*

What is the discriminant of $x^2 + x + 1$?

Proof. Suppose that $p = a^2 - ab + b^2$ for integers a and b. Then p doesn't divide either a or b (otherwise $p^2 \mid (a^2 - ab + b^2)$) and, hence, b is a unit in $\mathbb{F}_p$. Multiply both sides by $\left(b^{-1}\right)^2$ to obtain

$$p \left(b^{-1}\right)^2 = \left(-ab^{-1}\right)^2 + \left(-ab^{-1}\right) + 1,$$

so that $-a\left(b^{-1}\right)$ is a root of $x^2 + x + 1$ in $\mathbb{F}_p$.

Going the other way, suppose that there is a congruence class $[m] \in \mathbb{F}_p$ with

$$[m]^2 + [m] + 1 = 0 \quad \text{in } \mathbb{F}_p.$$

Then, in $\mathbb{Z}$, we have

$$m^2 + m + 1 \equiv 0 \bmod p;$$

that is,

$$p \mid (m^2 + m + 1).$$

Now move up to $\mathbb{Z}[\omega]$. We have

$$p \mid (m - \omega)(m - \omega^2),$$

or

$$p \mid (m - \omega)(m + 1 + \omega).$$

Check that $(m - \omega)(m - \omega^2) = m^2 + m + 1$.

But p doesn't divide either $m - \omega$ or $m + 1 + \omega$ in $\mathbb{Z}[\omega]$ (otherwise $\frac{m}{p} - \frac{1}{p}\omega$ or $\frac{m+1}{p} + \frac{1}{p}\omega$ would be an Eisenstein integer). Thus, by Euclid's Lemma, p is not prime in $\mathbb{Z}[\omega]$ and hence by Lemma 8.24, p is the norm of an Eisenstein integer $a + b\omega$; that is, $p = a^2 - ab + b^2$ for integers a and b. ∎

We get the next corollary.

Corollary 8.27. *A rational prime p factors in $\mathbb{Z}[\omega]$ if and only if $x^2 + x + 1$ has a root in $\mathbb{F}_p$.*

So, p factors in $\mathbb{Z}[\omega]$ if and only if $x^2 + x + 1$ factors in $\mathbb{F}_p[x]$.

Proof. Apply Proposition 8.26 and the Factor Theorem (Corollary 6.15). ∎

We summarize the chain of equivalent statements.

- p factors in $\mathbb{Z}[\omega]$.
- $p = N(z)$ for some z in $\mathbb{Z}[\omega]$.
- $p = a^2 - ab + b^2$ in $\mathbb{Z}$.
- $x^2 + x + 1$ has a root in $\mathbb{F}_p$.

Onward to a law of decomposition in $\mathbb{Z}[\omega]$. Numerical experiments (we hope you'll try some) suggest that if p is a rational prime and $p \equiv 1 \bmod 3$, then $x^2 + x + 1$ has a root in $\mathbb{F}_p$. The proof of Theorem 8.19 suggests a reason why.

Proposition 8.28. *If p is a prime and $p \equiv 1 \bmod 3$, then $x^2 + x + 1$ has a root in $\mathbb{F}_p$.*

Proof. Suppose that $p = 3k + 1$. Because p is prime, $a^{p-1} = 1$ for all non-zero a in $\mathbb{F}_p$ (Fermat's Little Theorem—Theorem 4.9). Hence, in $\mathbb{F}_p[x]$, we have the factorization

See Exercise 6.38 on page 263.

$$x^{p-1} - 1 = (x - 1)(x - 2)(x - 3) \cdots (x - (p - 1)). \tag{8.4}$$

But

$$
\begin{aligned}
x^{p-1} - 1 &= x^{3k} - 1 \\
&= \left(x^3\right)^k - 1 \\
&= (x^3 - 1)\left[\left(x^3\right)^{k-1} + \left(x^3\right)^{k-2} + \left(x^3\right)^{k-3} + \cdots + 1\right]
\end{aligned}
$$

Exercise 6.47 on page 269

$$
\begin{aligned}
&= (x^2 + x + 1)\left[(x - 1)\left(x^3\right)^{k-1} + \left(x^3\right)^{k-2} + \left(x^3\right)^{k-3} + \cdots + 1\right] \\
&= (x^2 + x + 1)h(x). \tag{8.5}
\end{aligned}
$$

Comparing Eqs. (8.4) and (8.5), the two factorizations of $x^{p-1} - 1$, and using unique factorization in $\mathbb{F}_p[x]$, we see that $x^2 + x + 1 = (x - \alpha)(x - \beta)$ for some $\alpha, \beta \in \mathbb{F}_p$; that is, $x^2 + x + 1$ has a root in $\mathbb{F}_p$. ∎

What about primes that are not congruent to 1 mod 3? One case is easily handled. Suppose that $p \equiv 2 \bmod 3$. By Proposition 8.26, $x^2 + x + 1$ has a root in $\mathbb{F}_p$ if and only if p can be written as $a^2 - ab + b^2$ for $a, b \in \mathbb{Z}$. But you can check that, for any choice of a and b, $a^2 - ab + b^2$ is never congruent to 2 mod 3 (just look at the possible congruence classes of a and b mod 3). Thus, p is inert; that is, p is prime in $\mathbb{Z}[\omega]$. There is only one more prime, namely 3, the prime congruent to 0 mod 3. And $x^2 + x + 1$ certainly has roots in $\mathbb{F}_3$, namely 0 and 1. Therefore, 3 must split; in fact,

$$3 = (2 + \omega)(2 + \omega^2) = (2 + \omega)(1 - \omega).$$

Recall that the units in $\mathbb{Z}[\omega]$ are ± 1, $\pm \omega$, and $\pm \omega^2$.

Are there other primes in $\mathbb{Z}[\omega]$ that are associate to their conjugates? That question is Exercise 8.18 on page 349.

In terms of ideals, $(3) = ((1 - \omega)^2) = (1 - \omega)^2$.

But the important thing is that the two factors on the right are associates. You can check that

$$-\omega^2(1 - \omega) = 2 + \omega.$$

So, our factorization of 3 can be written as

$$3 = -\omega^2(1 - \omega)^2,$$

and 3 is a ramified prime. Putting it all together, we have the law of decomposition in $\mathbb{Z}[\omega]$ as well as a description of all Eisenstein primes.

Theorem 8.29 (Law of Decomposition in Eisenstein Integers). *Every rational prime p decomposes in $\mathbb{Z}[\omega]$ in one of three ways.*

(1) *p splits into two conjugate prime factors if $p \equiv 1 \bmod 3$*

(2) *p is inert if $p \equiv 2 \bmod 3$*

(3) *3 ramifies into $-\omega^2(1 - \omega)^2$.*

Corollary 8.30 (Classification of the Eisenstein primes). *The primes* π *in* $\mathbb{Z}[\omega]$ *are of three types*:

(1) $\pi = a + b\omega$ *which lies above a rational prime congruent to* 1 mod 3; *in this case,* $N(\pi) = a^2 - ab + b^2$.

(2) *primes* p *in* $\mathbb{Z}$ *that are congruent to* 2 mod 3; *in this case,* $N(p) = p^2$.

(3) *the prime* $1 - \omega$ *and its associates; in this case,* $N(1 - \omega) = 3$.

The relation between the factorization of a prime p in $\mathbb{Z}[\omega]$ and the factorization of $x^2 + x + 1$ in $\mathbb{F}_p[x]$ can be used to factor Eisenstein integers.

Example 8.31. (i) Let $p = 31$. There are two roots of $x^2 + x + 1$ in $\mathbb{F}_{31}$, namely 5 and 25, and so $x^2 + x + 1 = (x - 5)(x - 25)$ in $\mathbb{F}_{31}[x]$. Lift this equation to $\mathbb{Z}[x]$:

$$x^2 + x + 1 = (x - 5)(x - 25) + 31(x - 4).$$

So, letting $x = \omega$, we have

$$(5 - \omega)(25 - \omega) = 31(4 - \omega).$$

Now $N(5 - \omega) = 31$, so $5 - \omega$ is a prime factor of 31, and the other is

$$\frac{25 - \omega}{4 - \omega} = 6 + \omega.$$

This example connects to Exercise 8.24 on page 349.

(ii) Let $p = 97$. There are two roots of $x^2 + x + 1$ in $\mathbb{F}_{97}$, namely 61 and 35. In fact,

$$x^2 + x + 1 = (x - 61)(x - 35) + 97(x - 22).$$

Letting $x = \omega$,

$$(61 - \omega)(35 - \omega) = 97(22 - \omega),$$

and so

$$\frac{(61 - \omega)(35 - \omega)}{22 - \omega} = 97.$$

Now, $N(22 - \omega) = 507 = 3 \cdot 13^2$; since $N(1 - \omega) = 3$ and $N(4 + \omega) = 13$, so (checking for unit factors), we have

$$22 - \omega = (1 - \omega)(4 + \omega)^2.$$

Some of these factors divide $61 - \omega$; the rest divide $35 - \omega$. We have

$$N(61 - \omega) = 3 \cdot 13 \cdot 97$$
$$N(35 - \omega) = 13 \cdot 97.$$

We can cancel the factor of 13 by dividing by $4 + \omega$; it's easier to work with $35 - \omega$:

$$\frac{35 - \omega}{4 + \omega} = \frac{(35 - \omega)(4 + \omega^2)}{13} = 8 - 3\omega.$$

Bingo: $N(8 - 3\omega) = 97$, so that

$$97 = (8 - 3\omega)(8 - 3\omega^2). \quad \blacktriangle$$

How to Think About It. Because $\mathbb{Z}[i]$ and $\mathbb{Z}[\omega]$ are commutative rings, we can construct quotient rings. And, since both rings are PIDs, they often look very similar to rings we have already met.

The result of this example will be useful in the next section.

Example 8.32. We investigate the quotient ring $R = \mathbb{Z}[\omega]/(\lambda)$, where

$$\lambda = 1 - \omega$$

is the prime lying over the rational (and ramified) prime 3. For any Eisenstein integer z, let's look at the remainder after dividing z by λ. Proposition 8.4 gives Eisenstein integers q and r such that

$$z = q\lambda + r, \quad \text{with } r = 0 \text{ or } N(r) < N(\lambda).$$

Now $N(\lambda) = 3$, so that $N(r)$ must be 0, 1, or 2. There are no Eisenstein integers of norm 2, because 2 is inert in $\mathbb{Z}[\omega]$. Hence $N(r)$ is 0 or 1. If $N(r) = 0$, then $r = 0$; if $N(r) = 1$, then r is a unit in $\mathbb{Z}[\omega]$. So, aside from 0, we need only investigate the six Eisenstein units. It turns out that each of these is congruent to 1 or -1 modulo λ:

If $r = 1$,   then $z = q\lambda + 1$ and $z \equiv 1 \bmod \lambda$.

If $r = -1$,   then $z = q\lambda - 1$ and $z \equiv -1 \bmod \lambda$.

If $r = \omega$,   then $z = q\lambda + \omega = (q-1)\lambda + 1$ and $z \equiv 1 \bmod \lambda$.

If $r = -\omega$,   then $z = q\lambda - \omega = (q+1)\lambda - 1$ and $z \equiv -1 \bmod \lambda$.

If $r = \omega^2$,   then $z = q\lambda + \omega^2 = (q-1-\omega)\lambda + 1$ and $z \equiv 1 \bmod \lambda$.

If $r = -\omega^2$, then $z = q\lambda - \omega^2 = (q+1+\omega)\lambda - 1$ and $z \equiv -1 \bmod \lambda$. ▲

So, every element of $\mathbb{Z}[\omega]$ is congruent mod λ to one of 0, 1, or -1. This suggests that $\mathbb{Z}[\omega]/(\lambda)$ is none other than our friend $\mathbb{F}_3$. And, in fact that's true.

Proposition 8.33. *If $\lambda = 1 - \omega$, then*

$$\mathbb{Z}[\omega]/(\lambda) \cong \mathbb{F}_3.$$

Proof. By Proposition 7.13, the quotient ring $\mathbb{Z}[\omega]/(\lambda)$ is a field, while Example 8.32 shows that the field has exactly 3 elements. Therefore, $\mathbb{Z}[\omega]/(\lambda) \cong \mathbb{F}_3$, for Corollary 7.40 says that two finite fields with the same number of elements are isomorphic. ■

The results in this section just scratch the surface, for life is more complicated; there are rings of cyclotomic integers that are not PIDs. We shall have more to say about this when we discuss the work of Kummer.

Further Results. The laws of decomposition for $\mathbb{Z}[i]$ (Theorem 8.21) and $\mathbb{Z}[\omega]$ (Theorem 8.29) show that the decomposition of a rational prime depends only on its congruence class modulo a fixed integer: 4 for $\mathbb{Z}[i]$ and 3 for $\mathbb{Z}[\omega]$. This theory was greatly generalized in the twentieth century to **Class Field Theory**, which determines laws of decomposition of primes in rings of cyclotomic integers, thereby bringing together under one roof many of the main ideas in modern algebra.

Exercises

8.18 * Which primes in $\mathbb{Z}[\omega]$ are associate to their conjugates?

8.19 For which primes p is $x^2 + x + 1$ a perfect square in $\mathbb{F}_p[x]$?

8.20 Working in $\mathbb{Z}[\omega]$, under what conditions are $a + b\omega$ and $b + a\omega$ associates?

8.21 In $\mathbb{Z}[\omega]$,

   (i)   What are all the associates of the prime $1 - \omega$?

   (ii) Show that $1 - \omega$ and $1 - \omega^2$ are associates.

   (iii) Write $(1 - \omega)(1 - \omega^2)$ as $a + b\omega$.

   (iv) What is the minimal polynomial of $1 - \omega$?

> Note that $x^2 + x + 1 = (x - \omega)(x - \omega^2)$. What happens if you put $x = 1$?

8.22 If z, w, v are elements of $\mathbb{Z}[\omega]$ (or $\mathbb{Z}[i]$), show that

   (i)   If $z \mid w$, then $\overline{z} \mid \overline{w}$.

   (ii) If $z \mid w$, then $N(z) \mid N(w)$ in $\mathbb{Z}$.

   (iii) If $z \equiv w \bmod v$ then $\overline{z} \equiv \overline{w} \bmod \overline{v}$.

8.23 Show that a rational prime p splits in $\mathbb{Z}[\omega]$ if and only if -3 is a square mod p.

8.24 Show there are isomorphisms of commutative rings,

   (i) $\mathbb{Z}[i] \cong \mathbb{Z}[x]/(x^2 + 1)$.

   (ii) $\mathbb{Z}[\omega] \cong \mathbb{Z}[x]/(x^2 + x + 1)$.

8.25 * Find all units u in $\mathbb{Z}[\omega]$ such that $u \equiv 1 \bmod 3$.

> Note that 3 is a unit times $(1 - \omega)^2$.

8.26 Factor into primes in $\mathbb{Z}[\omega]$.

   (i)   301   (ii)   307   (iii) $5 + 8\omega$

   (iv) $5 + \omega$ (v) $19 + 18\omega$ (vi) $39 + 55\omega$

   (vii) $61 - \omega$ (viii) $62 + 149\omega$ (ix) $87 - 62\omega$

8.27 Find the number of elements in

   (i) $\mathbb{Z}[\omega]/(2 + \omega)$ (ii) $\mathbb{Z}[\omega]/(4 - \omega)$

   (iii) $\mathbb{Z}[\omega]/(6 + \omega)$ (iv) $\mathbb{Z}[\omega]/(31)$

8.28 Take It Further. If z is an Eisenstein integer, show that

$$|\mathbb{Z}[\omega]/(z)| = N(z).$$

8.3   Fermat's Last Theorem for Exponent 3

The goal of this section is to prove Fermat's Last Theorem for exponent 3: there are no positive integers x, y, z satisfying $x^3 + y^3 = z^3$. The earliest proof is attributed to Euler [12] in 1770 (his proof has a gap that was eventually closed). We develop a different proof in this section that is a nice application of the arithmetic in $\mathbb{Z}[\omega]$.

How to Think About It.   The development of the proof is quite technical (we've polished it as much as we were able), but the essential idea is straightforward and has already been mentioned several times. It's based on the factorization of $x^3 + y^3$ in $\mathbb{Z}[\omega]$ (see Exercise 3.50 on page 115):

$$x^3 + y^3 = (x + y)(x + y\omega)(x + y\omega^2).$$

If there are positive integers x, y, z with $x^3 + y^3 = z^3$, then we'd have a factorization of z^3 in $\mathbb{Z}[\omega]$:

$$z^3 = (x + y)(x + y\omega)(x + y\omega^2).$$

The primes dividing z all show up with exponent at least 3 in z^3, and the idea is to show that this can't happen on the right-hand side. Heuristically, if the three factors on the right are relatively prime and none is divisible by the square of a prime, we're done. But it's not so easy, mainly because of some mischief caused by $\lambda = 1 - \omega$, the prime lying above 3. So, pull out your pencil again and follow along.

Preliminaries

Our development will often make use of a fact about $\mathbb{Z}[\omega]$ adapted from Corollary 6.37.

Proposition 8.34. *If x and y are rational integers that are relatively prime in $\mathbb{Z}$, then they are relatively prime in $\mathbb{Z}[\omega]$.*

See Exercise 8.22 on page 349.

Proof. If π is a prime in $\mathbb{Z}[\omega]$ dividing both x and y, then $N(\pi) \mid N(x)$ and $N(\pi) \mid N(y)$. That is, $N(\pi) \mid x^2$ and $N(\pi) \mid y^2$ (for both x and y lie in $\mathbb{Z}$). Now $N(\pi) \in \mathbb{Z}$; hence, if p is a prime factor of $N(\pi)$, then p is a common factor of x and y, a contradiction. ∎

The prime $\lambda = 1 - \omega$ will figure prominently in the story. In Theorem 8.29, we saw that $\lambda \mid 3$ and, in fact,

$$3 = -\omega^2 \lambda^2. \tag{8.6}$$

That λ lies above 3 implies that a rational integer divisible by λ in $\mathbb{Z}[\omega]$ is divisible by 3 in $\mathbb{Z}$. The next lemma explains the ubiquity of λ in the forthcoming proofs.

Lemma 8.35. *If x is a rational integer, then $3 \mid x$ in $\mathbb{Z}$ if and only if $\lambda \mid x$ in $\mathbb{Z}[\omega]$.*

Proof. If $3 \mid x$ in $\mathbb{Z}$, then Eq. (8.6) shows that $\lambda \mid 3$ in $\mathbb{Z}[\omega]$; hence, $\lambda \mid x$ in $\mathbb{Z}[\omega]$.

Conversely, if $\lambda \mid x$, then Exercise 8.22 on page 349 shows that $N(\lambda) \mid N(x)$ in $\mathbb{Z}$. But $N(\lambda) = 3$ and $N(x) = x^2$, so that $3 \mid x^2$ in $\mathbb{Z}$ and hence Euclid's Lemma gives $3 \mid x$ in $\mathbb{Z}$. ∎

In Example 8.32, we saw that every element in $\mathbb{Z}[\omega]$ is congruent mod λ to 0, 1, or -1. We'll often need to know "how congruent" an Eisenstein integer α is to one of these; that is, whether α is divisible by a power of λ. We introduce notation to capture this idea.

Definition. Define a function $\nu \colon \mathbb{Z}[\omega] - \{0\} \to \mathbb{N}$ as follows: if $z \in \mathbb{Z}[\omega]$ is nonzero and $n \geq 0$ is the largest integer with $\lambda^n \mid z$, then $\nu(z) = n$. We call ν the **valuation**.

Thus, $\nu(z)$ is the exponent of the highest power of λ dividing z:

$$\nu(z) = n \text{ if and only if } z = \lambda^n z' \text{ and } \lambda \nmid z'.$$

Some treatments define $\nu(0)$ to be ∞, but we won't do that here. Also, a valuation can be defined in an analogous way for any prime q in a UFD; just replace λ by q.

Put another way, $\lambda^{\nu(z)} \mid z$ and $\lambda^{\nu(z)+1} \nmid z$.

For example, $\nu(\lambda) = 1$, $\nu(\omega) = 0$, and $\nu(3) = 2$; in Example 8.31, we saw that $\nu(61 - \omega) = 1$. Indeed, $\nu(u) = 0$ for every unit u.

The valuation ν enjoys some properties that come from the properties of exponentiation. The next proposition reminds us of Exercise 2.15 on page 59.

Proposition 8.36. *If z, w are nonzero elements of $\mathbb{Z}[\omega]$, then*

(i) $\nu(zw) = \nu(z) + \nu(w)$.

(ii) *If n is a nonnegative integer, then $\nu(z^n) = n\,\nu(z)$.*

(iii) $\nu(z \pm w) \geq \min\{\nu(z), \nu(w)\}$ *and if $\nu(z) \neq \nu(w)$ then*

$$\nu(z \pm w) = \min\{\nu(z), \nu(w)\}.$$

Proof. This is Exercise 8.30 on page 358. ∎

How to Think About It. Most proofs of the theorem for exponent 3 are broken into two parts: the *first case* in which 3 doesn't divide x, y, or z, and the *second case* in which 3 does divide one of them. We'll follow this program and treat the two cases in turn. There are many proofs in the literature; our proof of the first case is not the easiest (see Exercise 8.31 on page 358 for a fairly simple alternative approach), but we choose it because it generalizes to a proof of the first case for any odd prime exponent p when $\mathbb{Z}[\zeta_p]$ has unique factorization (see Chapter 1 of [36] for the details). Our proof of the second case is based on the development in Chapter 17 of [17].

The First Case

The main result of this section is that there are no positive integers x, y, z with $\gcd(x, y) = 1$ and $3 \nmid xyz$ such that

$$x^3 + y^3 = z^3.$$

Assuming x and y are relatively prime is no loss in generality: a prime factor of x and y is a prime factor q of z, both sides can be divided by q^3, preserving the relationship; hence, infinite descent would apply. The proof will be by contradiction, and it will depend on the following lemma.

Lemma 8.37. *If x, y, and z are integers such that $3 \nmid xyz$, $\gcd(x, y) = 1$, and $x^3 + y^3 = z^3$, then the Eisenstein integers*

$$x + y, \quad x + \omega y, \quad x + \omega^2 y$$

are pairwise relatively prime in $\mathbb{Z}[\omega]$.

As we said on page 349, this relatively prime condition leads fairly directly to the desired proof.

Proof. Suppose that π is a prime in $\mathbb{Z}[\omega]$ that divides two of the three integers, say

$$\pi \mid x + \omega^i y \quad \text{and} \quad \pi \mid x + \omega^j y,$$

where $0 \leq i < j \leq 2$. Then π divides the difference

$$(x + \omega^i y) - (x + \omega^j y) = \omega^i y \left(1 - \omega^{j-i}\right)$$

But, by Exercise 8.21(ii) on page 349, $1 - \omega^{j-i}$ is an associate of $\lambda = 1 - \omega$, so that π divides

$$uy \left(1 - \omega\right),$$

where u is some unit in $\mathbb{Z}[\omega]$. Hence, by Euclid's Lemma, $\pi \mid y$ or $\pi = 1 - \omega = \lambda$. Similarly,

$$\pi \mid \omega^j \left(x + \omega^i y\right) - \omega^i \left(x + \omega^j y\right) = \omega^i x \left(\omega^{j-i} - 1\right).$$

And, because $\omega^{j-i} - 1$ and $1 - \omega$ are associates, we have $\pi \mid x$ or $\pi = 1 - \omega = \lambda$. Hence, if $\pi \neq \lambda$, then $\pi \mid x$ and $\pi \mid y$. This implies that x and y have a common factor in $\mathbb{Z}[\omega]$; thus, by Proposition 8.34, they have a common factor in $\mathbb{Z}$, contradicting the assumption that x and y are relatively prime. So $\pi = \lambda$.

Because $x + y - (x + \omega^i y) = y(1 - \omega^i) \equiv 0 \bmod \lambda$,

$$x + y \equiv x + \omega^i y \bmod \lambda.$$

We are assuming that $x + \omega^i y \equiv 0 \bmod \lambda$; thus, Lemma 8.35 implies that $x + y \equiv 0 \bmod 3$ in $\mathbb{Z}$. But then,

$$\begin{aligned}
z^3 &= x^3 + y^3 \\
&\equiv x + y \bmod 3 \quad \text{(Fermat's Little Theorem)} \\
&\equiv 0 \bmod 3;
\end{aligned}$$

that is, $3 \mid z$, which contradicts the hypothesis $3 \nmid xyz$. ∎

The hard work is done.

Proposition 8.38 (First Case for Exponent 3). *There are no positive integers* x, y, z *with* $\gcd(x, y) = 1$ *and* $3 \nmid xyz$ *such that*

$$x^3 + y^3 = z^3. \tag{8.7}$$

Proof. Suppose, on the contrary, that we have positive integers x, y, z as in the statement. Factoring the left-hand side of Eq. (8.7), we have

$$(x + y)(x + \omega y)(x + \omega^2 y) = z^3.$$

Lemma 8.37 guarantees that the three factors on the left-hand side are relatively prime. Hence, by unique factorization in $\mathbb{Z}[\omega]$, each is a unit times a cube in that ring (if β is a prime divisor of any factor, say $\beta \mid x + \omega y$, then $\beta \mid z^3$ and, by Euclid's Lemma, $\beta \mid z$. Hence, $\beta^3 \mid (x + \omega y)$: there exists an Eisenstein integer s such that

$$x + \omega y = \pm \omega^i s^3,$$

where $\pm \omega^i$ is one of the six units in $\mathbb{Z}[\omega]$ (i is 0, 1, or 2). We want to look at the equation mod 3. Suppose that $s = a + b\omega$ with $a, b \in \mathbb{Z}$. Then

$$\begin{aligned}
s^3 &= a^3 + 3a^2 b\omega + 3ab^2 \omega + b^3 \omega^3 \\
&= a^3 + 3a^2 b\omega + 3ab^2 \omega + b^3 \\
&\equiv a^3 + b^3 \bmod 3;
\end{aligned}$$

hence,

$$x + \omega y \equiv \pm \omega^i n \text{ mod } 3, \qquad (8.8)$$

Recall that $\alpha \equiv \beta$ mod 3 means that there is $\delta \in \mathbb{Z}[\omega]$ with $3\delta = \alpha - \beta$.

where n is a rational integer.

It follows (see Exercise 8.22 on page 349) that

$$x + \overline{\omega} y \equiv \pm \overline{\omega}^i n \text{ mod } 3.$$

But $\overline{\omega} = \omega^{-1}$, so that

$$x + \omega^{-1} y \equiv \pm \omega^{-i} n \text{ mod } 3. \qquad (8.9)$$

Eqs. (8.8) and (8.9) can be rewritten as:

$$\omega^{-i} (x + \omega y) \equiv \pm n \text{ mod } 3$$
$$\omega^i (x + \omega^{-1} y) \equiv \pm n \text{ mod } 3;$$

hence,

$$\omega^{-i} (x + \omega y) \equiv \omega^i (x + \omega^{-1} y) \text{ mod } 3.$$

Multiplying by ω^i gives

$$x + \omega y \equiv \omega^{2i} (x + \omega^{-1} y) \text{ mod } 3,$$

and so

$$x + \omega y - \omega^{2i} x - \omega^{2i-1} y \equiv 0 \text{ mod } 3. \qquad (8.10)$$

We claim, for each possible value of i, namely 0, 1, or 2, that Eq. (8.10) leads to a contradiction.

(i) $i = 0$:   Eq. (8.10) becomes

$$x + \omega y - x - \omega^{-1} y \equiv 0 \text{ mod } 3;$$

that is,

$$\left(\omega - \frac{1}{\omega} \right) y \equiv 0 \text{ mod } 3.$$

Multiplying both sides by ω^2 gives

$$(1 - \omega) y \equiv 0 \text{ mod } 3;$$

that is, there is some $\alpha \in \mathbb{Z}[\omega]$ with $\lambda y = 3\alpha$. But $3 = -\omega^2 \lambda^2$, by Eq. (8.6), so canceling λ gives $y = -\omega^2 \lambda \alpha$. Thus, $\lambda \mid y$ in $\mathbb{Z}[\omega]$, so that $3 \mid y$ in $\mathbb{Z}$, by Lemma 8.35. This contradicts the hypothesis $3 \nmid xyz$.

(ii) $i = 1$:   Eq. (8.10) becomes

$$x + \omega y - \omega^2 x - \omega y \equiv 0 \text{ mod } 3.$$

Thus, the ωy's drop out, and there is $\alpha \in \mathbb{Z}[\omega]$ with $x(1 - \omega^2) = 3\alpha$. But $1 - \omega^2 = i\lambda$, by Exercise 8.21 on page 349, and so $x i \lambda = -\omega^2 \lambda^2 \alpha$. Hence, $\lambda \mid x$ in $\mathbb{Z}[\omega]$, and Lemma 8.35 gives $3 \mid x$ in $\mathbb{Z}$, another contradiction.

(iii) $i = 2$: Eq. (8.10) becomes

$$x + \omega y - \omega^4 x - \omega^3 y \equiv 0 \bmod 3.$$

The left-hand side simplifies to $x - \omega x = \lambda x$, because $\omega^3 = 1$. As in parts (i) and (ii), this leads to $3 \mid xyz$, which is a contradiction.

We conclude that there is no solution to $x^3 + y^3 = z^3$ of the desired type. ∎

Gauss's Proof of the Second Case

Gauss gave an elegant proof of the second case of Fermat's Last Theorem for exponent 3, and we'll present it here. It turns out to be convenient to prove a more general result. The object of this section is to prove the following theorem.

This is another example where it's easier to do things in more generality. The reason for introducing u comes from the fact that we allow x, y, and z to be elements of $\mathbb{Z}[\omega]$, so arithmetic statements are true up to unit factors.

There are no Eisenstein integers u, x, y, z with $xyz \neq 0$, u a unit, and 3 a factor of exactly one of x, y, z, such that

$$x^3 + y^3 = uz^3.$$

The proof, which will use infinite descent, is a consequence of several lemmas and propositions. To start with, we can assume, for u, x, y, z as in the statement, that x, y, and z are not all divisible in $\mathbb{Z}[\omega]$ by $\lambda = 1 - \omega$ (otherwise there's a contradiction, for 3 is a divisor in $\mathbb{Z}$ of x, y, and z, by Lemma 8.35). We'll first prove the theorem in case $\lambda \nmid xy$ but $\lambda \mid z$. Since x and y are interchangeable in the hypothesis, the remaining case is $\lambda \nmid yz$ but $\lambda \mid x$. We'll see that the theorem is an easy consequence of this.

Example 8.32 shows that every Eisenstein integer α is congruent mod λ to 0, 1, or -1. In particular, if $\lambda \nmid \alpha$, then

$$\alpha \equiv \pm 1 \bmod \lambda.$$

Gauss's proof requires a lemma that shows how an "extra λ" sneaks into the cube of this congruence.

Lemma 8.39. *If α is an Eisenstein integer for which $\lambda \nmid \alpha$, then*

$$\alpha^3 \equiv \pm 1 \bmod \lambda^4.$$

Proof. Let's first consider the case $\alpha \equiv 1 \bmod \lambda$; say

$$\alpha = 1 + \lambda \beta$$

for some $\beta \in \mathbb{Z}[\omega]$. Substitute this into the usual factorization in $\mathbb{Z}[\omega]$:

$$\alpha^3 - 1 = (\alpha - 1)(\alpha - \omega)(\alpha - \omega^2).$$

Rewrite the first factor on the right-hand side: $\alpha - 1 = 1 + \lambda \beta - 1 = \lambda \beta$. Next, since $1 - \omega = \lambda$, we can rewrite the second factor:

$$\alpha - \omega = 1 + \lambda \beta - \omega = \lambda + \lambda \beta = \lambda(1 + \beta).$$

Now rewrite the third factor, using Exercise 8.21 on page 349, which says that $1 - \omega^2 = -\omega^2 \lambda$:

$$\alpha - \omega^2 = 1 + \lambda \beta - \omega^2 = \lambda \beta - \omega^2 \lambda = \lambda(\beta - \omega^2).$$

Therefore,

$$\alpha^3 - 1 = \lambda^3 \beta(1 + \beta)(\beta - \omega^2). \tag{8.11}$$

Example 8.32 shows that β is 0, -1, or 1 mod λ. In each of these cases, we'll see that there's an extra factor of λ in the expression on the right-hand side of Eq. (8.11). If $\beta \equiv 0$ mod λ for some $\beta' \in \mathbb{Z}[\omega]$, then $\beta = \lambda\beta'$, and the expression begins $\lambda^4\beta'$. If $\beta \equiv -1 + \lambda\beta'$, then the middle factor $1 + \beta$ equals $\lambda\beta'$, which contributes an extra λ. If $\beta = 1 + \lambda\beta'$, then Exercise 8.21 says that $1 - \omega^2 = -\omega^2\lambda$, and the last factor on the right-hand side becomes

$$\begin{aligned} \beta - \omega^2 &= 1 + \lambda\beta' - \omega^2 \\ &= 1 - \omega^2 + \lambda\beta' \\ &= -\omega^2\lambda + \lambda\beta' \\ &= \lambda(-\omega^2 + \beta'). \end{aligned}$$

Therefore, if $\alpha \equiv 1$ mod λ, then $\alpha^3 - 1$ is a multiple of λ^4; that is, $\alpha^3 \equiv 1$ mod λ^4.

The remaining case $\alpha \equiv -1$ mod λ is now easy. We have $-\alpha \equiv 1$ mod λ, so that

$$(-\alpha)^3 \equiv 1 \text{ mod } \lambda^4,$$

and so $\alpha^3 \equiv -1$ mod λ^4. ∎

Gauss used infinite descent on $\nu(z)$ and showed (as we will shortly) that if there was a solution to $x^3 + y^3 = uz^3$ of the desired type, then one could find another solution (x', y', z') of the same type with $\nu(z') < \nu(z)$. Iterating this process will eventually contradict the next lemma.

Lemma 8.40. *Suppose $x^3 + y^3 = uz^3$ for nonzero Eisenstein integers x, y, z. If $\lambda \nmid xy$ and $\lambda \mid z$, then $\lambda^2 \mid z$.*

Proof. Since $\lambda \nmid xy$, Euclid's Lemma in $\mathbb{Z}[\omega]$ says that $\lambda \nmid x$ and $\lambda \nmid y$, and so Lemma 8.39 applies to say that both x^3 and y^3 are congruent to ± 1 mod λ^4. Hence, reducing $x^3 + y^3 = uz^3$ mod λ^4 yields

$$(\pm 1) + (\pm 1) \equiv uz^3 \text{ mod } \lambda^4.$$

The left-hand side of these congruences is one of 0, 2, or -2. Since $\lambda \mid z$ and $\lambda \nmid 2$ (why?), we see that ± 2 are impossible. Thus, $0 \equiv uz^3$ mod λ^4, so that $\lambda^4 \mid z^3$ and $\nu(z^3) = 3\nu(z) \geq 4$. But $\nu(z)$ is an integer; hence, $\nu(z) \geq 2$ and $\lambda^2 \mid z$. ∎

Note that $\lambda \mid z$ implies that $\nu(z) \geq 1$.

Here's the main piece of the puzzle: the key step for infinite descent.

Proposition 8.41. *Suppose that u is a unit in $\mathbb{Z}[\omega]$ and $x^3 + y^3 = uz^3$ for Eisenstein integers x, y, z with $\lambda \nmid xy$ and $\lambda \mid z$. Then there exists a unit u' and $x', y', z' \in \mathbb{Z}[\omega]$ with $\lambda \nmid x'y'$ and $\nu(z') = \nu(z) - 1$, such that*

$$\left(x'\right)^3 + \left(y'\right)^3 = u'\left(z'\right)^3.$$

Before we dig into the proof, think about why this result, combined with Lemma 8.40, implies that there is no solution to $x^3 + y^3 = uz^3$ in Eisenstein integers with $\lambda \nmid xy$ and $\lambda \mid z$.

Proof. Given x, y, z as in the statement, we factor $x^3 + y^3$ to get

$$(x + y)(x + y\omega)(x + y\omega^2) = uz^3. \qquad (8.12)$$

Lemma 8.40 implies that $\lambda^2 \mid z$ so that $\nu(uz^3) \geq 6$. Hence at least one factor on the left-hand side of the above equation is divisible by λ^2. Because x, y, and z are Eisenstein integers, we can replace y by $y\omega$ or $y\omega^2$ without changing the equation or the claim of the proposition. Hence we can assume, without loss of generality, that $\lambda^2 \mid x + y$; that is, $\nu(x + y) \geq 2$.

Now,

> Recall the role of the factor $x + y$ in the proof of the first case.

$$x + y\omega = (x + y) - (1 - \omega)y = x + y - \lambda y$$

and, since $\lambda \nmid y$, we have $\nu(\lambda y) = 1$. Hence, $\nu(x + y\omega) = 1$, by Proposition 8.36(iii). Similarly, $\nu(x + y\omega^2) = 1$. So, applying ν to Eq. (8.12) and using Proposition 8.36, we have:

$$3\nu(z) = \nu(x + y) + \nu(x + y\omega) + \nu\left(x + y\omega^2\right)$$
$$= \nu(x + y) + 1 + 1,$$

so that

$$\nu(x + y) = 3\nu(z) - 2.$$

For convenience, let's call the right-hand side k:

$$k = 3\nu(z) - 2. \qquad (8.13)$$

The factors on the left-hand side of Eq. (8.12) are each divisible by λ. We claim that they can't have any other common factors. To see this, suppose that γ is a prime in $\mathbb{Z}[\omega]$, $\gamma \neq \lambda$. If γ divided $x + y$ and $x + y\omega$, then it would divide their difference, which is λy. By Euclid's Lemma, $\gamma \mid y$, but then $\gamma \mid x$, contradicting the fact that $\gcd(x, y) = 1$. Hence

$$\gcd(x + y, x + y\omega) = \lambda.$$

The same reasoning shows that the gcd of each of the other pairs of factors is λ.

Putting all this together, we have the following equation in $\mathbb{Z}[\omega]$:

$$\left(\frac{x + y}{\lambda^k}\right)\left(\frac{x + y\omega}{\lambda}\right)\left(\frac{x + y\omega^2}{\lambda}\right) = u\left(\frac{z}{\lambda^{\nu(z)}}\right)^3, \qquad (8.14)$$

where the three factors on the left-hand side are relatively prime [remember Eq. (8.13): $3\nu(z) = k + 2$].

Now invoke unique factorization in $\mathbb{Z}[\omega]$: the right-hand side of Eq. (8.14) is a cube, and the left-hand side is a product of three relatively prime factors (each having no factor λ). Hence, each is a cube and, more precisely, there are units u_1, u_2, u_3 and Eisenstein integers z_1, z_2, z_3 with

$$\frac{x + y}{\lambda^k} = u_1 z_1^3 \quad \text{and } \lambda \nmid z_1,$$

$$\frac{x + y\omega}{\lambda} = u_2 z_2^3 \quad \text{and } \lambda \nmid z_2,$$

$$\frac{x + y\omega^2}{\lambda} = u_3 z_3^3 \quad \text{and } \lambda \nmid z_3.$$

Clearing fractions, it follows that

$$x + y = u_1 \lambda^k z_1^3 \quad \text{where } \lambda \nmid z_1,$$
$$x + y\omega = u_2 \lambda z_2^3 \quad \text{where } \lambda \nmid z_2,$$
$$x + y\omega^2 = u_3 \lambda z_3^3 \quad \text{where } \lambda \nmid z_3.$$

Multiply the second of these equations by ω, the third by ω^2, and add them to the first to obtain

$$0 = u_1 \lambda^k z_1^3 + u_2 \omega \lambda z_2^3 + u_3 \omega^2 \lambda z_3^3.$$

Divide both sides by λ:

$$0 = u_1 \lambda^{k-1} z_1^3 + u_2 \omega z_2^3 + u_3 \omega z_3^3.$$

We're almost there. Letting $v_1 = -u_1$, $v_2 = u_2\omega$, and $v_3 = u_3\omega^2$, we have

$$v_2 z_2^3 + v_3 z_3^3 = v_1 \lambda^{k-1} z_1^3.$$

Note that v_1, v_2, and v_3 are all units.

Recalling that $k = \nu(z) - 2$, this can be written as

$$v_2 z_2^3 + v_3 z_3^3 = v_1 \left(\lambda^{\nu(z)-1} z_1 \right)^3, \quad \text{where } \lambda \nmid z_1 z_2 z_3. \tag{8.15}$$

Divide both sides by v_2, relabel everything, and we have the equation

$$\left(x' \right)^3 + v \left(y' \right)^3 = v' \left(z' \right)^3,$$

where $\lambda \nmid x'y'$, v and v' are units, and $\nu(z') = \nu(z) - 1$. Now λ^2 divides the right-hand side of the equation, by Lemma 8.40, so reducing the equation mod λ^2 yields

$$(\pm 1) + (\pm v) \equiv 0 \bmod \lambda^2.$$

Once again, trying all six Eisenstein units and all possible signs, you can check that $v = \pm 1$. Replacing y' by $-y'$ if necessary, we have

$$\left(x' \right)^3 + \left(y' \right)^3 = v' \left(z' \right)^3,$$

where $\lambda \nmid x'y'$ and $\nu(z') = \nu(z) - 1$, and this is what we wanted to show. ∎

Proposition 8.42. *There are no Eisenstein integers u, x, y, z with u a unit, $\lambda \nmid xy$, and $\lambda \mid z$, such that*

$$x^3 + y^3 = uz^3.$$

Proof. Suppose such elements u, x, y, z exist. Repeated use of Proposition 8.41 shows that there are elements u', x', y', z' with $\nu(z') < 2$. But Lemma 8.40 says that this is impossible.

Once again, we use infinite descent.

It remains to settle the case where $\lambda \mid yz$. If you've held on this long, there's a relatively simple finish: Given Eisenstein integers u, x, y, z with u a unit, $\lambda \mid x$, and $\lambda \nmid yz$ and

$$x^3 + y^3 = uz^3;$$

reduce mod λ^2 to obtain $\pm 1 \equiv u \bmod \lambda^2$. A check shows that $u = \pm 1$. But then

See Exercise 8.25 on page 349.

$$(\pm z)^3 + (-y)^3 = x^3,$$

and we can apply Proposition 8.41. ∎

This establishes Gauss's Theorem.

Theorem 8.43 (Gauss). *There are no Eisenstein integers u, x, y, z with u a unit, $xyz \neq 0$, and 3 a factor of exactly one of x, y, and z, such that*

$$x^3 + y^3 = uz^3.$$

Proof. Since $\lambda \mid 3$, the hypothesis in Proposition 8.42 that λ is a factor of exactly one of x, y, and z implies that 3 is a factor of exactly one of x, y, and z. ∎

After all this work, we have, as a simple corollary, what we wanted in the first place.

Theorem 8.44 (Fermat's Last Theorem for Exponent 3). *There are no positive integers x, y, z such that $x^3 + y^3 = z^3$.*

Proof. Since $\lambda \mid 3$, Proposition 8.38 and Theorem 8.43 (with $u = 1$) cover all the possible cases for x, y, z. ∎

Proving Fermat's Last Theorem for a given exponent n was split into two cases, as we have just seen for $n = 3$; the second case was also divided into two parts. The first case for all $n < 100$ was proved, around 1806, by Germain. In 1825, Legendre proved one part of the second case for $n = 5$, while Dirichlet proved the other part. In 1839, Lamé proved Fermat's Last Theorem for exponent $n = 7$. The level of difficulty increased with the exponent. It was not until Kummer that many exponents were completely settled simultaneously.

Exercises

8.29 Show that none of the six units u in $\mathbb{Z}[\omega]$ is congruent mod λ to 0, 2, or -2. (As usual, $\lambda = 1 - \omega$.)

8.30 Prove Proposition 8.36.

8.31 Without using Proposition 8.38, show that there are no integers x, y, z with $3 \nmid xyz$ such that $x^3 + y^3 = z^3$ mod 9. This exercise gives an alternative proof of Proposition 8.38.

8.32 Show that there are no integers x, y, z with $5 \nmid xyz$ such that $x^5 + y^5 = z^5$ mod 25. This exercise implies Fermat's Last Theorem for exponent 5 in the case that $5 \nmid xyz$.

8.33 Are there any integers x, y, z with $7 \nmid xyz$ such that $x^7 + y^7 = z^7$ mod 49?

8.34 (i) Sketch the graph of $x^3 + y^3 = 1$.

(ii) Show that the only rational points on the graph are $(1, 0)$ and $(0, 1)$.

8.35 Take It Further. Let G be the graph of $x^3 + y^3 = 9$.

(i) Sketch G.

(ii) Find the equation of the line ℓ tangent to G at $(2, 1)$.

(iii) Find the intersection of ℓ and G.

(iv) Show that there are infinitely many triples of integers (x, y, z) such that

$$x^3 + y^3 = 9z^3.$$

8.4   Approaches to the General Case

Almost all attempts to prove there are no positive integers x, y, z satisfying $x^p + y^p = z^p$, where p is an odd prime, divided the problem in half. The *first case* assumes that $\gcd(x, y) = 1$ and $p \nmid xyz$; the *second case* assumes that exactly one of x, y, and z is divisible by p.

Our choice of proof for exponent 3 contains some of the main ingredients of a proof of the first case for any odd prime p, provided that the ring $\mathbb{Z}[\zeta_p]$, where $\zeta_p = \cos(2\pi/p) + i \sin(2\pi/p)$, is a UFD. Once again, this is based on the factorization in Exercise 3.50 on page 115:

$$x^p + y^p = (x + y)(x + \zeta_p y) \ldots (x + \zeta_p^{p-1} y) \qquad (8.16)$$

The basic idea is to use the fact, in a UFD, that if a product of relatively prime elements is a pth power, then each of its factors is also a pth power. The proof is more complicated for large p because, while $\mathbb{Z}[\zeta_3] = \mathbb{Z}[\omega]$ has only six units, the ring $\mathbb{Z}[\zeta_p]$ for $p > 3$ may have infinitely many units. As we saw in the proof of Proposition 8.38, much of the argument depends on a careful analysis of how units enter into the calculations.

The commutative rings $\mathbb{Z}[\zeta_p]$ are called **rings of cyclotomic integers**, and investigating them has played an important part of the story of Fermat's Last Theorem, well into the 20th century. We'll start this section with a brief sketch of arithmetic in $\mathbb{Z}[\zeta_p]$, pointing to some major results, perhaps without proof, that generalize results we've already established for $\mathbb{Z}[\omega]$.

After that, we'll sketch the work of Kummer that deals with the situation when unique factorization fails. While these efforts didn't lead him to a proof of Fermat's Last Theorem, they did lead to some ideas that have had real staying power in algebra. One of them is his introduction of ideals as an important structural component of a commutative ring (Kummer called them **divisors**), not merely as subsets that happen to arise, say in studying gcd's. Another important idea is that of **class number**, a measure of how far off $\mathbb{Z}[\zeta_p]$ is from having unique factorization.

Here is a biography of Kummer we have adapted from that given in the history archives of the School of Mathematics and Statistics of the University of St. Andrews in Scotland.

Ernst Eduard Kummer was born in Sorau, Prussia, in 1810. He entered the University of Halle in 1828 with the intention of studying Protestant theology, but he received mathematics teaching as part of his degree which was designed to provide a proper foundation to the study of philosophy. Kummer's mathematics lecturer H. F. Scherk inspired his interest in mathematics, and Kummer soon was studying mathematics as his main subject.

In 1831 Kummer was awarded a prize for a mathematical essay he wrote on a topic set by Scherk. In the same year he was awarded his certificate enabling him to teach in schools and, on the strength of his prize-winning essay, he was awarded a doctorate. In 1832, Kummer was appointed to a teaching post at the Gymnasium in Liegnitz, now Legnica in Poland. He held this post for ten years, where he taught mathematics and physics. Some of his pupils had great ability and, conversely, they were extremely fortunate to find a school teacher of Kummer's quality and ability to inspire. His two most famous pupils were Kronecker and Joachimsthal and, under Kummer's guidance, they began mathematical research while at school, as did Kummer himself. He published

Joachimsthal was famed for the high quality of his lectures. His colleagues in Berlin included many famous mathematicians such as Eisenstein, Dirichlet, Jacobi, Steiner, and Borchardt.

a paper on hypergeometric series in Crelle's Journal in 1836, which he sent to Jacobi, and this led to Jacobi, and later Dirichlet, corresponding with Kummer. In 1839, although still a school teacher, Kummer was elected to the Berlin Academy on Dirichlet's recommendation. Jacobi now realized that he had to find Kummer a university professorship.

In 1842, with strong support from Jacobi and Dirichlet, Kummer was appointed a full professor at the University of Breslau, now Wroclaw in Poland, where he began research in number theory. In 1855, Dirichlet left Berlin to succeed Gauss at Göttingen, and he recommended that Berlin offer the vacant chair to Kummer, which they did. The clarity and vividness of Kummer's presentations brought him great numbers of students—as many as 250 were counted at his lectures. Kummer's popularity as a professor was based not only on the clarity of his lectures but on his charm and sense of humor as well. Moreover, he was concerned for the well-being of his students and willingly aided them when material difficulties arose.

During Kummer's first period of mathematics, he worked on function theory. He extended Gauss's work on hypergeometric series, giving developments that are useful in the theory of differential equations. He was the first to compute the monodromy groups of these series. In 1843 Kummer, realizing that attempts to prove Fermat's Last Theorem broke down because the unique factorization of integers did not extend to other rings of complex numbers, attempted to restore the uniqueness of factorization by introducing "ideal" numbers. Not only has his work been most fundamental in work relating to Fermat's Last Theorem, since all later work was based on it for many years, but the concept of an ideal allowed ring theory, and much of abstract algebra, to develop. The Paris Academy of Sciences awarded Kummer the Grand Prize in 1857 for this work. Soon after, he was elected to membership of the Paris Academy of Sciences and then, in 1863, he was elected a Fellow of the Royal Society of London. Kummer received numerous other honors in his long career; he died in 1893.

Cyclotomic integers

We shall assume throughout this section that p is an odd prime and $\zeta = \zeta_p = \cos(2\pi/p) + i\sin(2\pi/p)$. Recall some facts about $\mathbb{Q}(\zeta)$.

(1) $\text{irr}(\zeta, \mathbb{Q}) = \Phi_p(x) = 1 + x + x^2 + \cdots + x^{p-2} + x^{p-1}$ (Theorem 6.68 and Exercise 7.31 on page 300).

(2) $\left[\mathbb{Q}(\zeta_p) : \mathbb{Q}\right] = p - 1$ (Exercise 7.32 on page 300).

(3) $x^p - 1 = (x - 1)(x - \zeta)(x - \zeta^2)\ldots(x - \zeta^{p-1})$ (Exercise 6.46(i) on page 269).

(4) $\mathbb{Q}(\zeta) \cong \mathbb{Q}[x]/\left(\Phi_p(x)\right)$ (Theorem 7.25(i)).

We recall Proposition 7.20(v), which we now state as a lemma for your convenience.

Lemma 8.45. *Let p be an odd prime and $\zeta = \zeta_p$ be a pth root of unity. A basis for $\mathbb{Q}(\zeta)$ as a vector space over $\mathbb{Q}$ is*

$$B = 1, \zeta, \zeta^2, \ldots, \zeta^{p-2}.$$

The ring $\mathbb{Z}[\zeta] \subseteq \mathbb{Q}(\zeta)$ is thus the set of all linear combinations $\sum_{i=0}^{p-2} a_i \zeta^i$ with $a_i \in \mathbb{Z}$. It shares many of the algebraic properties of the Gaussian and Eisenstein integers except, alas, it is not always a UFD (more about this in the next section). But there are analogs for the laws of decomposition that we developed in $\mathbb{Z}[i]$ and in $\mathbb{Z}[\omega]$. Recall, for example, that there is equality of ideals in $\mathbb{Z}[i]$:

$$(2) = (1 - i)^2,$$

and also in $\mathbb{Z}[\omega]$,

$$(3) = (1 - \omega)^2.$$

It turns out that the ideal (p) ramifies in $\mathbb{Z}[\zeta]$ in a similar way. Let's look into this.

Lemma 8.46. *If p is an odd prime and $\zeta = \zeta_p$ is a pth root of unity, then*

$$p = \prod_{k=1}^{p-1}(1 - \zeta^k).$$

Proof. Since $x^p - 1 = (x - 1)(x - \zeta)(x - \zeta^2) \ldots (x - \zeta^{p-1})$, we have

$$\frac{x^p - 1}{x - 1} = \prod_{k=1}^{p-1}(x - \zeta^k).$$

But $(x^p - 1)/(x - 1) = \Phi_p(x) = 1 + x + \cdots + x^{p-1}$, so that

$$1 + x + \cdots + x^{p-1} = \prod_{k=1}^{p-1}(x - \zeta^k).$$

Now put $x = 1$.   ∎

Lemma 8.46 gives a factorization of p in $\mathbb{Z}[\zeta]$ into p factors. Our next goal is to show that the factors are all associates.

Proposition 8.47. *If $s, t \in \mathbb{N}$ and $p \nmid st$, then $1 - \zeta^s$ and $1 - \zeta^t$ are associates in $\mathbb{Z}[\zeta]$.*

Proof. In the field $\mathbb{F}_p$, let $t = r^{-1}s$, so that $tr \equiv s \bmod p$. Then $\zeta^{tr} = \zeta^s$ (why?), and so

$$\frac{1 - \zeta^s}{1 - \zeta^t} = \frac{1 - \zeta^{tr}}{1 - \zeta^t}$$

$$= \frac{1 - (\zeta^t)^r}{1 - \zeta^t}$$

$$= 1 + \zeta^t + (\zeta^t)^2 + \cdots + (\zeta^t)^{r-1}.$$

Hence, $(1 - \zeta^t)/(1 - \zeta^s) \in \mathbb{Z}[\zeta]$, and

$$(1 - \zeta^t) \mid (1 - \zeta^s)$$

in $\mathbb{Z}[\zeta]$. A similar argument shows that $(1 - \zeta^s) \mid (1 - \zeta^t)$ in $\mathbb{Z}[\zeta]$. It follows from Proposition 6.6 that $1 - \zeta^s$ and $1 - \zeta^t$ are associates in $\mathbb{Z}[\zeta]$.   ∎

As an immediate consequence, we can generalize the fact that there is equality of ideals: $(3) = (1 - \omega)^2$.

The next result says that p ramifies in $\mathbb{Z}[\zeta]$.

Corollary 8.48. *In $\mathbb{Z}[\zeta]$, there is a unit u such that*

$$p = u(1 - \zeta)^{p-1},$$

which gives a factorization of ideals

$$(p) = (1 - \zeta)^{p-1}.$$

Proof. Lemma 8.46 shows, as elements of $\mathbb{Z}[\zeta]$, that

$$p = \prod_{k=1}^{p-1}(1 - \zeta^k) = (1 - \zeta)\left(1 - \zeta^2\right)\ldots\left(1 - \zeta^{p-1}\right). \qquad (8.17)$$

Proposition 8.47 shows that there is a unit u_k $(2 \le k \le p - 1)$ so that

$$1 - \zeta^k = u_k(1 - \zeta).$$

Factoring out the units from Eq. (8.17) and writing their product as u, we see that

$$p = u(1 - \zeta)^{p-1}.$$

Hence, we have equality of ideals in $\mathbb{Z}[\zeta]$:

$$(p) = (1 - \zeta)^{p-1}. \quad \blacksquare$$

Corollary 8.49. *If $s, t \in \mathbb{N}$ and $p \nmid st$, then*

$$\frac{1 - \zeta^s}{1 - \zeta^t}$$

is a unit in $\mathbb{Z}[\zeta]$.

Proof. Since $1 - \zeta^s$ and $1 - \zeta^t$ are associates, there is a unit u in $\mathbb{Z}[\zeta]$ with $1 - \zeta^s = u(1 - \zeta^t)$, so $(1 - \zeta^s)/(1 - \zeta^t) = u$. $\quad \blacksquare$

How to Think About It. Since $\zeta^p = 1$, every integer power of ζ occurs among

$$1, \quad \zeta, \quad \zeta^2, \quad \ldots, \quad \zeta^{p-1}.$$

In particular, if $1 \le s \le p - 1$, then $\zeta^{-s} = \zeta^{p-s}$. We can calculate in $\mathbb{Z}[\zeta]$ by calculating in

$$\mathbb{Z}[x]/(x^{p-1} + x^{p-2} + \cdots + 1).$$

This allows us to use a CAS to do calculations and then to translate to $\mathbb{Z}[\zeta]$ via the map $f(x) \mapsto f(\zeta)$.

There are other units in $\mathbb{Z}[\zeta]$ that are real numbers; Corollary 8.49 gives a way to produce them.

Proposition 8.50. *The real number*

$$\zeta + \zeta^{-1} = 2\cos\left(\frac{2\pi}{p}\right)$$

is a unit in $\mathbb{Z}[\zeta]$.

Proof. Exercise 6.46 on page 269 shows that, in $\mathbb{Z}[\zeta][x]$,

$$x^p + 1 = (x+1)(x+\zeta)(x+\zeta^2)\cdots(x+\zeta^{p-1}).$$

Put $x = \zeta^{-1}$ to find that

$$\left(\zeta^{-1}\right)^p + 1 = (\zeta^{-1}+1)(\zeta^{-1}+\zeta)(\zeta^{-1}+\zeta^2)\cdots(\zeta^{-1}+\zeta^{p-1}). \quad (8.18)$$

The second factor on the right-hand side of Eq. (8.18) is the focus of the proposition. The left-hand side is 2 (because $\left(\zeta^{-1}\right)^p = 1$). Finally, the last factor on the right-hand side is

$$\zeta^{-1} + \zeta^{p-1} = \frac{1}{\zeta} + \zeta^{p-1} = \frac{1+\zeta^p}{\zeta} = \frac{2}{\zeta} = 2\zeta^{p-1}.$$

Hence

$$2 = 2\zeta^{p-1}(\zeta^{-1}+1)(\zeta^{-1}+\zeta)(\zeta^{-1}+\zeta^2)\cdots(\zeta^{-1}+\zeta^{p-2}).$$

Dividing both sides by 2, we see that $\zeta^{-1} + \zeta$ is a unit; in fact, its inverse is:

$$1 = \left(\zeta^{-1}+\zeta\right)\left[\zeta^{p-1}(\zeta^{-1}+1)(\zeta^{-1}+\zeta^2)\cdots(\zeta^{-1}+\zeta^{p-2})\right].$$

(The last equation gives us other units besides $\zeta + \zeta^{-1}$.)   ■

Further results. This is just the beginning.

- Corollary 8.48 is a piece of a ***law of decomposition*** in $\mathbb{Z}[\zeta]$. In $\mathbb{Z}[\omega]$, rational primes either stay prime, split, or ramify (and 3 is the only ramified prime). In $\mathbb{Z}[\zeta_p]$ for $p > 3$, rational primes can decompose in other ways, but it's still true that the way a prime decomposes depends only on its congruence class mod p. Indeed, if q is a prime and f is the smallest integer such that $q^f \equiv 1 \bmod p$, then q spits into f prime factors in $\mathbb{Z}[\zeta]$. This lovely theory is detailed in [5] Chapter 3 and [36] Chapter 2.

  It follows that if $q \equiv 1 \bmod p$, then q "splits completely" into $p - 1$ factors in $\mathbb{Z}[\zeta]$. What does this say in $\mathbb{Z}[\omega]$ (when $p = 3$)?

- Corollary 8.49 and Proposition 8.50 show how to build units in $\mathbb{Z}[\zeta]$. This is a piece of a complete classification of units in cyclotomic integers: Kummer proved that every unit in $\mathbb{Z}[\zeta]$ is a product $\zeta^s \epsilon$ for some integer s, where $\epsilon \in R = \mathbb{Z}[\zeta + \zeta^{-1}]$ (for a proof, see [36], p.3). Since

$$\zeta + \zeta^{-1} = 2\cos(2\pi/p) \in \mathbb{R},$$

R is a subring of $\mathbb{R}$, and every unit in $\mathbb{Z}[\zeta]$ is the product of a power of ζ and a real unit of $\mathbb{Z}[\zeta]$. For $p = 3$, $\zeta = \omega = \frac{1}{2}\left(-1 + i\sqrt{3}\right)$ and

$$\zeta + \zeta^{-1} = -1.$$

Hence $R = \mathbb{Z}$, and every unit is a power of ω times a unit in $\mathbb{Z}$, namely ± 1; this recovers the result from Exercise 4.45 on page 165.

The results in this section set the stage for a proof of Fermat's Last Theorem, along the same lines as our proof of the theorem for exponent 3, for arbitrary prime exponents p, as long as $\mathbb{Z}[\zeta_p]$ has unique factorization. Kummer did exactly this, for both cases of the theorem (a detailed historical account is in [23]). As in the case $p = 3$, the key players are Eq. (8.16), the prime $\lambda = 1 - \zeta$, and the units $\zeta^s \epsilon$ where ϵ is a real unit in $\mathbb{Z}[\zeta]$. We leave the story here, pointing to [5] Chapter 3 for the rest of the technical details.

Exercises

8.36 As usual, let $\zeta = \cos(2\pi/p) + i \sin(2\pi/p)$, where p is a rational prime.

    (i) Show that $\left[\mathbb{Q}(\zeta) : \mathbb{Q}\left(\zeta + \zeta^{-1}\right)\right] = 2$, and find

$$\mathrm{irr}\left(\zeta, \mathbb{Q}\left(\zeta + \zeta^{-1}\right)\right).$$

    (ii) What is $\left[\mathbb{Q}\left(\zeta + \zeta^{-1}\right) : \mathbb{Q}\right]$?

8.37 (i) Experiment with various values of p and calculate

$$\prod_{i=1}^{p-1} \left(1 + \zeta_p^i\right).$$

    (ii) Find a general formula (for any prime p) for

$$\prod_{i=1}^{p-1} \left(1 + \zeta_p^i\right).$$

8.38 For $1 \leq s \leq p - 1$ (p a prime), show that

$$\zeta_p^s + \zeta_p^{-s}$$

is a unit in $\mathbb{Z}[\zeta_p]$. Is $\zeta_p^s + \zeta_p^{-s}$ a real number?

8.39 In $\mathbb{Z}[x]/\left(\Phi_5(x)\right)$, calculate

    (i) $x^4 \left(x^4 + 1\right) \left(x^4 + x^2\right) \left(x^4 + x^3\right)$

    (ii) $\left(x + x^4\right) \left(x^3 + x^2\right)$

    (iii) $\left(x + x^4\right) (1 + x)$.

8.40 In $\mathbb{Z}[\zeta_5]$, calculate

    (i) $\zeta^4 \left(\zeta^4 + 1\right) \left(\zeta^4 + \zeta^2\right) \left(\zeta^4 + \zeta^3\right)$

    (ii) $\left(\zeta + \zeta^4\right) \left(\zeta^3 + \zeta^2\right)$

    (iii) $\left(\zeta + \zeta^4\right) (1 + \zeta)$.

8.41 Write $1 + \zeta_5$ as the product of a power of ζ_5 and an element of $\mathbb{Z}(\zeta_5 + \zeta_5^{-1})$.

8.42 Show that

$$\zeta_5^{-1} \frac{1 - \zeta_5^3}{1 - \zeta_5} = \frac{\sin(3\pi/5)}{\sin(\pi/5)}.$$

Kummer, Ideal Numbers, and Dedekind

It is natural to think that the rings $\mathbb{Z}[\zeta_p]$ are UFDs, as evidenced by the number of mathematicians in the 17th, 18th, and 19th centuries who assumed it. Indeed, it's true for all primes less than 23, but $\mathbb{Z}[\zeta_{23}]$ does not enjoy unique factorization ([23], p.7). How could so many not know this? It may seem that 23 is not that large, but the calculations in rings of cyclotomic integers are hefty, even with computers. Imagine the stamina required to calculate by hand with polynomials of degree 22 in ζ_{23}. Some of Kummer's tour-de-force calculations are recounted in [11] Chapter 4. The proof that unique factorization fails in $\mathbb{Z}[\zeta_{23}]$ is technical (again, see [11], Chapter 4, but the essential idea can be illustrated in the ring $R = \mathbb{Z}[\sqrt{-5}]$ This is a perfectly good commutative ring (of course, R is not a ring of cyclotomic integers):

$$R = \mathbb{Z}[\sqrt{-5}] \cong \mathbb{Z}[x]/\left(x^2 + 5\right).$$

If we let $\alpha = \sqrt{-5}$, then elements of R can be written as $a + b\alpha$ with $a, b \in \mathbb{Z}$. If $z = a + b\alpha$, then its complex conjugate is, as usual, $\overline{z} = a - b\alpha$. Just as in Gaussian and Eisenstein integers, we can take norms: $N(z) = z\overline{z}$, and we have

$$N(a + b\alpha) = (a + b\alpha)(a - b\alpha) = a^2 + 5b^2.$$

The usual properties of norm hold in R: it is multiplicative, the norm of a unit is 1, and conjugates have the same norm (Exercise 8.43 below).

There are two factorizations of 6 in R:

$$6 = 2 \cdot 3 = (1 + \sqrt{-5})(1 - \sqrt{-5}).$$

We claim that they are essentially different ways to factor 6 into primes. Let's see why.

Lemma 8.51. (i) *The rational integers* 2 *and* 3 *are prime in* $R = \mathbb{Z}[\sqrt{-5}]$.
(ii) $1 + \alpha$ *and* $1 - \alpha$ *are prime in* R.

Proof. (i) If $2 = zw$ for non-units z and w, then

$$4 = N(zw) = N(z) N(w).$$

By the Fundamental Theorem in $\mathbb{Z}$ (and the fact that neither z nor w is a unit, so that neither has norm 1), $N(z)$ would be a proper factor of 4, that is $N(z) = 2$. But 2 can't be written as $a^2 + 5b^2$. The proof for 3 uses exactly the same idea.

(ii) If $1 + \alpha = zw$ for non-units z and w, then

$$6 = N(1 + \alpha) = N(z) N(w).$$

By the Fundamental Theorem in $\mathbb{Z}$ (and the fact that neither z nor w is a unit), $N(z)$ would be a proper factor of 6, say $N(z) = 2$. But 2 can't be written as $a^2 + 5b^2$. The proof for $1 - \alpha$ uses exactly the same idea. ■

So, we have two factorizations of 6 into primes in R. We've seen in other rings that different-looking factorizations are really the same up to unit factors.

But that doesn't happen here because neither 2 nor 3 is associate to $1 + \alpha$, for neither has norm $6 = N(1 + \alpha)$. We have a problem!

Kummer was working on methods for factoring cyclotomic integers (not, as it turns out, towards a proof of Fermat's Last Theorem, but towards another, related question). He devised a way to think about our problem that actually shows up in elementary school when children think that 14×15 and 10×21 are different factorizations of 210. The students are not going far enough in their factorizations: if they write

$$14 = 2 \times 7 \quad \text{and} \quad 15 = 3 \times 5,$$

they see that the "other" factorization is just a rearrangement of the prime factors of 14×15:

$$10 = 2 \times 5 \quad \text{and} \quad 21 = 3 \times 7.$$

Now, our problem is different in the sense that we already have prime factorizations. But Kummer's idea was to imagine some "ghost factors" for each of $2, 3, 1 + \alpha$, and $1 - \alpha$, sort of "super primes" behind the scenes, that could be rearranged to produce the different factorizations. Kummer called these **ideal numbers** or *divisors*, and he imagined there was a further factorization into ideal numbers J_1, J_2, J_3, J_4:

$$2 = J_1 J_2$$
$$3 = J_3 J_4$$
$$1 + \alpha = J_1 J_3$$
$$1 - \alpha = J_2 J_4.$$

Kummer knew that no such J_i existed in R, but he was able to model these ghost factors, not as elements of R but as "lists" of elements, each list containing the non-associate divisors of $2, 3, 1 + \alpha$, and $1 - \alpha$. And he developed a theory extending R to a new system R' in which there was unique factorization into ideal numbers. Later, Dedekind refined Kummer's ideas, recasting *ideal numbers* into what we nowadays call **ideals**, a notion, as we've seen in this book, that has utility far beyond investigations into Fermat's Last Theorem. We'll use the contemporary notion of ideal to continue our story.

The basic idea is that products of elements are replaced by products of ideals. In a PID, nothing new is added, because there's a bijection between ring elements (up to associates) and principal ideals (Exercise 5.51(ii) on page 220). But rings that are not UFDs are not PIDs (Theorem 6.50), so there's a larger stash of ideals that can enter into factorizations.

Example 8.52. We've seen, in $R = \mathbb{Z}[\alpha]$, where $\alpha = \sqrt{-5}$, that

$$6 = 2 \cdot 3 = (1 + \alpha)(1 - \alpha).$$

The ghost factors that will resolve our problem are ideals in R generated by two elements:

$$J_1 = (2, 1 + \alpha) = \{2a + b(1 + \alpha) : a, b \in R\}$$
$$J_2 = (2, 1 - \alpha) = \{2a + b(1 - \alpha) : a, b \in R\}$$
$$J_3 = (3, 1 + \alpha) = \{3a + b(1 + \alpha) : a, b \in R\}$$
$$J_4 = (3, 1 - \alpha) = \{3a + b(1 - \alpha) : a, b \in R\}.$$

Actually, Kummer considered rings of cyclotomic integers. We're using R here just for the sake of example.

Exercises 5.51 and 5.52 on page 220 define the product of two ideals and develop the properties of the multiplication.

We claim that

$$(2) = J_1 J_2$$
$$(3) = J_3 J_4$$
$$(1 + \alpha) = J_1 J_3$$
$$(1 - \alpha) = J_2 J_4.$$

Note that these equations are equalities of ideals, not numbers.

The verifications all use the same method, so we'll carry it out for the first case only, leaving the rest for you as Exercise 8.45 below.

Let's show that $(2) = J_1 J_2$. Now the product of two ideals I and J is the set of all linear combinations of products rs where $r \in I$ and $s \in J$ (Exercise 5.51 on page 220). So, $J_1 J_2 = (2, 1 + \alpha)(2, 1 - \alpha)$ is the set of all linear combinations of the form (recall that $(1 - \alpha)(1 + \alpha) = 6$):

$$a(2 \cdot 2) + b\,(2(1 - \alpha)) + c\,(2(1 + \alpha)) + d(1 + \alpha)(1 - \alpha)$$
$$= 4a + 2b(1 - \alpha) + 2c(1 + \alpha) + 6d,$$

where $a, b, c, d \in R$. Well,

$$4a + 2b(1 - \alpha) + 2c(1 + \alpha) + 6d = 2\,[2a + b(1 - \alpha) + 2c(1 + \alpha) + 3d]\,,$$

so $J_1 J_2 \subseteq (2)$. And, if

$$(a, b, c, d) = (-1, 0, 0, 1),$$

we have

$$4a + 2b(1 - \alpha) + 2c(1 + \alpha) + 6d = 2,$$

so that $(2) \subseteq J_1 J_2$. Hence

$$(2) = J_1 J_2$$

as claimed. The other verifications follow in the same way.

Ah, but there's one glitch. What if one of the four ideals is (1), the unit ideal? If $J_1 = (1)$ for example, we'd have $(2) = (1 - \alpha)$, and we'd still have the same problem. But we can show that none of the J_i is the unit ideal. Let's show that $J_1 \neq (1)$—the arguments for the others are the same (Exercise 8.46 below).

Suppose, on the contrary, that $J_1 = (2, 1 + \alpha) = (1)$. Then there exist elements $r + s\alpha$ and $t + u\alpha$ in R, where $r, s, t, u \in \mathbb{Z}$, so that

$$1 = (r + s\alpha) \cdot 2 + (t + u\alpha)(1 + \alpha).$$

Multiply this out, using the fact that $\alpha^2 = -5$, and write the result as $x + y\alpha$ to obtain

$$1 = (2r + t - 5u) + (2s + t + u)\alpha.$$

It follows that

$$2r + t - 5u = 1$$
$$2s + t + u = 0.$$

Replace u by $-2s - t$ in the first equation to obtain

$$2r - 4t + 10u = 1.$$

Since the left-hand side is even, this is impossible. ▲

For rings of cyclotomic integers $\mathbb{Z}[\zeta_p]$, it turns out that this new kind of factorization into ideals is unique.

Kummer introduced another brilliant idea. Call two ideals I and J *equivalent* if there is a cyclotomic integer z so that

$$I = (z)J = \{zb : b \in J\}.$$

He was able to show that this gives an equivalence relation on nonzero ideals in $\mathbb{Z}[\zeta_p]$ (for symmetry, the set of all ideals must be enlarged by adding in certain subsets of $\text{Frac}(\mathbb{Z}[\zeta_p]) = \mathbb{Q}(\zeta_p)$ called **fractional ideals**). Most importantly, Kummer showed that this relation has only finitely many equivalence classes, and he called the number $h(p)$ of them the **class number** of $\mathbb{Z}[\zeta_p]$. If $\mathbb{Z}[\zeta_p]$ has class number 1, then all ideals are principal, there is unique factorization, and our proof of Fermat's Last Theorem can be refined to produce a proof for such exponents. In fact, Kummer generalized this, proving that if the class number $h(p)$ is not divisible by p, then there are no positive integer solutions to $a^p + b^p = c^p$. This was a monumental achievement. Kummer called primes p such that $p \nmid h(p)$ **regular primes**. For example, even though $\mathbb{Z}[\zeta_{23}]$ doesn't have unique factorization, 23 is a regular prime—$h(23) = 3$, and so Fermat's Last Theorem holds for it. Alas, there are irregular primes. The smallest is 37, and the next two are 59 and 67. Unfortunately, it is known that there are infinitely many irregular primes, and it's unknown whether there are infinitely many regular primes.

Let's now say a bit more about Kummer's ideal numbers (nowadays called *divisors*), but we view his idea through the eyes of Dedekind. Take a cyclotomic integer $a \in \mathbb{Z}[\zeta]$, and define its **divisor**

$$D(a) = \{z \in \mathbb{Z}[\zeta] : a \text{ is a divisor of } z\}.$$

Now $D(a)$ is closed under addition and multiplication by other cyclotomic integers; that is, if $z, z' \in D(a)$, then $z + z' \in D(a)$; if $z \in D(a)$ and $r \in \mathbb{Z}[\zeta]$, then $rz \in D(a)$. In other words, $D(a)$ is an ideal (in fact, a principal ideal) in precisely the sense we have been using the term in this book (and we see how natural the idea is when viewed in this context). The definition of divisor makes sense for any commutative ring R, not just for the rings $\mathbb{Z}[\zeta]$.

If $a, b \in R$, where R is a commutative ring, then $a \mid b$ if and only if $D(a) \supseteq D(b)$; thus, if R is a domain, then $D(a) = D(b)$ if and only if a and b are associates.

Now generalize the notion of divisor so that, instead of being a subset of a commutative ring R of the form $D(a)$ for some $a \in R$, it is a subset of R closed under addition and multiplication by elements of R; that is, let's replace $D(a)$, which is a principal ideal, by any ideal. Thus, if $a, b \in R$, then

$$D(a) + D(b) = \{z + w : z \in D(a) \text{ and } w \in D(b)\}$$

is a generalized divisor. If we denote $D(a) + D(b)$ by $D(c)$, pretending that generalized divisors are just ordinary divisors, then we cannot declare that c is an element of R. Thus, c is a "ghost" element. Of course, if R is a PID, then c is an element of R, but if R is not a PID, then c may be a creature of our imagination.

Consider the ring $R = \mathbb{Z}[\alpha]$ in Example 8.52, where $\alpha = \sqrt{-5}$. The factorizations of 6,

$$6 = 2 \cdot 3 \quad \text{and} \quad 6 = (1 + \alpha)(1 - \alpha),$$

involve four elements of R, each of which gives a divisor. As in the example, define

$$J_1 = D(2) + D(\alpha)$$
$$J_2 = D(2) + D(1 - \alpha)$$
$$J_3 = D(3) + D(1 + \alpha)$$
$$J_4 = D(3) + D(1 - \alpha).$$

We can pretend that there are ghosts c_i so that $J_i = D(c_i)$ for $i = 1, 2, 3, 4$.

To complete the story, we report that ghosts are primes: the ideals J_i can be shown to be prime ideals, using the notion of the norm of ideals. Moreover, one can prove that factorizations in terms of such ghosts are unique, using fractional ideals.

How to Think About It. One of the contributions of Fermat's Last Theorem to algebra is that it attracted mathematicians of the first order and, as they studied it, they enhanced the areas of mathematics impinging on it. For algebra in particular, it brought the idea of commutative rings, factorization, and unique factorization to the forefront. Kummer's recognition that unique factorization was not always present, and his restoration of it with his "ideal numbers," led Dedekind to introduce ideals into the study of rings. Dedekind's notion of ideal was taken up by Hilbert and then later by Emmy Noether. It is today one of the most fundamental ideas in modern algebra.

We have a confession to make. Our discussion in Chapter 6 explains parallels of the arithmetic of polynomials with coefficients in a field k with the arithmetic of integers by saying that both $k[x]$ and $\mathbb{Z}$ are PIDs. No doubt, our ancestors were aware of the analogy between these two systems, but viewing them in terms of ideals is a modern viewpoint, after Dedekind, dating from the 1920s. We wrote Chapter 6 using contemporary ideas because it unifies the exposition.

Richard Dedekind was born in 1831 in Braunschweig (in what is now Germany). He entered the University of Göttingen in 1850; it was a rather disappointing place to study mathematics at the time, for it had not yet become the vigorous research center it turned into soon afterwards. Gauss taught courses in mathematics, but mostly at an elementary level. Dedekind did his doctoral work under Gauss's supervision, receiving his doctorate in 1852; he was to be the last pupil of Gauss.

In 1854, both Riemann and Dedekind were awarded their habilitation degrees within a few weeks of each other. Dedekind was then qualified as a university teacher, and he began teaching at Göttingen. Gauss died in 1855, and Dirichlet was appointed to fill the vacant chair. This was an extremely important event for Dedekind, who found working with Dirichlet extremely profitable. He attended courses by Dirichlet, and they soon became close friends; the relationship was in many ways the making of Dedekind, whose mathematical interests took a new lease on life with their discussions. Around this time Dedekind studied the work of Galois, and he was the first to lecture on Galois theory when he taught a course on the topic at Göttingen.

In the spring of 1858, Dedekind was appointed to the Polytechnikum in Zurich. It was while he was thinking how to teach differential and integral calculus that the idea of a **Dedekind cut** came to him. His idea was that every real number r divides the rational numbers into two subsets, namely those greater than r and those less than r. Dedekind's brilliant idea was to represent the real numbers by such divisions of the rationals.

The Collegium Carolinum in Brunswick had been upgraded to the Brunswick Polytechnikum by the 1860s, and Dedekind was appointed there in 1862. He returned to his home town, remaining there for the rest of his life, retiring in 1894. Dedekind died in 1916.

Dedekind made a number of highly significant contributions to mathematics and his work would change the style of mathematics into what is familiar to us today. One remarkable piece of work was his redefinition of irrational numbers in terms of Dedekind cuts, as we mentioned above. His work in number theory, particularly in algebraic number fields, is of major importance. He edited Dirichlet's lectures, and it was in their third and fourth editions [8], published in 1879 and 1894, that Dedekind wrote supplements in which he introduced the notion of an ideal. Dedekind's work was quickly accepted, partly because of the clarity with which he presented his ideas.

Dedekind's brilliance consisted not only of the theorems and concepts that he studied but, because of his ability to formulate and express his ideas so clearly, his new style of mathematics has been a major influence on mathematicians ever since.

The full proof of Fermat's Last Theorem had to wait for much more powerful methods, developed in the latter half of the 20th century. More about this in the next chapter.

Exercises

8.43 Let $R = \mathbb{Z}[\sqrt{-5}]$ and let $N : R \to \mathbb{Z}$ be the norm map: $N(z) = z\,\bar{z}$. Show that

   (i) $N(zw) = N(z)\,N(w)$ for all $z, w \in R$.

   (ii) u is a unit in R if and only if $N(u) = 1$.

   (iii) If $z \in R$, $N(z) = N(\bar{z})$.

   (iv) If $a \in \mathbb{Z}$, $N(a) = a^2$.

8.44 Find all the units in $R = \mathbb{Z}[\sqrt{-5}]$.

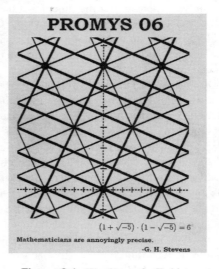

Figure 8.1. The front of a T-shirt.

8.45 Referring to Example 8.52, show that

$$(3) = J_3 J_4$$
$$(1 + \alpha) = J_1 J_3$$
$$(1 - \alpha) = J_2 J_4.$$

8.46 Referring to Example 8.52, show that none of J_2, J_3, J_4 is the unit ideal in R.

8.47 Referring to Example 8.52,

(i) The ideal generated by the norms of elements in J_1 is an ideal in $\mathbb{Z}$, and hence is principal. Find a generator for it.

(ii) Do the same for the other ideals J_i $(i = 2, 3, 4)$.

8.48 Take It Further. Figure 8.1 is the front of a T-shirt that illustrates that

$$2 \cdot 3 = (1 + \sqrt{-5})(1 + \sqrt{-5}).$$

Explain.

8.5 Connections: Counting Sums of Squares

This section investigates an extension of a question we asked in Section 8.2. You saw, in Corollary 8.20, that an odd rational prime can be written as a sum of two squares if and only if it is congruent to 1 mod 4. What about composite integers? For example, 15 can't be written as $a^2 + b^2$, but 65 can: $65 = 8^2 + 1^2$. In fact, 65 can be written as the sum of two squares in another way: $65 = 4^2 + 7^2$. This leads to the following question:

> *In how many ways can a positive integer be written as a sum of two squares?*

We should say "as a sum of two squares of nonnegative integers."

The surprising answer to this question was first discovered by Fermat. Just as we used the arithmetic of Eisenstein integers to prove Fermat's Last Theorem for exponent 3, we'll use the arithmetic of Gaussian integers to understand Fermat's discovery.

Before continuing, let's first consider $n = 5$. Now 5 is a sum of two squares: $5 = 2^2 + 1^2$. We recognize the norm of a Gaussian integer, for $5 = 2^2 + 1^2 = N(2+i)$. Is there another way to write 5 as a sum of two squares? Recall that 5 splits in $\mathbb{Z}[i]$ as $(2 + i)(2 - i)$, which suggests writing 5 as $N(2 - i)$; that is, $5 = 2^2 + (-1)^2$. If we agree, when we write $n = a^2 + b^2$, that both a and b are nonnegative, then we can ignore the second equation $5 = N(2 - i)$. Ah, but there's another way to write it as $N(a + bi)$ with both a, b nonnegative. While $2 - i$ doesn't have nonnegative real and imaginary parts, it is associate to $2 + i$, because

$$i(2 - i) = 1 + 2i;$$

and $2 + i$ and $1 + 2i$ are *not* associates (why?). So there are two bonafide non-associate Gaussian integers $a + bi$ with nonnegative a and b and norm 5. Let's agree, then, that 5 is a sum of two squares in *two* ways: $2^2 + 1^2$ and $1^2 + 2^2$. The following definition of a counting function makes sense.

Definition. The function $r \colon \mathbb{N} \to \mathbb{N}$ is defined on nonnegative integers by

$r(n) = $ the number of non-associate Gaussian integers of norm n.

Since we are interested in equations $n = a^2 + b^2$ in which a and b are nonnegative, it is reasonable to restrict our attention to non-associate Gaussian integers in the first quadrant. By Exercise 8.13 on page 343, every Gaussian integer is associate to exactly one Gaussian integer in the first quadrant of the complex plane, We've been using the term "first quadrant" throughout the book, often without properly defining it. We now insist that the positive x-axis is in the first quadrant but that the positive y-axis is not. The reason is, viewing $\mathbb{R}^2$ as $\mathbb{C}$, that we want to find a piece of the complex plane that contains one Gaussian integer from each class of associates.

Definition. The *first quadrant* of the complex plane is

$$Q_1 = \{z = a + bi \in \mathbb{C} : a > 0, b \geq 0\}.$$

The real number c is associate to ic (on the imaginary axis), and we don't include ic in the first quadrant. Also, the origin $0 = 0 + 0i$ does not lie in Q_1.

In light of these remarks, we modify the definition of $r(n)$ for the purpose at hand, counting only Gaussian integers in the first quadrant (as we wrote above, Exercise 8.13 says that two such integers are necessarily not associate).

$$r(n) = |\{z \in \mathbb{Z}[i] \mid N(z) = n \text{ and } z \in Q_1\}|.$$

It's worth calculating $r(n)$ to get a feel for what it is counting. High school students should enjoy working out some of these numbers by hand (there is no need to mention machinery of $\mathbb{Z}[i]$). Here are some values for them to check.

You can also check that $r(15625) = 7$ and, if you are ambitious, that $r(815730721) = 9$.

n	$r(n)$	n	$r(n)$	n	$r(n)$	n	$r(n)$	n	$r(n)$
1	1	11	0	21	0	31	0	41	2
2	1	12	0	22	0	32	1	42	0
3	0	13	2	23	0	33	0	43	0
4	1	14	0	24	0	34	2	44	0
5	2	15	0	25	3	35	0	45	2
6	0	16	1	26	2	36	1	46	0
7	0	17	2	27	0	37	2	47	0
8	1	18	1	28	0	38	0	48	0
9	1	19	0	29	2	39	0	49	1
10	2	20	2	30	0	40	2	50	3

Look for regularity in the table, make some conjectures, and try to prove them. For example, can you see anything that the values of n for which $r(n) = 0$ have in common?

Or, pick a few primes p, say $3, 5, 7, 11, 13$ and see what happens as you calculate the values of $r(p^k)$. What about products of two primes?

It's likely that Fermat did exactly these kinds of investigations—lots of purposeful numerical calculations—to arrive at an amazing result that we'll prove in this section.

Theorem 8.53 (Fermat). *The number of ways an integer n can be written as a sum of two squares is the excess of the number of divisors of n of the form $4k + 1$ over the number of divisors of n of the form $4k + 3$; that is, if*

$$A(n) = \text{the number of divisors of } n \text{ of form } 4k + 1$$

and

$$B(n) = \text{the number of divisors of } n \text{ of form } 4k + 3,$$

then

$$r(n) = A(n) - B(n).$$

The proof that we'll develop uses some new machinery as well as the law of decomposition for Gaussian integers. First, a few examples that show some of the delightful consequences of the theorem.

Example 8.54. (i) Consider $n = 65 = 13 \cdot 5$. Its divisors are

$$1, \ 5, \ 13, \ 65.$$

There are four divisors congruent to 1 mod 4 and none congruent to 3 mod 4, so $r(65) = 4$. Sure enough,

$$\begin{aligned}
65 &= 1 + 64 \\
&= 64 + 1 \\
&= 16 + 49 \\
&= 49 + 16.
\end{aligned}$$

(ii) Let $n = 21$. Its divisors are

$$1, \ 3, \ 7, \ 21.$$

There are two divisors that are 1 mod 4 and two that are 3 mod 4, so $r(21) = 0$. Thus, 21 is not a sum of two squares.

(iii) Let $n = 3^m$ for some integer m. Odd powers of 3 are congruent to 3 mod 4, while even powers are 1 mod 4. The divisors of 3^m are

$$1, \ 3, \ 3^2, \ 3^3, \ \ldots, \ 3^m.$$

It follows that

$$r(3^m) = \begin{cases} 0 & \text{if } m \text{ is odd} \\ 1 & \text{if } m \text{ is even.} \end{cases} \quad \blacktriangle$$

See Exercise 8.49 on page 377.

Corollary 8.55. *For any positive integer n, we have $A(n) \geq B(n)$; that is, n has at least as many divisors of the form $4k + 1$ as it has divisors of the form $4k + 3$.*

Proof. By Theorem 8.53, we have $A(n) - B(n) = r(n)$, and $r(n) \geq 0$. ∎

A Proof of Fermat's Theorem on Divisors

Let's now prove Theorem 8.53. Our proof requires a device that finds applications all over mathematics—a theory developed by Dirichlet that uses formal algebra to answer combinatorial questions in arithmetic.

Once again, it's time to pull out the pencil and paper.

Definition. A formal *Dirichlet series* is an expression of the form

$$\sum_{n=1}^{\infty} \frac{a(n)}{n^s} = a(1) + \frac{a(2)}{2^s} + \frac{a(3)}{3^s} + \cdots,$$

where the $a(n)$ are complex numbers. (It will be useful to write $a(n)$ instead of the usual a_n.)

The word "formal" is important here—we think of these series as book-keeping devices keeping track of combinatorial or numerical data (as in Example 2.31). So, we don't worry about questions of convergence; we think of s simply as an *indeterminate* rather than as a variable that can be replaced by a real or complex number. This misses many of the wonderful analytic applications of such series, but it turns out that their formal algebraic properties are all we need for this discussion.

Dirichlet series are added and multiplied formally. Addition is done term by term:

$$\sum_{n=1}^{\infty} \frac{a(n)}{n^s} + \sum_{n=1}^{\infty} \frac{b(n)}{n^s} = \sum_{n=1}^{\infty} \frac{a(n) + b(n)}{n^s}.$$

Multiplication is also done term by term, but then one gathers up all terms with the same denominator. So, for example, if we're looking for $c(12)/12^s$ in

$$\sum_{n=1}^{\infty} \frac{a(n)}{n^s} \sum_{n=1}^{\infty} \frac{b(n)}{n^s} = \sum_{n=1}^{\infty} \frac{c(n)}{n^s},$$

then a denominator of 12^s could come only from the products

$$\frac{a(1)}{1^s} \cdot \frac{b(12)}{12^s}, \frac{a(2)}{2^s} \cdot \frac{b(6)}{6^s}, \frac{a(3)}{3^s} \cdot \frac{b(4)}{4^s}, \frac{a(4)}{4^s} \cdot \frac{b(3)}{3^s}, \frac{a(6)}{6^s} \cdot \frac{b(2)}{2^s}, \frac{a(12)}{12^s} \cdot \frac{b(1)}{1^s}.$$

In general, the coefficient $c(n)$ in Eq. (8.5) is given by

$$c(n) = \sum_{d \mid n} a(d) \cdot b\left(\frac{n}{d}\right), \tag{8.19}$$

where $\sum_{d \mid n}$ means that the sum is over the divisors of n.

The simplest Dirichlet series is the ***Riemann zeta function***:

$$\zeta(s) = \sum_{n=1}^{\infty} \frac{1}{n^s}.$$

Eq. (8.19) implies that if

$$\zeta(s) \sum_{n=1}^{\infty} \frac{a(n)}{n^s} = \sum_{n=1}^{\infty} \frac{c(n)}{n^s},$$

then

$$c(n) = \sum_{d \mid n} a(d). \tag{8.20}$$

Let's state this as a theorem.

Theorem 8.56. *If*

$$\zeta(s) \sum_{n=1}^{\infty} \frac{a(n)}{n^s} = \sum_{n=1}^{\infty} \frac{c(n)}{n^s},$$

then $c(n) = \sum_{d \mid n} a(d)$.

... [to] omit those parts of the subject, however, is like listening to a stereo broadcast of, say Beethoven's Ninth Symphony, using only the left audio channel. [39].

Dirichlet series are not formal power series, and multiplication is not the same as in $\mathbb{C}[[x]]$.

Actually, the *zeta function* usually means the function of a complex variable s that analytically continues this infinite series.

Eq. (8.20) is, as we'll see, extremely useful, and it's the reason for defining Dirichlet series.

Proof. Expand

$$\left(\sum_{n=1}^{\infty}\frac{1}{n^s}\right)\left(\sum_{n=1}^{\infty}\frac{a(n)}{n^s}\right)$$

and gather terms with the same denominator. ∎

Sometimes, the coefficients $a(n)$ have interesting properties. For example,

Definition. A function $a\colon \mathbb{N} \to \mathbb{C}$ is *strongly multiplicative* if, for all nonnegative integers m, n,

$$a(mn) = a(m)a(n).$$

(A function $a\colon \mathbb{N} \to \mathbb{C}$ is *multiplicative* if $a(mn) = a(m)a(n)$ whenever $\gcd(m, n) = 1$.)

When a is strongly multiplicative, the Dirichlet series with coefficients $a(n)$ has an alternate form that shows its connection with arithmetic.

Theorem 8.57. *If a is a strongly multiplicative function, then the Dirichlet series*

$$f(s) = \sum_{n=1}^{\infty}\frac{a(n)}{n^s}$$

has a product expansion

$$f(s) = \prod_{p}\left(\frac{1}{1 - \frac{a(p)}{p^s}}\right),$$

where the product is over all prime numbers p.

Proof. Each factor on the right side is a geometric series:

$$\frac{1}{1 - \frac{a(p)}{p^s}} = 1 + \left(\frac{a(p)}{p^s}\right) + \left(\frac{a(p)}{p^s}\right)^2 + \left(\frac{a(p)}{p^s}\right)^3 + \dots$$

$$= 1 + \left(\frac{a(p)}{p^s}\right) + \left(\frac{a(p^2)}{p^{2s}}\right) + \left(\frac{a(p^3)}{p^{3s}}\right) + \dots.$$

To be rigorous, we should put some restrictions on the values of $a(k)$ to ensure that the series converges.

Multiply these together (one for each prime) and you get the sum of every possible expression of the form

$$\frac{a(p_1^{e_1})a(p_2^{e_2})\dots a(p_r^{e_r})}{p_1^{e_1 s}p_2^{e_2 s}\dots p_r^{e_r s}} = \frac{a(p_1^{e_1}p_2^{e_2}\dots p_r^{e_r})}{\left(p_1^{e_1}p_2^{e_2}\dots p_r^{e_r}\right)^s}.$$

Since every $n \in \mathbb{Z}$ can be written in one and only one way as a product of powers of primes (the fundamental theorem again), this is the same as the sum

$$\sum_{n=1}^{\infty}\frac{a(n)}{n^s}. \quad \blacksquare$$

Example 8.58. (i) The constant function $a(n) = 1$ is strongly multiplicative, so the Riemann zeta function has a product expansion

$$\zeta(s) = \sum_{n=1}^{\infty} \frac{1}{n^s} = \prod_p \frac{1}{1 - \frac{1}{p^s}}.$$

(ii) Here's a multiplicative function that's connected to our work with Gaussian integers:

$$\chi(n) = \begin{cases} 1 & \text{if } n \equiv 1 \pmod 4 \\ -1 & \text{if } n \equiv 3 \pmod 4 \\ 0 & \text{if } n \text{ is even} \end{cases}.$$

*χ is called a **quadratic character**.*

You can check that χ is strongly multiplicative, and so

$$\sum_{n=1}^{\infty} \frac{\chi(n)}{n^s} = \prod_p \left(\frac{1}{1 - \frac{\chi(p)}{p^s}} \right). \quad \blacktriangle$$

Now, by Theorem 8.56, if

$$\zeta(s) \sum_{n=1}^{\infty} \frac{\chi(n)}{n^s} = \sum_{n=1}^{\infty} \frac{a(n)}{n^s} \quad \text{then} \quad a(n) = \sum_{d|n} \chi(d). \tag{8.21}$$

So, $a(n)$ is the excess of the number of divisors of n of the form $4k + 1$ over the number of divisors of n of the form $4k + 3$. Bingo: this is exactly the function that is the at heart of Theorem 8.53. The idea, then, is to form the Dirichlet series with coefficients $r(n)$ and to show that

$$\sum_{n=1}^{\infty} \frac{r(n)}{n^s} = \zeta(s) \sum_{n=1}^{\infty} \frac{\chi(n)}{n^s}.$$

To do this, we'll convert each of the sums to products. We already have done this in Example 8.58 for the sums on the right-hand side; for the left-hand side, we argue as follows.

Each term in the left-hand sum is a sum of unit fractions, and the number of such fractions is the number of Gaussian integers with given norm. For example, $3/25^s$ comes from

$$\frac{1}{N(3 + 4i)} + \frac{1}{N(4 + 3i)} + \frac{1}{N(5 + 0i)}.$$

Using this idea and the multiplicativity of N, we get a product formula for the left-hand side.

$$\sum_{n=1}^{\infty} \frac{r(n)}{n^s} = \sum_{\alpha \in Q_1} \frac{1}{(N(\alpha))^s}$$

$$= \prod_{\mathfrak{p} \in Q_1} \sum_{k=0}^{\infty} \frac{1}{\left((N(\mathfrak{p}))^k\right)^s} \quad \text{(use the fundamental theorem in } \mathbb{Z}[i])$$

$$= \prod_{\mathfrak{p} \in Q_1} \frac{1}{1 - \frac{1}{N(\mathfrak{p})^s}} \quad \text{(sum a geometric series)}.$$

Here, the product is over all Gaussian primes in the first quadrant. This is another example that is best understood by calculating a few coefficients by hand.

Now use Theorem 8.21 (the law of decomposition for $\mathbb{Z}[i]$). Every prime in Q_1 lies over one of these:

- the prime 2. There's only one in the first quadrant: $1 + i$, and $N(1 + i) = 2$.
- a prime p congruent to 1 mod 4. There are two for each such p—if

$$p = \pi \, \overline{\pi},$$

then both π and $\overline{\pi}$ have an associate in Q_1 (and they are different), and each has norm p.

- a prime p congruent to 3 mod 4. There's only one such prime in Q_1, because such a p is inert and $N(p) = p^2$.

So,

$$\sum_{n=1}^{\infty} \frac{r(n)}{n^s} = \prod_{z \in Q_1} \frac{1}{1 - \dfrac{1}{N(z)^s}} = \frac{1}{1 - \dfrac{1}{2^s}} \left(\prod_{\substack{p \equiv \\ 1 \bmod 4}} \frac{1}{1 - \dfrac{1}{p^s}} \right)^2 \left(\prod_{\substack{p \equiv \\ 3 \bmod 4}} \frac{1}{1 - \dfrac{1}{p^{2s}}} \right)$$

$$= \frac{1}{1 - \dfrac{1}{2^s}} \left(\prod_{\substack{p \equiv \\ 1 \bmod 4}} \frac{1}{1 - \dfrac{1}{p^s}} \right)^2 \left(\prod_{\substack{p \equiv \\ 3 \bmod 4}} \frac{1}{1 - \dfrac{1}{p^s}} \right) \left(\prod_{\substack{p \equiv \\ 3 \bmod 4}} \frac{1}{1 + \dfrac{1}{p^s}} \right)$$

$$= \frac{1}{1 - \dfrac{1}{2^s}} \left(\prod_{p \text{ odd}} \frac{1}{1 - \dfrac{1}{p^s}} \right) \left(\prod_{p \equiv 1 \bmod 4} \frac{1}{1 - \dfrac{1}{p^s}} \right) \left(\prod_{p \equiv 3 \bmod 4} \frac{1}{1 + \dfrac{1}{p^s}} \right)$$

$$= \zeta(s) \left(\prod_{p \equiv 1 \bmod 4} \frac{1}{1 - \dfrac{\chi(p)}{p^s}} \right) \left(\prod_{p \equiv 3 \bmod 4} \frac{1}{1 - \dfrac{\chi(p)}{p^s}} \right)$$

$$= \zeta(s) \left(\prod_{p \text{ odd}} \frac{1}{1 - \dfrac{\chi(p)}{p^s}} \right) = \zeta(s) \sum_{n=1}^{\infty} \frac{\chi(n)}{n^s}$$

$$= \sum_{n=1}^{\infty} \frac{a(n)}{n^s},$$

where, by Eq. (8.21),

$$a(n) = \sum_{d \mid n} \chi(d).$$

It follows that $r(n) = a(n)$, and we've proved Theorem 8.53.

Exercises

8.49 Suppose that $m \geq 1$ is an integer. Show that if p is a prime and $p \equiv 3 \bmod 4$,

$$r(p^m) = \begin{cases} 0 & \text{if } m \text{ is odd} \\ 1 & \text{if } m \text{ is even.} \end{cases}$$

8.50 Suppose that $m \geq 1$ is an integer. Show that if p is a prime and $p \equiv 1 \bmod 4$,

$$r\left(p^m\right) = m + 1.$$

8.51 Suppose that $m \geq 1$ is an integer. Show that

$$s\left(2^m\right) = 1.$$

8.52 Show that if a is a strongly multiplicative function, so is b, where b is defined by

$$b(n) = \sum_{d \mid n} a(d).$$

8.53 A ***multiplicative function*** is a function $a : \mathbb{N} \to \mathbb{C}$ so that $a(mn) = a(m)a(n)$ whenever $\gcd(m, n) = 1$.

  (i) Give an example of a multiplicative function that is not strongly multiplicative.

If a is strongly multiplicative, is b?

  (ii) Show that if a is a multiplicative function, so is b, where

$$b(n) = \sum_{d \mid n} a(d).$$

8.54 Show that, if $\gcd(m, n) = 1$, then

$$r(mn) = r(m)r(n).$$

8.55 Show that an integer can be written as a sum of two squares if and only if the primes in its prime factorization that are congruent to 3 mod 4 show up with even exponents.

8.56 **Take It Further.** Show that

$$\zeta(s) \sum_{n=1}^{\infty} \frac{\phi(n)}{n^s} = \zeta(s - 1),$$

where ϕ is the Euler ϕ-function.

8.57 **Take It Further.** Tabulations of r show some erratic behavior with no apparent pattern. When that happens with a function f, it's often useful to look at the asymptotic behavior of its average value:

It might be easier to see things if you use $4r$ instead of r, allowing Gaussian integers in all four quadrants.

$$\lim_{n \to \infty} \frac{1}{n} \sum_{k=1}^{n} f(k).$$

Investigate the asymptotic behavior of the average value of r.

9

Epilog

Attempts to resolve Fermat's Last Theorem have led to much modern algebra. There were many other areas of mathematical research in the seventeenth, eighteenth and nineteenth centuries, one of which was determining the roots of polynomials. Informally, a polynomial is *solvable by radicals* if its roots can be given by a formula generalizing the classical quadratic, cubic, and quartic formulas. In 1824, Abel proved that there are quintic polynomials that are not solvable by radicals and in 1828 he found a class of polynomials, of any degree, that are solvable by radicals. In 1830, Galois, the young wizard who was killed before his 21st birthday, characterized *all* the polynomials which are solvable by radicals, greatly generalizing Abel's theorem. Galois' brilliant idea was to exploit symmetry through his invention of *group theory*.

After a brief account of the lives of Abel and Galois, we will use ring theory to make the notion of solvability by radicals precise. This will enable us to understand the work of Abel and Galois showing why there is no generalization of the classical formulas to polynomials of higher degree. We will then introduce some group theory, not only because groups were the basic new idea in the study of polynomials, but because they are one of the essential ingredients in Wiles' proof of Fermat's Last Theorem in 1995. In fact, symmetry is an important fundamental idea arising throughout mathematics. In the last section, we will say a bit about Andrew Wiles and his proof of Fermat's Last Theorem.

9.1 Abel and Galois

Niels Abel was born in Frindöe, Norway, near Stavanger, in 1802. Norway was then suffering extreme poverty as a consequence of economic problems arising from European involvement in the Napoleonic wars. Abel's family was very poor, although things improved a little in 1816 when his father, a Protestant minister, became involved in politics (Norway, which had been part of Denmark, claimed independence, and then became a largely autonomous kingdom in a union with Sweden). The next year, Abel was sent to a school in Christiana (present day Oslo), but he was an ordinary student there with poor teachers (the best teachers having gone to the recently opened University of Christiana). But two years later, a new mathematics teacher, B. Holmboë, joined the school and inspired Abel to study mathematics. Holmboë was convinced Abel had great talent, and he encouraged him to read the works of contemporary masters. In 1820, Abel's father died; there was no money for Abel to complete his education nor to enter the University. But Holmboë continued his support, helping him to obtain a scholarship to enter the University in 1821; Abel graduated the following year.

So, an inspiring teacher helped set the course of modern algebra.

379

The importance of numerical examples can't be overestimated.

In 1821, while in his final year at the University of Christiana, Abel thought he had proved that quintic polynomials are solvable by radicals, and he submitted a paper to the Danish mathematician Degen for publication by the Royal Society of Copenhagen. Degen asked Abel to give a numerical example of his method and, in trying to do this, Abel discovered a mistake in his paper. Degen had also advised Abel to study elliptic integrals, and Abel wrote several important fundamental papers on the subject. In 1824, Abel returned to quintic polynomials, proving that the general quintic polynomial is not solvable by radicals.

In 1825, having now done brilliant work in two areas of mathematics, the Norwegian government gave Abel a scholarship to travel abroad. He went to Germany and France, hoping to meet eminent mathematicians, but Gauss was not interested in Abel's work on the quintic, and the mathematicians in Paris did not yet appreciate his remarkable theorems on elliptic functions. By 1827, Abel's health deteriorated, he was heavily in debt, and he returned home to Norway. In 1828, he briefly returned to polynomials, proving a theorem describing a class of polynomials (of any degree) that are solvable by radicals. By this time, Abel's fame had spread to all mathematical centers. Legendre saw the new ideas in papers of Abel and of Jacobi, and he wrote

> *Through these works you two (Abel and Jacobi) will be placed in the class of the foremost analysts of our times.*

Strong efforts were made to secure a suitable position for Abel by a group from the French Academy, who addressed King Bernadotte of Norway-Sweden; Crelle also worked to secure a professorship for him in Berlin. But it was too late. Abel died in 1829, at age 26.

An imprecise measure of Abel's influence on modern mathematics is the number of areas named after him: abelian groups, abelian varieties, abelian differentials, abelian integrals, abelian categories, abelian extensions, abelian number fields, abelian functions. The Niels Henrik Abel Memorial Fund was established in 2002, and the Norwegian Academy of Science and Letters awards the Abel Prize for outstanding scientific work.

Évariste Galois was born in Bourg La Reine, near Paris, in 1811. France, and especially Paris, was then in the throes of great political and social change as a consequence of the French Revolution in 1789, the Napoleonic era 1799–1815, the restoration of the French monarchy with King Louis XVIII in 1815, his overthrow by King Charles X in 1824, and another revolution in 1830.

In April 1829, Galois' first mathematics paper (on continued fractions) was published; he was then 17 years old. In May and June, he submitted articles on the algebraic solution of equations to Cauchy at the Academy of Science. Cauchy advised him to rewrite his article, and Galois submitted *On the condition that an equation be solvable by radicals* in February 1830. The paper was sent to Fourier, the secretary of the Academy, to be considered for the Grand Prize in mathematics. But Fourier died in April 1830, Galois' paper was never subsequently found, and so it was never considered for the prize. July 1830 saw another revolution. King Charles X fled France, and there was rioting in the streets of Paris. Later that year, Galois (now age 19) was arrested for making threats against the king at a public dinner, but he was acquitted. Galois was invited by Poisson to submit a third version of his memoir on equations to the Academy, and he did so in January 1831. On July 14, Galois was arrested again. While in prison he received a rejection of his memoir. Poisson reported that "His argument is neither sufficiently clear nor sufficiently developed to allow us to judge its rigor.... There is no good way of deciding whether a given polynomial ... is solvable." He did, however, encourage Galois to publish a more complete account of his work.

In March 1832, a cholera epidemic swept Paris, and prisoners were transferred to boarding houses. The prisoner Galois was moved to a pension, where he apparently fell in love with Stéphanie-Félice du Motel, the daughter of the resident physician. After he was released on April 29, Galois exchanged letters with Stéphanie, and it is clear that she tried to distance herself from the affair. Galois fought a duel on May 30, the reason for the duel not being clear but certainly linked with Stéphanie. A note in the margin of the manuscript that Galois wrote the night before the duel reads, "There is something to complete in this proof. I do not have the time." It is this note which has led to the legend that he spent his last night writing out all he knew about group theory (but this story appears to have been exaggerated). Galois was mortally wounded in the duel, and he died the next day, only 20 years old. His funeral was the focus of a Republican rally, and riots lasting several days followed.

According to Galois' wish, his friend Chevalier and Galois' brother Alfred copied Galois' mathematical papers and sent them to Gauss, Jacobi, and others. No record exists of any comment these men may have made. Eventually, the papers reached Liouville who, in September 1843, announced to the Academy that he had found in Galois' papers a concise solution "... as correct as it is deep, of this lovely problem: given an irreducible equation of prime degree, decide whether or not it is solvable by radicals." Liouville published these papers of Galois in his Journal in 1846. What Galois outlined in these papers is called *Galois Theory* today.

The following quotation is from the Epilog of Tignol's book [33].

> *After the publication of Galois' memoir by Liouville, its importance dawned upon the mathematical world, and it was eventually realized that Galois had discovered a mathematical gem much more valuable than any hypothetical external characterization of solvable equations. After all, the problem of solving equations by radicals was utterly artificial. It had focused the efforts of several generations of brilliant mathematicians because it displayed some strange, puzzling phenomena. It contained something mysterious, profoundly appealing. Galois had taken the pith out of the problem, by showing that the difficulty of an equation was related to the ambiguity of its roots and pointing out how this ambiguity could be measured by means of a group. He had thus set the theory of equations and, indeed, the whole subject of algebra, on a completely different track.*

We have chosen Fermat's Last Theorem as an organizing principle of this book, but an interesting abstract algebra text could be written centered on group theory and roots of polynomials.

9.2 Solvability by Radicals

Informally, a polynomial is *solvable by radicals* if there is a formula for its roots that generalizes the classical quadratic, cubic, and quartic formulas. Let us now examine the classical formulas to make this rather vague idea more precise.

How to Think About It. Even though much of what we shall say applies to polynomials over any field, the reader may assume all fields coming up are subfields of the complex numbers $\mathbb{C}$. We point out, however, that some familiar results may not be true for all fields. For example, the quadratic formula doesn't hold in $k[x]$ when k has characteristic 2 (for $\frac{1}{2}$ doesn't make sense in k); similarly, neither the cubic formula nor the quartic formula holds in $k[x]$ when k has characteristic either 2 or 3.

Definition. A field extension K/k is a *pure extension* if $K = k(u)$, where $u^n \in k$ for some $n \geq 1$.

In more detail, $K = k(u)$, where u is a root of $x^n - a$ for some $a \in k$; that is, $u = \sqrt[n]{a}$, and so we are adjoining an nth root of a to k. But there are several nth roots of a in $\mathbb{C}$, namely

$$\sqrt[n]{a}, \quad \zeta \sqrt[n]{a}, \quad \ldots, \quad \zeta^{n-1} \sqrt[n]{a},$$

where $\zeta = \zeta_n = \cos(2\pi/n) + i \sin(2\pi/n)$ is a primitive nth root of unity. To avoid having to decide which one to adjoin, let's adjoin all of them by adjoining any one of them together with all the nth roots of unity. This is reasonable, for we are seeking formulas for roots of polynomial equations that involve square roots, cube roots, etc., and roots of numbers appear explicitly in the classical formulas.

Let's consider the classical formulas for polynomials of small degree, for we'll see that they give rise to a sequence of pure extensions.

Quadratics

If $f(x) = x^2 + bx + c$, then the quadratic formula gives its roots as

$$\frac{1}{2}\left(-b \pm \sqrt{b^2 - 4c}\right).$$

Let $k = \mathbb{Q}(b, c)$. Define $K_1 = k(u)$, where $u = \sqrt{b^2 - 4c}$. Then K_1 is a pure extension, for $u^2 \in k$. Moreover, the quadratic formula implies that K_1 is the splitting field of f.

Cubics

Let $f(X) = X^3 + bX^2 + cX + d$, and let $k = \mathbb{Q}(b, c, d)$. The change of variable $X = x - \frac{1}{3}b$ yields a new polynomial $\widetilde{f}(x) = x^3 + qx + r \in k[x]$ having the same splitting field E (for if u is a root of $\widetilde{f}$, then $u - \frac{1}{3}b$ is a root of f); it follows that $\widetilde{f}$ is solvable by radicals if and only if f is. The cubic formula gives the roots of $\widetilde{f}$ as

$$g + h, \quad \omega g + \omega^2 h, \quad \text{and} \quad \omega^2 g + \omega h,$$

where $g^3 = \frac{1}{2}\left(-r + \sqrt{R}\right)$, $h = -q/3g$, $R = r^2 + \frac{4}{27}q^3$, and ω is a primitive cube root of unity. Because of the constraint $gh = -\frac{1}{3}q$, each choice of $g = \sqrt[3]{\frac{1}{2}(-r + \sqrt{R})}$ has a "mate," namely $h = -q/(3g)$, $-q/(3\omega g) = \omega^2 h$, and $-q/(3\omega^2 g) = \omega h$.

Define $K_1 = k(\sqrt{R})$, and define $K_2 = K_1(g)$, where $g^3 = \frac{1}{2}(-r + \sqrt{R})$. The cubic formula shows that K_2 contains the root $g + h$ of $\tilde{f}$, where $h = -q/3g$. Finally, define $K_3 = K_2(\omega)$, where $\omega^3 = 1$. The other roots of $\tilde{f}$ are $\omega g + \omega^2 h$ and $\omega^2 g + \omega h$, both of which lie in K_3, and so $E \subseteq K_3$.

Thus, a sequence of pure extensions seems to capture the notion we are seeking. We now give the formal definition of *solvability by radicals*, after which we will show that all polynomials of degree ≤ 4 are solvable by radicals.

Definition. A *radical extension* of a field k is a field extension K/k for which there exists a tower of field extensions

$$k = K_0 \subseteq K_1 \subseteq \cdots \subseteq K_t = K,$$

where K_i / K_{i-1} is a pure extension for all $i \geq 1$.

A polynomial $f(x) \in k[x]$ is *solvable by radicals* if there is a splitting field E/k and a radical extension K/k with $E \subseteq K$.

Quadratics are solvable by radicals, and the cubic formula shows that every cubic $f(x) = x^3 + qx + r \in \mathbb{Q}[x]$ is solvable by radicals: a radical extension containing a splitting field of f is

$$\mathbb{Q}(q, r) = K_0 \subseteq K_1 = K_0(\omega) \subseteq K_2 = K_1(\sqrt{R}) \subseteq K_3 = K_2(g),$$

where we are using the notation in the cubic formula.

Why do we say in the definition of solvable by radicals, that $E \subseteq K$ instead of $E = K$? That is, why don't we say that some splitting field is a radical extension? The answer is that it isn't. Consider the following theorem, due to Hölder.

Theorem 9.1 (Casus Irreducibilis). *Let $f(x) \in \mathbb{Q}[x]$ be an irreducible cubic having three real roots. If $E \subseteq \mathbb{C}$ is the splitting field of f and K is a radical extension of $\mathbb{Q}$ containing E, then $K \subsetneq \mathbb{R}$.*

Proof. [25], p. 217. ∎

If $f(x) \in \mathbb{Q}[x]$ is an irreducible cubic all of whose roots are real ($f(x) = 3x^3 - 3x + 1$ is such a cubic, by Example 6.58), then its splitting field $E \subseteq \mathbb{R}$. We have just seen that f is solvable by radicals, so there is a radical extension $K/\mathbb{Q}$ with $E \subseteq K$. But the Casus Irreducibilis says that $K \subsetneq \mathbb{R}$. Therefore, $E \neq K$ (because $E \subseteq \mathbb{R}$); that is, the splitting field of f is not itself a radical extension.

Here is a more remarkable consequence of the Casus Irreducibilis. In down-to-earth language, it says that any formula for the roots of an irreducible cubic in $\mathbb{Q}[x]$ having all roots real requires the presence of complex numbers! After all, the formula involves a cube root of unity. In other words, it is impossible to "simplify" the cubic formula to eliminate i; we must use complex numbers to find real roots! How would this have played in Piazza San Marco?

We now show that quartic polynomials are solvable by radicals.

Proposition 9.2. *Every polynomial $f(X) = X^4 + bX^3 + cX^2 + dX + e \in \mathbb{Q}[x]$ is solvable by radicals.*

Proof. Let $k = \mathbb{Q}(b, c, d, e)$. The change of variable $X = x - \frac{1}{4}b$ yields a new polynomial $\widetilde{f}(x) = x^4 + qx^2 + rx + s \in k[x]$; moreover, the splitting field E of f is equal to the splitting field of $\widetilde{f}$, for if u is a root of $\widetilde{f}$, then $u - \frac{1}{4}b$ is a root of f. Factor $\widetilde{f}$ in $\mathbb{C}[x]$:

$$\widetilde{f}(x) = x^4 + qx^2 + rx + s = (x^2 + jx + \ell)(x^2 - jx + m),$$

and determine j, ℓ, and m. Now j^2 is a root of the ***resolvent cubic***:

$$(j^2)^3 + 2q(j^2)^2 + (q^2 - 4s)j^2 - r^2.$$

The cubic formula gives j^2, from which we can determine m and ℓ, and hence the roots of the quartic.

Define a radical extension

$$k = K_0 \subseteq K_1 \subseteq K_2 \subseteq K_3,$$

as in the cubic case, so that $j^2 \in K_3$. Define $K_4 = K_3(j)$ (so that $\ell, m \in K_4$). Finally, define $K_5 = K_4\left(\sqrt{j^2 - 4\ell}\right)$ and $K_6 = K_5\left(\sqrt{j^2 - 4m}\right)$, giving roots of the quadratic factors $x^2 + jx + \ell$ and $x^2 - jx + m$ of $\widetilde{f}$. The quartic formula gives $E \subseteq K_6$. Therefore, f is solvable by radicals. ∎

Example 9.3. $f(x) = x^5 - 1 \in \mathbb{Q}[x]$ is solvable by radicals. We know that $f(x) = (x - 1)h(x)$, where h is a quartic. But we have just seen that quartics are solvable by radicals. (Actually, Gauss proved that $x^n - 1$ is solvable by radicals for all $n \geq 1$, and this led to his construction of the regular 17-gon by ruler and compass.) ▲

We have just seen that quadratics, cubics, and quartics in $\mathbb{Q}[x]$ are solvable by radicals. Conversely, let $f(x) \in \mathbb{Q}[x]$ be a polynomial of any degree, and let $E/\mathbb{Q}$ be a splitting field. If f is solvable by radicals, we claim that there is a formula that expresses its roots in terms of its coefficients. Suppose that

$\mathbb{Q}$ can be replaced by any field of characteristic 0.

$$\mathbb{Q} = K_0 \subseteq K_1 \subseteq \cdots \subseteq K_t$$

is a radical extension with $E \subseteq K_t$. Let z be a root of f. Now $z \in K_t = K_{t-1}(u)$, where u is an mth root of some element $\alpha \in K_{t-1}$; hence, z can be expressed in terms of u and K_{t-1}; that is, z can be expressed in terms of $\sqrt[m]{\alpha}$ and K_{t-1}. But $K_{t-1} = K_{t-2}(v)$, where some power of v lies in K_{t-2}. Hence, z can be expressed in terms of u, v, and K_{t-2}. Ultimately, z is expressed by a formula analogous to the classical formulas. Therefore, *solvability by radicals* has now been translated into the language of fields.

9.3 Symmetry

Recognizing and exploiting symmetry is an important ingredient in geometry, algebra, number theory, and, indeed, in all of mathematics.

Here is the basic idea: an object is *symmetric* if, when you transform it in a certain way, you get the same object back. For example, what do we mean when we say that an isosceles triangle Δ is symmetric? Figure 9.1 shows $\Delta = \Delta ABC$ with its base $\overline{AB}$ on the x-axis and with the y-axis being the

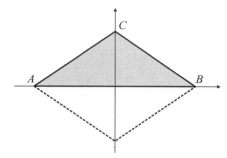

Figure 9.1. Isosceles triangle $\triangle ABC$.

Reflection in the y-axis is the function $(x, y) \mapsto (-x, y)$.
Reflection in the x-axis is the function $(x, y) \mapsto (x, -y)$.

perpendicular-bisector of $\overline{AB}$. Close your eyes; pretend that the y-axis is a mirror, and let $\triangle$ be reflected in the y-axis (so that the vertices A and B are interchanged); open your eyes. You cannot tell that $\triangle$ has been reflected; that is, $\triangle$ is symmetric in the y-axis. On the other hand, if $\triangle$ were reflected in the x-axis, then it would be obvious, once your eyes are reopened, that a reflection had taken place; that is, $\triangle$ is not symmetric in the x-axis.

Here is a non-geometric example: the polynomial $f(x, y) = x^3 + y^3 - xy$ is symmetric because, if you transform it by interchanging x and y, you get the same polynomial back. Another example arises from $g(x) = x^6 - x^2 + 3$. This polynomial is symmetric because $g(-x) = g(x)$; this symmetry induces symmetry of the graph of $g(x)$ in the y-axis, for $(-x, y)$ lies on the graph if and only if (x, y) does.

The transformations involved in defining symmetry are usually permutations.

Definition. A *permutation* of a set X is a bijection $\alpha \colon X \to X$.

Here is a precise definition of symmetry in geometry.

Definition. An *isometry* of the plane is a function $\varphi \colon \mathbb{R}^2 \to \mathbb{R}^2$ that is *distance preserving*: for all points $P = (a, b)$ and $Q = (c, d)$ in $\mathbb{R}^2$,

Now you see how important the Pythagorean Theorem really is; it allows us to define distance.

$$\|\varphi(P) - \varphi(Q)\| = \|P - Q\|,$$

where $\|P - Q\| = \sqrt{(a - c)^2 + (b - d)^2}$ is the distance from P to Q.

A *symmetry* of a subset Ω of the plane is an isometry σ with $\sigma(\Omega) = \Omega$ (by definition, $\sigma(\Omega) = \{\sigma(\omega) : \omega \in \Omega\}$).

It is clear that every isometry is an injection, for if $P \neq Q$, then $\|P - Q\| \neq 0$, hence $\|\sigma(P) - \sigma(Q)\| \neq 0$, and $\sigma(P) \neq \sigma(Q)$. It is also true (but harder to prove) that isometries are surjections ([26], p. 141). Thus, isometries are bijections; that is, they are permutations of the plane.

Some figures have more symmetries than others. Consider the triangles in Figure 9.2. The first, equilateral, triangle has six symmetries: rotations by 120°, 240°, 360° = 0° about its center, and reflections in each of the three angle bisectors. The second, isosceles, triangle has only two symmetries, the identity isometry and the reflection in the angle bisector, while the scalene triangle has only one symmetry, the identity isometry. A circle has infinitely many symmetries (for example, all rotations about its center).

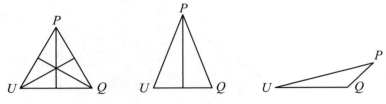

Figure 9.2. Triangles.

We now introduce symmetry in an algebraic setting.

Definition. An *automorphism* of a commutative ring R is an isomorphism $\sigma: R \to R$. Given a field extension E/k, an automorphism $\sigma: E \to E$ *fixes* k if $\sigma(a) = a$ for every $a \in k$.

The following theorem should remind you of Theorem 3.12 (which is the special case when $E/k = \mathbb{C}/\mathbb{R}$ and σ is complex conjugation). Of course, automorphisms are certain kinds of permutations.

Theorem 9.4. *Let k be a field, let $f(x) \in k[x]$, and let E/k be a splitting field of f. If $\sigma: E \to E$ is an automorphism fixing k, then σ permutes the set Ω of all the roots of f.*

Proof. Let $f(x) = a_0 + a_1 x + \cdots + a_n x^n$, where $a_i \in k$ for all i. If $u \in E$ is a root of f, then $f(u) = 0$ and

$$
\begin{aligned}
0 &= \sigma(f(u)) \\
&= \sigma(a_0 + a_1 u + \cdots + a_n u^n) \\
&= \sigma(a_0) + \sigma(a_1)\sigma(u) + \cdots + \sigma(a_n)\sigma(u^n) \\
&= a_0 + a_1\sigma(u) + \cdots + a_n\sigma(u)^n \\
&= f(\sigma(u)).
\end{aligned}
$$

Therefore, $\sigma(u)$ is also a root of f, so that $\sigma(\Omega) \subseteq \Omega$; that is, $\operatorname{im}(\sigma|\Omega) \subseteq \Omega$ and the restriction $\sigma|\Omega$ is a function $\Omega \to \Omega$. But $\sigma|\Omega$ is an injection, because σ is, and the Pigeonhole Principle, Exercise A.11 on page 419, says that it is a permutation. ∎

The following definition, due to E. Artin around 1930, modernizes and simplifies Galois' original definition given 100 years earlier (it is equivalent to Galois' definition).

Definition. If k is a field, $f(x) \in k[x]$, and E/k is a splitting field of f, then the *Galois group* of f is

$$\operatorname{Gal}(f) = \{\text{automorphisms } \sigma: E \to E \text{ fixing } k\}.$$

Just as some triangles are more symmetric than others, some polynomials are more symmetric than others. For example, consider $f(x) = x^2 - 2$ and $g(x) = x^2 - 9$, where we consider both polynomials as lying in $\mathbb{Q}[x]$. The splitting field of f is $E = \mathbb{Q}(\sqrt{2})$, and there is an automorphism $\sigma: E \to E$ that interchanges the roots $\sqrt{2}$ and $-\sqrt{2}$, namely $\sigma: a + b\sqrt{2} \mapsto a - b\sqrt{2}$. On the other hand, the splitting field of g is $\mathbb{Q}$, for both 3 and -3 lie in $\mathbb{Q}$, and so $\text{Gal}(g)$ consists only of the identity permutation.

The astute reader may have noticed that $\text{Gal}(f)$ really depends only on the fields k and E; two polynomials in $k[x]$ having the same splitting field have the same Galois group. For this reason, we usually write

$$\text{Gal}(f) = \text{Gal}(E/k).$$

Example 9.5. We show that not every permutation of the roots of a polynomial f is the restriction of some automorphism $\sigma \in \text{Gal}(f)$. We saw in Example 3.7 that the roots of

$$f(x) = x^4 - 10x^2 + 1 \in \mathbb{Q}[x]$$

are

$$\alpha = \sqrt{2} + \sqrt{3}, \quad \beta = \sqrt{2} - \sqrt{3}, \quad \gamma = -\sqrt{2} + \sqrt{3}, \quad \delta = -\sqrt{2} - \sqrt{3}.$$

Let $E/\mathbb{Q}$ be a field extension containing these four roots, and let π be the permutation that interchanges β and γ and fixes the other two roots:

$$\pi(\alpha) = \alpha, \quad \pi(\beta) = \gamma, \quad \pi(\gamma) = \beta, \quad \pi(\delta) = \delta.$$

In E, we have $\alpha - \beta = 2\sqrt{3}$. Suppose there is an automorphism σ of E with $\sigma|\{\alpha, \beta, \gamma, \delta\} = \pi$. Then $\sigma(\alpha - \beta) = \sigma(2\sqrt{3})$, and

$$\sigma(\alpha - \beta) = \sigma(\alpha) - \sigma(\beta) = \pi(\alpha) - \pi(\beta) = \alpha - \gamma = 2\sqrt{2}.$$

Hence, $\sigma(2\sqrt{3}) = 2\sqrt{2}$. Square both sides:

$$\sigma(2\sqrt{3})^2 = (2\sqrt{2})^2 = 8.$$

The left-hand side is $\sigma(2\sqrt{3})^2 = \sigma\big((2\sqrt{3})^2\big) = \sigma(12) = 12$, and this is a contradiction. Therefore, $\pi \notin \text{Gal}(f)$. ▲

An important class of symmetric polynomials are the ***elementary symmetric polynomials*** in n variables $\alpha_1, \ldots, \alpha_n$. For two variables, the elementary symmetric polynomials are $\alpha_1 + \alpha_2$ and $\alpha_1\alpha_3$. For three variables, they are

$$s_1 = \alpha_1 + \alpha_2 + \alpha_3,$$
$$s_2 = \alpha_1\alpha_2 + \alpha_1\alpha_3 + \alpha_2\alpha_3,$$
$$s_3 = \alpha_1\alpha_2\alpha_3$$

and, for n variables, s_i is the sum of all products of the α_i, taken i at a time. We've met these before in the context of roots of polynomials. The coefficients of a monic polynomial are the elementary symmetric polynomials of its roots, with alternating sign, so that, for example,

$$(x - \alpha_1)(x - \alpha_2)(x - \alpha_3)$$
$$= x^3 - (\alpha_1 + \alpha_2 + \alpha_3)x^2 + (\alpha_1\alpha_2 + \alpha_1\alpha_3 + \alpha_2\alpha_3)x - \alpha_1\alpha_2\alpha_3.$$

It turns out that every symmetric polynomial in n variables can be expressed as a polynomial in the elementary symmetric polynomials s_i (see [2], Chapter IIG). For example,

$$\alpha_1^2 + \alpha_2^2 + \alpha_3^2 = (\alpha_1 + \alpha_2 + \alpha_3)^2 - 2(\alpha_1\alpha_2 + \alpha_1\alpha_3 + \alpha_1\alpha_3)$$
$$= s_1^2 - 2s_2.$$

Example 9.6. The elementary symmetric polynomials can be used to give alternate derivations of the quadratic and cubic formulas (Exercises 9.1 and 9.3 below). Let's sketch a derivation of the cubic formula along these lines.

Assume our cubic has been reduced and is, as before, of the form

$$x^3 + qx + r.$$

Suppose further that we let its roots be α_1, α_2 and α_3. Then we know that

$$\alpha_1 + \alpha_2 + \alpha_3 = 0,$$
$$\alpha_1\alpha_2 + \alpha_1\alpha_3 + \alpha_2\alpha_3 = q,$$
$$\alpha_1\alpha_2\alpha_3 = -r.$$

As usual, $\omega = \frac{1}{2}\left(-1 + i\sqrt{3}\right)$.

Form the two expressions s and u:

$$s = \alpha_1 + \alpha_2\omega + \alpha_3\omega^2$$
$$u = \alpha_1 + \alpha_2\omega^2 + \alpha_3\omega.$$

So, we have three expressions in the roots

$$0 = \alpha_1 + \alpha_2 + \alpha_3$$
$$s = \alpha_1 + \alpha_2\omega + \alpha_3\omega^2$$
$$u = \alpha_1 + \alpha_2\omega^2 + \alpha_3\omega.$$

Adding the equations, we see that $s + u = 3\alpha_1$. Hence, if s and u can be expressed in terms of q and r, then α_1 can be so expressed (and, by symmetry, the other roots can be expressed in terms of q and r). Experimenting with a CAS or by hand, we find that

$$su = \alpha_1^2 + \alpha_2^2 + \alpha_3^2 - \alpha_1\alpha_2 - \alpha_1\alpha_3 - \alpha_2\alpha_3$$
$$= (\alpha_1 + \alpha_2 + \alpha_3)^2 - 3(\alpha_1\alpha_2 + \alpha_1\alpha_3 + \alpha_2\alpha_3)$$
$$= 0 - 3q = -3q.$$

A CAS is a great help here (See Appendix A.6).

Expanding $s^3 + u^3$ and factoring the result, we get

$$s^3 + u^3$$
$$= -(\alpha_1 + \alpha_2 - 2\alpha_3)(\alpha_1 + \alpha_3 - 2\alpha_2)(\alpha_2 + \alpha_3 - 2\alpha_1)$$
$$= -(\alpha_1 + \alpha_2 + \alpha_3 - 3\alpha_3)(\alpha_1 + \alpha_2 + \alpha_3 - 3\alpha_2)(\alpha_1 + \alpha_2 + \alpha_3 - 3\alpha_1)$$
$$= -(-3\alpha_3)(-3\alpha_2)(-3\alpha_1)$$
$$= 27\alpha_1\alpha_2\alpha_3 = -27r.$$

From $su = -3q$, we get $s^3u^3 = -27q^3$. Coupled with $s^3 + u^3 = -27r$, we see that s^3 and u^3 are roots of the quadratic polynomial

$$x^2 + 27rx - 27q^3.$$

We can solve this for s^3 and u^3, take cube roots, and recover α_1, leading to Cardano's formula (Exercise 9.3 below).   ▲

Exercises

Exercise 9.1 shows how to derive the quadratic formula without completing the square.

9.1 * Suppose the roots of $x^2 + bx + c$ are α and β. Find, without using the quadratic formula, an expression for and $\alpha - \beta$ in terms of $\alpha + \beta$ and $\alpha\beta$. Use it and the fact that $\alpha + \beta = -b$ to find α in terms of b and c.

9.2 Find the roots of the cubic from page 83,

$$x^3 - 18x - 35,$$

using the method of Example 9.6.

9.3 * Finish the derivation of the cubic formula outlined in Example 9.6.

9.4 Groups

Galois invented *groups* to exploit symmetry. Our purpose here is only to display Galois' ideas in enough detail so that Theorem 9.16 below is plausible; we wish to dispel some of the mystery that would arise if we merely cited the ultimate result (you can follow the proofs in [26], Chapter 5).

Commutative rings are sets with two binary operations; a group is a set having only one binary operation. Permutations, as any functions from a set X to itself, can be composed and, as we show in Appendix A.1, composition equips the family of all permutations of X with a a binary operation. This viewpoint begets a kind of algebra, called *group theory*.

Definition. A *group* is a nonempty set G with a binary operation

$$*: G \times G \to G,$$

where $*: (a, b) \mapsto a * b$, satisfying the following properties:

(i) $(a * b) * c = a * (b * c)$ for all $a, b, c \in G$,

(ii) there is $e \in G$ with $e * a = a = a * e$ for all $a \in G$,

(iii) for all $a \in G$, there is $a' \in G$ with $a' * a = e = a * a'$.

The element e is called the *identity* of G, and the element a' is called the *inverse* of a (the inverse of a is usually denoted by a^{-1}).

It is not difficult to prove, for groups as for commutative rings, that the identity element is unique (if $e' * a = a = a * e'$ for all $a \in G$, then $e' = e$), and the inverse of every element is unique (if $a'' * a = e = a * a''$, then $a'' = a'$).

Example 9.7. Theorem A.12 in Appendix A.1 shows that S_X, the family of all the permutations of a nonempty set X, is a group with composition as its binary operation. In the special case when $X = \{1, 2, \ldots, n\}$, denote S_X by

$$S_n,$$

and call it the *symmetric group* on n letters.   ▲

Example 9.8. Just because we call a Galois group a group doesn't make it so. Recall that the Galois group $\mathrm{Gal}(E/k)$ of a field extension E/k consists of all the automorphisms σ of E that fix k. We now show that $\mathrm{Gal}(E/k)$ with binary operation composition is a group.

If $\sigma, \tau \in \mathrm{Gal}(E/k)$, then their composite $\tau\sigma$ is an automorphism of E fixing k; that is, $\tau\sigma \in \mathrm{Gal}(E/k)$, so that composition is a binary operation on $\mathrm{Gal}(E/k)$. Proposition A.5 says that composition of functions is always associative. The identity $1_E : E \to E$ is an automorphism fixing k, and it is a routine calculation to show that if $\sigma \in \mathrm{Gal}(E/k)$, then its inverse σ^{-1} also lies in $\mathrm{Gal}(E/k)$. In particular, if $f(x) \in k[x]$ and E/k is a splitting field of f, then $\mathrm{Gal}(f) = \mathrm{Gal}(E/k)$ is a group. We have assigned a group to every polynomial. ▲

Example 9.9. The set $G = \mathrm{GL}(2, \mathbb{R})$ of all nonsingular 2×2 matrices with real entries is a group with binary operation matrix multiplication μ. First, we have $\mu : G \times G \to G$ because the product of two nonsingular matrices is also nonsingular. Matrix multiplication is associative, and the identity matrix $\begin{bmatrix} 1 & 0 \\ 0 & 1 \end{bmatrix}$ is the identity element. Finally, nonsingular matrices have inverses (this is the definition of nonsingular!), and so G is a group. ▲

How to Think About It. We warn the reader that new terms are going to be introduced at a furious pace. You need not digest everything; if a new idea seems only a little reasonable, continue reading nevertheless. One way to keep your head above water is to see that definitions and constructions for groups (*subgroups, homomorphisms, kernels, normal subgroups, quotient groups*) are parallel to what we have already done for commutative rings (subrings, homomorphisms, kernels, ideals, quotient rings). Your reward will be a better appreciation of the beautiful results of Abel and Galois.

Definition. A *subgroup* of a group G is a nonempty subset $S \subseteq G$ such that $s, t \in S$ implies $s * t \in S$ and $s \in S$ implies $s^{-1} \in S$.

Subgroups $S \subseteq G$ are themselves groups, for they satisfy the axioms in the definition. In particular, since S is not empty, there is some $s \in S$; by definition, its inverse s^{-1} also lies in S, and so $e = s * s^{-1}$ lies in S.

Food for Thought. If Ω is a set, we may view any subgroup of S_Ω, the group of all permutations of Ω, as symmetries of it. The notion of symmetry depends on the permutation group: isometries of the plane are one kind of symmetry; another kind arises from the group of all homeomorphisms of the plane; yet another arises from the group of all nonsingular linear transformations. This observation is the basis of Klein's **Erlanger Programm**, which classifies different types of geometries according to which geometric properties of figures are left invariant.

Multiplication in a commutative ring R is, by definition, commutative: if $a, b \in R$, then $ab = ba$. But multiplication in a group need not be commutative: $a * b$ and $b * a$ may be different. For example, composition in the symmetric group S_3 is not commutative: define $\sigma, \tau \in S_3$ by $\sigma(1) = 2$, $\sigma(2) = 1$, $\sigma(3) = 3$, and $\tau(1) = 1$, $\tau(2) = 3$, $\tau(3) = 2$. It is easy to see that $\sigma \circ \tau \neq \tau \circ \sigma$ (for $\sigma\tau(1) = \sigma(1) = 2$ and $\tau\sigma(1) = \tau(2) = 3$).

You also know that the product of two matrices depends on which is written first: $AB \neq BA$ is possible. Hence, the group $\mathrm{GL}(2, \mathbb{R})$ is not commutative.

Definition. A group G is *abelian* if $a * b = b * a$ for all $a, b \in G$.

Abel's 1828 theorem says, in modern language, that a polynomial with an abelian Galois group is solvable by radicals; this is why abelian groups are so-called. From now on, we shall simplify notation by writing the product of two group elements as ab instead of by $a * b$ and the identity as 1 instead of by e.

Proposition 9.10. *If k is a field, $a \in k$, and k contains all the nth roots of unity, $\zeta, \zeta^2, \ldots, \zeta^n = 1$, then the Galois group of $f(x) = x^n - a$ is abelian.*

Sketch of proof. Since k contains all the nth roots of unity,

$$E = k\left(\omega^i \beta : \beta^n = a \text{ and } 1 \leq i \leq n\right)$$

is a splitting field of f. Any automorphism σ of E must permute the roots, and so $\sigma(\omega^i \beta) = \omega^j \beta$ for some j depending on i. Similarly, if τ is another automorphism, then $\tau(\omega^j \beta) = \omega^\ell \beta$. It follows that both $\sigma\tau$ and $\tau\sigma$ send $\omega^i \beta$ to $\omega^{j+\ell} \beta = \omega^{\ell+j} \beta$; from this fact it is not hard to see that $\mathrm{Gal}(E/k)$ is an abelian group. ∎

There is always a *primitive* nth root of unity; that is, an element ζ with $\zeta^n = 1$ such that every nth root of unity is a power of ζ. In particular, a primitive complex nth root of unity is $e^{2\pi i/n} = \cos(2\pi/n) + i \sin(2\pi/n)$.

Galois was able to translate a polynomial being solvable by radicals into a certain property of its Galois group by constructing analogs for groups of the constructions we have done in earlier chapters for commutative rings.

Definition. If G and H are groups, then a ***homomorphism*** is a function $\varphi \colon G \to H$ such that, for all $a, b \in G$,

$$\varphi(ab) = \varphi(a)\varphi(b).$$

Of course, abstract algebra did not exist in Galois' time. In particular, rings, fields, and homomorphisms were not in anyone's vocabulary; nor were groups.

An ***isomorphism*** is a homomorphism that is a bijection. If there is an isomorphism $\varphi \colon G \to H$, then we say that G and H are ***isomorphic*** and we write $G \cong H$.

We can be more precise. If groups are denoted by $(G, *)$ and $(H, \circ)$, where $*$ and $\circ$ are binary operations, then a homomorphism $\varphi \colon G \to H$ is a function for which

$$\varphi(a * b) = \varphi(a) \circ \varphi(b).$$

It is easy to see that if φ is a homomorphism, then $\varphi(1) = e$, where 1 is the identity of G and e is the identity of H; moreover, for each $a \in G$, we have $\varphi(a^{-1}) = \varphi(a)^{-1}$ (the latter being the inverse of the element $\varphi(a)$ in H). If $a, b \in G$ commute, then $ab = ba$. Hence, if $\varphi \colon G \to H$ is a homomorphism, then

$$\varphi(a)\varphi(b) = \varphi(ab) = \varphi(ba) = \varphi(b)\varphi(a).$$

It follows that if G is abelian and φ is an isomorphism, then H is abelian.

Every polynomial determines a group of symmetries of its roots.

Theorem 9.11. *If a polynomial $f(x) \in k[x]$ has n roots, then its Galois group $\mathrm{Gal}(E/k)$ is isomorphic to a subgroup of the symmetric group S_n.*

Proof. By Theorem 9.4, elements of $\mathrm{Gal}(E/k)$ permute the roots of f. Now see [26], p. 454. ∎

If X and Y are sets of n elements, then $S_X \cong S_n \cong S_Y$. Thus, groups don't care if you are permuting n numbers, n roots, or n monkeys.

Definition. The *kernel* of a homomorphism $\varphi \colon G \to H$ of groups is

$$\ker \varphi = \{a \in G : \varphi(a) = 1\},$$

where 1 is the identity element of H. The *image* of φ is

$$\operatorname{im} \varphi = \{h \in H : h = \varphi(g) \text{ for some } g \in G\}.$$

If $\varphi \colon G \to H$ is a homomorphism, then $\ker \varphi$ is a subgroup of G and $\operatorname{im} \varphi$ is a subgroup of H.

Just as the kernel of a ring homomorphism has special properties—it is an ideal—so, too, is the kernel of a group homomorphism special. If $a \in \ker \varphi$ and $b \in G$, then

$$\begin{aligned}
\varphi(bab^{-1}) &= \varphi(b)\varphi(a)\varphi(b^{-1}) \\
&= \varphi(b)1\varphi(b)^{-1} \\
&= \varphi(b)\varphi(b)^{-1} = 1.
\end{aligned}$$

Definition. A subgroup N of a group G is a *normal subgroup* if, for each $a \in N$, we have $bab^{-1} \in N$ for every $b \in G$.

Thus, kernels of homomorphisms are always normal subgroups. In an abelian group, every subgroup is normal but, in general, there are subgroups that are not normal. For example, if σ is the permutation of $X = \{1, 2, 3\}$ that interchanges 1 and 2 and fixes 3, then $S = \{e, \sigma\}$ is a subgroup of the symmetric group S_3. But S is not a normal subgroup of S_3, for if τ is the permutation that fixes 1 and interchanges 2 and 3, then $\tau\sigma\tau^{-1}(1) = \tau\sigma(1) = \tau(2) = 3$. Hence, $\tau\sigma\tau^{-1} \neq e$ and $\tau\sigma\tau^{-1} \neq \sigma$; that is, $\sigma \in S$ but $\tau\sigma\tau^{-1} \notin S$.

Just as we used ideals to construct quotient rings, we can use normal subgroups to construct *quotient groups*. If N is a subgroup of a group G, define certain subsets, called *cosets* of N in G, as follows: if $a \in G$, then

$$aN = \{as : s \in N\} \subseteq G.$$

The family of all cosets of N is denoted by

$$G/N = \{\text{all cosets } aN : a \in G\}.$$

When N is a normal subgroup, then G/N is a group if we define a binary operation by

$$aN * bN = abN$$

(normality of N is needed to prove that this multiplication is well-defined: if $a'N = aN$ and $b'N = bN$, then $a'b'N = abN$). The group G/N is called the *quotient group*.

There is an isomorphism theorem for groups analogous to the isomorphism theorem for commutative rings.

Theorem 9.12 (First Isomorphism Theorem). *If $\varphi \colon G \to H$ is a group homomorphism, then $\operatorname{im} \varphi$ is a subgroup of H, $N = \ker \varphi$ is a normal subgroup of G, and there is an isomorphism $\Phi \colon G/N \to \operatorname{im} \varphi$ given by $\Phi \colon aN \mapsto \varphi(a)$.*

Sketch of proof. Adapt the proof of the First Isomorphism Theorem for commutative rings. ■

How to Think About It. Without a doubt, this section contains too much new material; there's too much to digest. Fortunately, you have seen analogs of these definitions for commutative rings so, at least, they sound familiar. You can now sympathize with the members of the Academy in Paris in 1830 as they struggled, without benefit of ever having seen any abstract algebra at all, to read such things in the paper Galois submitted to them!

Let us now see why normal subgroups are important for polynomials. If $k \subseteq B \subseteq E$, then $\mathrm{Gal}(E/B)$ is a subset of $\mathrm{Gal}(E/k)$:

$$\mathrm{Gal}(E/B) = \{\sigma \in \mathrm{Gal}(E/k) : \sigma(B) = B\}.$$

It is easy to check that $\mathrm{Gal}(E/B)$ is a subgroup of $\mathrm{Gal}(E/k)$.

Theorem 9.13. *Let $k \subseteq B \subseteq E$ be a tower of fields, where E/k is a splitting field of some $f(x) \in k[x]$. If B is the splitting field of some $g(x) \in k[x]$, then $\mathrm{Gal}(E/B)$ is a normal subgroup of $\mathrm{Gal}(E/k)$, and*

$$\mathrm{Gal}(E/k)/\mathrm{Gal}(E/B) \cong \mathrm{Gal}(B/k).$$

Sketch of proof. Define $\varphi \colon \mathrm{Gal}(E/k) \to \mathrm{Gal}(E/B)$ by $\varphi \colon \sigma \mapsto \sigma|B$. The restriction $\sigma|B$ does send B to itself, since B is a splitting field (so automorphisms permute the roots of g, by Theorem 9.4), and so $\sigma|B \in \mathrm{Gal}(B/k)$. By Theorem 9.12, the First Isomorphism Theorem, it suffices to find $\mathrm{im}\,\varphi$ and $\ker \varphi$. It is obvious that $\mathrm{Gal}(E/B) \subseteq \ker \varphi$, and a short calculation gives equality. We can prove that φ is surjective and hence $\mathrm{im}\,\varphi = \mathrm{Gal}(B/k)$ (the proof of surjectivity ([26] p. 455) is not straightforward). ■

The converse of Theorem 9.13 is true: the ***Fundamental Theorem of Galois Theory*** says, in part, that if N is a normal subgroup of $\mathrm{Gal}(E/k)$, then there is a subfield

$$B = \{\alpha \in E : \sigma(\alpha) = \alpha \text{ for all } \sigma \in N\} \subseteq E$$

that is a splitting field of some polynomial in $k[x]$.

How to Think About It.

The subgroups get smaller as the field extensions get bigger. If $K \subseteq L \subseteq E$, then $\mathrm{Gal}(E/L) \subseteq \mathrm{Gal}(E/K)$: if σ is an automorphism of E that fixes everything in L, then surely σ fixes everything in $K \subseteq L$.

Lemma 9.14. *Let $K \subseteq L \subseteq E$ be a tower of fields, where K contains all roots of unity. If L/K is a pure extension, then $\mathrm{Gal}(E/L)$ is a normal subgroup of $\mathrm{Gal}(E/K)$ and the quotient group $\mathrm{Gal}(E/K)/\mathrm{Gal}(E/L)$ is abelian.*

Sketch of proof. The field extension L/K is a splitting field (because the subfield K contains all needed roots of unity), and so Theorem 9.12 gives

It is not true that that if A is a normal subgroup of B and B is a normal subgroup of C, then A is a normal subgroup of C.

Gal(E/L) a normal subgroup of Gal(E/K). By Proposition 9.10, the quotient group is abelian.   ∎

We conclude the story by applying Lemma 9.14 to each pure extension K_i/K_{i-1} in a tower of a radical extension.

Lemma 9.15. *Let k be a field containing all roots of unity. If $f(x) \in k[x]$ is solvable by radicals, then there is a chain of subgroups*

$$G_0 = \text{Gal}(K_t/k) \supseteq G_1 \supseteq G_2 \supseteq \cdots \supseteq G_t = \{1\},$$

where each G_{i+1} is a normal subgroup of G_i and each quotient group G_i/G_{i+1} is abelian.

This lemma suggests the following definition.

Definition. A group G is ***solvable*** if there is a chain of subgroups

$$G = G_0 \supseteq G_1 \supseteq G_2 \supseteq \cdots \supseteq G_t = \{1\}$$

where each G_{i+1} is a normal subgroup of G_i and each quotient group G_i/G_{i+1} is abelian.

Clearly, every abelian group is solvable—take $G_1 = \{1\}$. It is shown in [26], p. 466, that the symmetric group S_4 and all its subgroups are solvable groups, but that S_5 is not solvable.

Using these ideas, Galois proved the following beautiful theorem.

Theorem 9.16 (Galois). *Let k be a field and $f(x) \in k[x]$. If f is solvable by radicals, then its Galois group is a solvable group. If k has characteristic 0, then the converse is true.*

Proof. [25], p. 189 and p. 208.   ∎

Galois' Theorem explains why the classical theorems hold for polynomials of degree ≤ 4.

Corollary 9.17. *If k is a field of characteristic 0 and $f(x) \in k[x]$ has degree ≤ 4, then f is solvable by radicals.*

Proof. Since deg(f) ≤ 4, Theorem 9.11 says that Gal(f) is (isomorphic to) a subgroup of S_4 and, hence, it is a solvable group. Theorem 9.16 now says that f is solvable by radicals.   ∎

Finally, the next theorem explains why degree 5 was so troublesome.

Ruffini's name occurs here because he published a proof of this result in 1799. Although his ideas were correct, there were gaps in his proof, and it was not accepted by his contemporaries.

Corollary 9.18 (Abel–Ruffini). *The general polynomial of degree 5 is not solvable by radicals.*

Proof. The Galois group of the general quintic $f(x) \in \mathbb{Q}[x]$ is S_5 ([26], p. 468), which is not a solvable group, and so Galois' Theorem says that f is not solvable by radicals.   ∎

Here is an explicit numerical example. The quintic

$$f(x) = x^5 - 4x + 2 \in \mathbb{Q}[x]$$

(see Figure 9.3) is not solvable by radicals because its Galois group is S_5 ([26], p. 469).

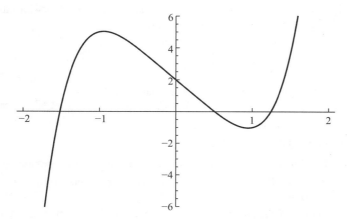

Figure 9.3. $f(x) = x^5 - 4x + 2$.

Corollary 9.18 is often misquoted. It says the general quintic is not solvable by radicals: there is no formula involving only addition, subtraction, multiplication, division, and extraction of roots that expresses the roots of the general quintic polynomial in terms of its coefficients. But it doesn't say that roots of quintics cannot be found. There are other kinds of formulas; for example, *Newton's method* gives the roots as $\lim_{n \to \infty} x_n$, where $x_{n+1} = x_n - f(x_n)/f'(x_n)$. Thus, it is not accurate to say that there is no formula finding the roots of a quintic polynomial.

Exercises

9.4 Prove that every subgroup of an abelian group is abelian.

9.5 Let $f(x), g(x) \in \mathbb{Q}[x]$ be solvable by radicals.

(i) Show that $f(x)g(x)$ is also solvable by radicals.

(ii) Give an example showing that $f(x) + g(x)$ need not be solvable by radicals.

9.6 Assuming that $x^n - 1$ is solvable by radicals, prove that $x^n - a$ is solvable by radicals, where $a \in \mathbb{Q}$.

9.7 Prove that S_3 is a solvable group and that it is not abelian.

9.8 Recall Exercise 1.56 on page 35: if $m \geq 2$ is an integer, $\gcd(k, m) = 1$, and $\gcd(k', m) = 1$, then $\gcd(kk', m) = 1$.
Prove that

$$U_m = \{[k] \in \mathbb{Z}_m : \gcd(k, m) = 1\}$$

is a group under multiplication.

9.9 If k is a field, prove that $k^\times = \{a \in k : a \neq 0\}$ is a group under multiplication.

9.10 If R is a commutative ring, prove that R is an abelian group under addition. (Note that 0 is the identity element and that $-a$ is the (additive) inverse of a.)

9.11 Let k be a field.

(i) Prove that $k^{\times}$ is an abelian group under multiplication, where $k^{\times}$ denotes the set of nonzero elements of k.

(ii) Prove that $GL_2(k)$, the set of all 2×2 nonsingular matrices with entries in k, is a group under matrix multiplication.

$GL_2(k)$ is called the *General Linear group*, and $SL_2(k)$ is called the *Special Linear group*.

(iii) Prove that the determinant function,

$$\det\colon GL_2(k) \to k^{\times},$$

is a surjective homomorphism of groups.

(iv) Prove that $\ker(\det) = SL_2(k)$, the set of all 2×2 matrices over k having determinant 1.

(v) Prove that $GL_2(k)/SL_2(k) \cong k^{\times}$.

9.12 (i) Prove that $\mathbb{R}$ is an abelian group with addition as binary operation.

(ii) Prove that $\mathbb{Q}$ is an abelian group with addition as binary operation; indeed, it is a subgroup of $\mathbb{R}$.

(iii) Let $\mathbb{R}^{>}$ be the group of positive real numbers. Show that $\mathbb{R}^{>}$ is a group with addition as as binary operation.

The "laws of exponents" from high school algebra preview the results of Exercise 9.13.

9.13 (i) Prove that $\exp\colon \mathbb{R} \to \mathbb{R}^{>}$, defined by $a \mapsto e^{a}$, is a group homomorphism.

(ii) Prove that $\log\colon \mathbb{R}^{>} \to \mathbb{R}$, defined by $b \mapsto \log b$, is a group homomorphism.

(iii) Prove that $\exp$ is an isomorphism by showing that its inverse is log.

9.14 (i) Prove that $\mathbb{R}^{>}$, the set of all positive real numbers, is an abelian group with multiplication as binary operation, and prove that $\mathbb{Q}^{>}$, the set of all positive rational numbers, is a subgroup of $\mathbb{R}^{>}$.

(ii) Prove that $\mathbb{Z}[x]$ is an abelian group under addition.

(iii) Use the Fundamental Theorem of Arithmetic to prove that the additive group $\mathbb{Z}[x]$ is isomorphic to the multiplicative group $\mathbb{Q}^{>}$ of all positive rational numbers.

Hint. Define $\varphi\colon \mathbb{Z}[x] \to \mathbb{Q}^{>}$ by

$$\varphi\colon e_0 + e_1 x + \cdots + e_n x^n \mapsto p_0^{e_0} p_1^{e_1} \cdots p_n^{e_n},$$

where $p_0 = 2$, $p_1 = 3$, $p_2 = 5, \ldots$ is the list of all primes.

9.5 Wiles and Fermat's Last Theorem

Andrew Wiles proved Fermat's Last Theorem in 1995: Modular elliptic curves and Fermat's last theorem, *Ann. Math.* (2) 141 (1995), pp. 443-551. He has said,

> I was a ten year old and one day I happened to be looking in my local public library and I found a book on maths and it told a bit about the history of this problem and I, a ten year old, could understand it. From that moment I tried to solve it myself; it was such a challenge, such a beautiful problem. This problem was Fermat's Last Theorem.

and

*There's no other problem that will mean the same to me. I had this very
rare privilege of being able to pursue in my adult life what had been my
childhood dream. I know it's a rare privilege, but I know if one can do
this it's more rewarding than anything one can imagine.*

Andrew Wiles was born in Cambridge, England in 1953. He was awarded
a doctorate in 1980 from the University of Cambridge, then spent a year in
Bonn before joining the Institute for Advanced Study in Princeton. In 1982,
he was appointed Professor at Princeton. Around 1985, Wiles learned that the
Shimura–Taniyama–Weil conjecture about **elliptic curves**, if true, would imply Fermat's Last Theorem (we will say more about this below). Wiles was
able to prove a special case of this conjecture (the full conjecture was proved
in 2001) which was strong enough to give Fermat's Last Theorem.

In 1994, Wiles was appointed Eugene Higgins Professor of Mathematics at
Princeton. Wiles received many honors for his outstanding work. For example, he was awarded the Schock Prize in Mathematics from the Royal Swedish
Academy of Sciences, the Prix Fermat from Université Paul Sabatier, the Wolf
Prize in Mathematics from the Wolf Foundation in Israel, and the Cole Prize
from the American Mathematical Society. He was elected a member of the
National Academy of Sciences of the United States, receiving its mathematics
prize, and Andrew Wiles became "Sir Andrew Wiles" when he was knighted
by the Queen of England. In 1998, not being eligible for a Fields medal (the
mathematics prize equivalent to a Nobel prize) because he was over forty
years of age, the International Mathematical Union presented him with a silver
plaque at the International Congress of Mathematicians.

Elliptic Integrals and Elliptic Functions

The context of Wiles' proof of Fermat's Last Theorem is that of elliptic curves,
an area with an interesting history. Leibniz, one of the founders of calculus, posed the problem of determining which integrals could be expressed in
"closed form;" that is, as linear combinations of familiar functions such as rational functions, exponentials, logarithms, trigonometric functions, and their
inverse functions. One of the first integrals that could not be so expressed (although the proof of this fact, by Liouville, waited until 1833) is the arclength
of an ellipse. If $f(x, y) = 0$ is the equation of a curve in the plane, then its
arclength is given in terms of the indefinite integral

$$\int \sqrt{1 + (dy/dx)^2}\, dx.$$

Consider the ellipse with equation

$$\frac{x^2}{a^2} + \frac{y^2}{b^2} = 1,$$

where $a > b > 0$. We have $y = b\sqrt{1 - (x^2/a^2)}$, so that

$$\frac{dy}{dx} = \frac{-bx}{a^2\sqrt{1 - (x^2/a^2)}},$$

and the arclength integral is

$$\frac{1}{a} \int \sqrt{\frac{a^4 - (a^2 - b^2)x^2}{a^2 - x^2}}\, dx.$$

We refer the reader to
the books of Siegel [30],
Silverman–Tate [31], and
Stillwell [32], as well
as to the article by M.
Rosen, Abel's Theorem
on the Lemniscate, *Amer.
Math. Monthly* 88 (1981),
pp. 387–395, for further
details of this discussion.

The *eccentricity* of the ellipse is

$$E = \sqrt{1 - (b/a)^2}.$$

Make the substitution $x = a \sin \theta$, so that $\cos \theta = \frac{1}{a}\sqrt{a^2 - x^2}$ and $dx = a \cos \theta \, d\theta$, to obtain

$$\frac{1}{a} \int \sqrt{\frac{a^4 - (a^2 - b^2)x^2}{a^2 - x^2}} \, dx = a \int \sqrt{1 - \left(\frac{a^2 - b^2}{a^2}\right) \sin^2 \theta} \, d\theta$$

$$= a \int \sqrt{1 - E^2 \sin^2 \theta} \, d\theta.$$

Finally, we rewrite the last integral using the tangent half-angle formula $t = \tan(\theta/2)$ in Chapter 1 (so that $d\theta = 2dt/(1 + t^2)$ and $\sin \theta = 2t/(1 + t^2)$). We obtain

$$a \int \sqrt{1 - E^2 \sin^2 \theta} \, d\theta = 2a \int \frac{\sqrt{g(t)}}{(1 + t^2)^2} \, dt,$$

where $g(t) = t^4 + (2 - 4E^2)t^2 + 1$. Thus, if $R(x, y)$ is the rational function in two variables,

$$R(x, y) = \frac{y}{(1 + x^2)^2},$$

then the arclength of an ellipse has the form

$$2a \int R\big(t, \sqrt{g(t)}\big) \, dt,$$

where $g(t)$ is a quartic polynomial. A similar integral arises from the arclength of the hyperbola $x^2/a^2 - y^2/b^2 = 1$.

Definition. An *elliptic integral* is an indefinite integral of the form

$$\int R\big(t, \sqrt{g(t)}\big) \, dt,$$

where $R(x, y)$ is a rational function and $g(t)$ is either a cubic or a quartic polynomial having no repeated roots.

The substitution $t = 1/u$ transforms the cubic integrand

$$\frac{dt}{\sqrt{(t-a)(t-b)(t-c)}}$$

into the quartic

$$\frac{-du}{\sqrt{u(1-ua)(1-ub)(i-uc)}}.$$

These integrals are so called because, as we have just seen, the arclength of an ellipse was one of the first examples of them. Another example of an elliptic integral, studied by Jacob Bernoulli in 1679, arises from computing the arclength of a spiral. In 1694, James Bernoulli examined the shape an elastic rod takes if its ends are compressed; he found the resulting curve to be the lemniscate $r^2 = \cos 2\theta$; see Figure 9.4 (there are eight mathematicians in the Bernoulli family, in the seventeenth and eighteenth centuries, listed in the MacTutor History of Mathematics Archive). Recall that the arclength of a curve $r = f(\theta)$ in polar coordinates is

$$\int \sqrt{1 + r^2(d\theta/dr)^2} \, dr.$$

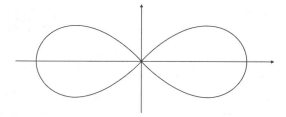

Figure 9.4. Lemniscate $r^2 = \cos 2\theta$.

If $r^2 = \cos 2\theta$, then

$$1 + r^2(d\theta/dr)^2 = 1 + \frac{r^4}{\sin^2 2\theta}$$

$$= 1 + \frac{r^4}{1 - \cos^2 2\theta}$$

$$= 1 + \frac{r^4}{1 - r^4}$$

$$= \frac{1}{1 - r^4}.$$

Therefore, the arclength of the lemniscate is $\displaystyle\int \frac{dr}{\sqrt{1 - r^4}}$.

Yet another example,

$$\int \frac{dt}{\sqrt{(1 - t^2)(1 - at^2)}},$$

arises when calculating the period of oscillation of a simple pendulum. Computing the electrical capacity of an ellipsoid with equation $x^2/a^2 + y^2/b^2 + z^2/c^2 = 1$ involves the integral

$$\int \frac{dt}{\sqrt{(a^2 + t)(b^2 + t)(c^2 + t)}},$$

an elliptic integral involving a cubic.

Since there are interesting elliptic integrals and they are difficult to evaluate, they were the subject of much investigation. In 1718, Fagnano proved a *Duplication Formula*:

$$2 \int_0^u \frac{dt}{\sqrt{1 - t^4}} = \int_0^{Q(u)} \frac{dt}{\sqrt{1 - t^4}},$$

where $Q(u) = 2u\sqrt{1 - u^4}/(1 + u^4)$. In proving this, Fagnano inverted $I(x) = \int_0^x 1/\sqrt{1 - t^4}\, dt$, getting the inverse function $I^{-1}(x) = \sqrt{2}x/\sqrt{1 + x^4}$. In 1751, Euler generalized the duplication formula of Fagnano, obtaining an *Addition Theorem*:

$$\int_0^u \frac{dt}{\sqrt{1 - t^4}} + \int_0^v \frac{dt}{\sqrt{1 - t^4}} = \int_0^{P(u,v)} \frac{dt}{\sqrt{1 - t^4}},$$

where $P(u, v) = \left(u\sqrt{1 - v^4} + v\sqrt{1 - u^4}\right)/(1 + u^2v^2)$ (so that $P(u, u) = Q(u)$). Euler further generalized this by replacing the integrand by $1/\sqrt{p(t)}$, where $p(t)$ is any quartic polynomial.

In 1797, Gauss considered the elliptic integrals $\int_0^u \frac{dt}{\sqrt{1-t^3}}$ and $\int_0^u \frac{dt}{\sqrt{1-t^4}}$. He saw an analogy (as, most likely, did Fagnano and Euler) with

$$\sin^{-1} u = \int_0^u \frac{dt}{\sqrt{1-t^2}},$$

and he inverted many elliptic integrals; after all, $\sin x$ is the inverse function of $\sin^{-1} x$. Nowadays, inverse functions of elliptic integrals are called ***elliptic functions***. Just as $\sin x$ is *periodic*, that is, $\sin(x + 2\pi) = \sin x$ for all x, so, too, are elliptic functions f; there is some number p with $f(x + p) = f(x)$ for all x. Gauss then studied *complex* elliptic integrals $I(z) = \int_0^z \frac{d\zeta}{\sqrt{g(\zeta)}}$; their inverse functions $f(z) = I^{-1}(z)$ are called elliptic functions of a complex variable. Gauss saw that complex elliptic functions are ***doubly periodic***: there are (noncollinear) complex numbers p and q with

$$f(z + mp + nq) = f(z)$$

for all complex z and all $m, n \in \mathbb{Z}$. This fact has important geometric consequences, both for elliptic functions and for complex variables in general. Alas, Gauss never published these ideas, and they became known only later.

In 1823, Abel investigated elliptic functions, rediscovered many of Gauss's theorems, and proved new beautiful results about them. For example, just as Gauss had found all n for which one can divide the circle into n equal arcs using ruler and compass ($n = 2^m p_1 \cdots p_k$, where $m \geq 0$ and the p_i are distinct primes of the form $2^{2^t} + 1$), Abel obtained the same result (for exactly the same n) for the lemniscate. At the same time, Jacobi began his investigations of elliptic functions, further explaining and generalizing work of Euler by introducing *theta functions* and *modular curves*.

Congruent Numbers Revisited

The search for congruent numbers can be viewed as the search for "generalized Pythagorean triples"—right triangles with rational side-lengths and integer area. Recall from Chapter 1 that a *congruent number* is a positive integer n that arises from asking which integers are areas of right triangles having rational side-lengths; that is, there are positive rational numbers a, b, and c such that

$$a^2 + b^2 = c^2 \quad \text{and}$$
$$\tfrac{1}{2}ab = n.$$

Since $ab = 2n > 0$, we have $a \neq 0$ and $b \neq 0$. It follows that $c \neq 0$, too.

Let's loosen the constraints a bit and allow a, b, and c to be negative rational numbers as well. We'd like to replace the two equations in four unknowns with a simpler set of constraints. We'll see that the solution can be realized as the search for rational points on a polynomial curve.

In Theorem 1.9, we reduced the defining pair (for $n = 2$) to a degree 4 equation in three variables. We'll do a little better here.

We now turn the pair of defining equations into a single equation in two variables. The equation $a^2 + b^2 = c^2$ can be written as

$$b^2 = c^2 - a^2 = (c - a)(c + a).$$

Let $k = c - a$, so that we have

$$b^2 = k(c + a).$$

Since $c = k + a$, this is the same as

$$b^2 = k(k + 2a) = k^2 + 2ak,$$

or

$$2ak = b^2 - k^2.$$

Since $a = 2b/n$, this is equivalent to

$$\frac{4nk}{b} = b^2 - k^2,$$

or

$$4nk = b^3 - k^2 b.$$

This is beginning to look like a cubic in b. To homogenize it, multiply both sides by $\left(\frac{n}{k}\right)^3$ to get

$$\frac{4n^4}{k^2} = \left(\frac{bn}{k}\right)^3 - n^2 \left(\frac{bn}{k}\right)$$

or

$$\left(\frac{2n^2}{k}\right)^2 = \left(\frac{bn}{k}\right)^3 - n^2 \left(\frac{bn}{k}\right);$$

remembering that $k = c - a$, we have a single cubic equation satisfied by a, b, c, and n:

$$\left(\frac{2n^2}{c-a}\right)^2 = \left(\frac{bn}{c-a}\right)^3 - n^2 \left(\frac{bn}{c-a}\right).$$

We're assuming that $c - a \neq 0$, or else $b = 0$.

This shows that if $a^2 + b^2 = c^2$ and $ab = 2n$, then

$$\left(\frac{bn}{c-a}, \frac{2n^2}{c-a}\right)$$

is a rational point on the graph of $y^2 = x^3 - n^2 x$ (the graph of $y^2 = x^3 - 25x$ is shown in Figure 9.5).

Except for sign changes and points on the x-axis, the correspondence goes both ways.

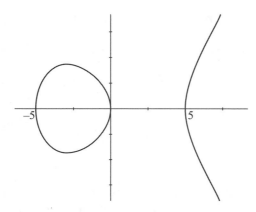

Figure 9.5. $y^2 = x^3 - 25x$

Theorem 9.19. *Let n be a positive integer. There's a bijection between triples of rational numbers* (a, b, c) *satisfying*

$$a^2 + b^2 = c^2 \quad and \quad ab = 2n$$

and rational points on the graph of $y^2 = x^3 - n^2 x$ *with* $y \neq 0$.

Proof. The calculation preceding the statement of the theorem shows that a triple produces such a point on the graph. Going the other way, if (x, y) is a point on the graph with $y \neq 0$, we can solve the system

$$x = \frac{bn}{c - a}$$
$$y = \frac{2n^2}{c - a}$$
$$c^2 = a^2 + b^2$$

for a, b, and c, either by hand or CAS (see Figure 9.6) to find

$$a = \frac{x^2 - n^2}{y}$$
$$b = \frac{2nx}{y}$$
$$c = \frac{x^2 + n^2}{y}.$$

It is easily checked that this produces a triple of rational numbers of the desired type. ■

$$\mathrm{solve}\left(\left\{x = \frac{b \cdot n}{c - a}, y = \frac{2 \cdot n^2}{c - a}, c^2 = a^2 + b^2\right\} \{a, b, c\}\right)$$

$$\frac{x^2 - c \cdot y - n^2}{y} \neq 0 \text{ and } a = \frac{x^2 - n^2}{y} \text{ and } b = \frac{2 \cdot n \cdot x}{y} \text{ and } c = \frac{x^2 + n^2}{y}$$

Figure 9.6. Solving for a, b, and c.

The first part of what the CAS returns (for $(x^2 - cy - n^2)/y \neq 0$) combined with the value given for c, just say that $n \neq 0$.

Let's state the conversion formulas explicitly.

Corollary 9.20. *The bijection guaranteed by* Theorem 9.19 *is given explicitly by*

$$(a, b, c) \mapsto \left(\frac{bn}{c - a}, \frac{2n^2}{c - a}\right) \quad and \quad (x, y) \mapsto \left(\frac{x^2 - n^2}{y}, \frac{2nx}{y}, \frac{x^2 + n^2}{y}\right)$$

Example 9.21. The correspondence between rational right triangles with integer area and cubic curves allows us to generate infinitely many congruent triangles with the same area from a given such triangle.

For example, on page 18, we saw that there are two rational right triangles with area 5. One comes from a scaled copy of $\triangle(9, 40, 41)$ whose area is $5 \cdot 6^2$. To find the second one, it would take a very long time (even with a computer) to find $\triangle(2420640, 2307361, 3344161)$ whose area is $5 \cdot 747348^2$. But we can use an idea related to the "sweeping lines" method of Diophantus. The rational right triangle with side-lengths $\left(\frac{3}{2}, \frac{20}{3}, \frac{41}{6}\right)$ corresponds, via the formulas in Corollary 9.20, to the point $P = \left(\frac{25}{4}, \frac{75}{8}\right)$ on the curve C defined by

$$y^2 = x^3 - 25x.$$

The idea is to take a line tangent to C through P; it intersects C in a second point P', which is also rational because P is (see Exercise 9.16 below). From this new rational point, we can build a new right triangle.

A CAS (or at least a calculator) is a very useful tool for these calculations.

From $y^2 = x^3 - 25x$, we have, using implicit differentiation that

$$\frac{dy}{dx} = \frac{3x^2 - 25}{2y}.$$

Using this, we find that the slope of the tangent to C at P is 59/12, and hence the tangent line to C at P has equation

$$y = \frac{59}{12}\left(x - \frac{25}{4}\right) + \frac{75}{8}.$$

Solving the system (see Figure 9.7)

$$y^2 = x^3 - 25x$$
$$y = \frac{59}{12}\left(x - \frac{25}{4}\right) + \frac{75}{8},$$

we get P and

$$P' = \left(\frac{1681}{144}, \frac{62279}{1728}\right).$$

$$\text{solve}\left(x^3 - 25 \cdot x = y^2 \text{ and } y = \frac{59}{12}\left(x - \frac{25}{4}\right) + \frac{75}{8}, \{x, y\}\right)$$

$$x = \frac{25}{4} \text{ and } y = \frac{75}{8} \text{ or } x = \frac{1681}{144} \text{ and } y = \frac{62279}{1728}$$

Figure 9.7. P and P'

Finally, using Corollary 9.20 again, we recover (a, b, c) from P':

$$a = \frac{1519}{492}, \quad b = \frac{4920}{1519}, \quad c = \frac{334416}{747348}.$$

These are the side-lengths of the triangle on page 18.   ▲

Exercises

9.15 Show that there are no rational points (x, y) with $y \neq 0$ on the graph of

(i) $y^2 = x^3 - x$

(ii) $y^2 = x^3 - 4x$

9.16 Show that a cubic equation with rational coefficients and two rational roots has, in fact, three rational roots.

9.17 Find a third rational right triangle with area 5, different from the two we found in Example 9.21.

Elliptic Curves

The curves defined by the equation in Theorem 9.19,

$$y^2 = \text{a cubic polynomial in } x$$

show up all across mathematics. We just saw how they can be used to generate congruent numbers.

Before that, we saw that the integral defining arcsine, $\int dt / \sqrt{1 - t^2}$, suggested studying elliptic functions, the inverse functions of elliptic integrals. Just as the unit circle is parametrized by sine and cosine (it consists of the points $(\sin \theta, \cos \theta)$), Gauss, Abel, and Jacobi considered curves parametrized by elliptic functions; that is, curves consisting of the points $(f(u), f'(u))$, where f is an elliptic function (cosine is the derivative of sine). What sort of curves are these? Expand the integrand of an elliptic integral as a power series (since it has a denominator, the series begins with a negative power), and then integrate term by term. There results a differential equation involving $x = f$ and $y = f'$, which turns out to be a cubic in two variables (see [9], pp. 17–19). After some manipulations, one obtains a **Weierstrass normal form** for the points (x, y) on the curve $y^2 = ax^3 + bx^2 + cx + d$ (there is another, simpler, Weierstrass normal form, $y^2 = 4x^3 - g_2 x - g_3$, where g_2, g_3 are constants).

This definition is not quite accurate, for an elliptic curve is really an equivalence class of such curves.

Definition. An *elliptic curve* over a field k is a curve $C \subseteq k^2$ with equation

$$y^2 = g(x),$$

where $g(x) = ax^3 + bx^2 + cx + d \in k[x]$ has no repeated roots.

Curves over $\mathbb{C}$ are two-dimensional surfaces when viewed over $\mathbb{R}$.

The most interesting elliptic curves are over $\mathbb{C}$ (for complex variables) or over $\mathbb{Q}$ (for number theory), while elliptic curves over finite fields $\mathbb{F}_q$ give rise to public access codes that are more secure than the RSA codes we discussed in Chapter 4.

Elliptic functions and elliptic curves, whose humble origins are in arclength problems, occur in analysis, geometry, and complex variables. In the previous subsection, we saw that congruent numbers lead to rational points on elliptic curves. More generally, let's now see the connection with number theory and with Fermat's Last Theorem in particular.

Definition. A *Diophantine equation* is an equation $f(x_1, \ldots, x_m) = 0$, where $f(x_1, \ldots, x_m) \in \mathbb{Q}[x_1, \ldots, x_m]$ is a polynomial in several variables having rational coefficients.

A **solution** to a Diophantine equation $f(x_1, \dots, x_m) = 0$ is an m-tuple $(q_1, \dots, q_m) \in \mathbb{Q}^m$ for which $f(q_1, \dots, q_m) = 0$; an **integer solution** is a solution in $\mathbb{Z}^m$.

For example, $x^n + y^n - 1 = 0$ is a Diophantine equation. Rational solutions $(u, v) = (a/c, b/c)$ give rise, by clearing denominators, to integer solutions of the Diophantine equation $x^n + y^n = z^n$. Of course, this example arises from Fermat's Last Theorem.

A curve in the plane is the locus of solutions to an equation $f(x, y) = 0$. Let's focus on polynomials $f(x, y) \in \mathbb{Q}[x, y]$; that is, on Diophantine equations of two variables. A **rational point** (u, v) on the curve $f(x, y) = 0$ is a geometric way of viewing a solution to the Diophantine equation $f(x, y) = 0$. The method of Diophantus classifies Pythagorean triples by intersecting the unit circle $x^2 + y^2 = 1$ with lines through the rational point $(-1, 0)$, thereby parametrizing the circle with rational functions, and then finding the rational points on the circle that correspond to Pythagorean triples. We saw, in Chapter 1, that it is worthwhile to generalize the method by replacing the unit circle by conic sections.

Now pass from conic sections, curves in the plane corresponding to quadratic polynomials $f(x, y) \in \mathbb{R}[x, y]$, to cubic polynomials of two variables. For example, rational points on $x^3 + y^3 = 1$ correspond to integer solutions of $a^3 + b^3 = c^3$, so that the truth of Fermat's Last Theorem for $n = 3$ says that the curve $x^3 + y^3 = 1$ has no rational points.

Diophantus also studied cubic curves. There was no analytic geometry in his day, and geometry was not explicit in his results. However, both Fermat and, later, Newton believed that geometry explains Diophantus's method of finding solutions of cubic Diophantine equations $f(x, y) = 0$; indeed, Newton called it the *chord–tangent construction*. Just as lines usually intersect a conic section in two points, lines usually intersect cubic curves in three points. If $y = mx + h$ is the equation of a line L, then L meets the curve in points $(x, mx + h)$ for which $f(x, mx + h) = 0$. But $f(x, mx + h)$ is a cubic, and so it has three roots (if we admit complex numbers). In particular, given rational points P, Q on a cubic curve C, say $P = (a, b)$ and $Q = (c, d)$, then the slope $(d - b)/(c - a)$ of the line L they determine is rational. If C has equation $f(x, y) = 0$, where $f(x, y) \in \mathbb{Q}[x, y]$, then L intersects C in a third point, which is also a rational point. Thus, the *chord* joining rational points P and Q on C determines a third rational point on C (see Figure 9.8); denote the third point by

$$P * Q.$$

If we think of the tangent line T to C at P as a limit of chords through P, then it is natural to consider where T meets C. The slope of T is also rational: if

$$A(x)y^3 + B(x)y^2 + C(x)y + D(x) = 0,$$

where $A, B, C, D \in \mathbb{Q}[x]$ have degrees, respectively, $0, 1, 2, 3$, then implicit differentiation gives

$$y'(x, y) = -\frac{B'y^2 + C'y + D'}{3Ay^2 + 2By + C};$$

since the coefficients of A, B, C, D are rational and $P = (a, b)$ is a rational point, the slope $y'(a, b)$ of T is rational. It follows easily that if T meets C,

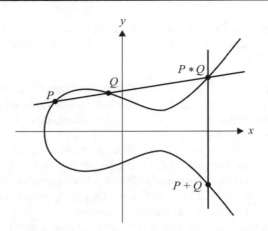

Figure 9.8. Adding points on a cubic.

then the point of intersection is another rational point; denote such a point by

$$P * P$$

(thus, the tangent line T intersects C in another rational point, say Q, and the two points P, Q determine a third rational point).

If we are considering cubic curves C in the plane, then it is possible that a line meets C in only one point, not three (a cubic in $\mathbb{R}[x]$ always has a real root, but its other roots may not be real). To make all work smoothly, we enlarge the plane to the *Riemann sphere* $\widehat{\mathbb{C}} = \mathbb{R}^2 \cup \{\infty\}$, where we regard ∞ as a point "at infinity." We agree that lines through ∞ are precisely the vertical lines; we declare that ∞ lies on every cubic C and that ∞ is a rational point on C. Given two points P, Q on the curve, the line they determine meets the curve in a third rational point $P * Q$. Define $P + Q$ to be the intersection of C with the vertical line V through $P * Q$; that is, V is the line determined by the two rational points ∞ and $P * Q$ (see Figure 9.8). The wonderful discovery is that this allows us to "add" points P, Q on elliptic curves (indeed, the set of all rational points on an elliptic curve is an abelian group under this binary operation). In particular, if C is the elliptic curve arising from the lemniscate (or any of the elliptic functions considered by Euler), then the limits of integration in Euler's Addition Theorem are given by the chord–tangent construction: for example,

This is a good reason to consider cubics over the complex numbers.

$$\int_0^P \frac{dt}{\sqrt{1 - t^4}} + \int_0^Q \frac{dt}{\sqrt{1 - t^4}} = \int_0^{P+Q} \frac{dt}{\sqrt{1 - t^4}}.$$

As we have seen, congruent numbers n arising from rational side-lengths (a, b, c) of a right triangle correspond to rational points on the elliptic curve $y^2 = x^3 - n^2 x$. The binary operation shows how to construct new congruent numbers from given ones. The importance of this operation is illustrated by Theorem 9.19.

What has this discussion to do with Fermat's Last Theorem? The abelian group of rational points on elliptic curves, an example of ***complex multiplication***, is only the beginning of deep connections between Diophantine equations and elliptic curves. The following account by the number theorist Andrew Granville summarizes the recent history.

*It all begun in 1955, with a question posed by the Japanese mathematician Yutaka Taniyama: Could one explain the properties of elliptic curves, equations of the form $y^2 = x^3 + ax + b$ with a and b given whole numbers, in terms of a few well-chosen curves? That is, is there some very special class of equations that in some way encapsulate everything there is to know about our elliptic curves? Taniyama was fairly specific about these very special curves (the so-called **modular curves**) and, in 1968, André Weil, one of the leading mathematicians of the twentieth century, made explicit which modular curve should describe which elliptic curve. In 1971, the first significant proven evidence in favor of this abstract understanding of equations was given by Goro Shimura at Princeton University, who showed that it works for a very special class of equations. This somewhat esoteric proposed approach to understanding elliptic curves is now known as the Shimura–Taniyama–Weil conjecture. There the matter stood until 1986, when Gerhard Frey made the most surprising and innovative link between this very abstract conjecture and Fermat's Last Theorem. What he realized was that if $c^n = a^n + b^n$, then it seemed unlikely that one could understand the equation $y^2 = x(x - a^n)(x + b^n)$ in the way proposed by Taniyama. It took deep and difficult reasoning by Jean-Pierre Serre and Ken Ribet to strengthen Frey's original concept to the point that a counterexample to Fermat's Last Theorem would directly contradict the Shimura–Taniyama–Weil conjecture.*

This is the point where Wiles enters the picture. Wiles drew together a vast array of techniques to attack this question. Motivated by extraordinary new methods of Victor Kolyvagin and Barry Mazur, Wiles established the Shimura–Taniyama–Weil conjecture for an important class of examples, including those relevant to proving Fermat's Last Theorem. His work can be viewed as a blend of arithmetic and geometry, and has its origins way back in Diophantus's Arithmetic. However he employs the latest ideas from a score of different fields, from the theories of L-functions, group schemes, crystalline cohomology, Galois representations, modular forms, deformation theory, Gorenstein rings, Euler systems and many others. He uses, in an essential way, concepts due to many mathematicians from around the world who were thinking about very different questions.

The work of Wiles is a tour de force, and will stand as one of the scientific achievements of the century. His work is not to be seen in isolation, but rather as the culmination of much recent thinking in many directions. Wiles' proof, starting from scratch, would surely be over a thousand pages long.

The story of this important discovery is a tribute to the deeper and more abstruse levels of abstract understanding that mathematicians have long claimed is essential. Many of us, while hailing Wiles' magnificent achievement, yearn for Fermat to have been correct, and for the truly marvellous, and presumably comparatively straightforward, proof to be recovered.

We refer the reader to [31] for more about elliptic curves and Diophantine equations. We also recommend the expository article of Cox, Introduction to Fermat's Last Theorem, *Amer. Math. Monthly* 101 (1994), pp. 3–14, for more details.

Appendices

A.1 Functions

Pick up any calculus book; somewhere near the beginning is a definition of *function* which reads something like this: a ***function*** f is a rule that assigns to each element a in a set A exactly one element, denoted by $f(a)$, in a set B. Actually, this isn't too bad. The spirit is right: f is dynamic; it is like a machine, whose input consists of elements of A and whose output consists of certain elements of B. The sets A and B may be made up of numbers, but they don't have to be.

There is a slight notational surprise. We are used to writing a function, not as f, but as $f(x)$. For example, integrals are written

$$\int f(x)\, dx.$$

Logically, one notation for a function, say f, and another for its value at a point a in A, say $f(a)$, does make sense. However, some notation is grand-fathered in. For example, we will continue to write polynomials as $f(x) = a_n x^n + a_{n-1} x^{n-1} + \cdots + a_0$, trigonometric functions as $\sin x$ and $\cos x$, and the exponential as e^x (but some authors denote the exponential function by exp). Still, the simpler notation f is usually a good idea.

One problem we have with the calculus definition of function involves the word *rule*. To see why this causes problems, we ask when two functions are equal. If f is the function $f(x) = x^2 + 2x + 1$ and g is the function $g(x) = (x + 1)^2$, is $f = g$? We usually think of a rule as a recipe, a set of directions. With this understanding, f and g are surely different: $f(5) = 25 + 10 + 1$ and $g(5) = 6^2$. These are different recipes, but note that both cook the same dish: for example, $f(5) = 36 = g(5)$.

A second problem with the calculus definition is just what is allowed to be a rule. Must a rule be a formula? If so, then $f(x)$, defined by

$$f(x) = \begin{cases} 1 & \text{if } x \text{ is rational} \\ 0 & \text{if } x \text{ is irrational,} \end{cases}$$

is not a function. Or is it? The simplest way to deal with these problems is to avoid the imprecise word *rule*.

If A is a set, then we write

$$a \in A,$$

which abbreviates "a belongs to A" or "a is an element of A."

We can define *ordered pair* from scratch: see Exercise A.2 on page 418. Cartesian product could then be defined by induction on n.

Definition. If $A_1, A_2, \ldots, A_n$ are sets, their **cartesian product** is

$$A_1 \times A_2 \times \cdots \times A_n = \{(a_1, a_2, \ldots, a_n) : a_i \in A_i \text{ for all } i\}.$$

An element $(a_1, a_2) \in A_1 \times A_2$ is called an **ordered pair**, and $(a_1, a_2) = (a_1', a_2')$ if and only if $a_1 = a_1'$ and $a_2 = a_2'$. More generally, two n-tuples $(a_1, a_2, \ldots, a_n)$ and $(a_1', a_2', \ldots, a_n')$ are **equal** if $a_i = a_i'$ for all subscripts i.

We now review *subsets* and *equality*, for functions $f : A \to B$ are subsets of $A \times B$. We say that U is a **subset** of V or U is **contained** in V, denoted by

$$U \subseteq V,$$

if, for all $u \in U$, we have $u \in V$. Formally: $(\forall u)(u \in U \Rightarrow u \in V)$.

Two subsets U and V of a set X are **equal**, that is,

$$U = V,$$

if they are comprised of exactly the same elements: thus, $U = V$ if and only if $U \subseteq V$ and $V \subseteq U$. This obvious remark is important because many proofs of equality of subsets break into two parts, each part showing that one subset is contained in the other. For example, let

$$U = \{x \in \mathbb{R} : x \geq 0\} \text{ and } V = \{x \in \mathbb{R} : \text{there exists } y \in \mathbb{R} \text{ with } x = y^2\}.$$

Now $U \subseteq V$ because $x = (\sqrt{x})^2 \in V$, while $V \subseteq U$ because $y^2 \geq 0$ for every real number y (if $y < 0$, then $y = -a$ for $a > 0$ and $y^2 = a^2$). Hence, $U = V$.

If U is a **proper subset** of V; that is, if $U \subseteq V$ but $U \neq V$, then we write

$$U \subsetneqq V.$$

An **empty set**, denoted by $\varnothing$, is a set with no elements. Given a set X, it is always true that $\varnothing \subseteq X$. To see this, observe that the negation

$$(\exists u)(u \in \varnothing \Rightarrow u \notin V)$$

is false, for there is no $u \in \varnothing$. The empty set is unique: if $\varnothing'$ is also an empty set, then $\varnothing \subseteq \varnothing'$ and $\varnothing' \subseteq \varnothing$, so that $\varnothing = \varnothing'$. There is only one empty set.

Informally, a function *is* what we usually call its *graph*.

Definition. Let A and B be sets. A **function** $f : A \to B$ is a subset $f \subseteq A \times B$ such that, for each $a \in A$, there is a unique $b \in B$ with $(a, b) \in f$. The set A is called its **domain**, and the set B is called its **target**.

If f is a function and $(a, b) \in f$, then we write $f(a) = b$ and we call b the **value** of f at a. Define the **image** (or *range*) of f, denoted by im f, to be the subset of B consisting of all the values of f.

When we say that A is the domain of a function $f : A \to B$, we mean that $f(a)$ is defined for *every* $a \in A$. Thus, the reciprocal $f(x) = 1/(x - 1)$ is not a function $\mathbb{R} \to \mathbb{R}$, but it is a function $\mathbb{R}' \to \mathbb{R}$, where $\mathbb{R}'$ denotes the set of all real numbers not equal to 1.

The second problem above—is $f : \mathbb{R} \to \mathbb{R}$ a function, where $f(x) = 1$ if x is rational and $f(x) = 0$ if x is irrational—can now be resolved; yes, f is a function:

$$f = \{(x, 1) : x \text{ is rational}\} \cup \{(x, 0) : x \text{ is irrational}\} \subseteq \mathbb{R} \times \mathbb{R}.$$

Let's look at more examples before resolving the first problem arising from the imprecise term *rule*.

Example A.1. (i) The squaring function $f: \mathbb{R} \to \mathbb{R}$, given by $f(x) = x^2$, is the parabola consisting of all points in the plane $\mathbb{R} \times \mathbb{R}$ of the form (a, a^2). It satisfies the definition, and so f is a function (if f wasn't a bona fide function, we would change the definition!).

(ii) If A and B are sets and $b_0 \in B$, then the **constant function** at b_0 is the function $f: A \to B$ defined by $f(a) = b_0$ for all $a \in A$ (when $A = \mathbb{R} = B$, then the graph of a constant function is a horizontal line).

(iii) For a set A, the **identity function** $1_A: A \to A$ is the function consisting of all (a, a) (the *diagonal* of $A \times A$), and $1_A(a) = a$ for all $a \in A$.

(iv) The usual functions appearing in calculus are also functions according to the definition just given. For example, the domain of $\sin x$ is $\mathbb{R}$, its target is usually $\mathbb{R}$, and its image is the closed interval $[-1, 1]$. ▲

How to Think About It. A function $f: A \to B$ is the subset of $A \times B$ consisting of all ordered pairs $(a, f(a))$ (this subset *is* the function but, informally, it is usually called its **graph**). In order to maintain the spirit of a function being dynamic, we often use the notation

$$f: a \mapsto b$$

instead of $f(a) = b$. For example, we may write the squaring function as $f: a \mapsto a^2$ instead of $f(a) = a^2$. We often say that f *sends* a to $f(a)$.

Let's return to our first complaint about rules; when are two functions equal?

Proposition A.2. *Let $f: A \to B$ and $g: A \to B$ be functions. Then $f = g$ if and only if $f(a) = g(a)$ for every $a \in A$.*

Proof. Assume that $f = g$. Functions are subsets of $A \times B$, and so $f = g$ means that each of f and g is a subset of the other. If $a \in A$, then $(a, f(a)) \in f$; since $f = g$, we have $(a, f(a)) \in g$. But there is only one ordered pair in g with first coordinate a, namely, $(a, g(a))$, because the definition of function says that g gives a *unique* value to a. Therefore, $(a, f(a)) = (a, g(a))$, and equality of ordered pairs gives $f(a) = g(a)$, as desired.

Conversely, assume that $f(a) = g(a)$ for every $a \in A$. To see that $f = g$, it suffices to show that $f \subseteq g$ and $g \subseteq f$. Each element of f has the form $(a, f(a))$. Since $f(a) = g(a)$, we have $(a, f(a)) = (a, g(a))$ and hence $(a, f(a)) \in g$. Therefore, $f \subseteq g$. The reverse inclusion $g \subseteq f$ is proved similarly. Therefore, $f = g$. ∎

Proposition A.2 resolves the first problem arising from the term *rule*: if $f, g: \mathbb{R} \to \mathbb{R}$ are given by $f(x) = x^2 + 2x + 1$ and $g(x) = (x + 1)^2$, then $f = g$ because $f(a) = g(a)$ for every number a.

Let us clarify another point. Can functions $f: A \to B$ and $g: A' \to B'$ be equal? Here is the commonly accepted usage.

Definition. Functions $f: A \rightarrow B$ and $g: A' \rightarrow B'$ are **equal** if $A = A'$, $B = B'$, and $f(a) = g(a)$ for all $a \in A$.

Thus, a function $f: A \rightarrow B$ has three ingredients—its domain A, its target B, and its graph—and we are saying that functions are equal if and only if they have the same domains, the same targets, and the same graphs.

It is plain that the domain and the graph are essential parts of a function; why should we care about the target of a function when its image is more important? As a practical matter, when first defining a function, one usually doesn't know its image. For example, what's the image of $f: \mathbb{R} \rightarrow \mathbb{R}$, defined by

$$f(x) = \log\left(\frac{1 + |x|^{e^{-x}}}{\sqrt[5]{x^2 + \cos^2 x}}\right) - \int_0^x \frac{dt}{\sqrt[7]{1 + t^6}} ?$$

We must analyze f to find its image, and this is no small task. But if targets have to be images, then we couldn't even write down $f: \mathbb{R} \rightarrow \mathbb{R}$ without having first found the image of f. Thus, targets are convenient to use.

See the discussion on page 437 for a more sophisticated reason why targets are important.

If A is a subset of a set B, the **inclusion** $i: A \rightarrow B$ is the function given by $i(a) = a$ for all $a \in A$; that is, i is the subset of $A \times B$ consisting of all (a, a) with $a \in A$. If S is a proper subset of a set A, then the inclusion $i: S \rightarrow A$ is not the identity function 1_S because its target is A, not S; it is not the identity function 1_A because its domain is S, not A.

Instead of saying that the values of a function f are unique, we sometimes say that f is **single-valued** or that it is **well-defined**. For example, if $\mathbb{R}^{\geq}$ denotes the set of nonnegative reals, then $\sqrt{\ } : \mathbb{R}^{\geq} \rightarrow \mathbb{R}^{\geq}$ is a function because we agree that $\sqrt{a} \geq 0$ for every nonnegative number a. On the other hand, $g(a) = \pm\sqrt{a}$ is not single-valued, and hence it is not a function. The simplest way to verify whether an alleged function f is single-valued is to phrase uniqueness of values as an implication:

$$\text{if } a = a', \text{ then } f(a) = f(a').$$

For example, consider the addition function $\alpha: \mathbb{R} \times \mathbb{R} \rightarrow \mathbb{R}$. To say that α is well-defined is to say that if $(u, v) = (u', v')$ in $\mathbb{R} \times \mathbb{R}$, then $\alpha(u, v) = \alpha(u', v')$; that is, if $u = u'$ and $v = v'$, then $u + v = u' + v'$. This is usually called the **Law of Substitution**.

Another example is addition of fractions. We define

$$\frac{a}{b} + \frac{c}{d} = \frac{ad + bc}{bd}.$$

But fractions have many names. If $a/b = a'/b'$ and $c/d = c'/d'$, is $(ad + bc)/bd = (a'd' + b'c')/b'd'$? We verified that this formula does not depend on the choices of names of the fractions on page 193. On the other hand, the operation

$$\frac{a}{b} * \frac{c}{d} = \frac{a + c}{bd}$$

is not well-defined: $\frac{1}{2} = \frac{2}{4}$, but $\frac{1}{2} * \frac{3}{4} = \frac{4}{8}$, while $\frac{2}{4} * \frac{3}{4} = \frac{5}{16} \neq \frac{4}{8}$.

There is a name for functions whose image is equal to the whole target.

Definition. A function $f: A \to B$ is *surjective* (or *onto* or a *surjection*) if

$$\text{im } f = B.$$

Thus, f is surjective if, for each $b \in B$, there is some $a \in A$ (depending on b) with $b = f(a)$.

Example A.3. (i) The identity function $1_A: A \to A$ is a surjection.

(ii) The sine function $\mathbb{R} \to \mathbb{R}$ is not surjective, for its image is $[-1, 1]$, a proper subset of its target $\mathbb{R}$. The function $s: \mathbb{R} \to [-1, 1]$, defined by $s(x) = \sin x$, is surjective.

(iii) The functions $x^2: \mathbb{R} \to \mathbb{R}$ and $e^x: \mathbb{R} \to \mathbb{R}$ have target $\mathbb{R}$. Now im x^2 consists of the nonnegative reals and im e^x consists of the positive reals, so that neither x^2 nor e^x is surjective.

(iv) Let $f: \mathbb{R} \to \mathbb{R}$ be defined by

$$f(a) = 6a + 4.$$

To see whether f is a surjection, we ask whether every $b \in \mathbb{R}$ has the form $b = f(a)$ for some a; that is, given b, can we find a so that

$$6a + 4 = b?$$

Since $a = \frac{1}{6}(b - 4)$, this equation can always be solved for a, and so f is a surjection.

(v) Let $f: \mathbb{R} - \{\frac{3}{2}\} \to \mathbb{R}$ be defined by

$$f(a) = \frac{6a + 4}{2a - 3}.$$

To see whether f is a surjection, we seek, given b, a solution a: can we solve

$$b = f(a) = \frac{6a + 4}{2a - 3}?$$

This leads to the equation $a(6 - 2b) = -3b - 4$, which can be solved for a if $6 - 2b \neq 0$ (note that $(-3b - 4)/(6 - 2b) \neq 3/2$). On the other hand, it suggests that there is no solution when $b = 3$ and, indeed, there is not: if $(6a + 4)/(2a - 3) = 3$, cross multiplying gives the false equation $6a + 4 = 6a - 9$. Thus, $3 \notin \text{im } f$, and f is not a surjection (in fact, $\text{im } f = \mathbb{R} - \{3\}$). ▲

The following definition gives another important property a function may have.

Definition. A function $f: A \to B$ is *injective* (or *one-to-one* or an *injection*) if, whenever a and a' are distinct elements of A, then $f(a) \neq f(a')$. Equivalently, the contrapositive states that f is injective if, for every pair $a, a' \in A$, we have

$$f(a) = f(a') \text{ implies } a = a'.$$

Being injective is the converse of being single-valued: f is single-valued if $a = a'$ implies $f(a) = f(a')$; f is injective if $f(a) = f(a')$ implies $a = a'$.

Most functions are neither injective nor surjective. For example, the squaring function $f: \mathbb{R} \to \mathbb{R}$, defined by $f(x) = x^2$, is neither.

Example A.4.   (i) The identity function $1_A: A \to A$ is injective.

(ii) Let $f: \mathbb{R} - \{\frac{3}{2}\} \to \mathbb{R}$ be defined by

$$f(a) = \frac{6a + 4}{2a - 3}.$$

To check whether f is injective, suppose that $f(a) = f(b)$:

$$\frac{6a + 4}{2a - 3} = \frac{6b + 4}{2b - 3}.$$

Cross multiplying yields

$$12ab + 8b - 18a - 12 = 12ab + 8a - 18b - 12,$$

which simplifies to $26a = 26b$ and hence $a = b$. We conclude that f is injective. (We saw, in Example A.3, that f is not surjective.)

(iii) Consider $f: \mathbb{R} \to \mathbb{R}$, given by $f(x) = x^2 - 2x - 3$. If we try to check whether f is an injection by looking at the consequences of $f(a) = f(b)$, as in part (ii), we arrive at the equation $a^2 - 2a = b^2 - 2b$; it is not instantly clear whether this forces $a = b$. Instead, we seek the roots of f, which are 3 and -1. It follows that f is not injective, for $f(3) = 0 = f(-1)$; that is, there are two distinct numbers having the same value.   ▲

Sometimes there is a way of combining two functions to form another function, their *composite*.

Definition. If $f: A \to B$ and $g: B \to C$ are functions (the target of f is the domain of g), then their ***composite***, denoted by $g \circ f$, is the function $A \to C$ given by

$$g \circ f: a \mapsto g(f(a));$$

that is, first evaluate f on a, and then evaluate g on $f(a)$.

We usually abbreviate the notation for composites in the text, writing gf instead of $g \circ f$, but we shall always write $g \circ f$ in this Appendix.

Composition is thus a two-step process: $a \mapsto f(a) \mapsto g(f(a))$. For example, the function $h: \mathbb{R} \to \mathbb{R}$, defined by $h(x) = e^{\cos x}$, is the composite $g \circ f$, where $f(x) = \cos x$ and $g(x) = e^x$. This factorization is plain as soon as one tries to evaluate, say $h(\pi)$; one must first evaluate $f(\pi) = \cos \pi = -1$ and then evaluate

$$h(\pi) = g(f(\pi)) = g(-1) = e^{-1}.$$

The chain rule in calculus is a formula that computes the derivative $(g \circ f)'$ in terms of g' and f':

$$(g \circ f)'(x) = (g' \circ f)(x) \cdot f'(x) = g'(f(x)) \cdot f'(x).$$

If $f: A \to B$ is a function, and if S is a subset of A, then the **restriction** of f to S is the function $f|S: S \to B$, defined by $(f|S)(s) = f(s)$ for all $s \in S$. It is easy to see that if $i: S \to A$ is the inclusion, then $f|S = f \circ i$; that is, the functions $f|S$ and $f \circ i$ have the same domain, the same target, and the same graph (see Exercise A.4 on page 419).

If $f: \mathbb{N} \to \mathbb{N}$ and $g: \mathbb{N} \to \mathbb{R}$ are functions, then $g \circ f: \mathbb{N} \to \mathbb{R}$ is defined, but $f \circ g$ is not defined (for $\text{target}(g) = \mathbb{R} \neq \mathbb{N} = \text{domain}(f)$). Even when $f: A \to B$ and $g: B \to A$, so that both composites $g \circ f$ and $f \circ g$ are defined, they need not be equal. For example, define $f, g: \mathbb{N} \to \mathbb{N}$ by $f: n \mapsto n^2$ and $g: n \mapsto 3n$; then $g \circ f: 2 \mapsto g(4) = 12$ and $f \circ g: 2 \mapsto f(6) = 36$. Hence, $g \circ f \neq f \circ g$.

Given a set A, let

$$A^A = \{\text{all functions } A \to A\}.$$

The composite of two functions in A^A is always defined, and it is, again, a function in A^A. As we have just seen, composition is not **commutative**; that is, $f \circ g$ and $g \circ f$ need not be equal. Let us now show that composition is always **associative**.

Proposition A.5. *Composition of functions is associative*: *given functions* $f: A \to B$, $g: B \to C$, *and* $h: C \to D$, *then*

$$h \circ (g \circ f) = (h \circ g) \circ f.$$

Proof. We show that the value of either composite on an element $a \in A$ is just $d = h(g(f(a)))$. If $a \in A$, then

$$h \circ (g \circ f): a \mapsto (g \circ f)(a) = g(f(a)) \mapsto h(g(f(a))) = d,$$

and

$$(h \circ g) \circ f: a \mapsto f(a) \mapsto (h \circ g)(f(a)) = h(g(f(a))) = d.$$

Since both are functions $A \to D$, it follows from Proposition A.2 that the composites are equal. ∎

In light of this proposition, we need not write parentheses: the notation $h \circ g \circ f$ is unambiguous.

Suppose that $f: A \to B$ and $g: C \to D$ are functions. If $B \subseteq C$, then some authors define the composite $h: A \to D$ by $h(a) = g(f(a))$. We do not allow composition if $B \neq C$. However, we can define h as the composite $h = g \circ i \circ f$, where $i: B \to C$ is the inclusion.

The next result implies that the identity function 1_A behaves for composition in A^A just as the number one does for multiplication of numbers.

Proposition A.6. *If* $f: A \to B$, *then* $1_B \circ f = f = f \circ 1_A$.

Proof. If $a \in A$, then

$$1_B \circ f: a \mapsto f(a) \mapsto f(a)$$

and

$$f \circ 1_A: a \mapsto a \mapsto f(a). \quad \blacksquare$$

Are there "reciprocals" in A^A; that is, are there any functions $f : A \to A$ for which there is $g \in A^A$ with $f \circ g = 1_A$ and $g \circ f = 1_A$? The following discussion will allow us to answer this question.

Definition. A function $f : A \to B$ is **bijective** (or a *one-one correspondence* or a **bijection**) if it is both injective and surjective.

Example A.7.   (i) Identity functions are always bijections.

(ii) Let $X = \{1, 2, 3\}$ and define $f : X \to X$ by

$$f(1) = 2, \quad f(2) = 3, \quad f(3) = 1.$$

It is easy to see that f is a bijection.   ▲

We can draw a picture of a function $f : X \to Y$ in the special case when X and Y are finite sets (see Figure A.1). Let $X = \{1, 2, 3, 4, 5\}$, let $Y = \{a, b, c, d, e\}$, and define $f : X \to Y$ by

$$f(1) = b \qquad f(2) = e \qquad f(3) = a \qquad f(4) = b \qquad f(5) = c.$$

Now f is not injective, because $f(1) = b = f(4)$, and f is not surjective,

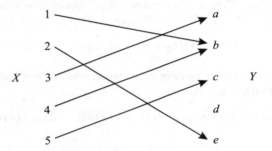

Figure A.1. Picture of a function.

because there is no $x \in X$ with $f(x) = d$. Can we reverse the arrows to get a function $g : Y \to X$? There are two reasons why we can't. First, there is no arrow going to d, and so $g(d)$ is not defined. Second, what is $g(b)$? Is it 1 or is it 4? The first problem is that the domain of g is not all of Y, and it arises because f is not surjective; the second problem is that g is not single-valued, and it arises because f is not injective (this reflects the fact that being single-valued is the converse of being injective). Neither problem arises when f is a bijection.

Definition. A function $f : X \to Y$ is **invertible** if there exists a function $g : Y \to X$, called its **inverse**, with both composites $g \circ f$ and $f \circ g$ being identity functions.

We do not say that every function f is invertible; on the contrary, we have just given two reasons why a function may not have an inverse. Notice that if an inverse function g does exist, then it "reverses the arrows" in Figure A.1. If $f(a) = y$, then there is an arrow from a to y. Now $g \circ f$ being the identity says that $a = (g \circ f)(a) = g(f(a)) = g(y)$; therefore $g : y \mapsto a$, and so the picture of g is obtained from the picture of f by reversing arrows. If f twists something, then its inverse g untwists it.

Lemma A.8. *If* $f: X \to Y$ *and* $g: Y \to X$ *are functions such that* $g \circ f = 1_X$, *then* f *is injective and* g *is surjective.*

Proof. Suppose that $f(a) = f(a')$. Apply g to obtain $g(f(a)) = g(f(a'))$; that is, $a = a'$ (because $g(f(a)) = a$), and so f is injective. If $x \in X$, then $x = g(f(x))$, so that $x \in \operatorname{im} g$; hence g is surjective. ∎

Proposition A.9. *A function* $f: X \to Y$ *has an inverse* $g: Y \to X$ *if and only if it is a bijection.*

Proof. If f has an inverse g, then Lemma A.8 shows that f is injective and surjective, for both composites $g \circ f$ and $f \circ g$ are identities.

Assume that f is a bijection. Let $y \in Y$. Since f is surjective, there is some $a \in X$ with $f(a) = y$; since f is injective, a is unique. Defining $g(y) = a$ thus gives a (single-valued) function whose domain is Y (g merely "reverses arrows;" since $f(a) = y$, there is an arrow from a to y, and the reversed arrow goes from y to a). It is plain that g is the inverse of f; that is, $f(g(y)) = f(a) = y$ for all $y \in Y$ and $g(f(a)) = g(y) = a$ for all $x \in X$. ∎

Example A.10. If a is a real number, then ***multiplication by a*** is the function $\mu_a: \mathbb{R} \to \mathbb{R}$, defined by $r \mapsto ar$ for all $r \in \mathbb{R}$. If $a \neq 0$, then μ_a is a bijection; its inverse function is ***division by a***, namely, $\delta_a: \mathbb{R} \to \mathbb{R}$, defined by $r \mapsto \frac{1}{a}r$; of course, $\delta_a = \mu_{1/a}$. If $r \in \mathbb{R}$, then

$$\mu_a \circ \delta_a: r \mapsto \tfrac{1}{a}r \mapsto a\tfrac{1}{a}r = r;$$

hence, $\mu_a \circ \delta_a = 1_\mathbb{R}$. Similarly, $\delta_a \circ \mu_a = 1_\mathbb{R}$.

If $a = 0$, however, then $\mu_a = \mu_0$ is the constant function $\mu_0: r \mapsto 0$ for all $r \in \mathbb{R}$, which has no inverse function. ▲

Etymology. The inverse of a bijection f is denoted by f^{-1} (Exercise A.6 on page 419 says that a function cannot have two inverses). This is the same notation used for inverse trigonometric functions in calculus; for example, $\sin^{-1} x = \arcsin x$ satisfies $\sin(\arcsin(x)) = x$ and $\arcsin(\sin(x)) = x$. Of course, $\sin^{-1}$ does *not* denote the reciprocal $1/\sin x$, which is $\csc x$.

What are the domain and image of arcsin?

Example A.11. Here is an example of two functions $f, g: \mathbb{N} \to \mathbb{N}$ with one composite the identity, but with the other composite not the identity; thus, f and g are *not* inverse functions.

Define $f, g: \mathbb{N} \to \mathbb{N}$ as follows:

$$f(n) = n + 1;$$

$$g(n) = \begin{cases} 0 & \text{if } n = 0 \\ n - 1 & \text{if } n \geq 1. \end{cases}$$

The composite $g \circ f = 1_\mathbb{N}$, for $g(f(n)) = g(n+1) = n$ (because $n + 1 \geq 1$). On the other hand, $f \circ g \neq 1_\mathbb{N}$ because $f(g(0)) = f(0) = 1 \neq 0$. ▲

Two strategies are now available to determine whether a given function is a bijection: use the definitions of injective and surjective, or find an inverse

What are the domains and targets of log and exp?

function. For example, if $\mathbb{R}^>$ denotes the positive real numbers, let us show that the exponential function $f \colon \mathbb{R} \to \mathbb{R}^>$, defined by $f(x) = e^x = \sum x^n / n!$, is a bijection. A direct proof that f is injective would require showing that if $e^a = e^b$, then $a = b$; a direct proof showing that f is surjective would involve showing that every positive real number c has the form e^a for some a. It is simplest to prove these statements using the (natural) logarithm $g(y) = \log y$. The usual formulas $e^{\log y} = y$ and $\log e^x = x$ show that both composites $f \circ g$ and $g \circ f$ are identities, and so f and g are inverse functions. Therefore, f is a bijection, for it has an inverse.

The next theorem summarizes some results of this section. If X is a nonempty set, define the **symmetric group** on X:

$$S_X = \{\text{bijections } \sigma \colon X \to X\}.$$

Theorem A.12. *If X is a nonempty set, then composition $(f, g) \mapsto g \circ f$ is a function $S_X \times S_X \to S_X$ satisfying the following properties:*

(i) $(f \circ g) \circ h = f \circ (g \circ h)$ *for all* $f, g, h \in S_X$;

(ii) *there is* $1_X \in S_X$ *with* $1_X \circ f = f = f \circ 1_X$ *for all* $f \in S_X$;

(iii) *for all* $f \in S_X$, *there is* $f' \in S_X$ *with* $f' \circ f = 1_X = f \circ f'$.

Proof. Exercise A.12(iii) on page 420 says that the composite of two bijections is itself a bijection, and so composition has target S_X. Part (i) is Proposition A.5, part (ii) is Proposition A.6, and part (iii) is Proposition A.9. ∎

Exercises

A.1 True or false, with reasons.

(i) If $S \subseteq T$ and $T \subseteq X$, then $S \subseteq X$.

(ii) Any two functions $f \colon X \to Y$ and $g \colon Y \to Z$ have a composite $f \circ g \colon X \to Z$.

(iii) Any two functions $f \colon X \to Y$ and $g \colon Y \to Z$ have a composite $g \circ f \colon X \to Z$.

(iv) For every set X, we have $X \times \varnothing = \varnothing$.

(v) If $f \colon X \to Y$ and $j \colon \operatorname{im} f \to Y$ is the inclusion, then there is a surjection $g \colon X \to \operatorname{im} f$ with $f = j \circ g$.

(vi) If $f \colon X \to Y$ is a function for which there is a function $g \colon Y \to X$ with $f \circ g = 1_Y$, then f is a bijection.

(vii) The formula $f(a/b) = (a + b)(a - b)$ is a well-defined function $\mathbb{Q} \to \mathbb{Z}$.

(viii) If $f \colon \mathbb{N} \to \mathbb{N}$ is given by $f(n) = n + 1$ and $g \colon \mathbb{N} \to \mathbb{N}$ is given by $g(n) = n^2$, then the composite $g \circ f$ is $n \mapsto n^2(n + 1)$.

(ix) Complex conjugation $z = a + ib \mapsto \overline{z} = a - ib$ is a bijection $\mathbb{C} \to \mathbb{C}$.

Hint. (i) True. (ii) False. (iii) True. (iv) True. (v) True. (vi) False. (vii) False. (viii) False. (ix) True.

A.2 * Let A and B be sets, and let $a \in A$ and $b \in B$. Define their **ordered pair** as follows:

$$(a, b) = \{a, \{a, b\}\}.$$

If $a' \in A$ and $b' \in B$, prove that $(a', b') = (a, b)$ if and only if $a' = a$ and $b' = b$.

Hint. In any formal treatment, one is obliged to define new terms carefully. In particular, in set theory, one must discuss the membership relation $\in$. Does $x \in x$ make sense? If it does, is it ever true? One of the axioms constraining $\in$ is that the statement $a \in x \in a$ is always false.

A.3 Let $L = \{(x, x) : x \in \mathbb{R}\}$; thus, L is the line in the plane that passes through the origin and makes an angle of $45°$ with the x-axis.

(i) If $P = (a, b)$ is a point in the plane with $a \neq b$, prove that L is the perpendicular-bisector of the segment PP' having endpoints $P = (a, b)$ and $P' = (b, a)$.

Hint. You may use Lemma 3.16 and the fact that $(\frac{1}{2}(a + c), \frac{1}{2}(b + d))$ is the midpoint of the line segment having endpoints (a, b) and (c, d).

(ii) If $f : \mathbb{R} \to \mathbb{R}$ is a bijection whose graph consists of certain points (a, b) (of course, $b = f(a)$), prove that the graph of f^{-1} is

$$\{(b, a) : (a, b) \in f\}.$$

A.4 * Let X and Y be sets, and let $f : X \to Y$ be a function.

(i) If S is a subset of X, prove that the restriction $f | S$ is equal to the composite $f \circ i$, where $i : S \to X$ is the inclusion map.

Hint. Use the definition of equality of functions on page 412.

(ii) If im $f = A \subseteq Y$, prove that there exists a surjection $f' : X \to A$ with $f = j \circ f'$, where $j : A \to Y$ is the inclusion.

A.5 If $f : X \to Y$ has an inverse g, show that g is a bijection.

Hint. Does g have an inverse?

A.6 * Show that if $f : X \to Y$ is a bijection, then it has exactly one inverse.

A.7 Show that $f : \mathbb{R} \to \mathbb{R}$, defined by $f(x) = 3x + 5$, is injective and surjective, and find its inverse.

A.8 Determine whether $f : \mathbb{Q} \times \mathbb{Q} \to \mathbb{Q}$, given by

$$f(a/b, c/d) = (a + c)/(b + d)$$

is a function.

Hint. It isn't.

A.9 Let $X = \{x_1, \ldots, x_m\}$ and $Y = \{y_1, \ldots, y_n\}$ be finite sets, where the x_i are distinct and the y_j are distinct. Show that there is a bijection $f : X \to Y$ if and only if $|X| = |Y|$; that is, $m = n$.

Hint. If f is a bijection, there are m distinct elements $f(x_1), \ldots, f(x_m)$ in Y, and so $m \leq n$; using the bijection f^{-1} in place of f gives the reverse inequality $n \leq m$.

A.10 Suppose there are 11 pigeons, each sitting in some pigeonhole. If there are only 10 pigeonholes, prove that there is a hole containing more than one pigeon.

A.11 * **(Pigeonhole Principle)**. If X and Y are finite sets with the same number of elements, show that the following conditions are equivalent for a function $f : X \to Y$:

(i) f is injective   (ii) f is bijective   (iii) f is surjective.

Hint. If $A \subseteq X$ and $|A| = n = |X|$, then $A = X$; after all, how many elements are in X but not in A?

A.12 * Let $f: X \to Y$ and $g: Y \to Z$ be functions.

   (i) If both f and g are injective, prove that $g \circ f$ is injective.

   (ii) If both f and g are surjective, prove that $g \circ f$ is surjective.

   (iii) If both f and g are bijective, prove that $g \circ f$ is bijective.

   (iv) If $g \circ f$ is a bijection, prove that f is an injection and g is a surjection.

A.13 (i) If $f: (-\pi/2, \pi/2) \to \mathbb{R}$ is defined by $a \mapsto \tan a$, then f has an inverse function g; indeed, $g = \arctan$.

   Hint. Compute composites.

   (ii) Show that each of $\arcsin x$ and $\arccos x$ is an inverse function (of $\sin x$ and $\cos x$, respectively) as defined in this section. (Domains and targets must be chosen with care.)

A.2   Equivalence Relations

When fractions are first discussed in grammar school, students are told that $\frac{1}{3} = \frac{2}{6}$ because $1 \times 6 = 3 \times 2$; cross-multiplying makes it so! Don't believe your eyes that $1 \neq 2$ and $3 \neq 6$. Doesn't everyone see that $1 \times 6 = 6 = 3 \times 2$? Of course, a good teacher wouldn't just say this. Further explanation is required, and here it is. We begin with the general notion of *relation*.

Definition. Let X and Y be sets. A **relation from X to Y** is a subset R of $X \times Y$ (if $X = Y$, then we say that R is a **relation on** X). We usually write $x R y$ instead of $(x, y) \in R$.

Here is a concrete example. Certainly $\leq$ should be a relation on $\mathbb{R}$; to see that it is, define the subset

$$R = \{(x, y) \in \mathbb{R} \times \mathbb{R} : (x, y) \text{ lies on or above the line } y = x\}.$$

You should check that $(x, y) \in R$ if the second coordinate is bigger than the first. Thus, $x R y$ here coincides with the usual meaning $x \leq y$.

Example A.13.   (i) Every function $f: X \to Y$ is a relation from X to Y.

(ii) Equality is a relation on any set X; it is the **diagonal**

$$\{(x, x) : x \in X\} \subseteq X \times X.$$

(iii) For every natural number m, congruence mod m is a relation on $\mathbb{Z}$. Can you describe the subset of $\mathbb{Z} \times \mathbb{Z}$?

(iv) If $X = \{(a, b) \in \mathbb{Z} \times \mathbb{Z} : b \neq 0\}$, then cross multiplication defines a relation $\equiv$ on X by

$$(a, b) \equiv (c, d) \quad \text{if} \quad ad = bc. \quad \blacktriangle$$

Definition. A relation $x \equiv y$ on a set X is

 (i) **reflexive** if $x \equiv x$ for all $x \in X$,

 (ii) **symmetric** if $x \equiv y$ implies $y \equiv x$ for all $x, y \in X$,

 (iii) **transitive** if $x \equiv y$ and $y \equiv z$ imply $x \equiv z$ for all $x, y, z \in X$.

An **equivalence relation** on a set X is a relation on X that has all three properties: reflexivity, symmetry, and transitivity.

Example A.14.   (i) Ordinary equality is an equivalence relation on any set.

(ii) If $m \geq 0$, then Proposition 4.3 says that $x \equiv y \bmod m$ is an equivalence relation on $\mathbb{Z}$.

(iii) If I is an ideal in a commutative ring R, then Proposition 7.1 shows that congruence mod I is an equivalence relation on R.

(iv) We claim that cross multiplication is an equivalence relation on $X = \{(a, b) \in \mathbb{Z} \times \mathbb{Z} : b \neq 0\}$. Verification of reflexivity and symmetry is easy. For transitivity, assume that $(a, b) \equiv (c, d)$ and $(c, d) \equiv (e, f)$. Now $ad = bc$ gives $adf = bcf$, and $cf = de$ gives $bcf = bde$; thus, $adf = bde$. We may cancel the nonzero integer d to get $af = be$; that is, $(a, b) \equiv (e, f)$.

(v) In calculus, equivalence relations are implicit in the discussion of vectors. An *arrow* from a point P to a point Q can be denoted by the ordered pair (P, Q); call P its *foot* and Q its *head*. An equivalence relation on arrows can be defined by saying that $(P, Q) \equiv (P', Q')$ if the arrows have the same length and the same direction. More precisely, $(P, Q) \equiv (P', Q')$ if the quadrilateral obtained by joining P to P' and Q to Q' is a parallelogram (this definition is incomplete, for one must also relate collinear arrows as well as "degenerate" arrows (P, P)). The direction of an arrow from P to Q is important; if $P \neq Q$, then $(P, Q) \not\equiv (Q, P)$.   ▲

An equivalence relation on a set X yields a family of subsets of X.

Definition.   Let $\equiv$ be an equivalence relation on a nonempty set X. If $a \in X$, the *equivalence class* of a, denoted by $[a]$, is defined by

$$[a] = \{x \in X : x \equiv a\} \subseteq X.$$

We now display the equivalence classes arising from the equivalence relations in Example A.14.

Example A.15.   (i) Let $\equiv$ be equality on a set X. If $a \in X$, then $[a] = \{a\}$, the subset having only one element, namely, a. After all, if $x = a$, then x and a are equal!

(ii) Consider the relation of congruence mod m on $\mathbb{Z}$, and let $a \in \mathbb{Z}$. The congruence class of a, defined by

$$\{x \in \mathbb{Z} : x = a + km \text{ where } k \in \mathbb{Z}\},$$

is equal to the equivalence class of a, namely

$$[a] = \{x \in \mathbb{Z} : x \equiv a \bmod m\}.$$

(iii) If I is an ideal in a commutative ring R, then the equivalence class of an element $a \in R$ is the coset $a + I$.

(iv) The equivalence class of (a, b) under cross multiplication, where $a, b \in \mathbb{Z}$ and $b \neq 0$, is

$$[(a, b)] = \{(c, d) : ad = bc\}.$$

If we denote $[(a, b)]$ by a/b, then the equivalence class *is* precisely the fraction usually denoted by a/b. After all, it is plain that $(1, 3) \neq (2, 6)$, but $[(1, 3)] = [(2, 6)]$ because $1 \times 6 = 3 \times 2$; that is, $1/3 = 2/6$.

(v) An equivalence class $[(P, Q)]$ of arrows, as in Example A.14, is called a ***vector***; we denote it by $[(P, Q)] = \overrightarrow{PQ}$. ▲

It is instructive to compare rational numbers and vectors, for both are defined as equivalence classes. Every rational a/b has a "favorite" name—its expression in lowest terms; every vector has a favorite name—an arrow (O, Q) with its foot at the origin O. Although it is good to have familiar favorites, working with fractions in lowest terms is not always convenient; for example, even if both a/b and c/d are in lowest terms, their sum $(ad + bc)/bd$ and product ac/bd may not be. Similarly, it is not always best to think of vectors as arrows with foot at the origin. Vector addition is defined by the parallelogram law (see Figure A.2): $\overrightarrow{OP} + \overrightarrow{OQ} = \overrightarrow{OR}$, where O, P, Q, and R are the vertices of a parallelogram. But $\overrightarrow{OQ} = \overrightarrow{PR}$, because $(O, Q) \equiv (P, R)$, and it is more natural to write $\overrightarrow{OP} + \overrightarrow{OQ} = \overrightarrow{OP} + \overrightarrow{PR} = \overrightarrow{OR}$.

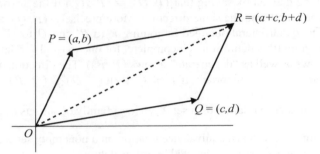

Figure A.2. Parallelogram Law.

The next lemma says that we can replace equivalence by honest equality at the cost of replacing elements by their equivalence classes.

Lemma A.16. *If $\equiv$ is an equivalence relation on a set X, then $x \equiv y$ if and only if $[x] = [y]$.*

Proof. Assume that $x \equiv y$. If $z \in [x]$, then $z \equiv x$, and so transitivity gives $z \equiv y$; hence $[x] \subseteq [y]$. By symmetry, $y \equiv x$, and this gives the reverse inclusion $[y] \subseteq [x]$. Thus, $[x] = [y]$.

Conversely, if $[x] = [y]$, then $x \in [x]$, by reflexivity, and so $x \in [x] = [y]$. Therefore, $x \equiv y$. ■

Here is a set-theoretic idea, *partitions*, that we'll see is intimately involved with equivalence relations.

Definition. Subsets A and B of a set X are ***disjoint*** if $A \cap B = \varnothing$; that is, no $x \in X$ lies in both A and B. A family $\mathcal{P}$ of subsets of a set X is called ***pairwise disjoint*** if, for all $A, B \in \mathcal{P}$, either $A = B$ or $A \cap B = \varnothing$.

A ***partition*** of a set X is a family $\mathcal{P}$ of nonempty pairwise disjoint subsets, called ***blocks***, whose union is X.

We are now going to prove that equivalence relations and partitions are merely different ways of viewing the same thing.

Proposition A.17. *If $\equiv$ is an equivalence relation on a nonempty set X, then the equivalence classes form a partition of X. Conversely, given a partition $\mathcal{P}$ of X, there is an equivalence relation on X whose equivalence classes are the blocks in $\mathcal{P}$.*

Proof. Assume that an equivalence relation $\equiv$ on X is given. Each $x \in X$ lies in the equivalence class $[x]$ because $\equiv$ is reflexive; it follows that the equivalence classes are nonempty subsets whose union is X. To prove pairwise disjointness, assume that $a \in [x] \cap [y]$, so that $a \equiv x$ and $a \equiv y$. By symmetry, $x \equiv a$, and so transitivity gives $x \equiv y$. Therefore, $[x] = [y]$, by Lemma A.16, and so the equivalence classes form a partition of X.

Conversely, let $\mathcal{P}$ be a partition of X. If $x, y \in X$, define $x \equiv y$ if there is $A \in \mathcal{P}$ with $x \in A$ and $y \in A$. It is plain that $\equiv$ is reflexive and symmetric. To see that $\equiv$ is transitive, assume that $x \equiv y$ and $y \equiv z$; that is, there are $A, B \in \mathcal{P}$ with $x, y \in A$ and $y, z \in B$. Since $y \in A \cap B$, pairwise disjointness gives $A = B$ and so $x, z \in A$; that is, $x \equiv z$. We have shown that $\equiv$ is an equivalence relation.

It remains to show that the equivalence classes are the blocks in $\mathcal{P}$. If $x \in X$, then $x \in A$ for some $A \in \mathcal{P}$. By the definition of $\equiv$, if $y \in A$, then $y \equiv x$ and $y \in [x]$; hence, $A \subseteq [x]$. For the reverse inclusion, let $z \in [x]$, so that $z \equiv x$. There is some B with $x \in B$ and $z \in B$; thus, $x \in A \cap B$. By pairwise disjointness, $A = B$, so that $z \in A$, and $[x] \subseteq A$. Hence, $[x] = A$. ■

Corollary A.18. *If $\equiv$ is an equivalence relation on a set X and $a, b \in X$, then $[a] \cap [b] \neq \varnothing$ implies $[a] = [b]$.*

Example A.19.   (i) If $\equiv$ is the identity relation on a set X, then the blocks are the one-point subsets of X.

(ii) Let $X = [0, 2\pi]$, and define the partition of X whose blocks are $\{0, 2\pi\}$ and the singletons $\{x\}$, where $0 < x < 2\pi$. This partition identifies the endpoints of the interval (and nothing else), and so we may regard this as a construction of a circle.   ▲

Exercises

A.14 Let $X = \{\text{rock}, \text{paper}, \text{scissors}\}$. Recall the game whose rules are: paper dominates rock, rock dominates scissors, and scissors dominates paper. Draw a subset of $X \times X$ showing that domination is a relation on X.

A.15 Which of the following relations are equivalence relations? State your reasons.

   (i) The relation $\leq$ on $\mathbb{R}$.

   (ii) The relation R on $\mathbb{Z}$ given by $m \, R \, n$ if $m - n$ is odd.

   (iii) The relation R on $\mathbb{Z}$ given by $m \, R \, n$ if $m - n$ is even.

   (iv) The relation on a group of people of having a common friend.

A.16 Let $f \colon X \to Y$ be a function. Define a relation on X by $x \equiv x'$ if $f(x) = f(x')$. Prove that $\equiv$ is an equivalence relation. If $x \in X$ and $f(x) = y$, the equivalence class $[x]$ is denoted by $f^{-1}(y)$; it is called the **fiber** over y.

A.17 (i) Find the error in the following argument that claims to prove that a symmetric and transitive relation R on a set X must be reflexive; that is, R is an equivalence relation on X. If $x \in X$ and xRy, then symmetry gives yRx and transitivity gives xRx.

Hint. What is y?

(ii) Give an example of a symmetric and transitive relation on the closed unit interval $X = [0, 1]$ that is not reflexive.

A.3   Vector Spaces

Linear algebra is the study of vector spaces and their homomorphisms (namely, linear transformations), with applications to systems of linear equations. We assume that most readers have had a course involving matrices with real entries. Such courses deal mainly with computational aspects of the subject, such as Gaussian elimination, finding inverses, and determinants, but here we do not emphasize this important aspect of linear algebra. Instead, we focus on vector spaces with only a few words about linear transformations.

Introductory linear algebra courses begin with vector spaces whose scalars are real numbers but, toward the end of the course, scalars are allowed to be complex numbers. The instructor usually says that the results about vector spaces over $\mathbb{R}$ hold, more generally, for vector spaces over $\mathbb{C}$. This hand-waving bothers most students. We are now going to generalize the definition of vector space so that scalars may belong to any field k, and we will prove that the usual theorems about vector spaces over $\mathbb{R}$ do, in fact, hold not only for vector spaces over $\mathbb{C}$ but for vector spaces over k. In particular, they hold for vector spaces over $\mathbb{Q}$ or over $\mathbb{F}_q$.

The first definitions do not change when we allow more general scalars.

Definition. If k is a field, then a **vector space over k** is a set V equipped with **addition** $V \times V \to V$, denoted by $(u, v) \mapsto u + v$, that satisfies

(i) $(u + v) + w = u + (v + w)$ for all $u, v, w \in V$,

(ii) there is $0 \in V$ with $0 + v = v$ for all $v \in V$,

(iii) for each $v \in V$, there is $-v \in V$ with $-v + v = 0$,

(iv) $u + v = v + u$ for all $u, v \in V$;

and **scalar multiplication** $k \times V \to V$, denoted by $(a, v) \mapsto av$, that satisfies, for all $a, b, 1 \in k$ and all $u, v \in V$,

(i) $a(u + v) = au + av$,

(ii) $(a + b)v = av + bv$,

(iii) $(ab)v = a(bv)$,

(iv) $1v = v$.

The elements of V are called **vectors** and the elements of k are called **scalars**. It is not difficult to prove that the vector $-v$ in the third axiom of addition is equal to the scalar product $(-1)v$.

Etymology. The word *vector* comes from the Latin word meaning "to carry;" vectors in Euclidean space carry the data of length and direction. The word *scalar* comes from regarding $v \mapsto av$ as a change of scale. The terms *scale* and *scalar* come from the Latin word meaning "ladder," for the rungs of a ladder are evenly spaced.

Example A.20. (i) Euclidean space $V = \mathbb{R}^n$ is a vector space over $\mathbb{R}$. Vectors are n-tuples $(a_1, \ldots, a_n)$, where $a_i \in \mathbb{R}$ for all i. Picture a vector v as an arrow from the origin to the point having coordinates $(a_1, \ldots, a_n)$. Addition is given by

$$(a_1, \ldots, a_n) + (b_1, \ldots, b_n) = (a_1 + b_1, \ldots, a_n + b_n);$$

geometrically, the sum of two vectors is described by the *parallelogram law* (see Figure A.2 on page 422).

Scalar multiplication is given by

$$av = a(a_1, \ldots, a_n) = (aa_1, \ldots, aa_n).$$

Scalar multiplication $v \mapsto av$ "stretches" v by a factor $|a|$, reversing its direction when a is negative (we put quotes around *stretches* because av is shorter than v when $|a| < 1$).

(ii) The example in part (i) can be generalized. If k is any field, define $V = k^n$, the set of all $n \times 1$ column vectors $v = (a_1, \ldots, a_n)$, where $a_i \in k$ for all i. Addition is given by

$$(a_1, \ldots, a_n) + (b_1, \ldots, b_n) = (a_1 + b_1, \ldots, a_n + b_n),$$

and scalar multiplication is given by

$$av = a(a_1, \ldots, a_n) = (aa_1, \ldots, aa_n).$$

(iii) The polynomial ring $R = k[x]$, where k is a field, is another example of a vector space over k. Vectors are polynomials f, scalars are elements $a \in k$, and scalar multiplication gives the polynomial af; that is, if

$$f = b_n x^n + \cdots + b_1 x + b_0,$$

then

$$af = ab_n x^n + \cdots + ab_1 x + ab_0.$$

Thus, the polynomial ring $k[x]$ is a vector space over k.

(iv) Let R be a commutative ring and let k be a subring; if k is a field, then R can be viewed as a vector space over k. Regard the elements of R as vectors and the elements of k as scalars; define scalar multiplication av, where $a \in k$ and $v \in R$, to be the given product of two elements in R. The axioms in the definition of vector space are just particular cases of axioms holding in the commutative ring R. For example, if a field k is a subfield of a larger field E, then E is a vector space over k; in particular, $\mathbb{C}$ is a vector space over $\mathbb{R}$, and it is also a vector space over $\mathbb{Q}$.

(v) The set $C[0, 1]$ of all continuous real-valued functions on the closed interval $[0, 1]$ is a vector space over $\mathbb{R}$ with the usual operations: if $f, g \in C[0, 1]$ and $c \in \mathbb{R}$, then

$$f + g : a \mapsto f(a) + g(a)$$
$$cf : a \mapsto cf(a). \quad \blacktriangle$$

Informally, a *subspace* of a vector space V is a nonempty subset of V that is closed under addition and scalar multiplication in V.

Definition. If V is a vector space over a field k, then a **subspace** of V is a subset U of V such that

(i) $0 \in U$,

(ii) $u, u' \in U$ imply $u + u' \in U$,

(iii) $u \in U$ and $a \in k$ imply $au \in U$.

Every subspace U of a vector space V is itself a vector space. For example, since $u_1 + (u_2 + u_3) = (u_1 + u_2) + u_3$ holds for all vectors $u_1, u_2, u_3 \in V$, it holds, in particular, for all vectors $u_1, u_2, u_3 \in U$.

Example A.21. (i) The extreme cases $U = V$ and $U = \{0\}$ (where $\{0\}$ denotes the subset consisting of the zero vector alone) are always subspaces of a vector space. A subspace $U \subseteq V$ with $U \neq V$ is called a **proper subspace** of V; we may write $U \subsetneq V$ to denote U being a proper subspace of V.

(ii) If $v = (a_1, \ldots, a_n)$ is a nonzero vector in $\mathbb{R}^n$, then the line through the origin,

$$\ell = \{av : a \in \mathbb{R}\},$$

is a subspace of $\mathbb{R}^n$.

Similarly, a plane through the origin consists of all vectors of the form $av_1 + bv_2$, where v_1, v_2 is a fixed pair of noncollinear vectors, and a, b vary over $\mathbb{R}$. It is easy to check that planes through the origin are subspaces of $\mathbb{R}^n$.

(iii) If k is a field, then a **homogeneous linear system over** k of m equations in n unknowns is a set of equations

$$a_{11}x_1 + \cdots + a_{1n}x_n = 0$$
$$a_{21}x_1 + \cdots + a_{2n}x_n = 0$$
$$\vdots \qquad\qquad \vdots$$
$$a_{m1}x_1 + \cdots + a_{mn}x_n = 0,$$

where $a_{ji} \in k$. A **solution** of this system is an $n \times 1$ column vector $c = (c_1, \ldots, c_n) \in k^n$, where $\sum_i a_{ji}c_i = 0$ for all j; a solution $(c_1, \ldots, c_n)$ is **nontrivial** if some $c_i \neq 0$. The set of all solutions forms a subspace of k^n, called the **solution space** (or **nullspace**) of the system. Using matrices, we can say this more succinctly: if $A = [a_{ij}]$ is the $m \times n$ *coefficient matrix*, then the linear system is $Ax = 0$ and a solution is an $n \times 1$ column vector c for which $Ac = 0$.

In particular, we can solve systems of linear equations over $\mathbb{F}_p$, where p is a prime. This says that we can treat a not necessarily homogeneous system of congruences mod p just as one treats an ordinary system of equations.

For example, the system of congruences

$$3x - 2y + z \equiv 1 \bmod 7$$
$$x + y - 2z \equiv 0 \bmod 7$$
$$-x + 2y + z \equiv 4 \bmod 7$$

can be regarded as a system of equations over the field $\mathbb{F}_7$. The system can be solved just as in high school, for inverses mod 7 are now known: $[2][4] = [1]$; $[3][5] = [1]$; $[6][6] = [1]$. The solution is

$$(x, y, z) = ([5], [4], [1]). \quad \blacktriangle$$

Bases and Dimension

The key observation in getting the "right" definition of *dimension* is to understand why $\mathbb{R}^3$ is 3-dimensional. Every vector (x, y, z) is a linear combination of the three vectors $e_1 = (1, 0, 0)$, $e_2 = (0, 1, 0)$, and $e_3 = (0, 0, 1)$; that is,

$$(x, y, z) = xe_1 + ye_2 + ze_3.$$

It is not so important that every vector is a linear combination of these specific vectors; what is important is that there are three of them, for it turns out that three is the smallest number of vectors with this property; that is, one cannot find two vectors $u = (a, b, c)$ and $u' = (a', b', c')$ with every vector a linear combination of u and u'.

Definition. A *list* in a vector space V is an ordered set $X = v_1, \ldots, v_n$ of vectors in V.

More precisely, we are saying that there is some $n \geq 1$ and a function

$$\varphi: \{1, 2, \ldots, n\} \to V,$$

with $\varphi(i) = v_i$ for all i. Thus, the subset im φ is **ordered** in the sense that there is a first vector v_1, a second vector v_2, and so forth. A vector may appear several times on a list; that is, φ need not be injective.

Definition. Let V be a vector space over a field k. A *k-linear combination* of a list $v_1, \ldots, v_n$ in V is a vector v of the form

$$v = a_1 v_1 + \cdots + a_n v_n,$$

where $a_i \in k$ for all i.

We often write *linear combination* instead of k-linear combination if it is clear where the scalar coefficients live.

Definition. If $X = v_1, \ldots, v_m$ is a list in a vector space V, then

$$\text{Span} \langle X \rangle = \langle v_1, \ldots, v_m \rangle,$$

the set of all the k-linear combinations of $v_1, \ldots, v_m$, is called the *subspace spanned by X*. We also say that $v_1, \ldots, v_m$ *spans* $\text{Span}\langle v_1, \ldots, v_m \rangle$.

It is easy to check that $\text{Span} \langle v_1, \ldots, v_m \rangle$ is, indeed, a subspace.

Lemma A.22. *Let V be a vector space over a field k.*

(i) *Every intersection of subspaces of V is itself a subspace.*

(ii) *If $X = v_1, \ldots, v_m$ is a list in V, then the intersection of all the subspaces of V containing X is $\text{Span} \langle v_1, \ldots, v_m \rangle$, the subspace spanned by $v_1, \ldots, v_m$, and so $\text{Span} \langle v_1, \ldots, v_m \rangle$ is the **smallest subspace** of V containing X.*

Proof. Part (i) is routine. For part (ii), let $X = \{v_1, \ldots, v_m\}$ and let $\mathcal{S}$ denote the family of all the subspaces of V containing X; we claim that

$$\bigcap_{S \in \mathcal{S}} S = \text{Span} \langle v_1, \ldots, v_m \rangle.$$

The inclusion $\subseteq$ is clear, because $S = \text{Span} \langle v_1, \ldots, v_m \rangle \in \mathcal{S}$. For the reverse inclusion, note that if $S \in \mathcal{S}$, then S contains $v_1, \ldots, v_m$, and so it contains the set of all k-linear combination of $v_1, \ldots, v_m$, namely, Span $\langle v_1, \ldots, v_m \rangle$. ∎

The next observation is important.

Corollary A.23. *The subspace spanned by a list $X = v_1, \ldots, v_m$ does not depend on the ordering of the vectors, but only on the set of vectors themselves.*

Proof. This follows from part (ii) of Lemma A.22. ∎

See Exercise A.24 on page 440 to see other properties of a list that do not depend on the ordering of its vectors.

If $X = \varnothing$, then Span $\langle X \rangle = \bigcap_{S \in \mathcal{S}} S$, where $\mathcal{S}$ is the family of *all* the subspaces of V containing X. Now $\{0\} \subseteq$ Span $\langle \varnothing \rangle = \bigcap_{S \in \mathcal{S}} S$, for $\{0\}$ is contained in every subspace S of V. For the reverse inclusion, one of the subspaces S of V occurring in the intersection is $\{0\}$ itself, and so Span $\langle \varnothing \rangle = \bigcap_{S \subseteq V} S \subseteq \{0\}$. Therefore, Span $\langle \varnothing \rangle = \{0\}$.

Example A.24.   (i) Let $V = \mathbb{R}^2$, let $e_1 = (1, 0)$, and let $e_2 = (0, 1)$. Now $V = \text{Span} \langle e_1, e_2 \rangle$, for if $v = (a, b) \in V$, then

$$\begin{aligned} v &= (a, 0) + (0, b) \\ &= a(1, 0) + b(0, 1) \\ &= ae_1 + be_2 \in \text{Span} \langle e_1, e_2 \rangle. \end{aligned}$$

(ii) If k is a field and $V = k^n$, define e_i as the $n \times 1$ column vector having 1 in the ith coordinate and 0s elsewhere. The reader may adapt the argument in part (i) to show that $e_1, \ldots, e_n$ spans k^n.

(iii) A vector space V need not be spanned by a finite list. For example, let $V = k[x]$, and suppose that $X = f_1(x), \ldots, f_m(x)$ is a finite list in V. If d is the largest degree of any of the f_i, then every (nonzero) k-linear combination of $f_1, \ldots, f_m$ has degree at most d. Thus, x^{d+1} is not a k-linear combination of vectors in X, and so X does not span $k[x]$. ▲

The following definition makes sense even though we have not yet defined *dimension*.

Definition. A vector space V is called *finite-dimensional* if it is spanned by a finite list; otherwise, V is called *infinite-dimensional*.

Part (ii) of Example A.24 shows that k^n is finite-dimensional, while part (iii) shows that $k[x]$ is infinite-dimensional. Now $\mathbb{C}$ is a vector space over $\mathbb{R}$, and it is finite-dimensional ($\mathbb{C}$ is spanned by $1, i$); by Example A.20(iii), both $\mathbb{R}$ and $\mathbb{C}$ are vector spaces over $\mathbb{Q}$ (each can be shown to be infinite-dimensional).

If a subspace U of a vector space V is finite-dimensional, then there is a list $v_1, v_2, \ldots, v_m$ that spans U. But there are many such lists: if u is a vector in U, then the extended list $v_1, v_2, \ldots, v_m, u$ also spans U. Let us, therefore, seek a *shortest* list that spans U.

Notation. If $v_1, \ldots, v_m$ is a list, then $v_1, \ldots, \widehat{v_i} \ldots, v_m$ is the shorter list with v_i deleted.

Proposition A.25. *If V is a vector space over a field k, then the following conditions on a list $v_1, \ldots, v_m$ spanning V are equivalent:*

(i) *$v_1, \ldots, v_m$ is not a shortest spanning list*

(ii) *some v_i is in the subspace spanned by the others*

(iii) *there are scalars $a_1, \ldots, a_m$, not all zero, with*

$$\sum_{\ell=1}^{m} a_\ell v_\ell = 0.$$

Proof. (i) $\Rightarrow$ (ii). If $v_1, \ldots, v_m$ is not a shortest spanning list, then one of the vectors, say v_i, can be thrown out, and the shorter list still spans. Hence, $v_i \in \mathrm{Span}\,\langle v_1, \ldots, \widehat{v_i}, \ldots, v_m \rangle$.

(ii) $\Rightarrow$ (iii). If $v_i = \sum_{j \neq i} c_j v_j$, define $a_i = -1 \neq 0$ and $a_j = c_j$ for all $j \neq i$.

(iii) $\Rightarrow$ (i). The given equation implies that one term, say $a_i v_i$, is nonzero. Since k is a field, a_i^{-1} exists, and

$$v_i = \left(-a_i^{-1} \right) \sum_{j \neq i} a_j v_j. \tag{A.1}$$

Deleting v_i gives a shorter list that still spans V: write any $v \in V$ as a linear combination of all the v_j (including v_i); now substitute the expression Eq. (A.1) for v_i and collect terms. ∎

We now give a name to lists described in Proposition A.25.

Definition. A list $X = v_1, \ldots, v_m$ in a vector space V is ***linearly dependent*** if there are scalars $a_1, \ldots, a_m$, not all zero, with $\sum_{\ell=1}^{m} a_\ell v_\ell = 0$; otherwise, X is called ***linearly independent***.

The empty set $\varnothing$ is defined to be linearly independent (we interpret $\varnothing$ as a list of length 0).

Example A.26. (i) A list $X = v_1, \ldots, v_m$ containing the zero vector is linearly dependent: if $v_j = 0$, then $\sum_i a_i v_i = 0$, where $a_j = 1$ and $a_i = 0$ for $i \neq j$.

(ii) A list v_1 of length 1 is linearly dependent if and only if $v_1 = 0$; hence, a list v_1 of length 1 is linearly independent if and only if $v_1 \neq 0$.

(iii) A list v_1, v_2 is linearly dependent if and only if one of the vectors is a scalar multiple of the other: if $a_1v_1 + a_2v_2 = 0$ and $a_1 \neq 0$, then $v_1 = -(a_2/a_1)v_2$. Conversely, if $v_2 = cv_1$, then $cv_1 - v_2 = 0$ and the list v_1, v_2 is linearly dependent (for the coefficient -1 of v_2 is nonzero).

(iv) If there is a repetition in the list $v_1, \ldots, v_m$ (that is, if $v_i = v_j$ for some $i \neq j$), then $v_1, \ldots, v_m$ is linearly dependent: define $c_i = 1$, $c_j = -1$, and all other $c = 0$. Therefore, if $v_1, \ldots, v_m$ is linearly independent, then all the vectors v_i are distinct.   ▲

Linear independence has been defined indirectly, as not being linearly dependent. Because of the importance of linear independence, let us define it directly.

Definition. A list $v_1, \ldots, v_m$ is ***linearly independent*** if, whenever a k-linear combination $\sum_{\ell=1}^{m} a_\ell v_\ell = 0$, then every $a_i = 0$.

It follows that every sublist of a linearly independent list is itself linearly independent (this is one reason for decreeing that $\varnothing$ be linearly independent).

Corollary A.27. *If $X = v_1, \ldots, v_m$ is a list spanning a vector space V, then X is a shortest spanning list if and only if X is linearly independent.*

Proof. These are just the contrapositives of (i) $\Rightarrow$ (iii) and (iii) $\Rightarrow$ (i) in Proposition A.25.   ■

We have arrived at the notion we have been seeking.

Definition. A ***basis*** of a vector space V is a linearly independent list that spans V.

Thus, bases are shortest spanning lists. Of course, all the vectors in a linearly independent list $v_1, \ldots, v_n$ are distinct, by Example A.26(iv).

Example A.28. In Example A.24(ii), we saw that $X = e_1, \ldots, e_n$ spans k^n, where e_i is the $n \times 1$ column vector having 1 in the ith coordinate and 0s elsewhere. We now show that X is linearly independent. If $0 = \sum_i c_i e_i$, then

$$
\begin{aligned}
c_1 e_1 &= & (c_1, 0, 0, \ldots, 0) \\
+ \, c_2 e_2 &= + & (0, c_2, 0, \ldots, 0) \\
\vdots & & \vdots \\
\underline{+ \, c_n e_n} &= + & \underline{(0, 0, 0, \ldots, c_n)} \\
0 &= & (c_1, c_2, \ldots, c_n).
\end{aligned}
$$

Hence, $c_i = 0$ for all i, X is linearly independent, and X is a basis; it is called the ***standard basis*** of k^n.   ▲

Proposition A.29. *A list $X = v_1, \ldots, v_n$ in a vector space V over a field k is a basis of V if and only if each $v \in V$ has a unique expression as a k-linear combination of the vectors in X.*

Proof. Since X is a basis, it spans V, and so each vector $v \in V$ is a k-linear combination: $v = \sum a_i v_i$. If also $v = \sum b_i v_i$, then $\sum (a_i - b_i) v_i = 0$, and linear independence gives $a_i = b_i$ for all i; that is, the expression is unique.

Conversely, existence of an expression shows that the list X spans V. Moreover, if $0 = \sum c_i v_i$ with not all $c_i = 0$, then the vector 0 does not have a unique expression as a linear combination of the v_i: a second expression is $0 = \sum a_i v_i$ with all $a_i = 0$. ∎

Definition. If $X = v_1, \ldots, v_n$ is a basis of a vector space V and $v \in V$, then Proposition A.29 says that there are unique scalars $a_1, \ldots, a_n$ with

$$v = \sum_{i=1}^{n} a_i v_i.$$

The n-tuple $(a_1, \ldots, a_n)$ is called the ***coordinate list*** of a vector $v \in V$ relative to the basis X.

If $v_1, \ldots, v_n$ is the standard basis of $V = k^n$, then this coordinate list coincides with the usual coordinate list.

How to Think About It. If $v_1, \ldots, v_n$ is a basis of a vector space V over a field k, then each vector $v \in V$ has a unique expression

$$v = a_1 v_1 + a_2 v_2 + \cdots + a_n v_n,$$

where $a_i \in k$ for all i. Since there is a first vector v_1, a second vector v_2, and so forth, the coefficients in this k-linear combination determine a unique n-tuple $(a_1, a_2, \ldots, a_n)$. Were a basis merely a subset of V and not a list (i.e., an ordered subset), then there would be $n!$ coordinate lists for every vector. But see Exercise A.24(iv) on page 440.

We are going to define the *dimension* of a vector space V to be the number of vectors in a basis. Two questions arise at once.

 (i) Does every vector space have a basis?

 (ii) Do all bases of a vector space have the same number of elements?

The first question is easy to answer; the second needs some thought.

Theorem A.30. *Every finite-dimensional vector space V has a basis.*

Proof. A finite spanning list X exists, since V is finite-dimensional. If X is linearly independent, it is a basis; if not, Proposition A.25 says that we can throw out some element from X, leaving a shorter spanning list, say X'. If X' is linearly independent, it is a basis; if not, we can throw out an element from X' leaving a shorter spanning sublist. Eventually, we arrive at a shortest spanning list, which is linearly independent, by Corollary A.27 and hence it is a basis. ∎

The definitions of spanning and linear independence can be extended to infinite lists in a vector space, and we can then prove that infinite-dimensional vector spaces also have bases. For example, it turns out that a basis of $k[x]$ is $1, x, x^2, \ldots, x^n, \ldots$.

We can now prove *invariance of dimension*, perhaps the most important result about vector spaces.

Lemma A.31. *Let $u_1, \ldots, u_n, v_1, \ldots, v_m$ be elements in a vector space V with $v_1, \ldots, v_m \in \mathrm{Span}\langle u_1, \ldots, u_n \rangle$. If $m > n$, then $v_1, \ldots, v_m$ is linearly dependent.*

Proof. The proof is by induction on $n \geq 1$.

Base Step. If $n = 1$, there are at least two vectors v_1, v_2, and $v_1 = a_1 u_1$ and $v_2 = a_2 u_1$. If $u_1 = 0$, then $v_1 = 0$ and the list of v's is linearly dependent (by Example A.26(i)). Suppose $u_1 \neq 0$. We may assume that $v_1 \neq 0$, or we are done; hence, $a_1 \neq 0$. Therefore, v_1, v_2 is linearly dependent, for $1 \cdot v_2 - a_2 a_1^{-1} v_1 = 0$ and hence the larger list $v_1, \ldots, v_m$ is linearly dependent.

Inductive Step. There are equations

$$v_i = a_{i1} u_1 + \cdots + a_{in} u_n$$

for $i = 1, \ldots, m$. We may assume that some $a_{i1} \neq 0$, otherwise $v_1, \ldots, v_m \in \langle u_2, \ldots, u_n \rangle$, and the inductive hypothesis applies. Changing notation if necessary (that is, by re-ordering the v's), we may assume that $a_{11} \neq 0$. For each $i \geq 2$, define

Write out the proof in the special case $m = 3$ and $n = 2$.

$$v_i' = v_i - a_{i1} a_{11}^{-1} v_1 \in \mathrm{Span}\langle u_2, \ldots, u_n \rangle.$$

Each v_i' is a linear combination of the u's, and the coefficient of u_1 is $a_{i1} - (a_{i1} a_{11}^{-1}) a_{11} = 0$. Since $m - 1 > n - 1$, the inductive hypothesis gives scalars $b_2, \ldots, b_m$, not all 0, with

$$b_2 v_2' + \cdots + b_m v_m' = 0.$$

Rewrite this equation using the definition of v_i':

$$\left(-\sum_{i \geq 2} b_i a_{i1} a_{11}^{-1} \right) v_1 + b_2 v_2 + \cdots + b_m v_m = 0.$$

Not all the coefficients are 0, and so $v_1, \ldots, v_m$ is linearly dependent. ∎

The following familiar fact illustrates the intimate relation between linear algebra and systems of linear equations.

Corollary A.32. *If a homogeneous system of linear equations over a field k has more unknowns than equations, then it has a non-trivial solution.*

Proof. Recall that an n-tuple $(\beta_1, \ldots, \beta_n)$ is a *solution* of a system

$$\alpha_{11} x_1 + \cdots + \alpha_{1n} x_n = 0$$

$$\vdots \qquad \vdots \qquad \vdots$$

$$\alpha_{m1} x_1 + \cdots + \alpha_{mn} x_n = 0$$

if $\alpha_{i1}\beta_1 + \cdots + \alpha_{in}\beta_n = 0$ for all i. In other words, if $c_1, \ldots, c_n$ are the columns of the $m \times n$ coefficient matrix $A = [\alpha_{ij}]$ (note that $c_i \in k^m$), then

$$\beta_1 c_1 + \cdots + \beta_n c_n = 0.$$

Now k^m can be spanned by m vectors (the standard basis, for example). Since $n > m$, by hypothesis, Lemma A.31 shows that the list $c_1, \ldots, c_n$ is linearly dependent; there are scalars $\gamma_1, \ldots, \gamma_n$, not all zero, with $\gamma_1 c_1 + \cdots + \gamma_n c_n = 0$. Therefore, $(\gamma_1, \ldots, \gamma_n)$ is a nontrivial solution of the system. $\blacksquare$

Theorem A.33 (Invariance of Dimension). *If* $X = x_1, \ldots, x_n$ *and* $Y = y_1, \ldots, y_m$ *are bases of a vector space* V, *then* $m = n$.

Proof. If $m \neq n$, then either $n < m$ or $m < n$. In the first case, $y_1, \ldots, y_m \in$ Span$\langle x_1, \ldots, x_n \rangle$, because X spans V, and Lemma A.31 gives Y linearly dependent, a contradiction. A similar contradiction arises if $m < n$, and so we must have $m = n$. $\blacksquare$

It is now permissible to make the following definition, for all bases of a vector space have the same size.

Definition. If V is a finite-dimensional vector space over a field k, then its **dimension**, denoted by $\dim_k(V)$ or by $\dim(V)$, is the number of elements in a basis of V.

Corollary A.34. *Let* k *be a finite field with* q *elements. If* V *is an* n-*dimensional vector space over* k, *then* $|V| = q^n$.

Proof. If $v_1, \ldots, v_n$ is a basis of V, then every $v \in V$ has a unique expression

$$v = c_1 v_1 + \cdots + c_n v_n,$$

where $c_i \in k$ for all i. There are q choices for each c_i, and so there are q^n vectors in V. $\blacksquare$

Example A.35. (i) Example A.28 shows that k^n has dimension n, which agrees with our intuition when $k = \mathbb{R}$: the plane $\mathbb{R} \times \mathbb{R}$ is 2-dimensional, and $\mathbb{R}^3$ is 3-dimensional!

(ii) If $V = \{0\}$, then $\dim(V) = 0$, for there are no elements in its basis $\varnothing$. (This is another good reason for defining $\varnothing$ to be linearly independent.)

(iii) Let I be a finite set with n elements. Define

$$k^I = \{\text{functions } f: I \to k\}.$$

Now k^I is a vector space if we define addition $f + f'$ to be

$$f + f': i \mapsto f(i) + f'(i)$$

and scalar multiplication af, for $a \in k$ and $f: I \to k$, by

$$af: i \mapsto af(i)$$

(see Exercise A.18(i) on page 439). It is easy to check that the set of n functions of the form f_i, where $i \in I$, defined by

$$f_i(j) = \begin{cases} 1 & \text{if } j = i \\ 0 & \text{if } j \neq i \end{cases}$$

form a basis, and so $\dim(k^I) = n = |I|$.

This is not a new example: an n-tuple $(a_1, \ldots, a_n)$ is really a function $f : \{1, \ldots, n\} \to k$ with $f(i) = a_i$ for all i. Thus, the functions f_i comprise the standard basis. ▲

Definition. A *longest* (or *maximal*) linearly independent list $X = u_1, \ldots, u_m$ is a linearly independent list for which there is no vector $v \in V$ such that $u_1, \ldots, u_m, v$ is linearly independent.

Lemma A.36. *If V is a finite-dimensional vector space, then a longest linearly independent list $X = v_1, \ldots, v_n$ is a basis of V.*

Proof. If $v_1, \ldots, v_n$ is not a basis, then it does not span V, for this list is linearly independent. Thus, there is $w \in V$ with $w \notin \text{Span} \langle v_1, \ldots, v_n \rangle$. But the longer list $v_1, \ldots, v_n, w$ is linearly independent, by Proposition A.25, contradicting X being a longest linearly independent list. ■

The converse of Lemma A.36 is true; bases are longest linearly independent lists. This follows from the next proposition, which is quite useful in its own right.

Proposition A.37. *Let V be an n-dimensional vector space. If $Z = u_1, \ldots, u_m$ is a linearly independent list in V, where $m < n$, then Z can be extended to a basis; that is, there are vectors $v_{m+1}, \ldots, v_n \in V$ such that $u_1, \ldots, u_m, v_{m+1}, \ldots, v_n$ is a basis of V.*

Proof. If the linearly independent list Z does not span V, there is $v_{m+1} \in V$ with $v_{m+1} \notin \text{Span} \langle u_1, \ldots, u_m \rangle$, and the longer list $u_1, \ldots, u_m, v_{m+1}$ is linearly independent, by Proposition A.25. If $u_1, \ldots, u_m, v_{m+1}$ does not span V, there is $v_{m+2} \in V$ with $v_{m+2} \notin \text{Span} \langle u_1, \ldots, u_m, v_{m+1} \rangle$. Since $\dim(V) = n$, Lemma A.31 says that the length of these lists can never exceed n, and so this process of adjoining elements $v_{m+1}, v_{m+2}, \ldots$ must stop. But the only reason a list stops is that it spans V; hence, it is a basis. ■

Corollary A.38. *Let V be an n-dimensional vector space. Then a list in V is a basis if and only if it is a longest linearly independent list.*

Proof. Lemma A.36 shows that longest linearly independent lists are bases. Conversely, if X is a basis, it must be a longest linearly independent list: otherwise, Proposition A.37 says we could lengthen X to obtain a basis of V which is too long. ■

We now paraphrase Lemma A.31.

Corollary A.39. *If $\dim(V) = n$, then a list of $n + 1$ or more vectors is linearly dependent.*

Proof. Otherwise, the list could be extended to a basis having too many elements. ∎

Corollary A.40. *Let V be a vector space with* $\dim(V) = n$.

(i) *A list of n vectors that spans V must be linearly independent.*

(ii) *Any linearly independent list of n vectors must span V.*

In either case, the list is a basis of V.

Proof. (i) Were the list linearly dependent, then it could be shortened to give a basis, and this basis is too small.

(ii) If the list does not span, then it could be lengthened to give a basis, and this basis is too large. ∎

Corollary A.41. *Let U be a subspace of a vector space V of dimension n.*

(i) *U is finite-dimensional.*

(ii) $\dim(U) \leq \dim(V)$.

(iii) *If* $\dim(U) = \dim(V)$, *then* $U = V$.

Proof. (i) Take $u_1 \in U$. If $U = \text{Span}\langle u_1 \rangle$, then U is finite-dimensional. If $U \neq \text{Span}\langle u_1 \rangle$, there is $u_2 \notin \text{Span}\langle u_1 \rangle$. By Proposition A.25, u_1, u_2 is linearly independent. If $U = \text{Span}\langle u_1, u_2 \rangle$, we are done; if not, there is $u_3 \notin \text{Span}\langle u_1, u_2 \rangle$, and the list u_1, u_2, u_3 is linearly independent. This process cannot be repeated $n + 1$ times, for then $u_1, \ldots, u_{n+1}$ would be a linearly independent list in $U \subseteq V$, contradicting Corollary A.39.

(ii) A basis of U is linearly independent, and so it can be extended to a basis of V.

(iii) If $\dim(U) = \dim(V)$, then a basis of U is already a basis of V (otherwise it could be extended to a basis of V that would be too large). ∎

Linear Transformations

Linear transformations are homomorphisms of vector spaces; they are really much more important than vector spaces, but vector spaces are needed in order to define them, and bases of vector spaces are needed to describe them by matrices. (You are surely familiar with the next definition, at least for $k = \mathbb{R}$.)

Definition. Let V and W be vector spaces over a field k. A *linear transformation* is a function $T \colon V \to W$ such that, for all vectors $v, v' \in V$ and scalars $a \in k$, we have

(i) $T(v + v') = T(v) + T(v')$

(ii) $T(av) = aT(v)$.

It follows by induction on $n \geq 1$ that linear transformations preserve linear combinations:

$$T(a_1 v_1 + \cdots + a_n v_n) = a_1 T(v_1) + \cdots + a_n T(v_n).$$

You've certainly seen many examples of linear transformations in a linear algebra course. Here are a few more.

Example A.42. (i) If A is an $m \times n$ matrix, then $x \mapsto Ax$, where x is an $n \times 1$ column vector, is a linear transformation $k^n \to k^m$.

(ii) If we regard the complex numbers $\mathbb{C}$ as a 2-dimensional vector space over $\mathbb{R}$, then complex conjugation $T: z \mapsto \bar{z}$ is a linear transformation.

(iii) Let $V = k[x]$, where k is a field. If $a \in k$, then evaluation $e_a: k[x] \to k$ is a linear transformation (we can view k as a 1-dimensional vector space over itself).

(iv) Integration $f \mapsto \int_0^1 f(x)\, dx$ is a linear transformation $C[0,1] \to \mathbb{R}$ (see Example A.20(v)). ▲

We now associate matrices to linear transformations.

Theorem A.43. *Let V and W be vector spaces over a field k. If $v_1, \ldots, v_n$ is a basis of V and $w_1, \ldots, w_n$ is a list of elements in W (possibly with repetitions), then there exists a unique linear transformation $T: V \to W$ with $T(v_j) = w_j$ for all j.*

Proof. By Proposition A.29, every vector $v \in V$ has a unique expression as a linear combination of basis vectors:

$$v = a_1 v_1 + \cdots + a_n v_n.$$

Therefore, there is a well-defined function $T: V \to W$ with $T(v_j) = w_j$ for all j, namely

$$T(a_1 v_1 + \cdots + a_n v_n) = a_1 w_1 + \cdots + a_n w_n.$$

It is routine to check that T is a linear transformation. If $v' = a_1' v_1 + \cdots + a_n' v_n$, then

$$v + v' = (a_1 + a_1')v_1 + \cdots + (a_n + a_n')v_n,$$

and

$$\begin{aligned}
T(v + v') &= (a_1 + a_1')w_1 + \cdots + (a_n + a_n')w_n \\
&= \big(a_1 w_1 + \cdots + a_n w_n\big) + \big(a_1' w_1 + \cdots + a_n' w_n\big) \\
&= T(v) + T(v').
\end{aligned}$$

Similarly,

$$\begin{aligned}
T(av) &= T\big(a(a_1 v_1 + \cdots + a_n v_n)\big) \\
&= T(aa_1 v_1 + \cdots + aa_n v_n) \\
&= aa_1 w_1 + \cdots + aa_n w_n \\
&= a(a_1 w_1 + \cdots + a_n w_n) \\
&= aT(v).
\end{aligned}$$

To prove uniqueness, suppose that $S: V \to W$ is a linear transformation with $S(v_j) = w_j$ for all j. Since S preserves linear combinations,

$$\begin{aligned}
S(a_1 v_1 + \cdots + a_n v_n) &= a_1 S(v_1) + \cdots + a_n S(v_n) \\
&= a_1 w_1 + \cdots + a_n w_n \\
&= T(a_1 v_1 + \cdots + a_n v_n),
\end{aligned}$$

and so $S = T$. ■

Definition. Let $T: V \to W$ be a linear transformation. Given bases $v_1, \ldots, v_n$ and $w_1, \ldots, w_m$ of V and W, respectively, each $T(v_j)$ is a linear combination of the w's:

$$T(v_j) = a_{1j}w_1 + \cdots + a_{mj}w_m.$$

The $m \times n$ matrix $A = [a_{ij}]$ whose jth column is $a_{1j}, \ldots, a_{mj}$, the coordinate list of $T(v_j)$ with respect to the w's, is called the ***matrix associated to*** T.

Example A.44. As in Example A.42(ii), view $\mathbb{C}$ as $\mathbb{R}^2$. A basis is $1, i$; that is, $(1, 0), (0, 1)$. Since $\bar{1} = 1$ and $\bar{i} = -i = (0, -1)$, the matrix of complex conjugation relative to the basis is

$$A = \begin{bmatrix} 1 & 0 \\ 0 & -1 \end{bmatrix}. \quad \blacktriangle$$

Of course, the matrix A associated to a linear transformation $T: V \to W$ depends on the choices of bases of V and of W.

The next theorem shows why the notation $a_{1j}w_1 + \cdots + a_{mj}w_m$ is chosen instead of $a_{j1}w_1 + \cdots + a_{jm}w_m$.

Theorem A.45. *If $T: k^n \to k^m$ is a linear transformation, then*

$$T(v) = Av,$$

where A is the matrix associated to T from the standard bases of k^n and k^m, v is an $n \times 1$ column vector, and Av is matrix multiplication.

Proof. If A is an $m \times n$ matrix and v_j is the $n \times 1$ column vector whose jth entry is 1 and whose other entries are 0, then Av_j is the jth column of A. Thus, $Av_j = T(v_j)$ for all j, and so $Av = T(v)$ for all $v \in k^n$, by Exercise A.29 on page 441. $\blacksquare$

In Appendix A.1, we defined functions to be equal if they have the same domain, same target, and same graph. It is natural to require the same domain and the same graph, but why should we care about the target? The coming discussion gives a (persuasive) reason why targets are important.

Definition. Let V be a vector space over k, and regard k as a 1-dimensional vector space over itself. A ***functional*** on V is a linear transformation $f: V \to k$. The ***dual space*** V^* of a vector space V is the set of all functionals on V.

It is shown in Example A.42 that evaluation and integration give rise to functionals.

Proposition A.46. *Let $T: V \to W$ be a linear transformation, where V and W are vector spaces over a field k.*

(i) *V^* is a vector space over k.*

(ii) *If $f \in W^*$, then the composite $f \circ T$ is in V^*.*

(iii) *The function $T^*\colon W^* \to V^*$, given by $T^*\colon f \mapsto f \circ T$, is a linear transformation.*

Proof.   (i) This is Exercise A.30 on page 441.

(ii) Since $T\colon V \to W$ and $f\colon W \to k$ are linear transformations, so is their composite $f \circ T\colon V \to k$; that is, $f \circ T$ is a functional on V.

(iii) That $f \mapsto f \circ T$ is a linear transformation $W^* \to V^*$ follows easily from the formulas $(f + g) \circ T = f \circ T + g \circ T$ and $(cf) \circ T = c(f \circ T)$, where $f, g \in W^*$ and $c \in k$ (note that cf is a functional on W, for W^* is a vector space). ■

Use Proposition A.2 to prove $(f + g) \circ T = f \circ T + g \circ T$: just evaluate both sides on $v \in V$.

Proposition A.47. *If $v_1, \ldots, v_n$ is a basis of a vector space V over a field k, then there are functionals $v_j^*\colon V \to k$, for each j, with*

$$v_j^*(v_i) = \begin{cases} 1 & \text{if } i = j \\ 0 & \text{if } i \neq j, \end{cases}$$

*and $v_1^*, \ldots, v_n^*$ is a basis of V^* (it is called the **dual basis**).*

Proof. By Theorem A.43, it suffices to prescribe the values of v_j^* on a basis of V.

Linear Independence: If $\sum_j c_j v_j^* = 0$, then $\sum_j c_j v_j^*(v) = 0$ for all $v \in V$. But $\sum_j c_j v_j^*(v_j) = c_j$, so that all the coefficients c_j are 0 and, hence, $v_1^*, \ldots, v_n^*$ is linearly independent.

Spanning: If $g \in V^*$, then $g(v_j) = d_j \in k$ for all j. But $g = \sum_j d_j v_j^*$, for both sides send each v_j to d_j. Thus, g is a linear combination of $v_1^*, \ldots, v_n^*$. ■

Corollary A.48. *If V is an n-dimensional vector space, then*

$$\dim(V^*) = n = \dim(V).$$

Proof. A basis of V and its dual basis have the same number of elements. ■

If $T\colon V \to W$ is a linear transformation, what is a matrix associated to the linear transformation $T^*\colon W^* \to V^*$?

Lemma A.49. *Let $v_1, \ldots, v_n$ be a basis of a vector space V over k. If $g \in V^*$, then $g = d_1 v_1^* + \cdots + d_n v_n^*$, where $d_j = g(v_j)$ for all j. Therefore, the coordinate list of g relative to the dual basis $v_1^*, \ldots, v_n^*$ of V^* is $d_1, \ldots, d_n$.*

Proof. We saw this in the proof of Proposition A.47, when showing that the dual basis spans V^*. ■

The next result shows that dual spaces are intimately related to transposing matrices. If $A = [a_{ij}]$ is an $m \times n$ matrix, then its **transpose** $A^\top$ is the $n \times m$ matrix $[a_{ji}]$ whose ij entry is a_{ji}. In words, for each i, the ith row $a_{i1}, \ldots, a_{in}$ of A is the ith column of $A^\top$ (and, necessarily, each jth column of A is the jth row of $A^\top$).

Proposition A.50. *If $T: V \to W$ is a linear transformation and A is the matrix of T arising from bases $v_1, \ldots, v_n$ of V and $w_1, \ldots, w_m$ of W, then the matrix of $T^*: W^* \to V^*$ arising from the dual bases is the transpose $A^\top$ of A.*

Proof. Let B be the matrix associated to $T^*: W^* \to V^*$. The recipe for constructing B says that if w_i^* is a basis element, then the ith column of B is the coordinate list of $T^*(w_i^*)$. Let's unwind this. First, $T^*(w_i^*) = w_i^* \circ T$. Second, the coordinate list of $w_i^* \circ T$ is obtained by writing it as a linear combination of $v_1^*, \ldots, v_n^*$: Lemma A.49 does this by computing $\left(w_i^* \circ T\right)(v_j)$ for all j. Now

$$\left(w_i^* \circ T\right)(v_j) = w_i^*(T(v_j)) = w_i^* \left(a_{1j} w_1 + \cdots + a_{mj} w_m\right) = a_{ij}.$$

Thus, the ith column of B is $a_{i1}, \ldots, a_{im}$; that is, the ith column of B is the ith row of A. In other words, $B = A^\top$. ∎

If $T: V \to W$ is a linear transformation, then the domain of T^* is W^*, which depends on the target of T. Suppose that W is a subspace of a vector space U; let $i: W \to U$ be the inclusion. Now $S = i \circ T: V \to U$ is also a linear transformation. The transformations T and S have the same domain, namely V, and the same graph (for $T(v) = S(v)$ for all $v \in V$); they differ only in their targets. Now $T^*: W^* \to V^*$, while $S^*: U^* \to V^*$. Since T^* and S^* have different domains, they are certainly different functions, for we have agreed that the domain of a function is a necessary ingredient of its definition. We conclude that S and T are distinct; that is, if you like transposes of matrices, then you must admit that targets are essential ingredients of functions.

Exercises

A.18 (i) * If k is a field, $c \in k$, and $f : k \to k$ is a function, define a new function $cf : k \to k$ by $a \mapsto cf(a)$. With this definition of scalar multiplication, prove that the commutative ring k^k of all functions on k is a vector space over k (see Example A.35(iii)).

(ii) Prove that $\mathrm{Poly}(k)$, the set of all polynomial functions $k \to k$, is a subspace of k^k.

A.19 If the only subspaces of a vector space V are $\{0\}$ and V itself, prove that $\dim(V) \le 1$.

A.20 Prove, in the presence of all the other axioms in the definition of vector space, that the commutative law for vector addition is redundant; that is, if V satisfies all the other axioms, then $u + v = v + u$ for all $u, v \in V$.

Hint. If $u, v \in V$, evaluate $-[(-v) + (-u)]$ in two ways.

A.21 If V is a vector space over $\mathbb{F}_2$ and if $v_1 \neq v_2$ are nonzero vectors in V, prove that v_1, v_2 is linearly independent. Is this true for vector spaces over any other field?

A.22 Prove that the columns of an $m \times n$ matrix A over a field k are linearly dependent in k^m if and only if the homogeneous system $Ax = 0$ has a nontrivial solution.

A.23 Prove that the list of polynomials $1, x, x^2, x^3, \ldots, x^{100}$ is a linearly independent list in $k[x]$, where k is a field.

A.24 * Let $X = v_1, \ldots, v_n$ be a list in a vector space V, and let $Y = y_1, \ldots, y_n$ be a permutation of $v_1, \ldots, v_n$.

  (i) Prove that X spans V if and only if Y spans V.

  (ii) Prove that X is linearly independent if and only if Y is linearly independent.

  (iii) Prove that X is a basis of V if and only if Y is a basis of V.

  (iv) Conclude that spanning, being linearly independent, or being a basis are properties of a subset of vectors, not merely of a list of vectors. See Corollary A.23.

A.25 It is shown in analytic geometry that if ℓ_1 and ℓ_2 are lines with slopes m_1 and m_2, respectively, then ℓ_1 and ℓ_2 are perpendicular if and only if $m_1 m_2 = -1$. If

$$\ell_i = \{\alpha v_i + u_i : \alpha \in \mathbb{R}\},$$

for $i = 1, 2$, prove that $m_1 m_2 = -1$ if and only if the dot product $v_1 \cdot v_2 = 0$. (Since both lines have slopes, neither is vertical.) See Lemma 3.16.

A.26 (i) In calculus, a *line in space passing through a point u* is defined as

$$\{u + \alpha w : \alpha \in \mathbb{R}\} \subset \mathbb{R}^3,$$

where w is a fixed nonzero vector. Show that every line through the origin is a one-dimensional subspace of $\mathbb{R}^3$.

  (ii) In calculus, a *plane in space passing through a point u* is defined as the subset

$$\{v \in \mathbb{R}^3 : (v - u) \cdot n = 0\} \subset \mathbb{R}^3,$$

where $n \neq 0$ is a fixed *normal vector*. Prove that a plane through the origin is a two-dimensional subspace of $\mathbb{R}^3$.

    If the origin $(0, 0, 0)$ lies on a plane H, then $u = 0$ and

$$H = \{v = (x, y, z) \in \mathbb{R}^3 : v \cdot n = 0\},$$

where $n = (\alpha, \beta, \gamma)$ is a (nonzero) normal vector; that is, H is the set of all vectors orthogonal to n.

A.27 If U and W are subspaces of a vector space V, define

$$U + W = \{u + w : u \in U \text{ and } w \in W\}.$$

  (i) Prove that $U + W$ is a subspace of V.

  (ii) If U and U' are subspaces of a finite-dimensional vector space V, prove that

$$\dim(U) + \dim(U') = \dim(U \cap U') + \dim(U + U').$$

Hint. Extend a basis of $U \cap U'$ to a basis of U and to a basis of U'.

A.28 If U and W are vector spaces over a field k, define their **direct sum** to be the set of all ordered pairs,

$$U \oplus W = \{(u, w) : u \in U \text{ and } w \in W\},$$

with addition

$$(u, w) + (u', w') = (u + u', w + w')$$

and scalar multiplication

$$\alpha(u, w) = (\alpha u, \alpha w).$$

(i) Show that $U \oplus W$ is a vector space.

(ii) If U and W are finite-dimensional vector spaces over a field k, prove that

$$\dim(U \oplus W) = \dim(U) + \dim(W).$$

A.29 * Let $S, T : V \to W$ be linear transformations, where V and W are vector spaces over a field k. Prove that if there is a basis $v_1, \ldots, v_n$ of V for which $S(v_j) = T(v_j)$ for all j, then $S = T$.

A.30 * Prove that the dual space V^* of a vector space V over a field k is a vector space over k.

A.4 Inequalities

Many properties of inequality follow from a few basic properties. Denote the set of all positive real numbers by P (we do not regard 0 as positive). We assume the set P satisfies

Recall that $\mathbb{N}$ is the set of all nonnegative integers, so that $\mathbb{N} = P \cup \{0\}$.

(i) $a, b \in P$ implies $a + b \in P$

(ii) $a, b \in P$ implies $ab \in P$

(iii) **Trichotomy** : If a is a number, then exactly one of the following is true:

$$a \in P, \quad a = 0, \quad -a \in P.$$

The first two properties say that P is **closed** under addition and multiplication. We now define inequality.

Definition. Given real numbers a and b, we say that a is **less than** b, written $a < b$, if $b - a \in P$; we say that a is **less than or equal to** b, written $a \leq b$, if $b - a \in \mathbb{N}$; that is, $a < b$ or $a = b$.

Thus, a is **positive** if $0 < a$ (that is, $a \in P$), and a is **negative** if $a < 0$ (that is, $-a \in P$).

Other notation: if $a < b$, we may write $b > a$ and, if $b \leq a$, we may write $a \geq b$.

Just to complete the picture, $a > b$ means $b < a$ (and $a \geq b$ means $b \leq a$).

Here are some standard properties of inequality.

Proposition A.51. *Let a, b, B be real numbers with $b < B$.*

(i) *If $a > 0$, then $ab < aB$; if $a < 0$, then $ab > aB$.*

(ii) *If $a > 0$ and $b < 0$, then $ab < 0$.*

(iii) *If $b > 0$, then $b^{-1} > 0$; if $b < 0$, then $b^{-1} < 0$.*

(iv) $a + b < a + B$ *and* $a - b > a - B$.

(v) *If c, d are positive, then $d < c$ if and only if $d^{-1} > c^{-1}$.*

Proof. We prove the first three parts; the last two proofs are similar and appear in Exercise A.31 below.

(i) By definition, $b < B$ means that $B - b \in P$.

- Suppose that $a > 0$; that is, $a \in P$. To show that $ab < aB$, we must show that $aB - ab = a(B - b) \in P$, and this follows from Property (i) of P.

- If $a < 0$, then $-a \in P$. Therefore, $(-a)(B - b) \in P$, and so

$$(-a)(B - b) = (-1)a(B - b) = a(b - B) \in P.$$

(ii) The first part says that if $b < B$ and both sides are multiplied by a positive number, then the sense of the inequality stays the same. So, if $B < 0$, then $aB < a \cdot 0 = 0$.

(iii) Suppose that $b > 0$. If $b^{-1} < 0$, then

$$1 = b \cdot b^{-1} < b \cdot 0 = 0,$$

a contradiction. If $b^{-1} = 0$ then

$$1 = bb^{-1} = b \cdot 0 = 0,$$

another contradiction. Hence, Trichotomy gives $b^{-1} > 0$.   ∎

Exercises

A.31 * Prove parts (iv) and (v) of Proposition A.51.

A.32 Prove, or disprove and salvage if possible. Suppose a, b, c, and d are real numbers.

> "Disprove" here means "give a concrete counterexample." "Salvage" means "add a hypotheis to make it true."

  (i) If $a < b$, then $a^2 < b^2$.

 (ii) If $a^2 < b^2$, then $a < b$.

(iii) If $a < b$ and $c < d$, then $ac < bd$.

(iv) If $a^3 > 0$, then $a > 0$.

A.33 Does $\mathbb{C}$ have a subset P' like P; that is, P' is closed under addition and multiplication, and it satisfies Trichotomy?

A.5   Generalized Associativity

Recall that a ***set with a binary operation*** is a set G equipped with a function $G \times G \to G$; we denote the value of the function by $(a, b) \mapsto a * b$. Examples of such sets are the real numbers and the complex numbers, each of which is usually viewed as having two binary operations: addition $(a, b) \mapsto a + b$ and multiplication $(a, b) \mapsto ab$. More generally, every commutative ring has binary operations addition and multiplication. Another example is given by $G = X^X$, the family of all functions from a set X to itself: composition of functions is a binary operation on G.

> We do not assume that $*$ is commutative; that is, $a * b \neq b * a$ is allowed.

The adjective *binary* means two: two elements $a, b \in G$ are combined to produce the element $a * b \in G$. But it is often necessary to combine more than two elements: for example, we may have to multiply several numbers. The binary operations in the examples cited above are ***associative***; we can combine three elements unambiguously. If $a, b, c \in G$, then

$$a * (b * c) = (a * b) * c.$$

Since we are told only how to combine two elements, there is a choice when confronted with three elements: first combine b and c, obtaining $b * c$, and then combine this new element with a to get $a * (b * c)$, or first get $a * b$ and then combine it with c to get $(a * b) * c$. Associativity says that either choice yields the same element of G. Thus, there is no confusion in writing $a * b * c$ without parentheses. In contrast, subtraction is not associative, for it is not clear whether $a - b - c$ means $(a - b) - c$ or $a - (b - c)$, and these may be different: $9 - (5 - 3) = 7$ while $(9 - 5) - 3 = 1$.

Suppose we want to combine more than three elements; must we assume more complicated identities? Consider powers of real numbers, for example. Is it obvious that $a^3 a^2 = (a[aa^2]) a$? The remarkable fact is: assuming we don't need parentheses for three factors, we don't need parentheses for more than three factors. To make all concrete, we now call a binary operation *multiplication*, and we simplify notation by omitting $*$ and writing ab instead of $a * b$.

Definition. Let G be a set with a binary operation; an ***expression in*** G is an n-tuple $(a_1, a_2, \ldots, a_n) \in G \times \cdots \times G$ that is rewritten as $a_1 a_2 \cdots a_n$; we call the a_i ***factors*** of the expression.

An expression yields many elements of G by the following procedure. Choose two adjacent a's, multiply them, and obtain an expression with $n - 1$ factors: the new product just formed and $n - 2$ original factors. In the shorter new expression, choose two adjacent factors (either an original pair or an original one together with the new product from the first step) and multiply them. Repeat this procedure until there is a ***penultimate*** expression having only two factors; multiply them and obtain an element of G that we call an ***ultimate product***. For example, consider the expression $abcd$. We may first multiply ab, obtaining $(ab)cd$, an expression with three factors, namely, ab, c, d. We may now choose either the pair c, d or the pair ab, c; in either case, multiply them to obtain expressions having two factors: ab, cd, or $(ab)c, d$. The two factors in the last expressions can now be multiplied to give two ultimate products from $abcd$, namely $(ab)(cd)$ and $((ab)c)d$. Other ultimate products derived from the expression $abcd$ arise from multiplying bc or cd as the first step. It is not obvious whether the ultimate products from a given expression are equal.

Definition. Let G be a set with a binary operation. An expression $a_1 a_2 \cdots a_n$ in G ***needs no parentheses*** if all its ultimate products are equal elements of G.

Theorem A.52 (Generalized Associativity). *If G is a set with an associative binary operation, then every expression $a_1 a_2 \cdots a_n$ in G needs no parentheses.*

Proof. The proof is by induction on $n \geq 3$. The base step holds because the operation is associative. For the inductive step, consider two ultimate products U and V obtained from a given expression $a_1 a_2 \cdots a_n$ after two series of choices:

$$U = (a_1 \cdots a_i)(a_{i+1} \cdots a_n) \quad \text{and} \quad V = (a_1 \cdots a_j)(a_{j+1} \cdots a_n);$$

the parentheses indicate the penultimate products displaying the last two factors that multiply to give U and V, respectively; there are many parentheses inside each of the shorter expressions. We may assume that $i \leq j$. Since each of the four expressions in parentheses has fewer than n factors, the inductive hypothesis says that each of them needs no parentheses. It follows that $U = V$ if $i = j$. If $i < j$, then the inductive hypothesis allows the first expression to be rewritten as

$$U = (a_1 \cdots a_i) ([a_{i+1} \cdots a_j][a_{j+1} \cdots a_n])$$

and the second to be rewritten as

$$V = \left([a_1 \cdots a_i][a_{i+1} \cdots a_j]\right)(a_{j+1} \cdots a_n),$$

where each of the expressions $a_1 \cdots a_i$, $a_{i+1} \cdots a_j$, and $a_{j+1} \cdots a_n$ needs no parentheses. Thus, the three expressions yield unique elements A, B, and C in G, respectively. The first expression gives $U = A(BC)$ in G, the second gives $V = (AB)C$ in G, and so $U = V$ in G, by associativity. ∎

Corollary A.53. *If G is a set with a binary operation, $a \in G$, and $m, n \geq 1$, then*

$$a^{m+n} = a^m a^n \quad and \quad (a^m)^n = a^{mn}.$$

Proof. In the first case, both elements arise from the expression having $m + n$ factors each equal to a; in the second case, both elements arise from the expression having mn factors each equal to a. ∎

A.6   A Cyclotomic Integer Calculator

Several times in the previous chapters, we've advised you to use a CAS. Some uses are simply to reduce the computational overhead of algebraic calculations, such as the expansion in Lagrange interpolation on page 272. For this, you can use the CAS "right out of the box:" all the functionality you need is built in with commands like `expand` or `simplify`.

Other applications require programming that uses specific syntax for the CAS in use. A good example is the formula in Example 6.61 on page 265. The recursive formula for Φ_n can be implemented in almost any CAS, but the details for how to get a product over the divisors of an integer (especially if the product is in the denominator of an expression) can either be trivial, if the functionality is built-in, or extremely tricky to implement if it is not. There are many CAS environments, so it would be of little use to include actual programs here.

Computer Algebra is used regularly in many high school classrooms, implemented either on handheld devices or with tablet apps.

What we can do in this Appendix is point out how to use Proposition 7.20 on page 290 to model $\mathbb{Q}(\alpha)$, where α is algebraic over $\mathbb{Q}$ and its minimal polynomial $p = \mathrm{irr}(\alpha, \mathbb{Q})$ is known. The essential piece of that Proposition is that

$$\mathbb{Q}(\alpha) \cong \mathbb{Q}[x]/(p(x));$$

so, as long as your CAS can find the remainder when one polynomial is divided by another, you can use it to perform "modular arithmetic" with polynomials in $\mathbb{Q}(\alpha)$.

If p is a prime in $\mathbb{Z}$ and $\zeta = \cos(2\pi/p) + i \sin(2\pi/p)$, we know that

$$\mathrm{irr}(\zeta, \mathbb{Q}) = 1 + x + x^2 + \cdots + x^{p-1}.$$

This is easily implemented in a CAS with something like

```
Phi(x,p):= sum(x^k,k,0,p-1)
```

Suppose that your CAS command for polynomial remainder is `pmod`. For example,

```
pmod(x^3+4x^2-3x+1,x^2+1)
```

returns $-4x - 3$, the remainder when $x^3 + 4x^2 - 3x + 1$ is divided by $x^2 + 1$.

The two functions, Phi and pmod, allow us to calculate in $\mathbb{Q}(\zeta)$. Let's look at some examples.

Eisenstein Integers

Arithmetic with complex numbers is built into most CAS environments, so that you can do calculations with Gaussian integers right away. Arithmetic with Eisenstein integers isn't usually built in, but you can build a model of $\mathbb{Z}[\omega]$ by thinking of an Eisenstein integer as a congruence class mod $x^2 + x + 1$:

```
cl(f) := pmod (f,phi(x,3))
```

or even

```
 cl(f) := pmod (f, x^2+x+1).
```

Addition and multiplication of classes are defined as in Chapter 7:

```
add(f,g) = cl(f+g)
```

```
mult(f,g) = cl(fg).
```

So, now we can compute: to find, for example, $3\omega^5 - \omega^2 + 1$, you want the class of $3x^5 - x^2 + 1 \bmod x^2 + x + 1$

```
cl (3x^5-x^2+1)
> -2x-1
```

And, sure enough,

$$3\omega^5 - \omega^2 + 1 = -1 - 2\omega.$$

Your model can do generic calculations, giving the rules for addition and multiplication in $\mathbb{Z}[\omega]$:

```
add(a+b*x,c+d*x)
> a+c+(b+d)*x
```

```
mult(a+b*x,c+d*x)
> a*c-b*d + (a*d+b*c-b*d)*x
```

You can generate Eisenstein triples by squaring Eisenstein integers:

```
mult(3+2*x,3+2*x)
> 5+8*x
```

```
mult(5+x,5+x)
```

```
>  24+9*x

mult(4+3*x,4+3*x)
>  7+15*x
```

Symmetric Polynomials

In Example 9.6 on page 388, we derived the cubic formula via symmetric polynomials. There, we defined

$$s = \alpha_1 + \alpha_2\omega + \alpha_3\omega^2$$
$$u = \alpha_1 + \alpha_2\omega^2 + \alpha_3\omega,$$

and we saw that $s + u = 3\alpha$.

We also claimed that

$$s^3 + u^3 = 27\alpha_1\alpha_2\alpha_3.$$

Our CAS Eisenstein calculator can help. Replacing α, β, γ by a, b, c, we have

```
add((a+b*x+c*x^2)^3,(a+b*x^2+c*x)^3)
>  2*a^3-3*a^2*b+a^2*c-3*a*b^2-12*a**b*c-3*a*c^2+2*
   b^3-3*b^2*c-3*b*c^2+2*c^3

factor(2*a^3-3*a^2*b+a^2*c-3*a*b^2-12*a**b*c-3*a*
       c^2+2*b^3-3*b^2*c-3*b*c^2+2*c^3)
>  (a+b-2*c)*(a-2*b+c)*(2*a-b-c)
```

This can be written (with an eye to symmetric polynomials) as

$$(a + b + c - 3c)(a + b + c - 3b)(3a - (a + b + c)).$$

Since $a + b + c = 0$, this is exactly what we claimed in Example 9.6.

Algebra with Periods

One last example: in Section 7.3, we outlined Gauss's construction of the regular 17-gon with ruler and compass. Central to that is the specification of "periods" of various lengths, listed on page 323. They are constructed according to the formula on page 325: if $ef = 16$, the periods of length f are given by

$$\eta_{e,k} = \sum_{j=0}^{f-1} \zeta^{3^{k+je}}, \qquad 0 \le k < e.$$

mod is the CAS built-in "mod" function.

A CAS model follows the syntax pretty closely:

```
n(e,k):=sum(x^(mod(3^(k+e*j),17)),j,0,(16/e)-1).
```

Now change cl so that it gives the congruence class mod $\Phi_1 7(x)$:

```
cl(f) := pmod (f, phi(x,17)),
```

and we can calculate with the classes of the periods:

```
n(2,0)
> x^16+x^15+x^13+x^9+x^8+x^)+x^2+x

n(2,1)
> x^14+x^12+x^11+x^10+x^7+x^6+x^5+x^3

cl( add (n(2,0), n(2, 1)) )
>-1

cl( mult (n(2,0), n(2, 1)) )
> -4
```

So, as we claimed on page 323, $\eta_{2,0}$ and $\eta_{2,1}$ are roots of

$$x^2 + x - 4.$$

Exercises

A.34 Find a polynomial in $\mathbb{Q}[x]$ that has roots $\eta_{0,k}, \eta_{1,k}, \eta_{3,k}, \eta_{4,k}$.

References

[1] Apostol, Tom M., *Introduction to Analytic Number Theory*, Undergraduate Texts in Mathematics, Springer-Verlag, New York-Heidelberg, 1976.

[2] Artin, E., *Galois Theory*, Edwards Brothers, Ann Arbor, 1948.

[3] Baker, A., *Transcendental Number Theory*, 2d ed., Cambridge University Press, Cambridge, 1990.

[4] Barbeau, E. J., *Polynomials*, Springer-Verlag, New York, 1989.

[5] Borevich, Z. I., and Shafarevich, I. R., *Number Theory*, Academic Press, New York, 1966.

[6] Cajori, F., *A History of Mathematical Notation*, Open Court, 1928; Dover reprint, Mineola NY, 1993.

[7] Cuoco, A., *Mathematical Connections*, Classroom Resource Materials, MAA, Washington, 2005.

[8] Dirichlet, P. G. L., *Lectures on Number Theory*, Supplements by R. Dedekind. Translated from the 1863 German original and with an introduction by John Stillwell. History of Mathematics, 16. American Mathematical Society, Providence; London Mathematical Society, London, 1999.

[9] Du Val, P., *Elliptic Functions and Elliptic Curves*, London Mathematical Society Lecture Note Series 9, Cambridge University Press, London, 1973.

[10] EDC, *The CME Project*, Pearson, Boston, 2013

[11] Edwards, H. M., *Fermat's Last Theorem. A Genetic Introduction to Number Theory*, Graduate Texts in Mathematics vol. 50, Springer-Verlag, New York, 1977.

[12] Euler, L., *Elements of Algebra*, Translated from the German by John Hewlett. Reprint of the 1840 edition. With an introduction by C. Truesdell, Springer-Verlag, New York, 1984.

[13] Flannery, S., *In Code. A Mathematical Journey*, Reprint of the 2000 original, Workman Publishing, New York, 2001.

[14] Gauss, C. F., *Disquisitiones Arithmeticae*, Arthur A. Clarke (trans), Yale University Press, New Haven, 1966.

[15] Hadlock, C. R., *Field Theory and Its Classical Problems*, Carus Mathematical Monographs 19, Mathematical Association of America, Washington, 1978.

[16] Heath, T. L., *The Thirteen Books of Euclid's Elements*, Cambridge University Press, 1926; Dover reprint, Mineola NY, 1956.

[17] Ireland, K. and Rosen, M., *A Classical Introduction to Modern Number Theory*, Springer-Verlag, New York, 1982.

[18] Koblitz, N., *A Course in Number Theory and Cryptography*, Springer-Verlag, New York, 1987.

[19] ——, *Introduction to Elliptic Curves and Modular Forms*, Springer Verlag, New York, 1993.

[20] Loomis, E. S., *The Pythagorean Proposition*, Edwards Brothers., Ann Arbor, 1940.

[21] Montgomery, S., and Ralston, E. W., *Selected Papers in Algebra*, Raymond W. Brink Selected Mathematical Papers, Vol. 3, Mathematical Association of America, Washington, 1977.

[22] Needham, T., *Visual Complex Analysis*, Clarendon Press, Oxford, 1997.

[23] Ribenboim, P., *Thirteen Lectures on Fermat's Last Theorem*, Springer–Verlag, New York, 1979.

[24] Rosen, K. H., *Elementary Number Theory and Its Applications*, 4th ed., Addison-Wesley, Reading, MA, 2000.

[25] Rotman, J. J., *Advanced Modern Algebra*, 2d ed., Graduate Studies in Mathematics vol. 114, American Mathematical Society, Providence, 2010.

[26] ——, *A First Course in Abstract Algebra*, 3d ed., Prentice Hall, Upper Saddle River NJ, 2006.

[27] ——, *Galois Theory*, 2d ed., Springer-Verlag, New York, 1998.

[28] ——, *Journey into Mathematics*, Prentice Hall, Upper Saddle River NJ, 1998; Dover reprint, Mineola NY, 2007.

[29] Samuel, P., *Algebraic Theory of Numbers*, Houghton-Mifflin, Boston, 1992.

[30] Siegel, C. L., *Topics in Complex Function Theory* Vol. I, *Elliptic Functions and Uniformization Theory*, Wiley, New York, 1969.

[31] Silverman, J. H., and Tate, J., *Rational Points on Elliptic Curves*, Springer–Verlag, New York, 1992.

[32] Stillwell, J., *Mathematics and Its History*, 3d ed., Springer, New York, 2010.

[33] Tignol, J.-P., *Galois' Theory of Equations*, World Scientific Publishing, Singapore, 2001.

[34] van der Waerden, B. L., *Science Awakening*, John Wiley, New York, 1963.

[35] ——, *Geometry and Algebra in Ancient Civilizations*, Springer–Verlag, New York, 1983.

[36] Washington, L.C., *Introduction to Cyclotomic Fields*, Springer, New York, 1982.

[37] Weil, A., *Number Theory: An Approach Through History. From Hammurapi to Legendre*, Birkhäuser, Boston, 1984.

[38] Weyl, H., *Algebraic Theory of Numbers*, 6th printing, Princeton University Press, Princeton, 1971; Princeton Landmarks in Mathematics and Physics Series, 1998.

[39] Wilf, H., *Generatingfunctionology*, Academic Press, New York, 1994.

Index

Abel, Niels H., 379, 391, 394, 400
abelian group, 390
Addition Theorem, 399
adjoining to field, 294
Adleman, Leonard M., 150
al-Khwarizmi, 240
algebraic
    element, 293
    extension field, 293
algebraic closure, 254
algebraically closed, 301
anagram, 70
antanairesis, 31
Archimedes, 321
Aristarchus, 3
Arithmetic–Geometric Mean Inequal-
    ity, 6
Artin, Emil, 386
associate, 234
associated polynomial function, 204
associativity, 37
    generalized, 443
atom, 227
automorphism, 386

b-adic digits, 139
Babylonian method, 5
base b, 139
basis
    standard, 430
    vector space, 430
Bernoulli, Jacob, 398
Bernoulli, James, 398
Bhaskara I, 149
bijective, bijection, 416
binary operation, 155
Binomial Theorem
    in $\mathbb{Z}$, 66
    in commutative ring, 160
Boolean ring, 167, 223

Braunfeld, Peter, 53

calendar
    formula
        Conway, John H., 175
        Gregorian, 173
        Julian, 169
        Mayan, 144
cancellation law, 192
Cardano, Girolamo, 81
cartesian product, 410
CAS = Computer Algebra System, 142
casting out 9s, 137
castle problem, 90
Casus Irreducibilis, 383
century year, 170
characteristic function, 209
characteristic of field, 288
Ch'in Chiu-shao = Qin Jiushao, 90,
    146
Chinese Remainder Theorem, 142
chord–tangent construction, 405
closed under operation, 155
coconuts, 148
code (see RSA code), 150
coefficients, 197
commensurable, 28
common divisor, 24
    in $\mathbb{Z}$, 24
    polynomials, 243
common multiple
    in $\mathbb{Z}$, 55
    polynomials, 253
common year, 169
commutative, 37
commutative ring, 156
    formal power series, 200
    polynomial ring
        one variable, 200
        several variables, 205

compass, 309
complement, 168
completing the square, 2, 23
complex conjugate, 96
complex exponential, 108
complex number, 92
    absolute value, 99
    argument, 100
    conjugate, 96
    exponential form, 109
    imaginary part, 92
    modulus, 99
    norm, 116
    polar form, 101
    real part, 92
composite
    functions, 414
    number, 22
congruence class, 154, 421
congruent mod I, 278
congruent mod m, 132, 270
congruent numbers, 400
constant function, 411
constant polynomial, 201
constant term, 201
constructible, 311
    number, 312
    point, 311
    subfield, 313
contrapositive, 26
convex, 48
Conway, John H., 175
coordinate list, 431
coset
    ideal, 279
    subgroup, 392
Cramer's Rule, 164
Crelle, August L., 380
cross multiplication, 193
cubic formula, 84
cubic polynomial, 382
cyclotomic integers, 157
cyclotomic polynomial, 265

day, 169
de Moivre's Theorem, 107
    exponential form, 109
de Moivre, Abraham, 107
De Morgan, Augustus, 169
decimal expansion, 177
Dedekind, Richard, 218, 366, 369

degree, 198
    extension field, 291
derivative, 202
Descartes, René, 82
diagonal, 420
dimension, 433
Diophantine equation, 404
Diophantus, 5, 8, 12, 27, 405
direct product, 221
direct sum of vector spaces, 440
Dirichlet, J. P. G. Lejeune, 153, 358, 360
discriminant, 91
disjoint, 422
distance preserving, 385
distributive, 37
divides, 21
Division Algorithm
    integers, 23
    polynomials, 237
divisor, 21
    commutative ring, 233
    proper, 257
divisors, 368
domain, 192
    Euclidean domain, 333
    of function, 410
doomsday, 174
double angle formula, 107
dual basis, 438
dual space, 437

Eisenstein integers, 120
Eisenstein Criterion, 267
Eisenstein triple, 121
Eisenstein, F. Gotthold M., 120, 267
Elements, 2, 20
elliptic curve, 404
elliptic function, 400
elliptic integral, 398
empty set, 410
equality of functions, 412
equality of sets, 410
equivalence class, 421
equivalence relation, 420
Erlanger Programm, 390
etymology
    abelian, 391
    algebra, 240
    algorithm, 240
    arithmetic, xv

binomial, 63
calculus, 197
calendar, 173
casting out 9s, 137
coefficients, 198
corollary, 11
cubic, 202
degree
    field extension, 291
    polynomial, 198
geometry, xv
golden ratio, 76
homomorphism, 207
hypotenuse, 4
ideal, 218, 366
isomorphism, 207
lemma, 11
linear, 202
mathematics, xv
modulo, 132
monomial, 63
power, 51
proof, 11
proposition, 10
quadratic, 202
radical, 240
ring, 157
root, 239
scalar, 425
September, 171
theorem, 11
vector, 424
Euclid, 2, 20, 23
    Elements, 20
Euclid's Lemma
    integers, 25
    PID, 256
    polynomials, 248
Euclidean Algorithm I
    integers, 32
    polynomials, 249
Euclidean Algorithm II
    integers, 32
    polynomials, 250
Eudoxus, 28, 308
Euler ϕ-function, 111, 232
Euler, Leonhard, 108, 111, 131, 326, 330, 399
evaluation homomorphism, 215
extension field, 291
    algebraic, 293

degree, 291
finite, 291
pure, 382
radical, 383

Factor Theorem, 240
factorial, 51
Fagnano, Giulio, 399
Fermat prime, 325
Fermat's $n = 4$ Theorem, 15
Fermat's Last Theorem, 14
Fermat's Little Theorem, 136
Fermat's Theorem on Divisors, 372
Fermat's Two-Square Theorem, 342
Fermat, Pierre de, 14
Ferrari, Ludovico, 87
Fibonacci sequence, 75
Fibonacci, Leonardo, 19, 75
field, 163
    algebraically closed, 301
    extension, 291
    finite, 305
    prime, 293
    rational functions, 205
    splitting, 301
finite extension field, 291
finite-dimensional, 428
first day of month, 173
first form of induction, 57
First Isomorphism Theorem
    commutative rings, 282
    groups, 392
first quadrant, 372
fixes, 386
formal power series, 197
fraction field, 194
Frey, Gerhard, 407
function, 410
    bijective, 416
    constant, 411
    identity, 411
    inclusion, 412
    injective, 413
    restriction, 415
    surjective, 413
functional, 437
Fundamental Theorem
    Algebra, 105
    Arithmetic, 54
    Galois Theory, 393

Galois field, 306
Galois group, 386
Galois, Évariste, 305, 306, 380, 394
Gauss's Lemma, 260
Gauss, Carl F., 50, 258, 268, 321, 380, 384, 400
Gauss-Wantzel theorem, 325
Gaussian integers, 7, 119, 157
gcd, 24, 243, 255
Generalized Associativity, 443
Germain, Marie-Sophie, 358
golden ratio, 76
googol, 140
Granville, Andrew J., 406
graph of function, 411
greatest common divisor, 24, 243, 255
greatest integer function, 29
Gregorian calendar, 170
group, 389
    abelian, 390
    quotient, 392
    solvable, 394
    symmetric, 389

Heath, Thomas L., 2, 20
Helikon, 308
Heron triangle, 14
Hilbert, David, 157, 369
hockey stick, 68
Hölder, Otto L., 383
Holmes, Sherlock, 176
homomorphism, 207
    group, 391
Hume, James, 82
Hungerbühler, Norbert, 320
hyperbolic cosine, 6

ideal, 218
    generated by X, 168
    maximal, 286
    prime, 286
    principal, 218
    product, 220
    proper, 218
    sum, 220
identity element
    commutative ring, 156
    group, 389
identity function, 411
image
    function, 410

homomorphism, 217
inclusion, 412
Inclusion–Exclusion, 230
independent list
    longest, 434
indeterminate, 200
indirect proof, 27
induction
    base step, 47
    first form, 47
    inductive hypothesis, 47
    inductive step, 47, 57
    strong, 57
inductive reasoning, 45
inequalities, 441
inert, 342
infinite descent, 12
infinite-dimensional, 428
injective, injection, 413
integers mod m, 154
integers $\mathbb{Z}$, 21
    rational, 339
integers, cyclotomic, 359
integration, 42
invariance of dimension, 433
inverse element
    additive, 37
    group, 389
    multiplicative, 37, 156
inverse function, 416
$\mathrm{irr}(z, k)$, 296
irreducible in commutative ring, 234
isometry, 385
isomorphism, 207
    group, 391

Jacobi, Carl G. J., 360, 380, 400
Joachimsthal, Ferdinand, 359
Julian calendar, 169

k-linear combination, 427
Kaplansky, Irving, 236
kernel, 217
    group, 392
Klein, Felix C., 390
Kolyvagin, Victor, 407
Kronecker, Leopold, 28, 300, 359
Kummer, Eduard, 218, 358, 359

Lagrange Interpolation, 272
Lagrange, Joseph-Louis, 272

Lamé, Gabriel, 32, 358
lattice point, 13, 35
Law of Decomposition
    Eisenstein integes, 346
    Gaussian integers, 343
Law of Substitution, 37, 155, 412
Laws of Exponents, 52
leading coefficient, 198
leap year, 169
least common multiple
    in $\mathbb{Z}$, 55
    polynomials, 253
Least Integer Axiom, 21
Legendre, Adrien-Marie, 358
Leibniz, Gottfried W., 68, 397
Levi ben Gershon, 52
lies above, 340
Lindemann, C. L. Ferdinand von, 293
linear combination, 218
    in $\mathbb{Z}$, 24
    vector space, 427
linear transformation, 435
linearly dependent, 429
linearly independent, 429, 430
Liouville, Joseph, 381, 397
list, 427
longest independent list, 434
lowest terms
    in $\mathbb{Q}$, 26
    rational functions, 248

Mascheroni, Lorenzo, 320
matrix associated to linear transformation, 437
Maurolico, Francesco, 52
maximal ideal, 286
Mayan calendar, 144
Mazur, Barry, 407
minimal polynomial, 296
Möbius, August F., 263
modulus, 132
Mohr, Georg, 320
monic polynomial, 201
Moore, Eliakim H., 306
Motzkin, Theodore S., 335
multiple, 21
multiplicative function, 378
    strongly multiplicative function, 375
multiplicity, 254

natural map, 282
natural numbers $\mathbb{N}$, 21
n choose r, 63
negative, 37, 156
Newton, Isaac, 197, 405
Nine Chapters on the Mathematical Art, 3, 6
Noether, Emmy, 369
norm, 116
normal subgroup, 392

one-one correspondence
    see bijective, 416
one-to-one
    see injective, 413
onto (function)
    see surjective, 413
order of power series, 203
order of unit, 165
ordered pair, 410
Oresme, Nicole, 45
origin, 310
Oughtred, William, 82

pairwise disjoint, 422
parallelogram law, 93, 425
parity, 131
partition, 422
Pascal's triangle, 63
Pascal, Blaise, 52, 64
period, 179
permutation, 385
PID, 255
Pigeonhole Principle, 419
Plato, 308
Plimpton 322, 4
Pogrebishte, 176
pointwise addition, 157
pointwise multiplication, 157
polar form, 101
polynomial
    n variables, 205
    monic, 201
    one variable, 197
    reduced, 83
    splits, 301
    zero, 198
polynomial function, 204
polynomial ring
    one variable, 200
    several variables, 205

power series, 197
powers, 51
    commutative ring, 160
predecessor, 57
prime, 22
    inert, 342
    ramifies, 342
    rational, 339
    splits, 340
prime factorization, 54, 252
prime field, 293
prime ideal, 286
primes
    irregular, 368
    regular, 368
primitive element, 325
primitive Pythagorean triple, 27
primitive root of unity, 111, 264
principal ideal, 218
principal ideal domain, 255
private key, 150
Proclus, 20
product expansion, 375
proof
    by contradiction, 27
    indirect, 27
proper
    ideal, 218
    subset, 410
    subspace, 426
public key, 150
pure extension, 382
Pythagoras, 2
Pythagorean
    point, 10
    Theorem, 2
        converse, 7
    triple, 4
        primitive, 27
        similar, 9

Qin Jiushao, 90, 146
quadratic formula, 2
quartic formula, 87
quartic polynomial, 383
quotient, 23, 38
quotient group, 392
quotient ring, 281

radical extension, 383
Rahn, Johann H., 82

rational
    block, 179
    function, 205
    integer, 339
    line, 13
    period, 179
    point, 10
    prime, 339
    repeats, 179
    terminates, 179
Rational Root Theorem, 260
Recorde, Robert, 82
recurrence, 74
reduced polynomial, 83
reduction mod p, 260
reflexive relation, 420
relation, 420
relatively prime
    integers, 26
    polynomials, 248
remainder, 23
Remainder Theorem, 240
repeated roots, 263
resolvent cubic, 87, 384
restriction, 415
Ribet, Kenneth A., 407
Riemann zeta function, 374
Riemann, G. F. Bernhard, 369
ring, commutative, 156
Rivest, Ronald L., 150
root
    multiplicity, 254
    polynomial, 81, 239
root of unity, 111
    commutative ring, 165
    cube root, 82
    primitive, 111, 264
RSA code, 150
Ruffini, Paolo, 394
ruler, 309

scalar, 424
    multiplication, 424
Schönemann, Theodor, 267
Scherk, Heinrich F., 359
Scipione, 81
Serre, Jean-Pierre, 407
Shamir, Adi, 150
Shimura, Goro, 397, 407
similar, 9
Singer, Richard, 267

single-valued, 412
size function, 333
smallest subspace, 427
solution of linear system, 426
solution space, 426
solvable by radicals, 383
solvable group, 394
spans, 427
splits
    polynomial, 301
    prime, 340
splitting field, 301
squarefree integer, 34
standard basis, 430
straightedge (see ruler), 309
strong induction, 57
subfield, 167
    generated by X, 293
subgroup, 390
    normal, 392
subring, 166
    generated by X , 168
subset, 410
subspace, 426
    proper, 426
    smallest, 427
    spanned by X, 427
subtraction, 38
surjective, 413
surjective, surjection, 413
symmetric difference, 166
symmetric group, 389, 418
symmetric relation, 420
symmetry, 385

tangent half-angle formula, 43
Taniyama, Yutaka, 397, 407
target, 410
Tartaglia = Nicolo Fontana, 81
transcendental element, 293
transitive relation, 420
transpose, 438
triangular number, 68
trichotomy, 441
trigonometric identities, 42
triple angle formula, 110
Trotsky, Leon, 170
tzolkin calendar, 144

UFD, 258
Uncle Ben, 176

Uncle Charles, 174
unique, 23
unique factorization
    in $\mathbb{Z}$, 54
        polynomials, 252
unique factorization domain, 258
unit, 162
unit n-gon, 110
unit circle, 8
unit vector, 99

valuation, 350
value of function, 410
van der Waerden, Bartel L., 21, 28
Vandermonde's Identity, 73
Vandermonde, Alexandre, 73
vector space, 424
vectors, 424
Venn diagram, 222
Venn, John, 222
Viète, Francois, 82, 88

Wantzel, Pierre L., 318
Weierstrass, Karl, 404
Weil, André, 407
well-defined, 412
Well-Ordering Axiom, 21
Wessel, Caspar, 92
Widman, Johannes, 82
Wiles, Andrew J., 14, 396, 407
Williams, Kenneth S., 335
Wilson's Theorem, 263
Wilson, Jack C., 335
Wilson, John, 263

year, 169
    century year, 170
    common, 169
    leap year, 169

zero polynomial, 198
zero ring, 159
zero divisor, 192

About the Authors

Al Cuoco is Distinguished Scholar and Director of the Center for Mathematics Education at Education Development Center. He is lead author for The CME Project, a four-year NSF-funded high school curriculum, published by Pearson. He also co-directs Focus on Mathematics, a mathematics-science partnership that has established a mathematical community of mathematicians, teachers, and mathematics educators. The partnership evolved from his 25-year collaboration with Glenn Stevens (BU) on Boston University's PROMYS for Teachers, a professional development program for teachers based on the Ross program (an immersion experience in mathematics). Al taught high school mathematics to a wide range of students in the Woburn, Massachusetts public schools from 1969 until 1993. A student of Ralph Greenberg, Cuoco holds a Ph.D. from Brandeis, with a thesis and research in Iwasawa theory. In addition to this book, MAA published his *Mathematical Connections: a Companion for Teachers and Others*. But his favorite publication is a 1991 paper in the *American Mathematical Monthly*, described by his wife as an attempt to explain a number system that no one understands with a picture that no one can see.

Joseph Rotman was born in Chicago on May 26, 1934. He studied at the University of Chicago, receiving the degrees BA, MA, and Ph.D. there in 1954, 1956, and 1959, respectively; his thesis director was Irving Kaplansky.

Rotman has been on the faculty of the mathematics department of the University of Illinois at Urbana-Champaign since 1959, with the following ranks: Research Associate 1959–1961; Assistant Professor 1961–1963; Associate Professor 1963–1968; Professor 1968–2004; Professor Emeritus 2004–present. He has held the following visiting appointments: Queen Mary College, London, England 1965, 1985; Aarhus University, Denmark, Summer 1970; Hebrew University, Jerusalem, Israel 1970; University of Padua, Italy, 1972; Technion, Israel Institute of Technology and Hebrew University, Jerusalem (Lady Davis Professor), 1977–78; Tel Aviv University, Israel, 1982; Bar Ilan University, Israel, Summer 1982; Annual visiting lecture, South African Mathematical Society, 1985; Oxford University, England, 1990.

Professor Rotman was an editor of Proceedings of American Mathematical Society, 1970, 1971; managing editor, 1972, 1973.

Aside from writing research articles, mostly in algebra, he has written the following textbooks: *Group Theory* 1965, 1973, 1984, 1995; *Homological Algebra* 1970, 1979, 2009; *Algebraic Topology* 1988; *Galois Theory* 1990, 1998; *Journey into Mathematics* 1998, 2007; *First Course in Abstract Algebra* 1996, 2000, 2006; *Advanced Modern Algebra* 2002.